T0344974

5G System Design

5G System Design

Architectural and Functional Considerations and Long Term Research

Edited by

Patrick Marsch
Deutsche Bahn AG, Germany

Ömer Bulakçı
Huawei German Research Center (GRC), Germany

Olav Queseth
Ericsson Research, Sweden

Mauro Boldi
Telecom Italia, Italy

Registered Office(s)
John Wiley & Sons, Inc., 111 River Street, Hoboken, NJ 07030, USA
John Wiley & Sons Ltd, The Atrium, Southern Gate, Chichester, West Sussex, PO19 8SQ, UK

Editorial Office
The Atrium, Southern Gate, Chichester, West Sussex, PO19 8SQ, UK

For details of our global editorial offices, customer services, and more information about Wiley products visit us at www.wiley.com.

Wiley also publishes its books in a variety of electronic formats and by print-on-demand. Some content that appears in standard print versions of this book may not be available in other formats.

Library of Congress Cataloging-in-Publication data applied for

ISBN: 9781119425120

Cover Design: Wiley
Cover Images: ©cono0430/Shutterstock; ©Iaremenko Sergii/Shutterstock

Set in 10/12pt Warnock by SPi Global, Pondicherry, India

Printed in the UK

To my wife Ines and our kids Philipp and Daniel for their continuous love and support, and to the great team in Nokia Wrocław that I was privileged to work with in the past years.

Patrick Marsch

To my family for their continuous support and encouragement over the years and my big brother Mesut Bulakçı, MD for guiding me to the right career path at my early age.

Ömer Bulakçı

For Eskil, Ellen and Ester.

Olav Queseth

To the memory of my father Ivano.

Mauro Boldi

Contents

Part 2   5G System Architecture and E2E Enablers *79*

5   E2E Architecture *81*
Marco Gramaglia, Alexandros Kaloxylos, Panagiotis Spapis, Xavier Costa, Luis Miguel Contreras, Riccardo Trivisonno, Gerd Zimmermann, Antonio de la Oliva, Peter Rost and Patrick Marsch

8   Network Slicing   *181*

Alexandros Kaloxylos, Christian Mannweiler, Gerd Zimmermann, Marco Di Girolamo, Patrick Marsch, Jakob Belschner, Anna Tzanakaki, Riccardo Trivisonno, Ömer Bulakçı, Panagiotis Spapis, Peter Rost, Paul Arnold and Navid Nikaein

9   Security   *207*

Carolina Canales-Valenzuela, Madalina Baltatu, Luciana Costa, Kai Habel, Volker Jungnickel, Geza Koczian, Felix Ngobigha, Michael C. Parker, Muhammad Shuaib Siddiqui, Eleni Trouva and Stuart D. Walker

Part 4   Performance Evaluation and Implementation *451*

15   **Performance, Energy Efficiency and Techno-Economic Assessment** *453*
          Michał Maternia, Jose F. Monserrat, David Martín-Sacristán, Yong Wu, Changqing Yang, Mauro
          Boldi, Yu Bao, Frederic Pujol, Giuseppe Piro, Gennaro Boggia, Alessandro Grassi, Hans-Otto Scheck,
          Ioannis-Prodromos Belikaidis, Andreas Georgakopoulos, Katerina Demesticha and Panagiotis
          Demestichas

Contributor List

Adrian Kliks	Poznan University of Technology
Akira Matsunaga	KDDI
Alessandro Grassi	Politecnico di Bari
Alexandros Kaloxylos	University of Peloponnese
Ali Yaver	Nokia
Andreas Georgakopoulos	WINGS ICT Solutions
Anna Tzanakaki	University of Bristol
Antonio de la Oliva	Universidad Carlos III de Madrid
Athul Prasad	Nokia Bell Labs
Balázs Sonkoly	Budapest University of Technology and Economics
Caner Kilinc	Ericsson
Carolina Canales	Ericsson
Catherine Douillard	IMT Atlantique
Chan Zhou	Huawei German Research Center
Changqing Yang	Huawei
Charbel Abdel Nour	IMT Atlantique
Chen Xiaobei	IMT 2020
Chia-Yu Chang	EURECOM
Chris Pearson	5GAmericas
Christian Mannweiler	Nokia Bell Labs
Damiano Rapone	Telecom Italia
Daniel Calabuig	Universitat Politècnica de València
Daniel Camps Mur	I2CAT
Dario Sabella	Intel

David Garcia-Roger	Universitat Politècnica de València
David Gutierrez Estevez	Samsung
David Martín-Sacristán	Universitat Politècnica de València
Davide Sorbara	Telecom Italia
Didier Bourse	Nokia
Dieter Ferling	Nokia Bell Labs
Dimitra Simeonidou	University of Bristol
Dong Ku Kim	5G Forum Korea
Du Ho Kang	Ericsson
Eckhard Grass	IHP
Elena Trouva	NCSRD
Emmanouil Pateromichelakis	Huawei German Research Center
Evangelos Kosmatos	WINGS ICT Solutions
Felix Ngobigha	University of Essex
Fernando Sanchez Moya	Nokia
Frank Schaich	Nokia Bell Labs
Frederic Pujol	iDate
Fredrik Tillman	Ericsson
Gennaro Boggia	Politecnico di Bari
Gerd Zimmermann	Deutsche Telekom
Geza Koczian	University of Essex
Gian Michele Dell'Aera	Telecom Italia
Giorgio Calochira	Telecom Italia
Giovanna D'Aria	Telecom Italia
Giuseppe Piro	Politecnico di Bari
Hans Dieter Schotten	DFKI
Hans Otto Scheck	Nokia
Hao Lin	Orange
Hirsam El Shaer	Vodafone
Hitoshi Yoshino	Softbank
Hong Beom Jeon	Korea Telecom
Honglei Miao	Intel
Hua Wang	Keysight
Ioannis-Prodromos Belikaidis	WINGS ICT Solutions

Jakob Belschner	Deutsche Telekom
Jan Christoffersson	Ericsson
Jesús Gutiérrez	IHP
Ji Lianghai	University of Kaiserslautern
Jian Luo	Huawei German Research Center
Jinhyo Park	SK Telecom
Jose Alcaraz-Calero	UWS
Jose F. Monserrat	Universitat Politècnica de València
Josep M. Fabrega	CTTC
Josep Mangues	CTTC
Kai Habel	Fraunhofer HHI
Katerina Demesticha	WINGS ICT Solutions
Klaus Pedersen	Nokia Bell Labs
Kwang Taik Kim	Samsung
Laia Nadal	CTTC
László Toka	Budapest University of Technology and Economics
Leonardo Gomes Baltar	Intel
Luciana Costa	Telecom Italia
Luis M. Campoy	Telefónica
Luis Miguel Contreras Murillo	Telefónica
Madalina Baltatu	Telecom Italia
Malte Schellmann	Huawei German Research Center
Marco Caretti	Telecom Italia
Marco Di Girolamo	Hewlett Packard Enterprise
Marco Giordani	University of Padova
Marco Gramaglia	Universidad Carlos III de Madrid
Marco Mezzavilla	NYU
Maria Carmela De Gennaro	Magneti Marelli
Mark Doll	Nokia Bell Labs
Markos Anastasopoulos	University of Bristol
Marten Ericsson	Ericsson
Martin Kurras	Fraunhofer HHI
Mauro Boldi	Telecom Italia
Mehrdad Shariat	Samsung

Michael Färber	Intel
Michał Maternia	Nokia
Michela Svaluto Moreolo	CTTC
Mike Parker	University of Essex
Mikko Säily	Nokia Bell Labs
Milos Tesanovic	Samsung
Miquel Payaró	CTTC
Muhammad Shuaib Siddiqui	i2CAT
Nathan Gomes	University of Kent
Navid Nikaein	EURECOM
Nicolas Barati	NYU
Nikos Makris	UTH Nitlab
Nuno Pratas	Aalborg University
Nuria Molner	IMDEA Networks and Univ. Carlos III de Madrid
Nurul H. Mahmood	Aalborg University
Ömer Bulakçı	Huawei German Research Center
Olav Queseth	Ericsson
Pablo Serrano	Universidad Carlos III de Madrid
Panagiotis Demestichas	University of Piraeus
Panagiotis Spapis	Huawei German Research Center
Paris Flegkas	UTH Nitlab
Patrick Marsch	Nokia (now Deutsche Bahn)
Paul Arnold	Deutsche Telekom
Peter Rost	Nokia Bell Labs
Philippos Assimakopoulos	University of Kent
Qi Wang	UWS
Raffaele D'Errico	CEA-LETI
Raffaele de Peppe	Telecom Italia
Ramon Casellas	CTTC
Raúl Muñoz	CTTC
Rauno Ruismaki	Nokia Bell Labs
Raymond Knopp	EURECOM
Ricard Vilalta	CTTC
Riccardo Trivisonno	Huawei German Research Center

Salah El Ayoubi	Orange (now CentraleSupélec)
Samer Bazzi	Huawei German Research Center
Sandra Roger	Universitat Politècnica de València
Sergio Barberis	Telecom Italia
Shangbin Wu	Samsung
Shubhranshu	ITRI
Sinh L. H. Nguyen	Aalto University
Stavroula Vassaki	WINGS ICT Solutions
Stuart D. Walker	University of Essex
Sylvie Mayrargue	CEA-LETI
Takehiro Nakamura	NTT DOCOMO
Tao Chen	VTT
Tapio Rautio	VTT
Taylan Şahin	Huawei German Research Center and Technische Universität Berlin
Terje Tjelta	Telenor
Thanasis Korakis	UTH Nitlab
Thomas Rosowski	Deutsche Telekom
Thorsten Wild	Nokia Bell Labs
Tomasz Mach	Samsung
Tommy Svensson	Chalmers University of Technology
Valerio Frascolla	Intel
Victor Lopez Alvarez	Telefónica
Vladica Sark	IHP
Volker Jungnickel	Fraunhofer HHI
Wei Jiang	DFKI
Wu Yong	Huawei
Xavier Costa	NEC
Yang Yang	Intel
Yinan Qi	Samsung
Yu Bao	Orange
Yukihiko Okumura	NTT DOCOMO
Yves Bellego	Orange

Foreword 1

Digital technologies have a profound transformative impact on our societies and economies. The digital revolution opens the door to novel activities, applications and business cases that could not be envisaged before. As part of the Digital Single Market initiative launched as one of the ten priorities of the Juncker Commission, the European Commission proposed in 2016 an ambitious package of measures to foster the advent of a digital society and economy in Europe.

5G is an important pillar of this strategy. The European 5G vision co-created with a multiplicity of actors is fully aligned with our wider digitization strategy, as 5G is designed to support smart connectivity in domains as diverse as the automotive, healthcare, factories, energy or media sectors.

The stakes are high. 5G has the potential to open new B2B businesses, whilst operators are currently facing a stagnation of their revenues in Europe. Market estimates point at a potential of € 550 billion extra revenues in 2025 from vertical industries, adding to the classical broadband consumer markets. The "connectivity package" released in September 2016 by the European Commission has thus proposed an ambitious strategy for 5G in Europe. It includes a new connectivity strategy moving Europe in the Gigabit/s connectivity era, a reform of the telecom regulatory framework with specific spectrum and investment friendly measures, and a 5G Action Plan with a package of actions to put in place the right framework conditions for the launch of 5G in Europe in 2020.

European efforts are indeed key to keep abreast of a fierce global competition. The USA and South Korea have already announced the deployment of early versions of 5G technology in 2018. Japan plans 5G introductions in 2020, and China pursues a bold technological development plan. These pre-commercial initiatives are putting high pressure on the quick release of the required standards. In that context, it is imperative that the European 5G strategy targeting vertical markets gets quickly validated, both from a technology and business perspective. The 5G Action Plan consequently calls for early cross-industry and large-scale trials in Europe. These will be supported by the next phase of the 5G Public Private Partnership (5G PPP), a € 700 million Research and Innovation initiative launched in 2013 by the European Commission to materialize our bold ambitions in this domain.

The European 5G vision requires a versatile network platform that can adapt to demanding requirements of a multiplicity of business models, whilst current networks are more designed as "one size fits all" platforms. Moving towards 5G, deployment will largely piggyback on the results of previous phases of the 5G PPP, which have invested tremendous 5G research efforts in a multiplicity of domains, covering issues as diverse as new radio access technologies, network architectures with co-operation of a multiplicity of fixed or mobile access networks including satellites, operation of new spectrum in the millimeter wave ranges, network virtualization, redesign of the core network, applications of software techniques to network management, as typical examples.

This book presents the results of the research carried out in these multiple domains during the first phase (2014-16) of the 5G PPP. It shows an impressive set of technological achievements, unlocking many of the roadblocks on the road towards achieving the most demanding KPIs of 5G, such as data rates beyond 10 Gbit/s, latencies in the milliseconds range, or service creation and deployment within a few minutes. This work has also been instrumental in supporting the European industry to make informed choices for what concerns 5G standards and spectrum requirements and allocation.

I am grateful to all the colleagues who have shown undivided commitment to make 5G a reality in Europe, and I am sure that the readers will enjoy reading this book as a testimony to these efforts.

Khalil Rouhana, Deputy Director-General in DG CONNECT,
European Commission

Foreword 2

As chairman of the 5G Infrastructure Association board, it is with great pleasure that I see this book come to fruition, and I welcome the chance to add a personal message of support. I believe this is a timely and important work, which in hindsight will be seen as one of the key results from the 5G pre-standardization period.

Although not limited to only European research work, the major input to this book represents the key results from the 5G Public Private Partnership (5G-PPP) research programme. Within the 5G PPP programme, the 5G Infrastructure Association (5G IA) is the organization which represents European industry. The 5G IA is committed to the advancement of 5G in Europe and to building global consensus on 5G. To this aim, the Association brings together a global industry community of telecoms & digital actors, such as operators, manufacturers, research institutes, universities, verticals and SMEs. I believe this book may play a useful role for this 5G IA goal of advancement of 5G, by providing a definitive source for the current state of 5G research.

This book is timely because we are at a water shed in both the 5G PPP and in terms of 5G in general. Within the 5G PPP, we are at a time where most of the phase 1 projects have completed or are about to complete. This means that a lot of the fundamental 5G research has been completed and the focus of the programme in phase 2 will move more towards demonstration of trial systems and integration of vertical domains such as automotive, e-health and Industry 4.0. As such, it is a perfect time to document and disseminate the key results of those phase 1 5G PPP projects.

In terms of the broader view, 5G is also moving from the pre-standardization to post-standardization phase. At the time of writing this book, 3GPP has almost concluded on the so-called first drop of 5G technology, focusing on non-stand-alone operation of 5G in conjunction with (e)LTE. This first drop will of course not include the complete functionality, and it is important to point out that standardization will need many years to completely specify 5G. Nonetheless, some aspects of 5G will be finalized in the near future, and the influence of the 5G PPP projects on many of the design choices made both through direct research and creation of pre-standardization consensus should not be underestimated.

As well as giving an important snapshot of where we are in terms of 5G today and some clear guidance of where we think it should go in the future, I believe the depth and quality shown in this book is a clear validation for the vision and goals of the 5G PPP in general. There is still much work to do to make 5G a reality that lives up to the promised goals, and I believe this book is an important step on that journey.

Dr. Colin Willcock, Chairman of the Board,
5G Infrastructure Association

Acknowledgments

In the mid of 2015, the first phase of 5th Generation Public Private Partnership (5G PPP) projects kicked off, and the time until now has been hectic, but also very rewarding. During this period, 5G has moved from vision and concepts to technologies that are almost ready to be deployed, and we are glad and proud to have been part of this work.

This book is based on the outcome of 12 projects within the 5G PPP framework, as detailed in Section 1.2, and complemented by contributions from various additional 5G experts across the globe. We would like to thank all the contributors for the substantial effort and engagement invested into this book, despite the fact that the writing of the book collided with that of the final deliverables of most of the involved projects. In particular, we would like to express a big thank you to the main chapter editors for consolidating the often diverse viewpoints and terminologies used by different projects or entities into a coherent story. Knowing that many contributors have also spent their free time to finalize the book, and given that the work behind the development of new technologies like 5G is typically as demanding and time-consuming as it is rewarding and inspiring, we would also like to thank the families of the contributors for their continuous patience and support.

Naturally, we would like to thank the European Commission for funding the projects that have led to this book, and in particular Bernard Barani for his personal support of the book.

Beyond the researchers who have been directly involved in the projects, there are of course much more persons involved in our home organizations. We would hence like to thank all our colleagues in the mobile communications industry, research institutes and universities for inspiring discussions, the contribution of ideas, and the help on various tasks.

Dr. Bulakçı would also like to thank Wu Jianjun and Dr. Egon Schulz from Huawei for the support in preparation of this book.

Last but not least, we would like to thank Sandra Grayson, Louis Manoharan and Adalfin Jayasingh from Wiley for the pleasant collaboration and continuous support throughout the writing and production process of this book.

Patrick Marsch, Ömer Bulakçı, Olav Queseth and Mauro Boldi
On behalf of the book contributors

List of Abbreviations

Term	Meaning
3D	Three-Dimensional
3G	3rd Generation (*cellular communications*)
3GPP	3rd Generation Partnership Project
4G	4th Generation (*cellular communications*)
5G	5th Generation (*cellular communications*)
5G AP	5G Action Plan
5G IA	5G Infrastructure Association
5G PPP	5th Generation Public Private Partnership
5GAA	5G Automotive Association
5GC	5G Core Network
5GMF	Fifth Generation Mobile Communication Promotion Forum
5GPoAs	5G Points of Attachment
5GTTI	5G Trial & Testing Initiative
AAS	Active Antenna System
ABG	Alpha Beta Gamma
ABS	Almost Blank Subframes
ABNO	Applications-Based Network Operations
ACDM	Algebraic Channel Decomposition Multiplexing
ACK	Acknowledged
A-CPI	Application-Controller Plane Interface
ADB	Aggregated Data Bundle
ADC	Analog-to-Digital Converter
AE	Action Enforcer
AF	Application Function
AF-x	Adaptation Function–*number* x
AI	Air Interface
AIV	Air Interface Variant
AL	Aggregation level
AM	Acknowledged Mode
AMC	Adaptive Modulation and Coding
AMF	Access and Mobility Management Function

AN	Access Network
ANDSF	Access Network Discovery and Selection Function
AN-I	Access Network-Inner (*layer*)
ANN	Artificial Neural Networks
AN-O	Access Network-Outer (*layer*)
ANQP	Access Network Query Protocol
AP	Access Point
API	Application Programming Interface
APT	Average Power Tracking
AR	Augmented Reality
ARIB	Association of Radio Industries and Businesses
ARP	Allocation and Retention Priority
ARPU	Average Revenue Per User
ARQ	Automatic Repeat reQuest
AS	Access Stratum
ASIC	Application Specific Integrated Circuit
AS-PCE	Active Stateful Path Computation Element
ATIS	Alliance for Telecommunications Industry Solutions
AUSF	Authentication Server Functions
BB	BaseBand
BBU	BaseBand Unit
BCH	Reed Muller or Bose, Ray-Chaudhuri and Hocquenghem codes
BCJR	Bahl, Cocke, Jelinek and Raviv algorithm
BF	BeamForming
BF-OFDM	Block-Filtered OFDM
BH	BackHaul
BLER	Block Error Rate
BMS	Broadcast/Multi-cast Services
BPSK	Binary Phase Shift Keying
BRP	Beam Resource Pool
BS	Base Station
BSM	Basic Safety Message
BSR	Buffer State Reporting
BSS	Business Support System
B-TAG	Backbone VLAN Tag
BV	Bandwidth-Variable
BVT	Bandwidth-Variable Transponders
C/I	Carrier-to-Interference (*ratio*)
CA	Carrier Aggregation
CAGR	Compound Annual Growth Rate
CAM	Cooperative Awareness Messages
CAPEX	Capital Expenditures
CCDF	Complementary Cumulative Distribution Function
CCE	Control Channel Entity
CCH	Control Channel

CCNF	Common Control Network Functions
CCSA	China Communications Standards Association
CD	Code Division
CDF	Cumulative Distribution Function
CDMA	Code Division Multiple Access
CDN	Content Delivery Network
CH	Cluster Head
CI	Close-In
CIoT	Cellular Internet of Things
C-ITS	Cooperative-ITS
CLI	Cross-Link Interference
CMO	Control, Management and Orchestration
CMOS	Complementary Metal-Oxide Semiconductor
CN	Core Network
CoMP	Coordinated/Cooperative Multi-Point
COST	European Cooperation in Science and Technology
COTS	Commercial Off-The-Shelf
CP	Control Plane
CPE	Common Phase Error
CPM-19	Conference Preparatory Meeting in 2019
CP-OFDM	Cyclic Prefix OFDM
CPRI	Common Public Radio Interface
CPU	Central Processing Unit
CQI	Channel Quality Indicator
C-RAN or CRAN	Centralized/Cloud Radio Access Network
CRC	Cyclic Redundancy Check
C-RNTI	Cell Radio Network Temporary Identifier
CRS	Cell-specific Reference Symbols
CSI	Channel State Information
CSIT	Channel State Information (*transmitter side*)
CSMA/CA	Carrier Sense Multiple Access with Collision Avoidance
CSP	Communication Service Provider
CTC	Convolutional Turbo Codes
CTO	Chief Technology Officer
CU	Central Unit
D2D	Device-to-Device
D2N	Device-to-Network
DAC	Digital-to-Analog Converter
DAS	Distributed Antenna System
DBSCAN	Density-Based Spatial Clustering of Applications with Noise
DC	Dual Connectivity
DCC	Decentralized Congestion Control
DCI	Downlink Control Information
DCN	Dedicated Core Network
D-CPI	Data-Controller Plane Interface

DD-OFDM	Direct Detection OFDM
DDoS	Distributed Denial of Service
DECOR	Dedicated Core
DEI	Drop Eligible Indicator
DEN	Decentralized Environmental Notification
DFT	Digital Fourier Transform
DFT-s-OFDM	DFT-spread OFDM
DL	Downlink
DM	Decision-Maker
DMC	Dense Multipath Component
DMRS	DeModulation Reference Signal
DNS	Domain Name System
DoS	Denial of Service
DPD	Digital Pre-Distortion
DPDK	Data Plane Development Kit
DR	D2D Receiver
D-RAN	Distributed Radio Access Network
DRB	Data Radio Bearer
DRX	Discontinuous Reception
DSC	Decentralized Congestion Control Sensitivity Control
DSP	Digital Signal Processor
DSRC	Dedicated Short Range Communications
DT	Decision Tree
DTT	Digital Terrestrial Television
DTX	Discontinuous Transmission
DU	Distributed Unit
DVB	Digital Video Broadcasting
DWDM	Dense Wavelength Division Multiplexing
E2E	End-to-End
EATA	European Automotive-Telecom Alliance
EC	European Commission
ECDSA	Elliptic Curve Digital Signature Algorithm
EDCA	Enhanced Distributed Channel Access
eDECOR	evolved DECOR
eICIC	Enhanced Inter-Cell Interference Coordination
eIMTA	Enhanced Interference Mitigation and Traffic Adaptation
EIRP	Effective Isotropic Radiated Power
eLTE	enhanced Long-Term Evolution
EM	Element Management
eMBB	enhanced Mobile BroadBand
eMBMS	enhanced Mobile Broadband Multimedia Services
EMF	Electro-Magnetic Field
eNB	enhanced Node-B
EPC	Enhanced Packet Core
ePDCCH	enhanced PDCCH

EPS	Evolved Packet System
ET	Envelope Tracking
ETH	Ethernet
ETN	Edge Transport Nodes
ETSI	European Telecommunications Standards Institute
EU	European Union
E-UTRA	Evolved-UTRA
EVM	Error Vector Magnitude
F1	*Horizontal interface in the RAN*
F1-C	*Horizontal interface in the RAN (control plane)*
F1-U	*Horizontal interface in the RAN (user plane)*
FBMC	Filter Bank Multi-Carrier
FBR	Front-to-Back Ratio
FCAPS	Fault Configuration, Accounting, Performance, Security
FCC	Federal Communications Commission
FC-OFDM	Flexibly Configured – OFDM
FD	Full Duplex
FDD	Frequency Division Duplexing
FDM	Frequency Division Multiplexing
FDMA	Frequency Division Multiple Access
FDV	Frame-Delay Variation
FEC	Forward Error Correction
FeICIC	Further enhanced Inter-Cell Interference Coordination
FFT	Fast Fourier Transform
FH	Fronthaul
FI	Float Intercept
FIR	Finite Impulse Response
FN	False Negatives
FP	False Positives
FPGA	Field Programmable Gate Array
FQAM	Frequency Quadrature Amplitude Modulation
FS	Feature Selection
Fs-C	*Intra-RAN control plane interface*
FSPL	Free-Space Path Loss
Fs-U	*Intra-RAN user plane interface*
FTP	File Transfer Protocol
FTTA	Fiber-to-the-Antenna
GA	Genetic Algorithm
GAA	General Authorized Access
GaN	Gallium Nitride
GAN	Generic Access Network
GDB/GLDB	GeoLocation DataBase
GEPON	Gigabit Ethernet PON
GFDM	Generalized Frequency Division Multiplexing
GLOSA	Green Light Optimal Speed Advice

GMPLS	Generalized Multi-Protocol Label Switching
gNB	Gigabit (enhanced) Node-B
GOPS	Giga Operations
GPON	Gigabit PON
GPP	General Purpose Processing
GPRS	General Packet Radio Service
GPS	Global Positioning System
GSCM	Geometry-based Stochastic Channel Model
GSM	Global System for Mobile Communications
GTP	GPRS Tunnelling Protocol
GUI	Graphical User Interface
HARQ	Hybrid Automatic Repeat reQuest
HD	High Definition
HO	HandOver
HPA	High Power Amplifier
HSIC	Hybrid Self Interference Cancellation
HSPA	High-Speed Packet Access
HSS	Home Subscriber System
HSTD	Horizontal Security and Trust Domains
HTHP	High Tower High Power
HW	Hardware
I2I	Indoor-to-Indoor
IaaS	Infrastructure-as-a-Service
IAD	Interference-Aware Detection
IAN	Interference-as-Noise
IASD	Interference-Aware Successive Decoding
IAT	Inter-Arrival Time
IATN	Inter Area Transport Node
IBFD	In-Band Full-Duplex
ICI	Inter Carrier Interference
ICIC	Inter-Cell Interference Coordination
I-CPI	Intermediate-Controller Plane Interface
ICT	Information and Communications Technology
ID	Identifier
IDFT	Inverse DFT
IEEE	Institute of Electrical and Electronics Engineers
IETF	Internet Engineering Task Force
iFFT	inverse Fast Fourier Transform
IFOM	IP Flow Mobility
IIC	Industrial Internet Consortium
IIR	Infinite Impulse Response
IMS	IP Multimedia Subsystem
IMT	International Mobile Telecommunications
IMT2020 or IMT-2020	IMT for year 2020 and beyond

IMT-A	IMT-Advanced
INR	Interference-to-Noise Ratio
IoT	Internet of Things
IP	Internet Protocol
IPr	Intellectual Property
IPsec	Internet Protocol Security
IPv4	Internet Protocol version 4
IPv6	Internet Protocol version 6
IQ	Inphase and Quadrature Phase
IR	Incremental Redundancy
IR-HARQ	incremental redundancy HARQ
ISD	Inter-Site Distance
ISG	Industry Standard Group (in ETSI)
ISI	Inter Symbol Interference
ISM	Industrial, Scientific and Medical
ISO	International Organization for Standardization
IT	Information Technology
I-TAG	Backbone Service Instance Tag
ITS	Intelligent Transport Systems
ITU	International Telecommunication Union
ITU-R	ITU - Radiocommunication sector
JT	Joint transmission
KED	Knife Edge Diffraction
KPI	Key Performance Indicator
KSP	Known Symbol Padding
KTX	Korea Train eXpress
LAA	License(d)-Assisted Access
LBT	Listen-Before-Talk
LCP	Logical Channel Prioritization
LD	Linear Discriminant
LDM	Local Dynamic Map
LDPC	Low Density Parity Check
LIME	Linda in a Mobile Environment
LIPA	Local IP Access
LLC	Logical link control
LNA	Low Noise Amplifier
LO	Local Oscillator
LoRa	Long Range, low power technology
LOS	Line-of-Sight
LPWA	Low Power Wide Area
LSA	Licensed Shared Access
LSE	Least Squares Estimation
LSO	Lifecycle Service Orchestration
LTE(-A)	Long-Term Evolution (Advanced)
LWA	LTE-WLAN Aggregation

LWAAP	LTE-WLAN Aggregation Adaptation protocol
LWIP	LTE/Wi-Fi Radio Level Integration Using IPsec Tunnel
M2M	Mobile-to-Mobile
MAC	Medium Access Control
MANO	Management & Orchestration
MAP	Maximum a Posteriori
MAPCON	Multi-Access PDN Connectivity
MBB	Mobile BroadBand
MBHM	Map-Based Hybrid Modeling
MBMS	Multimedia Broadcast Multicast Services
MBSFN	Multicast-Broadcast Single Frequency Network
MC	Multi-Carrier
MCC	Mission Critical Communications
MCP	Measurement Configuration Profile
MCS	Modulation and Coding Scheme
MdO	Multi-domain Orchestrator
MD-PCE	Multi-Domain Path Computation Element
MEC	Mobile Edge Computing *or* Multi-Access Edge Computing
ME	Multi-Edge
ME-LDPC	Multi-Edge LDPC
MeNB	Master (enhanced) Node-B
METIS (-II)	Mobile and wireless communications Enablers for the Twenty-twenty Information Society (-II)
MIB	Master Information Block
MIC	Ministry of Internal Affairs and Communications, Japan
MIIT	Ministry of Industry and Information Technology
MIMO	Multiple-Input Multiple-Output
MIPS	Million Instructions Per Second
MISO	Multiple Input Single Output
ML	Machine Learning
MLD	Maximum Likelihood Decoding
MM	Mobility Management
MMC	Massive Machine Communications
MME	Mobility Management Entity
MMIMMO	Massive Multiple-Input Massive Multiple-Output
MMSE-IRC	Minimum Mean Squared Error – Interference Rejection Combining
mMTC	Massive Machine-Type Communications
mmW, mmWave	millimeter-Wave
MNO	Mobile Network Operator
MOCN	Multi-Operator Core Network
MOP	Multi-Objective Optimization
MORAN	Mobile Operator Radio Access Network
MP	Memory Polynomial
MP	Message Passing
MPLS	Multi Protocol Label Switching

MPLS-TP	MPLS Transport Profile
MP-TCP	Multi-Path Transmission Control Protocol
M-RRC	Master Radio Resource Control
MS	Min-Sum
MSE	Mean-Square-Error
MSIP	Ministry of Science, Information and Future Planning, South Korea
MT	Mobile Terminated
MTA	Multi-Tenancy Application
MTC	Machine-Type Communications
MTU	Maximum Transmission Unit
MU-MIMO	Multi User MIMO
MVNO	Mobile Virtual Network Operator
MWC	Mobile World Congress
N2	*Control plane interface betw. RAN and core network in 5G → NG-C*
N3	*User plane interface betw. RAN and core network in 5G → NG-U*
NACK	Negative Acknowledged
NAICS	Network Assisted Interference Cancellation and Suppression
NAS	Non Access Stratum
NB	NarrowBand
NBAP	Node-B Application Part
NBI	North Bound Interface
NBIFOM	Network-Based Internet Protocol Flow Mobility
NB-IoT	NarrowBand Internet of Things
NCO	Numerically Controlled Oscillator
NEF	Network Exposure Function
NF	Network Function
nFAPI	Network functional Application Programming Interface
NFV	Network Function Virtualization
NFVI	Network Function Virtualization Infrastructure
NFVO	NFV Orchestrator
NG	Next Generation
NG-C	*Control plane interface betw. RAN and core network in 5G → N2*
NGCO	New Generation Central Office
NGFI	Next Generation Fronthaul Interface
NGMN	Next Generation Mobile Networks
NGPoP	Next Generation Point of Presence
NG-RAN	Next Generation Radio Access Network
NG-U	*User plane interface betw. RAN and core network in 5G → N3*
NG-Uu	*Interface between network and device in 5G*
NHTSA	National Highway Traffic Safety Administration
NIC	Network Interface Controller
NLoS	Non-Line-of-Sight
NMSE	Normalized Mean Square Error
NN	Nomadic Node
NNSF	Network Node Selection Function

NOI	Notice of Inquiry
NPRM	Notice of Proposed Rulemaking
NPV	Net Present Value
NR	New Radio
NRA	National Regulatory Authority
NRF	Network Repository Function
NS	Network Service
NSA	Non-StandAlone
NSD	Network Service Descriptors
NS-FAID	Non-Surjective Finite Alphabet Iterative Decoder
NSI	Network Slice Instance
NSMF	Network Slice Management Function
NSSF	Network Slice Selection Function
NST	Network Slice Template
NYU	New York University
O2I	Outdoor-to-Indoor
O2O	Outdoor-to-Outdoor
OAI	Open Air Interface
OAM	Operations, Administration and Management/Maintenance
OBSAI	Open Base Station Architecture Initiative
OBU	On-Board Unit
OF	OpenFlow
OFDM	Orthogonal Frequency Division Multiplex
OFDMA	Orthogonal Frequency Division Multiple Access
OFNC	Open-Flow Network Controller
OLT	Optical Line Terminal
ONF	Open Networking Foundation
ONU	Optical Network Unit
OOB	Out Of Band
OPEX	Operational Expenditures
OPI	Offload Preference Indicator
OPS	Operations
OQAM	Offset Quadrature Amplitude Modulation
OS	Operating System
OSI	Open System Interconnection
OSM	Open Source Management and Orchestration
OSS	Operations Support System
OTN	Optical Transport Network
OTT	Over The Top
OvS	Open vSwitch
OVSDB	Open vSwitch Database Management Protocol
PA	Power Amplifier
PaaS	Platform-as-a-Service
PAE	Power Added Efficiency
PAL	Primary Access Licenses

PAM	Pulse-Amplitude Modulation
PAPR	Peak-to-Average Power Ratio
PBB	Provider Backbone Bridging
PBCH	Physical Broadcast CHannel
PC	Polar Codes
PCA	Principal Component Analysis
PCell	Primary Cell
PCF	Policy Control Function
PCHA	Personal Connected Health Alliance
PCI	Physical Cell ID
PCM	Parity Check Matrices
PCP	Priority Code Points
PCR	Pilot Contamination Regime
PCRF	Policy and Charging Rules Function
PDCA	Plan, Do, Check, Adjust
PDCCH	Physical Downlink Control CHannel
PDCP	Packet Data Convergence Protocol
PDN	Packet Data Network
PDP	Power Delay Profile
PDSCH	Physical Downlink Shared Channel
PDU	Protocol Data Unit
PDV	Packet Delay Variation
PER	Packet Error Rate
PGIA	Pre-emptive Geometrical Interference Analysis
PGW	Packet Data Network Gateway
PHY	Physical layer
PLL	Phase-Locked Loop
PLMN	Public Land Mobile Network
PLOS	Probability Loss
PLS	Physical Layer Security
PMI	Precoding Matrix Indicator
PMIP	Proxy Mobile IP
PMSE	Programme Making and Special Events
PN	Polyphase Network
PNF	Physical Network Function
P-OFDM	Pulse Shaped OFDM
PON	Passive Optical Network
PoP	Point of Presence
PPC	PolyPhase Components
PPDR	Public Protection and Disaster Relief
PPI	Power Preference Indicator
PPN	PolyPhase Network
PPP	Public Private Partnership
PRACH	Physical Random Access CHannel
PRB	Physical Resource Block

PRG	Precoding Resource Block Group
PS	Packet Switched
PSS	Primary Synchronization Signal
PTM	Point-To-Multi-point
PUCCH	Physical Upling Control CHannel
PUSCH	Physical Uplink Shared CHannel
PVM	Probe Vehicle Message
PVNO	Private Virtual Network Operator
PWS	Public Warning System
QAM	Quadrature Amplitude Modulation
QC	Quasi-Cyclic
QCI	QoS Class Identifier
QFI	QoS Flow Identity
QoE	Quality of Experience
QoS	Quality of Service
QPP	Quadratic Permutation Polynomial
RA	Random Access
RACH	Random Access Channel
RAM	Random Access Memory
RAN	Radio Access Network
RAT	Radio Access Technology
RAU	Radio Aggregation Unit
RB	Resource Block
RCC	Radio Cloud Center
RCLWI	RAN Controlled LTE WLAN Interworking
RE	Resource Element
REG	Resource Element Groups
REM	Radio Environment Map
RF	Radio Frequency
RFSIC	Radio Frequency Self Interference Cancelation
RIT	Radio Interface Technology
RITA	Research and Innovative Technologies Administration
RLC	Radio Link Control
RM	Resource Management
RMS	Root Mean Square
RNC	Radio Network Controller
RoE	Radio over Ethernet
RoHC	Robust Header Compression
ROI	Return on Investment
RRC	Radio Resource Control
RRH	Remote Radio Head
RRM	Radio Resource Management
RRU	Remote Radio Unit
RS	Reference Symbol
RSA	Roadside Alert

RSC	Resources Sharing Cluster
RSRP	Reference Signal Received Power
RSRQ	Reference Signal Received Quality
RTA	RAN Tracking Area
RTT	Round Trip Time
RU	Remote Unit
Rx	Receiver
SaaS	Software-as-a-Service
SAE	Society of Automotive Engineers
SAS	Spectrum Access System
SBI	Service-Based Interface
sBH	Self-BackHauling
S-BVT	Sliceable-Bandwidth Variable Transponders
SC	Single Carrier
SCell	Secondary Cell
SCF	Small Cell Forum
SC-FDMA	Single Carrier Frequency Division Multiple Access
SCM	Spatial Channel Model
SCME	Spatial Channel Model Extension
SCNF	Slice-specific Control Network Function
SD	Slice Differentiator
SDAP	Service Data Adaptation Protocol
SDMA	Space Division Multiple Access
SDM-C	Software-Defined Mobile Network Controller
SDM-X	Software-Defined Mobile Network Coordinator
SDN	Software-Defined Networking
SDR	Software-Defined Radio
SD-RAN	Software-Defined Radio Access Network
SDT	Small Data Transmission
SDU	Service Data Unit
SeNB	Secondary enhanced Node-B
SFBC	Spatial-Frequency Block Codes
SFN	Single Frequency Network
SGW	Serving Gateway
SHF	Super High Frequency
SI	Self Interference
SIB	System Information Block
SiC	Silicon Carbide
SIC	Self Interference Cancelation
SIM	Subscriber Identity Module
SIMO	Single Input Multiple Output
SINR	Signal to Interference plus Noise Ratio
SIPTO	Selected IP Traffic Offloading
SIW	Substrate Integrated Waveguide
SLA	Service Level Agreement

SMA	Service Monitoring & Analytics
SMF	Session Management Function
SMS	Spectrum Management System
SN	Sequence Number
SNR	Signal-to-Noise Ratio
S-NSSAI	Single Network Slice Selection Assistance Information
SOI	Silicon-On-Insulator
SON	Self-Organizing Network
SP	Strict Priority
SPAT	Signal Phase and Timing
SPM	Security Policy Manager
SPS	Semi-Persistent Scheduling
SPTP	Small Packet Transmit Procedure
SR	Scheduling Request
SRIT	Set of RITs
SRM	Slice Resource Manager
S-RRC	Secondary Radio Resource Control
SRS	Sounding Reference Signal
SSB	Synchronization Signal Block
SSS	Secondary Synchronization Signal
SST	Slice/Service Type
SU	Single User
SVD	Single Value Decomposition
SVM	Support Vector Machine
SW	Software
SWA	Software Architectures
SWCM	Sliding-Window Coded Modulation
SYNC	Synchronization
TA	Timing Advance
TAC	Transmit Access Control
TAU	Tracking Area Update
TBCC	Tail-Biting Convolutional Codes
TBS	Transport Block Size
TC	Turbo Code
TCP	Transmission Control Protocol
TD	Time Division
TDC	Transmit Datarate Control
TDD	Time Division Duplex
TDMA	Time Division Multiple Access
TE	Traffic Engineering
TG	Task Group
THR	Throughput
TM	Transverse Magnetic
TMC	Traffic Management Center
TMS	Transmission Mode Selection

TN	Transport Node
TNeg	True Negatives
TO	Timing Offset
TOSCA	Topology and Orchestration Specification for Cloud Applications
TP	True Positives
TPC	Transmit Power Control
TR	Technical Report
TRC	Transmit Rate Control
TRP	Transmission Reception Point
TS	Technical Specification
TSDSI	Telecommunications Standards Development Society
TSN	Time-Sensitive Networking
TSON	Time Shared Optical Network
TTA	Telecommunications Technology Association
TTC	Telecommunication Technology Committee
TTI	Transmission Time Interval
TV	Television
Tx	Transmitter
UC	Use Case
UCA	Use Customer Address
UDM	Unified Data Management
UDN	Ultra-Dense Network
UDP	User Datagram Protocol
UE	User Equipment
UEFA	Union of European Football Associations
UF-OFDM	Universal Filtered OFDM
UHD	Ultra High Definition
UHF	Ultra High Frequency
UK	United Kingdom
UL	Uplink
UM	Unacknowledged Mode
UMA	Unlicensed Mobile Access
UP	User Plane
UPF	User Plane Function
URLLC	Ultra-Reliable Low-Latency Communications
US	United States (*of America*)
USRP	Universal Software Radio Peripheral
UTRA	Universal Mobile Telecommunications System (UMTS) Terrestrial Radio Access
UW	Unique Word
V2I	Vehicle-to-Infrastructure
V2N	Vehicle-to-Network
V2P	Vehicle-to-Pedestrian
V2V	Vehicle-to-Vehicle
V2X	Vehicle-to-Everything

VCO	Voltage-Controlled Oscillator
VGA	Variable Gain Amplifier
VI	Virtualized Infrastructure
VIM	Virtual Infrastructure Manager
VLAN	Virtual Local Area Network
VLIW	Very Long Instruction Word
VNF	Virtual Network Function
VNFM	VNF Manager
VNN	Vehicular Nomadic Node
VoD	Verify-on-Demand
VoLTE	Voice over LTE
VPN	Virtual Private Networks
VR	Virtual Reality
vRAN/VRAN	Virtual Radio Access Network
VSF	Virtual Security Function
VSTD	Vertical Security and Trust Domains
WAN	Wide Area Network
WAS	Wireless Access System
WAVE	Wireless Access in Vehicular Environments
WBI	West Bound Interface
WCDMA	Wideband Code Division Multiple Access
WDM	Wavelength Division Multiplexing
WFQ	Weighted Fair Queueing
WI	Work Item
Wi-Fi	*Technology for wireless local area networking based on IEEE 802.11 standards*
WIM	WAN Infrastructure Manager
WIMAX	Worldwide Interoperability for Microwave Access
WINNER-II	Wireless world INitiative NEw Radio phase II
WLAN	Wireless Local Area Network
W-OFDM	Windowed OFDM
WP5D	ITU-R Work Package 5D
WRC	World Radiocommunication Conference
WRC-15	WRC in 2015
WRC-19	WRC in 2019
WRR	Weighted Round Robin
WSLM	Weighted Selective Mapping Technique
WSMP	WAVE Short Message Protocol
WT	WLAN Termination
X2	*Interface between base stations in LTE*
XaaS	Everything-as-a-Service
XGPON	10-Gigabit-capable passive optical network
xMBB	extreme Mobile BroadBand
Xn	*Interface between base stations in 5G*
ZC	Zadoff-Chu
ZP	Zero Padding
ZP-OFDM	Zero Padding OFDM
ZT	Zero Tail

Part 1

Introduction and Basics

1

Introduction and Motivation

Patrick Marsch[1], Ömer Bulakçı[2], Olav Queseth[3] and Mauro Boldi[4]

[1] *Nokia, Poland (now Deutsche Bahn, Germany)*
[2] *Huawei German Research Center, Germany*
[3] *Ericsson, Sweden*
[4] *Telecom Italia, Italy*

1.1 5th Generation Mobile and Wireless Communications

The 5th generation (5G) of mobile and wireless communications is expected to have a large impact on society and industry that will go far beyond the information and communications technology (ICT) field. On one hand, it will enable significantly increased peak data rates compared to previous cellular generations, and allow for high experienced data rates almost anytime and anywhere, to support enhanced mobile broadband (eMBB) services. While there is already a wide penetration of mobile broadband services today, 5G is expected to enable the next level of human connectivity and human-to-human or human-to-environment interaction, for instance with a pervasive usage of virtual or augmented reality [1], free-viewpoint video [2], and tele-presence.

On the other hand, 5G is expected to enable ultra-reliable low-latency communications (URLLC) and massive machine-type communications (mMTC), providing the grounds for the all-connected world of humans and objects. This will serve as a catalyst for developments or even disruptions in various other technologies and business fields beyond ICT, from the ICT perspective typically referred to as *vertical industries*, that can benefit from omnipresent mobile and wireless connectivity [3]. To name a few examples[1], it is expected that 5G will

- foster the 4th industrial revolution, also referred to as Industry 4.0 [4] or the Industrial Internet, by enabling reliability- and latency-critical communication between machines, or among machines and humans, in industrial environments;
- play a key role for the automotive sector and transportation in general, for instance allowing for advanced forms of collaborative driving and the protection of vulnerable road users [5], or increased efficiency in railroad transportation [6];

---

1 Note that more use case examples are described in Chapter 2 and in Section 17.3.

- enable the remote control of vehicles or machines in dangerous or inaccessible areas, as for instance in the fields of mining and construction [7];
- revolutionize health services, for instance through the possibility of wirelessly enabled smart pharmaceuticals or remote surgery with haptic feedback [8];
- accelerate and, in some cases, enable the adoption of solutions for so-called Smart Cities, improving the quality of life through better energy, environment and waste management, improved city transportation, etc. [9].

Ultimately, directly or indirectly through the stated impacts on vertical industries, 5G is likely to have a huge impact on the way of life and the societies in which we live [10].

The mentioned wide diversity of technology drivers and use cases is a unique characteristic of 5G in comparison to earlier generations of cellular communications, as illustrated in Figure 1-1. More precisely, previous generations have always been tailored towards one particular need and a particular business ecosystem, such as mobile broadband in the case of Long-Term Evolution (LTE), and have hence always been characterized by one monolithic system design. In contrast, 5G is from the very beginning associated with the need for multi-service and multi-tenancy support, as detailed in Section 5.2, and is commonly understood to comprise a variety of tightly integrated radio technologies, such as enhanced LTE (eLTE), Wi-Fi, and different variants of novel 5G radio interfaces that are tailored to different frequency bands, cell sizes or service needs.

Beyond the technology as such, 5G is also expected to imply an unprecedented change in the value chain of the mobile communications industry. Although a mobile-operator-centric ecosystem may

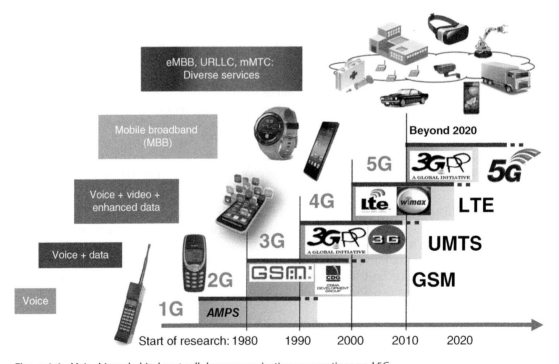

Figure 1-1. Main drivers behind past cellular communications generations and 5G.

prevail, a set of new players are deemed to enter the arena, such as enhanced connectivity providers, asset providers, data centre and relay providers, and partner service providers, as detailed in Section 2.6.

Clearly, the path to 5G is a well-beaten track by now. Early research on 5G started around 2010, and the first large-scale collaborations on 5G, such as METIS [11] and 5GNow [12] were launched in 2012. In the meanwhile, most geographical areas have launched initiatives and provided platforms for funded research or collaborative 5G trials, as detailed in Section 7.3. The International Telecommunications Union (ITU) has defined the requirements that 5G has to meet to be chosen as an official International Mobile Telecommunications 2020 (IMT-2020) technology [13], and published related evaluation guidelines [14]. On the way towards the fulfilment of the IMT-2020 framework, the standardization of an early phase of 5G by the 3rd Generation Partnership Project (3GPP) is in full swing [15], as summarized in the following section and detailed in Section 17.2.1. Further, 5G has now gained major public visibility through pre-commercial deployments alongside the Winter Olympics in South Korea, and will soon be showcased at further large-scale events such as the Summer Olympics in Tokyo in 2020 and the UEFA EURO 2020 soccer championship.

Nevertheless, even though 5G is moving full pace ahead towards first commercial deployments, there are still various design questions to be answered, and many topics are still open for longer-term research. This is in part due to the continuous acceleration of the 5G standardization timeline, requiring to set priorities and postpone parts of the original 5G vision to later, as detailed in the following section.

At this vital point in the 5G development timeline, this book aims not only to summarize the consensus that has already been reached in 3GPP and in research consortia, but also to elaborate on various design options and choices that are still to be made towards the complete 5G system, which is ultimately envisioned to respond to all the use cases and societal needs as listed before, and address or exceed the IMT-2020 requirements.

As a starting point to the book, Section 1.2 elaborates in more detail on the timing of the book w.r.t. the 5G developments in 3GPP and global initiatives. Section 1.3 stresses the exact scope of the 5G system design as covered in this book, and in particular puts this into perspective to what is currently covered in 3GPP Release 15 and likely covered in subsequent releases. Finally, Section 1.4 explains the approach pursued in writing this book, and introduces the structure and the following chapters of this book.

1.2 Timing of this Book and Global 5G Developments

At the time of the publication of this book, the Winter Olympic Games in South Korea are taking place, constituting the first large-scale pre-commercial 5G deployment connected to a major international event, and hence marking a major milestone in the 5G development.

Further, by the time the book appears, **3GPP** has likely just concluded the specification of the so-called *early drop* of New Radio (NR) [16], reflecting a subset of 5G functionalities that are just sufficient for very first commercial 5G deployments in so-called *non-stand-alone* (NSA) operation, i.e. where 5G radio is only used in conjunction with existing LTE technology, as detailed in Section 5.5.2. The full completion of 3GPP Release 15, often referred to as the Phase 1 of 5G, is expected for the second half of 2018, and will also include *stand-alone* (SA) operation [16]. More details on the 3GPP timeline can be found in Section 17.2.1.

Naturally, as the 5G standardization in 3GPP has been heavily accelerated to allow for very early commercial deployments, some prioritization had to be made w.r.t. the scope of the 5G system that is captured in Release 15. For instance, the discussion in 3GPP so far tends towards eMBB use cases, as most specific 5G deployment plans and related investments that have already been announced are related to eMBB, as visible in Section 17.3. In consequence, some design choices in 3GPP have so far been made with eMBB services in mind, leaving further modifications and optimizations for other service types for future study in upcoming releases. One example for such decisions is the choice of cyclic prefix based orthogonal frequency division multiplex (CP-OFDM) as the waveform for NR Release 15 [17][18], possibly enhanced with filtering that is transparent to the receiver. This approach is seen as suitable for eMBB as well as for several URLLC services, but it may not fully address the needs of some other specific URLLC and mMTC services or device-to-device (D2D) communications, as detailed in Sections 11.3 and 14.3. Another example is the choice of Low Density Parity Check (LDPC) codes and Polar codes for data and control channels in NR Release 15 [19], respectively, which has been accepted as a combination for eMBB, but which may not be the final choice for all service types envisioned for 5G, as detailed in Section 11.4. Again for the reason of speed, 3GPP is currently also putting most attention towards carrier frequencies below 40 GHz, i.e., not yet covering the full spectrum range up to 100 GHz envisioned in the longer term, see Section 3.4, which will be tackled in later releases.

However, one has to stress that 3GPP in general pursues the approach that whatever is introduced in early 5G releases has to be future-proof, or *forward-compatible*, i.e., it must not constitute a show-stopper for further developments in future releases. An example for this approach is the way how 3GPP handles self-backhauling, i.e., the usage of the same radio technology and spectrum for both backhaul and access links, as detailed in Section 7.4. While 3GPP will not be able to fully standardize this in Release 15, it ensures that the basic operation and essential features of NR that will also be needed for self-backhauling, such as flexible time division duplex (TDD), a minimization of always-on signals, asynchronous Hybrid Automated Repeat reQuest (HARQ), flexible scheduling time units, etc., are already covered well in Release 15. Based on this, the further standardization of self-backhauling, particularly covering higher-layer aspects in 3GPP RAN2 and RAN3, can then be taken up in Release 16.

Ultimately, 3GPP standardization is expected to take place in Releases 15 and 16 until 2020 [15], with the aim to submit a 5G system design to ITU, where NR, and NR in combination with enhanced LTE (eLTE), i.e. Release 15 and onwards, meet the IMT-2020 requirements [20][21]. The IMT process is covered in detail from a performance evaluation perspective in Section 15.2.1, and from an overall 5G deployment perspective in Section 17.2.2. Beyond the ITU submission, 5G standardization is naturally expected to continue further in Release 17 and beyond.

This book has been written at a point in time when most of the so-called Phase 1 of the **5G Public Private Partnership** (5G PPP) research projects have been concluded, and the Phase 2 has just started [22]. While Phase 1 has focused on 5G *concepts*, Phase 2 is dedicated to *platforms*, and Phase 3 to *trials*, as depicted in Figure 1-2. In fact, a big portion of this book is based on the output of the 5G PPP Phase 1 projects, in particular on the output of (in alphabetical order) [23]:

- **5G-Crosshaul**, which has developed a 5G integrated backhaul and fronthaul transport network enabling a flexible and software-defined reconfiguration of all networking elements in a multi-tenant and service-oriented unified management environment;
- **5GEx**, which has aimed at enabling the cross-domain orchestration of services over multiple administrations or over multi-domain single administrations;
- **5G-NORMA**, which has developed a novel, adaptive and future-proof 5G mobile network architecture, with an emphasis on multi-tenancy and multi-service support;

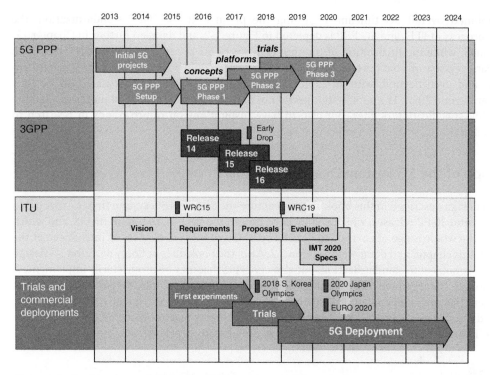

Figure 1-2. Combined overall 5G timeline of the mentioned different bodies.

- **5G-Xhaul**, which has developed a converged optical and wireless network solution able to flexibly connect small cells to the core network;
- **COHERENT**, which has developed a unified programmable control framework for coordination and flexible spectrum management in 5G heterogeneous access networks;
- **CHARISMA**, which has focused on an intelligent hierarchical routing and paravirtualized architecture uniting a devolved offload with an end-to-end security service chain via virtualized open access physical layer security;
- **FANTASTIC-5G**, which has developed a 5G flexible air interface for scalable service delivery, with a comprehensive PHY, MAC and RRM design;
- **Flex5GWare**, which has developed highly reconfigurable hardware and software platforms targeting both network elements and devices, and taking into account increased capacity, reduced energy footprint, as well as scalability and modularity for a smooth transition to 5G;
- **METIS-II**, which has developed an overall 5G RAN design, focusing on the efficient integration of evolved legacy and novel air interface variants (AIVs), and the support of network slicing;
- **mmMAGIC**, which has developed new RAN architecture concepts for millimeter-wave (mmWave) radio access technology, including its integration with lower frequency bands;
- **Selfnet**, which has developed an autonomic network management framework to achieve self-organizing capabilities in managing network infrastructures by automatically detecting and mitigating a range of common network problems; and finally
- **SPEED-5G**, which has investigated resource management techniques across technology 'silos', and medium access technologies to address densification in mostly unplanned environments.

The combined overall 5G timeline regarding the planned trials, 3GPP standardization, the IMT-2020 process of ITU, and 5G PPP is depicted in Figure 1-2, and detailed further in Chapter 17.

In a nutshell, while the finalization of the first features of 5G are ongoing these days, this book offers a clear overview of what the complete 5G system design could be at the end of the standardization phase, and even beyond, with an exploration of innovative features that may only be fully exploited far beyond 2020. The book is thus useful not only to have a clear understanding of what the current 3GPP specification defines, but also to have inspirations on future trends in research to further develop the 5G system and improve its performance.

1.3   Scope of the 5G System Described in this Book

The system design described in this book aims to capture the *complete* 5G system that is expected to exist after several 3GPP releases, which will meet or exceed the IMT-2020 requirements, and which will address the whole range of envisioned eMBB, URLLC and mMTC services as introduced at the beginning of this chapter and detailed in Section 2.2. Also, the book does not only describe 5G design aspects that are subject to standardization, but also concepts that may be proprietarily implemented, such as resource management (RM) strategies, orchestration frameworks, or general enablers of the 5G system that are independent of a particular standards release. Consequently, the book clearly goes beyond the scope of 3GPP NR Release 15, and covers aspects that are expected to be relevant in the Release 16 and 17 time frame, or further beyond, as illustrated in Figure 1-3.

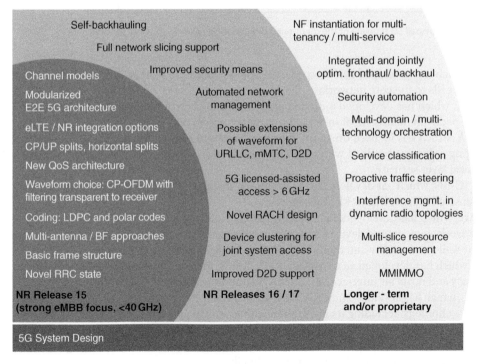

Figure 1-3. Illustration of the scope of the 5G system design covered in this book, in the form of few selected examples of the many topics covered in the book.

Just to provide some examples, for **NR Release 15** (including the "early drop"), the book covers all the early conclusions that have been drawn in 3GPP, for instance on:

- The extended channel models to be used for 5G (see Chapter 4);
- The overall modularized E2E 5G architecture that 3GPP has defined (Section 5.4.1), the various options for eLTE/NR integration (Section 5.5), and the forms of control/user plane (CP/UP) and horizontal RAN function splits that are envisioned (Section 6.6);
- The new QoS architecture that enables a dynamic mapping of so-called *QoS flows* to data radio bearers on RAN level (Sections 5.3.3 and 12.2.1);
- The waveform choice (Section 11.3), coding approaches (Section 11.4), multi-antenna and beam-forming support (Section 11.5) and basic frame structure (Section 11.6);
- The introduction of a new RRC state (Section 11.3) and related signalling optimizations.

As possible candidates for standardization in **NR Releases 16 or 17**, the book, for instance, covers:

- Self-backhauling, i.e., the usage of the same radio interface and spectrum for backhaul and access links (see Section 7.4);
- The extension of NR towards full network slicing support (Chapter 8);
- Improved security means and related architecture for 5G (Section 9.4);
- Automated network management and orchestration for 5G (Section 10.7);
- Possible extensions of waveforms for specific URLLC and mMTC services (Section 11.3) or better D2D support (Section 14.3);
- 5G licensed-assisted access (LAA) to enable NR operation in unlicensed bands, also above 6 GHz (Section 12.5.1);
- Novel Random Access CHannel (RACH) design for service prioritization already at initial access (Section 13.2);
- Device clustering for joint system access (Sections 13.2.6 and 13.4.2);
- Improved D2D support, e.g., through sidelink mobility management (Section 14.5).

Finally, the book also covers various concepts that are of further **longer-term nature**, and/or which could be **implemented proprietarily**, for instance:

- Network function instantiation for multi-tenancy and multi-service support (see Section 6.4.4);
- Integrated and jointly optimized fronthaul and backhaul (Section 7.6);
- Security automation (Section 9.4.6);
- Orchestration in multi-domain and multi-technology scenarios (Section 10.4);
- Machine-learning based service classification (Section 12.2);
- Proactive traffic steering that provides an early assessment of mmWave links to reduce link failures (Section 12.4.2);
- Interference management in dynamic radio topologies, for instance involving moving access nodes and related novel interference challenges (Section 12.5.1);
- Multi-slice resource management, based on real-time SLA monitoring and ensuring SLA fulfilment via slice-specific QoS enforcement (Section 12.6);
- Massive multiple-input massive multiple output (MMIMMO) involving a large number of antenna elements at both transmitter and receiver side (Section 11.5.4);
- Detailed hardware and software implementation considerations, based on flexible HW/SW partitioning (Chapter 16).

1.4   Approach and Structure of this Book

Several books on 5G have already been published. For instance, [24] and [10] have focused on identifying the main use cases for 5G and their requirements, as well as key technology components needed to address these. The authors of [25] have focused in particular on signal processing challenges related to 5G, for instance in the context of novel waveforms or massive multiple-input multiple-output (MIMO), while [26] takes a bit more critical stand on 5G, pointing out that continuous connectivity may be more relevant in the 5G era than ultra-high peak data rates in hotspots, and that many of the often claimed 5G capabilities are economically questionable. [27] views 5G from a R&D technical design perspective, with a particular focus on the physical layer, while [28] focuses on key protocols, network architectures and techniques considered for 5G The authors in [29] focus on mmWave and massive MIMO communications as specific technology components in 5G, while the authors in [30] delve into simulation and evaluation methodology for 5G, and [31] focuses on the specific usage of 5G for the Internet of Things.

This book differs from all mentioned publications in that it does not describe single 5G technology components, but rather captures the complete 5G system in its likely overall system design, i.e., covering all technology layers that are required to operate a complete 5G system. For this reason, the book does not contain chapters on typical 5G keywords such as massive MIMO, mmWave communications, or URLLC support, but instead describes the system from an overall architecture perspective and then layer-by-layer, inherently always covering all relevant components on each layer, and covering the support of all three main 5G service types stated before.

Further, this book is unique in that it is based on consolidated contributions from 158 authors from 54 companies, institutes or regional bodies, hence capturing the consensus on 5G that has already been obtained by key stakeholders, while also stressing the diversity of further system design concepts that have been raised, but not yet agreed, and which could hence appear in future 3GPP releases.

While this book is to a large extent based on the results of European Commission funded 5G PPP projects, as mentioned in Section 1.2, the fact that there are also many non-European partners involved in these projects ensures that the book does not only represent a purely European view. Further, various authors from outside Europe and outside the 5G PPP ecosystem have been invited to contribute to this book, for instance to Chapter 17 on the global deployment plans for 5G, to ensure that the book can legitimately claim to capture a global view on 5G.

This book is written such that it should be decently easily digestible for persons who are not yet familiar with cellular communications in general or with 5G, through detailed introductions and explanations of all covered topics, while also providing significant technical details for experts in the field. Naturally, a key challenge inherent to writing a book on a technology that is yet in the process of standardization, in particular a technology that is being as pushed and accelerated as 5G, is that certain technical details of the book may quickly become outdated. For instance, it is almost inevitable that there are aspects described in this book which are marked as "under discussion", which may have already been agreed upon or dropped by 3GPP by the time the book is published. For this reason, the book does not aim to meticulously capture the latest agreements in 3GPP, but rather explain general 5G design decisions from a more didactic perspective, also elaborating on the advantages and disadvantages of concepts that may have already been discarded in 3GPP, or which may be far further down the 5G horizon than what is currently covered in 3GPP. This way, the book is expected to also serve as a good *reference book* on cellular communication system design in general, irrespective of the specific road taken by 3GPP.

This book is structured into 4 parts, which are shortly introduced in the following:

Part 1 – Introduction and Basics

This part of the book sets the scene for the following parts, and in particular covers various basic aspects related to the expected 5G ecosystem and the spectrum usage in 5G, which are central to many 5G system design aspects discussed in the subsequent parts of the book. Beyond this introduction chapter, **Chapter 2**, for instance, covers the main service types and use cases typically considered for 5G, and elaborates on the related requirements and the expected transformation of the mobile network ecosystem in the context of 5G. **Chapter 3** ventures into spectrum usage in the 5G era, in particular stressing the need for different spectrum sharing forms, and the usage of diverse frequency bands from the sub-6 GHz regime up to 100 GHz, in order to address the diverse and stringent 5G requirements. **Chapter 4** then builds upon this and introduces the reader to the particular propagation challenges inherent in the usage of higher frequency bands in 5G, and the additional channel models that had to be introduced to be able to design and evaluate a 5G system appropriately.

Part 2 – 5G System Architecture and E2E Enablers

This largest part of the book then focuses on the architecture of the 5G system, and various required E2E enablers. Here, **Chapter 5** initially provides the big picture on the 5G E2E architecture, covering everything from the core network to transport network and radio access network (RAN), and introducing various general design principles, such as modularization, softwarization, network slicing and multi-tenancy. **Chapter 6** then focuses on the 5G RAN architecture, for instance discussing changes in the protocol stack w.r.t. 4G and the notion of service-specific protocol stack optimization and instantiation. It further covers RAN-based multi-connectivity among (e)LTE and 5G or within 5G, horizontal and vertical function splits in the RAN, and subsequent deployments. **Chapter 7** then delves into the same level of detail on the transport network architecture, explaining a possible holistic user plane and control plane design for the transport network as well as available transport technologies and specific overall concepts, such as self-backhauling. Based on the previous chapters, **Chapter 8** then takes an E2E perspective again and covers in detail the establishment and management of network slices, constituting E2E logical networks that are each operated to serve a particular business need. **Chapter 9** addresses a topic that is essential especially in the context of the many new use cases and business forms envisioned in the 5G era, namely that of security, by elaborating on the main attack vectors to be considered, security requirements, and possible security architecture to address these. Finally, **Chapter 10** elaborates on how an overall 5G system incorporating the aspects introduced in the previous chapters, and in particular based on software-defined networking (SDN) and network function virtualization (NFV), can be efficiently managed and orchestrated.

Part 3 – 5G Functional Design

This part of the book then delves into the details of the functional design of the system. More precisely, **Chapter 11** describes the lower part of the RAN protocol stack, namely the physical layer and Medium Access Control (MAC) layer, covering topics such as waveform design, coding, Hybrid Automatic Repeat reQuest (HARQ), frame design and massive MIMO. **Chapter 12** deals with traffic steering and resource management, which play a critical role to fulfil the stringent service and slice requirements envisioned for 5G in the context of highly heterogeneous networks. In particular, the chapter covers the classification of traffic, the fast steering of traffic to different radio interfaces, dynamic multi-service or multi-slice scheduling, interference management and RAN moderation. **Chapter 13** handles the control plane procedures for the access of user equipments (UEs) to the network, state handling and mobility, in particular covering novelties in 5G such as an extended Radio Resource Control (RRC) state machine and further means to reduce control plane latency in 5G and support a larger number of devices and diverse service requirements. Finally, **Chapter 14** delves into specific functionalities related to D2D and vehicular-to-anything (V2X) communications, also providing an in-depth background and implementation details on the usage of cellular technologies for Intelligent Transport Systems (ITS).

Part 4 – Performance Evaluation and Implementation

This part of the book finally focuses on vary practical aspects related to the development, implementation and roll-out of 5G technology. **Chapter 15**, for instance, focuses on evaluation methodology for 5G that allows to quantify the performance of key 5G design concepts long before any type of hardware and field implementation is available. Further, the chapter introduces the methodology and results related to the evaluation of 5G deployments from an energy efficiency and techno-economic perspective. Next, **Chapter 16** is dedicated to the implementation of 5G concepts and components from a hardware and software perspective, considering for instance the need for increased hardware versatility and the ability to operate with increasingly higher bandwidths and related data rates, especially at mmWave bands. The chapter explicitly also covers the notion of flexible hardware/software partitioning and contains a detailed study on practical virtualized RAN deployments for 5G. Finally, the book is concluded with **Chapter 17**, which presents the roadmap of the expected standardization and regulation activities towards a full 5G system deployment and covers trials and early commercialization plans in the three regions Europe, Americas and Asia.

References

1 Nunatak, White Paper, "Virtual and Augmented Reality", April 2016
2 Canon, Press Release, "Canon announces development of the Free Viewpoint Video System virtual camera system that creates an immersive viewing experience", Sept. 2017
3 European Commission, White Paper, "5G empowering vertical industries", April 2016
4 CGI, White Paper, "Industry 4.0: Making your business more competitive", 2017
5 5G Automotive Association, White Paper, "The Case for Cellular V2X for Safety and Cooperative Driving", Nov. 2016
6 CER, CIT, EIM and UIC, White Paper, "A Roadmap for Digital Railways", April 2016
7 ABI Research, "Remote Control in Construction Made Possible by 5G", Q3 2017
8 WWRF, White Paper, "A New Generation of e-Health Systems Powered by 5G", Dec. 2016
9 Accenture, White Paper, "How 5G Can Help Municipalities Become Vibrant Smart Cities", 2017
10 A. Osseiran, J. F. Monserrat and P. Marsch (editors), "5G Mobile and Wireless Communications Technology", Cambridge University Press, June 2016

11 FP7 METIS project, see http://www.metis2020.com

12 FP7 5GNow project, see http://5gnow.eu

13 ITU-R WP5D, M.2140, "Minimum requirements related to technical performance for IMT-2020 radio interface(s)", Nov. 2017

14 ITU-R WP5D, M.2412, "Guidelines for the evaluation of the radio interface technologies for IMT-2020", Nov. 2017

15 3GPP release overview, see http://www.3gpp.org/specifications/releases

16 3GPP RP-170794, "Work plan for Rel-15 New Radio access technology WI", NTT Docomo, March 2017

17 3GPP TR 38.802, "Study on new radio access technology physical layer aspects", V14.1.0, June 2017

18 3GPP TS 38.201, "NR; Physical layer; General description", V1.0.0, Sept. 2017

19 3GPP TS 38.212, "NR; Multiplexing and channel coding", V1.0.0, Sept. 2017

20 ITU-R WP5D, M.2411, "Requirements, evaluation criteria and submission templates for the development of IMT-2020", Nov. 2017

21 3GPP RP-172098, "3GPP submission towards IMT-2020", Sept. 2017

22 5G Public-Private Partnership, see https://5g-ppp.eu/

23 5G PPP Phase 1 projects, see https://5g-ppp.eu/5g-ppp-phase-1-projects/

24 J. Rodriguez (editor), "Fundamentals of 5G Mobile Networks", Wiley&Sons, 2015

25 F.-L. Luo and C. Zhang (editors), "Signal Processing for 5G: Algorithms and Implementations", Wiley&Sons, 2016

26 W. Webb, "The 5G Myth: When vision decoupled from reality", Webb Search, 2016

27 F, Hu (editor), "Opportunities in 5G Networks: A Research and Development Perspective", CRC Press, 2016

28 V.W.S. Wong, R. Schober, D. Wing Kwan Ng, L.-C. Wang, "Key Technologies for 5G Wireless Systems", Cambridge University Press, 2017

29 S. Mumtaz, J. Rodriguez and L. Dai (editors), "mmWave Massive MIMO: A Paradigm for 5G", Academic Press, 2017

30 Y. Yang, J. Xu and G. Shi (editors), "5G Wireless Systems: Simulation and Evaluation Techniques", Springer, 2017

31 V. Mohanan, R. Budiarto, I. Aldmour (editors), "Powering the Internet of Things With 5G Networks", IGI Global, 2017

2

Use Cases, Scenarios, and their Impact on the Mobile Network Ecosystem

Salah Eddine Elayoubi[1], Michał Maternia[2], Jose F. Monserrat[3], Frederic Pujol[4], Panagiotis Spapis[5], Valerio Frascolla[6] and Davide Sorbara[7]

[1] *Orange Labs, France (now CentraleSupélec, France)*
[2] *Nokia, Poland*
[3] *Universitat Politècnica de València, Spain*
[4] *iDATE, France*
[5] *Huawei German Research Center, Germany*
[6] *Intel, Germany*
[7] *Telecom Italia, Italy*
With contributions from Damiano Rapone[7] and Marco Caretti[7].

2.1   Introduction

This chapter delves in detail into the use cases (UCs) widely assumed to be addressed by the 5th generation (5G) wireless and mobile communications system, and the related requirements. In particular, this chapter takes into consideration and aggregates the requirements from different bodies like the International Telecommunication Union (ITU), Next Generation Mobile Networks (NGMN), and the 5G Public Private Partnership (5G PPP). The next part of the chapter is an analysis of the 5G ecosystem evolutions that are needed, and the novel value chains that can be expected for some UCs.

The chapter is structured as follows. The main service types considered for 5G are initially introduced in Section 2.2, before their detailed requirements are discussed in Section 2.3. Section 2.4 then presents key 5G UCs as considered by NGMN and different 5G PPP research projects, and Section 2.5 elaborates particularly in the UCs further discussed in specific parts of this book. Section 2.6 then delves into the likely ecosystem evolutions from a 5G mobile network perspective, with emerging value chains of mobile network operators (MNOs), before the chapter is summarized in Section 2.7.

5G System Design: Architectural and Functional Considerations and Long Term Research, First Edition.
Edited by Patrick Marsch, Ömer Bulakçı, Olav Queseth and Mauro Boldi.
© 2018 John Wiley & Sons Ltd. Published 2018 by John Wiley & Sons Ltd.

2.2 Main Service Types Considered for 5G

After several years of research and standardization on 5G wireless and mobile communications, there is broad consensus on the fact that 5G will not just be a simple evolution of 4G networks with new spectrum bands, higher spectral efficiencies and higher peak throughput, but also target new services and business models. In this respect, the main 5G service types typically considered are:

- **Enhanced mobile broadband (eMBB)**, related to human-centric and enhanced access to multimedia content, services and data with improved performance and increasingly seamless user experience. This service type, which can be seen as an evolution of the services nowadays provided by 4G networks, covers UCs with very different requirements, e.g. ranging from hotspot UCs characterized by a high user density, very high traffic capacity and low user mobility, to wide area coverage cases with medium to high user mobility, but the need for seamless radio coverage practically anywhere and anytime with visibly improved user data rates compared to today;
- **Ultra-reliable and low-latency communications (URLLC)**, related to UCs with stringent requirements for capabilities such as latency, reliability and availability. Examples include the wireless control of industrial manufacturing or production processes, remote medical surgery, distribution automation in a smart grid, transportation safety, etc. It is expected that URLLC services will provide a main part of the fundament for the 4th industrial revolution (often referred to as Industry 4.0) and have a substantial impact on industries far beyond the information and communication technology (ICT) industry;
- **Massive machine-type communications (mMTC)**, capturing services that are characterized by a very large number of connected devices typically transmitting a relatively low volume of non-delay-sensitive data. However, the key challenge is here that devices are usually required to be low-cost, and have a very long battery lifetime. Key examples for this service type would be logistics applications (e.g., involving the tracking of tagged objects), smart metering, or for instance agricultural applications where small, low-cost and low-power sensors are sprinkled over large areas to measure ground humidity, fertility, etc.

It is worth noting that these three service types have been considered quite early in the METIS project [1], under the names of extreme mobile broadband (xMBB, equivalent to eMBB), ultra-reliable machine-type communications (uMTC, equivalent to URLLC) and mMTC. They have also been adopted by ITU-R Working Party 5D (WP5D), who have recently issued the draft new recommendation "IMT Vision - Framework and overall objectives of the future development of IMT for 2020 and beyond" [2], where IMT stands for International Mobile Telecommunications.

It should further be stressed that many services envisioned in the 5G era cannot easily be mapped to one of the three main service types as listed above, as they combine the challenges and requirements related to multiple service types. As an example, augmented reality is expected to play a major role in the 5G era, where information is overlaid to the real environment for the purpose of education, safety, training or gaming, and which poses high requirements on both throughput and latency. Similarly, some Factory of the Future [3] related UCs foresee the wireless communication of items in a factory environment where both energy efficiency and latency play a strong role. Especially such compound use cases combining different types of requirements ultimately pose the strongest challenges towards the development of the 5G system.

It goes without saying that considering each service type, or even single UCs, separately and building a 5G network accordingly, one would likely end up with very different 5G system designs and architectures. However, only a common design that accommodates all three service types is seen as an economically and environmentally sustainable solution, as discussed in more detail in Sections 15.3 and 15.4 on energy efficiency and techno-economic assessment, respectively. In the following, we briefly present the groups of 5G UCs typically found in literature, which have been proposed as representative and specific embodiments of the three service types or mixtures thereof, with the main aim to understand the scenarios envisaged in the 2020-2030 time horizon and have a reference for the development of the 5G system. We first start, in the next section, by listing the detailed requirements of these main 5G UCs.

2.3 5G Service Requirements

Even if the qualitative requirements of the three main 5G service types can be roughly understood from their description, there is a need for defining them in quantitative terms. Towards this aim, the ITU-R has considered a set of parameters to be key capabilities of IMT-2020 [3]:

- **Peak data rate**, referring to the maximum achievable data rate under ideal conditions per user or device in bits per second. The minimum 5G requirements for peak data rate are 20 Gbps in the downlink (DL) and 10 Gbps in the uplink (UL);
- **Peak spectral efficiency**, defined as the maximum data rate under ideal conditions normalized by the channel bandwidth, in bps/Hz. The target set by ITU-R is 30 bps/Hz in the DL and 15 bps/Hz in the UL. The combination of this key performance indicator (KPI) and the aforementioned peak data rate requirement results in the need for 2-3 GHz of spectrum to meet the stated requirements;
- **User experienced data rate**, referring to the achievable data rate that is available ubiquitously across the coverage area to a mobile user or device in bits per second. This KPI corresponds to the 5% point of the cumulative distribution function (CDF) of the user throughput, and represents a kind of minimum user experience in the coverage area. This requirement is set by ITU-R to 100 Mbps in the DL and 50 Mbps in the UL;
- **5th percentile user spectral efficiency**, referring to the 5% point of the CDF of the user throughput normalized by the channel bandwidth in bps/Hz. The minimum requirements for this KPI depend on the test environments as follows:
  - Indoor Hotspot: 0.3 bps/Hz in the DL, 0.21 bps/Hz in the UL;
  - Dense Urban: 0.225 bps/Hz in the DL, 0.15 bps/Hz in the UL;
  - Rural: 0.12 bps/Hz in the DL, 0.045 bps/Hz in the UL.
- **Average spectral efficiency**, also known as spectrum efficiency and defined as the average data throughput per unit of spectrum resource and per cell in bps/Hz/cell. Again, the minimum requirements depend on the test environments as follows:
  - Indoor Hotspot: 9 bps/Hz/cell in the DL, 6.75 bps/Hz/cell in the UL;
  - Dense Urban: 7.8 bps/Hz/cell in the DL, 5.4 bps/Hz/cell in the UL;
  - Rural: 3.3 bps/Hz/cell in the DL, 1.6 bps/Hz/cell in the UL.
- **Area traffic capacity**, defined as the total traffic throughput served per geographic area in Mbps/m^2. ITU-R has defined this objective only for the indoor hotspot case, with a target of 10 Mbps/m^2 for the DL;
- **User plane latency**, given as the contribution of the radio network to the time from when the source sends a packet to when the destination receives it. The one-way end-to-end (E2E) latency requirement is set to 4 ms for eMBB services and 1 ms for URLLC;

- **Control plane latency**, reflecting the transition time from idle to active state. The objective is to make this transition in less than 20 ms;
- **Connection density**, corresponding to the total number of connected and/or accessible devices per unit area. ITU-R has specified a target of 1 000 000 devices per km^2 for mMTC services;
- **Energy efficiency**, on the network side referring to the quantity of information bits transmitted to or received from users, per unit of energy consumption of the RAN, and on the device side to the quantity of information bits per unit of energy consumption of the communication module, in both cases in bits/Joule. The specification given by ITU-R in this respect is that IMT-2020 air interfaces must have the capability to support a high sleep ratio and long sleep duration;
- **Reliability**, defined as the success probability of transmitting a data packet before a given deadline. The target is to transmit Medium Access Control (MAC) packets of 32 bytes in less than 1 ms in the cell edge of the dense urban test environment with 99.999% probability;
- **Mobility**, here defined as the maximum speed at which a defined quality of service (QoS) and seamless transfer between radio nodes which may belong to different layers and/or radio access technologies can be achieved. For the rural test environment, the normalized traffic channel link data rate at 500 km/h, reflecting the average user spectral efficiency, must be larger than 0.45 bps/Hz in the UL;
- **Mobility interruption time**, being the time during which the device cannot exchange data packets because of handover procedures. The minimum requirement for mobility interruption time is 0 ms, essentially meaning that a *make-before-break* paradigm has to be applied, i.e., the connection to the new cell has to be set up before the old one is dropped;
- **Bandwidth**, referring to the maximum aggregated system bandwidth. At least 100 MHz must be supported, but ITU-R encourages proponents to support bandwidths of more than 1 GHz.

The set of the eight most significant capabilities expected for IMT-2020 are shown in Figure 2-1 (a), in comparison with those of IMT-Advanced. Since the importance of the achieved capability values is not the same for all three service types, the comparison among the service types is additionally given in Figure 2-1 (b).

As of energy efficiency, it is considered as an overall design goal for the entire 5G system. For eMBB services, the energy consumption on the infrastructure side is very important, while device battery life is critical for mMTC services. The METIS-II project adopted the principle that the energy efficiency improvement in 5G should follow at least the capacity improvement [5], i.e., the overall energy consumption should be similar or ideally lower than that in existing networks [6] [7], despite the large traffic growth. Since the 5G system is expected to see several hundred times or even a thousand times the traffic of legacy systems, while having the same or less energy consumption, network energy efficiency consequently also has to increase by a factor of several hundred times or a thousand.

2.4 Use Cases Considered in NGMN and 5G PPP Projects

Several 5G PPP projects have proposed new scenarios for identifying the requirements of 5G. Similarly, other initiatives like NGMN, and standardization bodies like 3GPP and ITU-R, have captured the respective requirements so as to drive the research for handling the future demands. This process has resulted in a large number of UCs with diverse requirements. The METIS-II project has

(a)

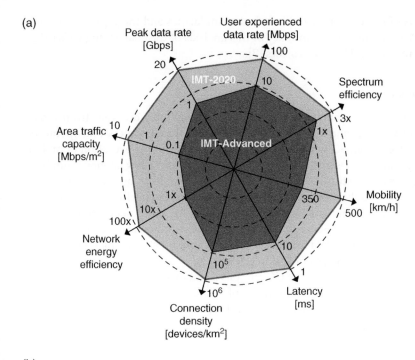

(b)

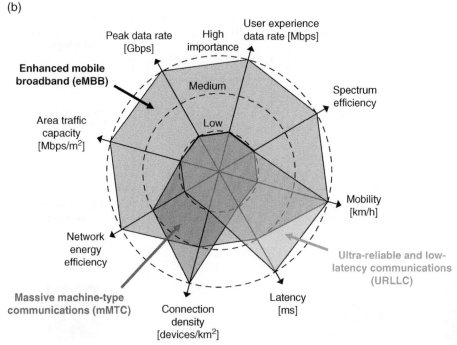

Figure 2-1. Key capabilities of IMT beyond 2020 [2]. a) Expected enhancements of IMT-2020 vs. IMT-Advanced. b) Importance of KPIs for different service types.

performed a detailed analysis of these in order to identify the similarities and the gaps between the already proposed UCs [4]. We present here a summary of this analysis of the challenging UCs originating from NGMN and from 5G PPP Phase 1 projects [7].

2.4.1 NGMN use Case Groups

According to NGMN [5], the business context beyond 2020 will be notably different from today, since it will have to handle the new UCs and business models driven by the customers' and operators' needs. According to the NGMN vision, 5G will have to support, apart from the evolution of mobile broadband, new UCs ranging from delay-sensitive video applications to ultra-low latency, from high speed entertainment applications in a vehicle to mobility for connected objects, and from best effort applications to reliable and ultra-reliable applications, for instance related to health and safety.

Thus, NGNM has performed a thorough analysis for capturing all the customers' and operators' needs. The analysis is based on 25 UCs for 5G grouped into eight UC families, as listed in Table 2-1 and illustrated in Figure 2-2. The UCs and UC families serve as an input for stipulating requirements and defining the building blocks of the 5G system design.

According to the NGMN 5G White Paper [5], the UC analysis is not exhaustive, though it provides a thorough and comprehensive analysis of the requirements of 5G. One can identify the key requirements and characteristics of each UC proposed by NGMN as listed in Table 2-1.

Broadband access in dense areas	Broadband access everywhere	Higher user mobility	Massive Internet of things
PERVASIVE VIDEO 	50+ MBPS EVERYWHERE 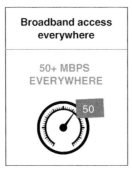	HIGH SPEED TRAIN 	SENSOR NETWORKS 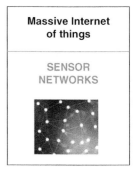
Extreme real-time communications	Lifeline communications	Ultra-reliable communications	Broadcast-like services
TACTILE INTERNET 	NATURAL DISASTER 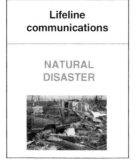	E-HEALTH SERVICES 	BROADCAST SERVICES 

Figure 2-2. UC families considered by NGMN with representative UCs [6].

Table 2-1. NGMN use case analysis by their characteristics and the dominant 5G service type, with H=high, L=low, and M=medium denoting the stringency of requirements.

UC	UC description UC name	UC requirements Number of devices	Mobility	Traffic type	Latency	Reliability	Availability	Service type eMBB	URLLC	mMTC
1	Pervasive video	H	L	Continuous	L	–	L	X	–	–
2	Smart office	H	No	Continuous	L	–	L	X	–	–
3	Operator cloud services	H	Yes	Continuous	M	–	L	X	–	–
4	HD video/photo sharing in stadiums or open air gatherings	H	No	Continuous	M	–	H	X	–	–
5	50+ Mbps everywhere	L	H	Continuous	H	–	H	X	–	–
6	Ultra-low-cost networks	L	M	Continuous	M	–	L	X	–	–
7	High-speed train	M	H	All types	L	–	L	X	–	–
8	Remote computing	L	H	Continuous	L	–	L	X	–	–
9	Moving hotspots	L	H	Bursty	L	–	H	X	–	–
10	3D connectivity, e.g. for aircrafts	L	H	H	L	–	H	X	–	–
11	Smart wearables	H	H	Periodic	L	–	H	–	–	X
12	Sensor networks	H	L	Periodic	L	–	H	–	–	X
13	Mobile video surveillance	H	H	Continuous	L	–	H	X	–	X
14	Tactile Internet	H	H	Various types	L	H	H	–	X	–
15	Natural disaster	H	L	Short messages	H	H	H	–	X	X
16	Automated traffic control and driving	L	H	All types	L	H	H	–	X	–
17	Collaborative robots: A control network for robots	L	No	Continuous	L	H	H	–	X	–
18	eHealth: extreme life critical	H	No/L	Short messages	L	H	H	–	X	–

(Continued)

Table 2-1. (Continued)

UC	UC description				UC requirements				Service type			
	UC name	Number of devices	Mobility	Traffic type	Latency	Reliability	Availability		eMBB	URLLC	mMTC	
19	Remote object manipulation, e.g. for remote surgery	L	L	Continuous	L	H	H		-	X	-	
20	3D connectivity, e.g. for drones	L	H	Continuous	L	H	H		-	X	-	
21	Public safety	L	L	Continuous	L	H	H		-	X	-	
22	News and information	H	H	All types	H	-	-		X	-	-	
23	Local broadcast-like services	H	L	All types	H	-	-		X	-	-	
24	Regional broadcast-like services	H	H	All types	H	-	-		X	-	-	
25	National broadcast-like services	H	H	All types	H	-	-		X	-	-	

2.4.2 Use Case Groups from 5G PPP Phase 1 Projects

Taking into consideration the rich literature of 5G UCs and scenarios including those of NGMN described before, 5G PPP Phase 1 projects have defined a set of UCs with the aim of evaluating the technological and architectural innovations developed in the projects. Without entering into the details of each project UC, we present here a grouping of these UCs and a mapping between these and the business cases identified in vertical industries.

Even if different 5G PPP projects have defined their own UCs, an in-depth analysis of these reveals strong similarities. This is because all 5G PPP projects agree on the three 5G service types listed in Section 2.2, and start in their UC definitions from the results of the METIS project, NGMN, ITU and other fora.

The UCs of 5G PPP Phase 1 projects can, ultimately, be classified into six families, as described in the 5G PPP White Paper on UCs and performance models [8] and detailed in Table 2-2.

This classification into UC families allows having a general idea on the individual UCs and their requirements, e.g., a UC belonging to the family "Future smart offices" is necessarily characterized by an indoor environment and very high user rates. However, this general classification does not reveal the detailed requirements of the UC, which may differ depending on the targeted application. Some UC families may feature enhanced diversity in terms of mixed requirements as well as mixed application environments, an example being the "Dense urban" UC family, where early 5G users could experience services demanding extreme data rates, such as virtual reality and ultra-high definition video in both indoor and outdoor environments, both requiring very high data rates but having heterogeneous latency requirements.

2.4.3 Mapping of the 5G-PPP Use Case Families to the Vertical Use Cases

While the 5G PPP projects have been intentionally mixing services with different requirements for the purpose of challenging the 5G RAN design, the 5G Infrastructure Association (5G IA), i.e. the private side of the 5G PPP including industry manufacturers, telecommunications operators, service providers and SMEs, has adopted a vertical industry driven approach in its business case definition,

Table 2-2. 5G PPP Phase 1 use case families.

Group	Description
Dense urban	Indoor and outdoor UCs, all in a dense urban environment
Broadband (50+ Mbps) everywhere	UCs that focus on suburban, rural environments and high speed trains
Connected vehicles	UCs containing URLLC and/or eMBB services related to vehicles, i.e. vehicle-to-vehicle (V2V) and/or vehicle-to-anything (V2X) applications
Future smart offices	UCs with very high data rates and low latency, indoor
Low-bandwidth Internet of Things (IoT)	UCs with a very large number of connected objects
Tactile Internet and automation	UCs with ultra-reliable communication and eMBB flavor

where each business case describes a specific vertical need and its requirements, as described in the 5G PPP White Paper on vertical requirements [8]. Table 2-3 illustrates the ambition of 5G PPP for a 5G network federating the needs of vertical industries.

Having a closer look at the business cases of Table 2-3, we can see that the 5G PPP UC families cover the requirements of most of them. Consequently, Table 2-4 highlights the relationship between the 8 NGMN UC families, the 6 5G PPP UC families and the main 5G service types.

Table 2-3. Vertical industry business cases.

Vertical Industry	Associated business cases	Corresponding 5G PPP use case families
Automotive	A1-Automated driving A2-Road safety and traffic efficiency services A3-Digitalization of transport and logistics A4-Intelligent navigation A5-Information society on the road A6-Nomadic nodes	Connected vehicles
eHealth	H1-Assets and interventions management in hospitals H2-Robotics (remote surgery, cloud service robotics for assisted living) H3-Remote monitoring of health or wellness data H4-Smarter medication	Dense urban (H3, H4) Broadband everywhere (H3, H4) IoT (H3) Tactile Internet (H2, H3)
Energy	E1-Grid access E2-Grid backhaul E3-Grid backbone	Dense urban (E1) Broadband everywhere (E3) IoT (E1) Tactile Internet (E2, E3)
Media & Entertainment	ME1-Ultra high fidelity media ME2-On-site live event experience ME3-User generated content & machine generated content ME4-Immersive and integrated media ME5-Cooperative media production ME6-Collaborative gaming	Dense urban (ME2, ME6) Broadband everywhere (ME1, ME3, ME4) Future smart offices (ME5)
Factories of the future	F1-Time-critical process optimization inside factory to support zero-defect manufacturing F2-Non time-critical optimizations inside factory to realize increased flexibility and eco-sustainability, and to increase operational efficiency F3-Remote maintenance and control optimizing the cost of operation while increasing uptime F4-Seamless intra-/inter-enterprise communication, allowing the monitoring of assets distributed in larger areas, the efficient coordination of cross value chain activities and the optimization of logistic flows F5-Connected goods, to facilitate the creation of new value added services	Dense urban (F2, F3, F4, F5) Broadband everywhere (F2, F4) IoT (F5) Tactile Internet (F1, F3)

Table 2-4. Relationship between the NGMN use case families, 5G PPP use case families and the three main 5G service types.

NGMN UC families	5G PPP UC families	5G service types
Broadband access in dense areas	Dense urban, Future smart offices	eMBB
Broadband access everywhere	Broadband (50+ Mbps) everywhere	
Broadcast-like services		
Higher user mobility		
Massive Internet of Things	Low-bandwidth IoT	mMTC
Extreme real-time communications	Tactile Internet and automation	URLLC
Lifeline communications		
Ultra-reliable communications	Connected vehicles	

2.5 Typical Use Cases Considered in this Book

Although the different chapters of this book focus on different aspects of the system design and do not necessarily investigate specific UCs, there are several UCs that are mostly represented in the book, in particular when it comes to performance evaluation, as covered in detail in Chapter 15. This section gives additional context and explanations for setting certain 5G KPI requirements in these representative UCs.

2.5.1 Dense Urban Information Society

Dense urban information society is a UC referring to the connectivity requirements of humans living in dense urban areas. This environment can host each of the 5G generic service types as defined in Section 2.2: high data rates of eMBB for both indoor and outdoor users, a massive number of mMTC transmissions (despite the limited area, the 3D distribution of mMTC devices pushes the overall number of communicating machines to the extreme), and the presence of URLLC, e.g., for vehicles. Such combination of services makes this environment critical when considering potential 5G solutions.

Evaluation results for dense urban information society presented in this book, e.g. in Section 15.2, focus on challenges of eMBB communication for human-generated and human-consumed traffic. eMBB users are located both indoors (following a 3D distribution) and outdoors. 5G should be able to provide public cloud services with expected user throughputs of up to 300 and 50 Mbps in DL and UL, respectively. In case of transmissions used by device-centric services, as for instance the communication between user equipments (UEs) or sensors, the required user throughput is in the range of 10 Mbps. Altogether, the 5G network is required to maintain those data rates for 95% of locations and time, for the users that on average generate a traffic volume of 500 GBytes per month. These assumptions lead to the overall traffic volume density of 750 and 125 Gbps/km^2 in the busy hour for DL and UL, respectively. Finally, the network should achieve this performance while taking into account cost and energy consumption. These expenses should be at the similar level as today's expenses for both infrastructure and broadband UE devices.

To efficiently cope with the uneven distribution of the traffic in dense urban environments, radio access sites are deployed in a heterogeneous network (HetNet) configuration. On one hand, an urban macro layer provides wide network coverage and caters for the edge users' experience and for the users on the move. To enable high data rates and relatively wide coverage, macro stations operate at a carrier frequency of, e.g., 3.5 GHz and are deployed every, e.g., 200 m, with antennas above the rooftop level. On the other hand, a small cell base station (BS) layer boosts available capacity over specific areas. To avoid heavy interference, small cell BSs are deployed with the minimum distance of 20 m between each other. They operate at millimetre-wave (mmWave) frequencies around 30 GHz and utilize a total system bandwidth of about 800 MHz. In contrary to macro BSs, antennas of small cells are located below the roof-top level, e.g., on the lamp post. Both cell types are expected to exploit massive antenna arrays.

2.5.2 Smart City

The main idea behind the Smart City concept is to exploit wireless communication of mMTC and IoT devices, to improve the overall quality of urban life, as also discussed in the context of early 5G trials in Section 17.3.1. This improvement can manifest in various ways, e.g., through a more efficient usage of utilities, better health and social care, or even faster public transport. To achieve this effect, low-cost and low energy consumption devices interact with each other or with city dwellers through applications running, e.g., directly in their own smartphones or in the cloud. As the legacy cellular systems were initially developed for broadband applications and the notion of Smart Cities only arose when the standard was already mature, 5G has the chance to provide a native support for this UC to fully address its expectations in a cost-efficient manner.

Although there are numerous applications related to Smart City concepts, out of which some are already implemented while new ones are constantly developed, there are certain challenges related to wireless communication that are common to the majority of appliances, and which 5G should address. Coverage, often characterized by the maximum coupling loss of the radio link, is one of such challenges. It is commonly associated with rural deployments, but the extensive penetration losses related to the attenuation or radio signal while propagating through building walls for indoor devices may be a crucial factor (e.g., for the case of a gas meter located in a basement). Coverage is also directly linked with the availability of a given service in an urban area, which is expected to be at the level of at least 99.9%. Another crucial metric is the energy efficient operation of Smart City units, as these are often located in isolated locations where battery exchange or recharge is difficult. To keep the costs at a low level, at least 10 years of energy efficient radio operations on a single 5 Wh battery should be possible, assuming sporadic data exchange. Low cost is also the driver for reduced complexity, as the Smart City devices are expected to be deployed in large volumes, which is also challenging for the radio network. The latter may in extreme cases for instance need to handle up to 1 million devices per km^2. Especially initial access solutions, as detailed in Section 13.2, are critical to meet aforementioned requirements.

2.5.3 Connected Cars

The connected cars UC facilitates safe and time-efficient journey by enabling URLLC services between the cars and their surrounding, as covered in detail in Chapter 14. The most critical KPIs that quantify the performance of such communication are ultra-high reliability and very low latency for the low payload messages exchanged for safety and efficiency reasons. Additionally, when driving in a car, bus or train, passengers are expecting the availability of remote services, despite the high mobility conditions. Such eMBB service may be used to provide entertainment or connectivity for humans on the move.

The performance assessment of the connected cars UC that is given in this book in Section 15.2 is based on an evaluation of URLLC only. As the safety of the passengers is at stake, an unpreceded level of reliability of transmissions is expected, with a specific target of 99.999%. This reliability is expected for low payload messages (up to 1600 Byte packets) that are exchanged periodically every 100 ms between connected cars.

Different environments are foreseen for testing the performance of the connected cars UC, and each one brings slightly different challenges. In a highway scenario, cars are moving at the speed of 140 km/h, using 3 lines in each direction. Network coverage is provided by rural macro BSs distributed with a distance of 1732 m between each other, and operating at a carrier frequency around 800 MHz with antennas located on high masts. The challenging factor here is the high velocity and related physical phenomena that deteriorate the error rate of the radio transmission. In an urban scenario, cars are moving at the maximum velocity of 60 km/h. However, the density of vehicles in proximity is much higher than in a freeway case. Network coverage is provided by urban macro BSs deployed with an inter-site-distance of 500 m, and with 10 MHz reserved for URLLC services at a carrier frequency around 2 GHz. In both highway and urban scenarios, a carrier frequency of 5.9 GHz is used for the sidelink communication between the vehicles, related to the dedicated Intelligent Transport Systems (ITS) bands that are defined in detail in Section 14.2.3.

2.5.4 Industry Automation

The industry automation UC is URLLC-related and refers to the Factories of the Future, as defined in more detail in Table 2-3. It involves direct device-to-device (D2D) communications between machines as well as access point to machine communications. The focus in this book is on URLLC services within the factory, whose requirements depend on the specific UC and range in terms of latency from 1 to 10 ms, in all cases requiring very high reliability. The traffic pattern also depends on the specific industrial UC, and is typically a mix of periodic and event-triggered traffic. The performance of specific concepts for network slicing is best done against requirements and assumptions of this UC, and hence an industry automation UC is also used as a detailed example for network slicing in Section 8.2.5.

2.5.5 Broadcast/Multicast Communications

In addition to the legacy broadcast services deployed today, e.g. TV, the fully mobile and connected society will need an efficient distribution of information from one source to many destinations [11], see also the video broadcasting scenario in Section 15.2. These services may distribute contents as done today, i.e. typically only using DL, but also provide an UL feedback channel for interactive services or acknowledgement information. Both real-time and non-real-time services are possible. Furthermore, such services are well suited to accommodate the needs of vertical industries. These services are characterized by having a wide distribution, in terms of either geographical distribution and/or a large address space, i.e., many end-users.

2.6 Envisioned Mobile Network Ecosystem Evolution

2.6.1 Current Mobile Network Ecosystem

The value chain of mobile networks is currently specialized into segments that include content-related services and applications, network infrastructure, integration services, access devices, and a multitude of sub-segments and niche applications. Figure 2-3 presents the current value chain of the mobile telecommunications industry. This value chain starts with the hardware providers that manufacture network equipment (i.e., BSs, network controllers, gateways, etc.) and user devices (i.e., mobile phones, smartphones, tablets, dongles). Software providers developing software enablers (middleware and applications), occupy the second position as they allow operating infrastructure and devices. Then come the facility and equipment managers (i.e., tower companies and urban furniture managers) that own assets which are useful for network coverage and capacity extensions. Note that MNOs are also subcontracting some of the network operation and management tasks to equipment vendors or specialized companies. MNOs are then in the middle of this value chain, intermediating between infrastructure players and content and service-related players. Among this latter group, we can cite content providers, over-the-top (OTT) players, especially those that provide telecommunication services (e.g., voice and video conferencing) and service providers that offer wireless services to end clients. Their services include voice calls (e.g., local, regional, national, and international), voice services like voice mail, caller ID, call waiting, call forwarding, and data services like SMS messaging, text alerts, Web browsing, e-mailing, streaming, etc., and mobile TV services. End users occupy the last position in this value chain. Note that this denomination covers a wide spectrum of customers as will be detailed in Section 2.6.3.

2.6.2 Identification of New Players and their Roles in 5G

In its 5G White Paper [5], NGMN describes new business models expected with 5G. New business roles described in this document make reference to asset providers, connectivity providers and partner service providers. In this section, we use the previous section on the identification of the current players and NGMN's input as a basis to identify new players and roles in the 5G field, as developed in [5]. Note that 5G IA produced a White Paper for Mobile World Congress 2017 [12] which also identified new business roles with 5G, and the related analysis converges to a large extent to that of [5].

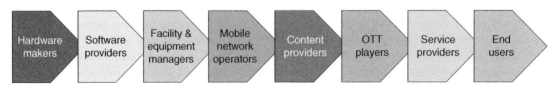

Figure 2-3. Current value chain of the mobile telecommunications industry.

2.6.2.1 Connectivity Providers

The business models associated to connectivity providers can be differentiated between "basic" and "enriched" models and are the following:

- **Basic connectivity providers**: In this model, only best effort IP connectivity is provided. This is the "dumb pipe" model for mobile operators, and we can include Wi-Fi access providers in the same model. In the years to come, we might see new players such as satellite service providers, low-power wide-area (LPWA) players, loons' players, etc. Wi-Fi first players, which use Wi-Fi as the primary connection option and switch to a mobile network only as a "backup solution", could play a bigger role in bundling other access networks as well (e.g., satellite, LPWA, loons). In the energy sector, an example of a basic connectivity provider is an evolution of the mobile virtual network operator (MVNO) concept called a private virtual network operator (PVNO). The long-term needs of energy grids are not fulfilled by existing mobile networks, leading the players of this sector to become MVNOs and to take full or partial control of a wireless network. The PVNO could control elements of the core network such as customer database and SIM cards. In countries like the Netherlands and France, utility companies have deployed cellular networks and were awarded spectrum for their own needs, as for instance in the 450 MHz band in the Netherlands;
- **Enhanced connectivity providers** could increase operator differentiation through network quality and configurability. Public safety players, new MVNOs providing machine-to-machine (M2M) enriched services (for vertical sectors, security purposes, etc.) could appear in this field. The broadcasting sector could also propose a new model called "tower overlay over 5G" (TOo5G), in which the broadcast operator would use its high-tower high-power (HTHP) infrastructure; the latter being already in place and serving for digital terrestrial television (DTT) services. This dedicated broadcasting infrastructure would provide broadcast and multicast services (such as video streaming) with lower transmission costs than in unicast mode, but the viability of this solution will depend on the 5G design choices, for instance related to the integration of digital video broadcasting (DVB) like air interfaces in 5G, and on the development of evolved multimedia broadcast multicast services (MBMS) solutions.

2.6.2.2 Asset Providers

The asset provider role covers both network sharing models and Anything-as-a-Service or Everything-as-a-Service (XaaS) models. With XaaS, everything can be accessed on demand via the cloud. XaaS gives a first sight at what would be the future of cloud services. Users have access to services remotely, whatever the device.

In addition to the Small-Cells-as-a-Service (SCaaS), XaaS asset provider models identified in the NGMN White Paper are Infrastructure-as-a-Service (IaaS), Platform-as-a-Service (PaaS) and Network-as-a-Service (NaaS). They should bring completely new business models in the 5G field.

In the IaaS model, hardware (e.g., servers, routers, etc.) and software elements, maintenance and backup means are managed by a third-party provider. These providers are able to provide dynamic scaling and policy-based services. They charge their customers on a subscription basis and can also take into account the amount of virtual machine space used. In the IaaS model, it is expected that Internet or traditional information technology (IT) companies such as IBM, HP, Google, etc., could become important players with 5G.

As of SCaaS, other parties can also provide it, and vendors are already entering this market. Municipalities or real estate owners can also jump into the business and monetize access to small cells. For example, small cells located in street furniture can be deployed almost anywhere and very close to the user.

In the PaaS model, applications are delivered over the Internet. Hardware and software tools are hosted by the infrastructure provider which provides applications to its customers. Internet and IT companies (e.g., Salesforce, Google, Microsoft, etc.) and telecom players will play a role here.

In the NaaS business model, network services are virtually delivered over the Internet thanks to the virtualization of network functions, as detailed in Section 10.2.2. This can be done on a monthly subscription or on a pay-per-use basis.

Network sharing represents another dimension for asset providers with real-time network sharing. One could for instance envision dynamic network sharing between commercial mobile networks and public safety networks. Capacity would be made available to commercial operators in absence of emergencies. Spectrum brokers could also play a role in the future and manage spectrum resources on behalf of mobile network operators in order to allow for real-time management of the spectrum.

2.6.2.3 New Players in Relation with RAN Evolution

With the expected development of cloud RAN architectures, new players such as content delivery network (CDN) providers or data center players could play a significant role, as listed in the following:

- Data center players could also operate baseband units (BBUs) in a centralized infrastructure, i.e. data centers, in the form of a large concentration of servers and databases. A limit to their possible investment in this field is the limited number of data centers, which are today only present in large cities;
- CDN players could provide services to mobile operators in supporting content hosting closer to the edge of the network. Today, Akamai dominates the market, followed by LimeLight and Jetstream;
- New players could offer both BBU hosting and management and CDN capabilities, and play a role in RAN sharing agreements;
- Relay owners could propose relays to extend coverage of a wireless network or to increase the area spectral efficiency, by means of shortening the radio path distance among end users and access nodes. The actor running and maintaining the relay could be an MNO, an end user that wants to provide enhanced performance in its specific area, or a third party like a restaurant owner interested in providing coverage enhancement based on a specific agreement with an MNO and the usage of the radio resources of the same.

2.6.2.4 Partner Service Providers

Disintermediation of the value chain provides opportunities to create innovative services. With its network, the MNO provides bandwidth to customers and evolves from the former pricing model (i.e., per minute, per volume, per data rate, etc.) to a value pricing model (i.e., involving various QoS metrics, availability, prioritization, latency, etc.).

In the partner service provider model, the MNO offer can be enriched by partners or, the other way round, the partner offer could be enriched by MNOs' capabilities and services.

- **MNO capabilities and offers enriched by partners**: In this model, the mobile operator still provides the service to the end user. As an example, collaboration with OTTs enables MNOs to differentiate their offers. In the coming years, payment solutions, content or integrated streaming solutions could be added by partners. In vertical industries (e.g., related to Factories of the Future), new players could provide data analysis on top of sensing & communications provided by the 5G operator;

- **Partner offers enriched by MNOs' capabilities and services**: In the second model, third party or OTTs are using an MNO's network and have a direct relationship with customers. Products such as smart body analyzer devices or smart wearables could use health monitoring features and connectivity provided by the MNO.

2.6.3 Evolution of the MNO-Centric Value Net

Having identified, in the previous section, the main actors and the interactions between these, the focus of this section is now on the MNOs, as they are expected to play a central role in 5G, as in previous generations. We construct the value net of these MNOs and its evolution with 5G, the aim being to identify their coopetition relations with the other actors.

The value net model has been elaborated by [13]. This model is a complementary approach to the value chain framework, but the analysis is more comprehensive, as the main players have to be integrated in four categories, namely customers, suppliers, competitors and complementors, following a vertical and horizontal dimension. Figure 2-4 gives the current value net of MNOs.

We first begin by defining customers, as this will give us a clear view about the positioning of the MNO. Two kinds of customers are identified: End users, either in the mass market (i.e., individuals), or other business customers (i.e., private companies or public administrations) are contracting with

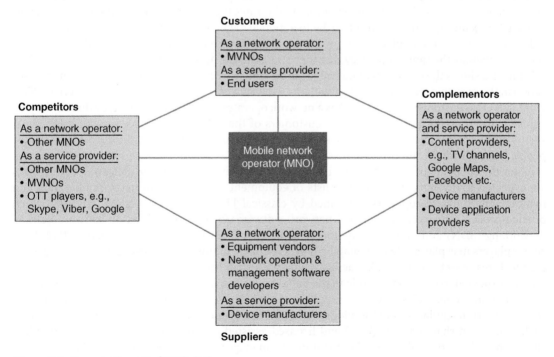

Figure 2-4. Current value net of MNOs [14].

the MNO as a service provider. MVNOs, however, are customers of the MNO, as they buy the right to use its network in order to serve their customers. When the MNO sells network access rights to MVNOs, it is behaving as a network operator. Based on this analysis, we can see that customer groups can be classified into two groups: customers of the MNO as a service provider, and customers of the MNO as a network operator. We will keep this classification for the rest of our analysis of the value net.

Next, we stay in the vertical dimension and identify suppliers. As a network operator, the MNO has as suppliers infrastructure vendors and network operation & management software makers, as identified in Figure 2-4. On the other hand, as a service provider, the MNO has as suppliers device manufacturers, as service providers usually buy devices from manufacturers and sell them at lower prices to end users.

Let us now move to the horizontal dimension and identify competitors. As a service provider, each MNO has as obvious competitors all other service providers, them being MNOs or MVNOs. OTT players, like Skype and Viber, are also seen as competitors of the MNO as a service provider, as they propose substitution services (e.g., voice, video conferences, etc.). As a network operator, the considered MNO has as competitors the other MNOs, as they offer the same services for MVNOs.

The most difficult task is to identify complementors whose presence incites customers to buy more services from the MNO. Obviously, content providers (e.g., online game developers, Google maps, TV channels, etc.) act as complementors, as people are more willing to buy mobile data access in order to benefit from their favorite contents everywhere. Device manufacturers are also complementors, as end users consider smartphones and tablets as valuable devices by themselves, and a smartphone or a tablet will be more useful with a wireless Internet connection. The device application developing industry is also a complementor, as the multitude of smartphone applications incites users to buy a smartphone and to subscribe to a mobile data connection. Note that we do not make a distinction between complementors of the MNO as service provider or as network operator, as they are generally the same in the way that they stimulate the need for network access.

Figure 2-5 shows the evolution of the value net of MNOs with 5G, based on the 5G player identification in the previous sections. We start with the evolution of the group of customers where PVNOs join MVNOs as customers of the MNO as a network operator, and where verticals, by directly buying connectivity to their customers, become customers of the MNO as a service provider. The same verticals become complementors as, by moving towards more connectivity, they provide needs for people (i.e., individuals and professionals) for 5G services.

As for the suppliers of the MNO, the increased heterogeneity and the virtualization of networks are expected to diversify their list. The lists of equipment vendors and of network operation and management software suppliers are joined by classical IT companies like IBM, HP, etc., which provide processing servers and virtual network software, for instance based on software-defined networking (SDN) and network function virtualization (NFV), as detailed in Section 10.2. Data center players may play a role in managing hostels of BBUs in this context, especially for cloud RAN architectures. Asset providers like facility managers, urban furniture managers and tower companies are expected to have a larger role in the deployment and the management of parts of the access network, reinforcing their position as suppliers of the MNO as a network operator. With the evolution of spectrum regulation and the allocation of new bands under innovative authorization schemes such as licensed shared access (LSA) and licensed-assisted access (LAA), detailed in Section 3.2, spectrum brokers could play a role in the future and manage spectrum resources on behalf of mobile network operators in order to allow real-time management of the spectrum. Finally, as a service

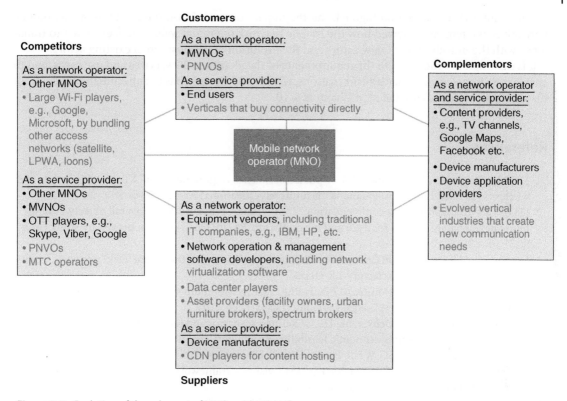

Figure 2-5. Evolution of the value net of MNOs with 5G [14].

provider, the MNO can make deals with CDN players for content hosting near end users at the network edge, making them suppliers with regards to its role as a service provider.

Finally, the advent of new LPWA networks and various access networks based on satellites and loons in addition to the increased integration of Wi-Fi evolutions within the 5G network introduce a variety of new competitors to the MNO in the RAN. A possible scenario, as discussed previously, is the emergence of large Wi-Fi players in the bundling of these various access networks. Regarding the service provider role of the MNO, PVNOs and MTC operators join MVNOs as competitors for offering services to end users.

2.7 Summary and Outlook

In this chapter, an overview of the envisioned main 5G service types eMBB, mMTC and URLLC and related requirements was presented, after which the 5G UCs from two main sources, namely NGMN and 5G PPP Phase 1 projects, were detailed. The synergies and commonalities between these UCs show a large consensus in the community on this topic. It was also shown how these UCs map to vertical needs, allowing 5G to reach a wide range of verticals, especially with mMTC and URLLC services.

The chapter then focused on highlighting the impact of the UCs on the evolution of the mobile network ecosystem, and showed how the mobile network operators' value net is expected to transform, with the introduction of new actors and the evolution of the position of existing actors.

It is clear that in order to fulfil the requirements of the identified 5G services and use cases, and to enable the discussed value net transformations, the 5G architecture has to substantially evolve from that of legacy systems, as detailed in Chapters 5-10.

References

1 FP7 METIS project, Deliverable D1.5, "Updated scenarios, requirements and KPIs for 5G mobile and wireless system with recommendations for future investigations", Apr. 2015

2 ITU-R WP5D, Draft New Recommendation, "IMT Vision - Framework and overall objectives of the future development of IMT for 2020 and beyond", Doc. R12-SG05-C-0199 (approved by Study Group SG5), June 2015

3 International Electrotechnical Commission, White Paper, "Factory of the Future", 2015, see www.iec.ch/whitepaper/pdf/iecWP-futurefactory-LR-en.pdf

4 ITU-R, Draft New Report, IMT-2020.TECH PERF REQ, "Minimum requirements related to technical performance for IMT-2020 radio interface(s)", Feb. 2017

5 5G PPP METIS-II project, Deliverable D1.1, "Refined scenarios and requirements, consolidated use cases, and qualitative techno-economic feasibility assessment", Jan. 2016

6 NGMN Alliance, "NGMN 5G White Paper", Feb. 2015

7 ITU-R, Recommendation ITU-R M.2083-0, "IMT vision – framework and overall objectives of the future development of IMT for 2020 and beyond", Sept. 2015

8 5G Public-Private Partnership (5G PPP), see https://5g-ppp.eu/

9 S. E. El Ayoubi and M. Maternia (editors), "5G PPP paper on use cases and performance models", May 2016

10 FP7 METIS project, Deliverable D6.6, "Final report on the METIS 5G system concept and technology roadmap", May 2015

11 J. Calabuig, J. F. Monserrat, and D. Gomez-Barquero, "5th generation mobile networks: A new opportunity for the convergence of mobile broadband and broadcast services", IEEE Communications Magazine, vol. 53, no. 2, pp. 198–205, Feb. 2015

12 S. E. El Ayoubi (editor), "5G innovations for new business opportunities", 5G IA White Paper for Mobile World Congress 2017, Feb. 2017

13 A. Brandenburger and B. Nalebuff, "Co-opetition", New York, Doubleday, 1996

14 F. Pujol, S. E. El Ayoubi, J. Markendahl and L. Salahaldin, "Mobile telecommunications ecosystem evolutions with 5G", Communications & Strategies, 109, 2016

3

Spectrum Usage and Management

Thomas Rosowski[1], Rauno Ruismaki[2], Luis M. Campoy[3], Giovanna D'Aria[4], Du Ho Kang[5] and Adrian Kliks[6]

[1] *Deutsche Telekom, Germany*
[2] *Nokia, Finland*
[3] *Telefónica, Spain*
[4] *Telecom Italia, Italy*
[5] *Ericsson, Sweden*
[6] *Poznan University of Technology, Poland*

3.1 Introduction

5th generation (5G) networks need to handle mobile data rates in the range from a few kbps up to several Gbps. The requirements w.r.t. the availability of wireless access and link reliability will also increase. Beside mobile broadband services, other utilizations like, e.g., automotive applications, smart grid or smart meter communications, manufacturing systems, and health care by electronic means are going to be incorporated into the 5G design for economies of scale, as pointed out in Chapter 2.

The total amount of spectrum below 6 GHz currently allocated for the mobile service and identified for International Mobile Telecommunications (IMT) in the International Telecommunication Union (ITU) Radio Regulations [1] is 1886 MHz (see Table 3-2). However, in individual countries, only parts of this spectrum are available or planned for mobile communications.

In principle, the capacity of mobile networks can be increased in three ways: a) through additional spectrum bands, b) through cell densification with the deployment of more access points, and c) by using advanced radio technologies obtaining higher spectral efficiency. Since cell densification and higher spectral efficiency alone are not sufficient to cope with the predicted extreme high mobile traffic for specific 5G usage scenarios, a significant amount of additional spectrum for mobile communications needs to be made available, preferably on a globally harmonized basis. For this reason, according to Resolution 238 in [1], a number of frequency bands between 24 GHz and 86 GHz (see Table 3-2) are under study for identification for 5G/IMT2020 at the World Radiocommunication Conference in 2019 (WRC-19).

Depending on the envisaged 5G use cases, spectrum is required in different frequency ranges: below 1 GHz, between 1 GHz and 6 GHz, and above 6 GHz. Consequently, 5G systems will need to be

5G System Design: Architectural and Functional Considerations and Long Term Research, First Edition.
Edited by Patrick Marsch, Ömer Bulakçı, Olav Queseth and Mauro Boldi.
© 2018 John Wiley & Sons Ltd. Published 2018 by John Wiley & Sons Ltd.

able to utilize various operational bandwidths in different deployment scenarios, in any frequency band ranging from below 1 GHz up to 100 GHz. Thus, a main challenge of spectrum management in future 5G networks is the integration of numerous frequency bands within a wide range of spectrum, and with differing spectrum access like, e.g., individual (licensed) or general (shared) authorization.

5G specifications need to support co-existence with legacy mobile technologies and flexible spectrum management to facilitate a smooth transition to 5G. Moreover, the principle of technology neutrality when authorizing spectrum usage provides operators with the flexibility to re-farm their spectrum holdings in order to allow for the evolution from existing to new technologies. For initial 5G deployments, the availability of new spectrum bands is required. This has been initiated in several regions and countries, for example in Europe with the adoption of a 5G Action Plan [2] and the nomination of 5G pioneer bands [3].

Exclusively licensed spectrum on a technology-neutral basis is essential to ensure a high quality of service, a good system performance, and the investment in network infrastructure needed for 5G. Shared spectrum and license-exempt spectrum can play a complementary role to increase capacity and user experience, while simultaneously allowing operators to guarantee a certain Quality of Service (QoS) in licensed spectrum [4].

In the following sections, a number of aspects related to spectrum utilization are considered in more detail. In Section 3.2, spectrum authorization schemes and usage scenarios relevant for 5G are introduced, as well as the spectrum usage requirements for these schemes and scenarios. Spectrum bandwidth demand for 5G is discussed in Section 3.3, by evaluating analysis tools, and elaborating on the impacts of 5G services and deployment scenarios on the spectrum demand estimation. Furthermore, a technology-agnostic approach for spectrum demand estimation is presented. Section 3.4 deals with frequency bands for 5G, depicting bands identified or under study for IMT, but also further potential candidate bands, as well as spectrum roadmaps for the 5G launch. In Section 3.5, spectrum usage aspects at high frequencies are analyzed, including propagation, coverage, deployment and co-existence. Evolutionary paths of dynamic spectrum management are discussed in Section 3.6, concluded by the introduction of a possible functional architecture. Finally, Section 3.7 gives a summary of this chapter and refers to studies above 86 GHz.

3.2 Spectrum Authorization and Usage Scenarios

In this section, spectrum authorization schemes and usage options for 5G are described, based on the findings of the work on spectrum aspects in the 5G Public Private Partnership (5G PPP) project METIS-II [5]. Furthermore, spectrum requirements for different 5G usage scenarios are considered.

3.2.1 Spectrum Authorization and Usage Options for 5G

Generally, the use of radio frequency spectrum can be authorized in two ways, first by "individual authorization" in the form of awarding licenses, and secondly by "general authorization", also referred to as license-exempt or unlicensed. In [6], four different user modes for the operation of 5G radio access systems have been defined, namely the "service dedicated user mode", the "exclusive user mode", the "Licensed Shared Access (LSA) user mode", and the "unlicensed user mode". The relationship between these user modes and the two authorization schemes is illustrated in the upper part of Figure 3-1 named "regulatory framework domain".

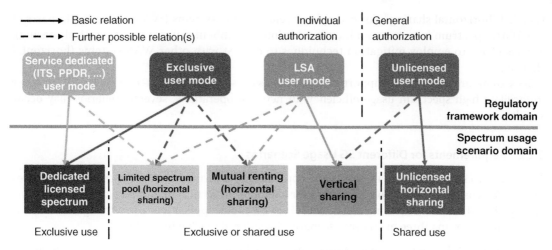

Figure 3-1. Concept for spectrum management and spectrum sharing [6].

Spectrum usage rights awarded by "individual authorization" are exclusive for the license holder at a given location and/or time. The "service dedicated user mode" refers to spectrum designated to services other than public mobile communications, which are indented to be integrated into the 5G ecosystem, for example Intelligent Transport Systems (ITS) or Public Protection and Disaster Relief (PPDR) applications. This spectrum is to be used only for dedicated services and applications. Spectrum designated to public mobile communications falls into the "exclusive user mode". In the "LSA user mode", a non-mobile communications license holder (incumbent) would share spectrum access rights with one (or more) LSA licensee(s), which can use the spectrum under defined conditions subject to an individual agreement and permission by the relevant regulatory authority. These three user modes can occur either in their basic form (see continuous lines in Figure 3-1) or as an evolution of current approaches in the form of "limited spectrum pool" or "mutual renting" (shown through dashed lines in Figure 3-1).

In the "limited spectrum pool" usage scenario, a limited number of known operators obtain authorizations to access a spectrum band dynamically. Mutual agreements between these licensees shall guarantee that in the long-term each participating operator has a predictable minimum value from the shared spectrum. In the "mutual renting" scenario, an operator would rent at least a part of its licensed spectrum resources to another operator, based on mutually agreed rules. Depending on the time period of the spectrum access, the spectrum usage scenarios "limited spectrum pool", "mutual renting" and "vertical sharing" may be considered as exclusive (static) or shared (dynamic) use.

Spectrum access and usage rights granted by general authorization are covered by the "unlicensed user mode", also known as license-exempt usage. This means that such users have no individual license, but the spectrum usage is subject to certain technical restrictions or conditions, for example limited transmission power or mitigation techniques like duty cycle or listen-before-talk. In this user mode, spectrum users cannot claim protection from interference by other users.

In the case when spectrum sharing takes place between systems of different priority, this is referred to as "vertical sharing", whereas spectrum sharing between systems of equal priority is called "horizontal sharing". For example, wireless access systems (WAS) operating in parts of the 5 GHz spectrum have to avoid interference into incumbent radar systems (vertical sharing), and also have to employ mitigation techniques to coexist with other WAS systems (horizontal sharing).

5G systems are expected to support all spectrum usage scenarios indicated in Figure 3-1, in order to facilitate high spectrum usage efficiency. In network operations, several scenarios may occur simultaneously.

3.2.2 Requirements for Different 5G Usage Scenarios

5G is going to support diverse use cases and applications, covering not only the traditional services for mobile subscribers, but also applications for a number of vertical industries like the automotive, energy, eHealth or manufacturing sector. All 5G use cases and applications can be assigned to one or more of the following three main usage scenarios introduced in Chapter 2 [7]:

1) **Enhanced Mobile Broadband (eMBB)**, addressing human-centric use cases for access to multi-media content and data services. This usage scenario embraces a number of use cases and deployment scenarios with quite diverging requirements. For example in a hotspot scenario, extreme high throughputs and low-latency communications are in the foreground, while for wide area coverage the customer Quality of Experience (QoE) with reliable and moderate data rates over the coverage area is in focus.
2) **Massive Machine-Type Communications (mMTC)**, characterized by wireless connectivity of billions of network-enabled devices with prioritization on wide area coverage and deep indoor penetration, typically transmitting non-delay-sensitive data at low rates. Usage scenarios for mMTC are for example smart cities, smart buildings, or sensor networks for farming and agriculture.
3) **Ultra-Reliable and Low Latency Communications (URLLC)**, having stringent requirements on latency and availability. Examples for URLLC are the wireless automation of production facilities, monitoring of critical infrastructures in a smart grid, remote medical surgery, remote robotics, the tactile Internet, or vehicular traffic efficiency and safety.

eMBB applications require a mixture of frequency bands including lower bands for coverage purposes as well as for low to medium data traffic, and higher bands with large contiguous bandwidths to deal with the expected extremely high traffic demand. Exclusive licensed spectrum is essential to guarantee coverage obligations and a minimum QoS for the customers. Spectrum authorized by other licensing regimes, for example LSA or unlicensed access, is a supplementary option to increase the overall spectrum availability.

mMTC applications mainly demand frequency spectrum below 6 GHz, and spectrum below 1 GHz is needed in particular for wide area coverage and reliable outdoor to indoor penetration. Therefore, exclusive licensed spectrum is the preferred option. However, also higher frequency bands and other licensing regimes might be considered, subject to the specific mMTC application requirements.

URLLC applications require high and reliable spectrum availability. Thus, licensed spectrum is considered most appropriate for these kinds of services. For communications for automotive safety and efficiency, see also Chapter 14, the frequency band 5875-5925 MHz harmonized for ITS is an option. Particularly for high-speed vehicles and in rural environments, spectrum below 1 GHz is well suited.

3.3   Spectrum Bandwidth Demand Determination

Radiocommunication networks are deployed over a specific geographical area to provide one or multiple services characterized by the offered QoS, using a chunk of spectrum according to the aforementioned authorization and usage options. Since spectrum is a finite resource, it is of paramount importance to determine the spectrum bandwidth demand for each radio service in order to be able to fulfil the service requirements.

3.3.1   Main Parameters for Spectrum Bandwidth Demand Estimations

The required bandwidth for any specific radiocommunication network, i.e. also the future 5G networks, greatly depends on the following three parameters:

1) **The targeted QoS**. This parameter may differ to a large extent, depending on the service provided by the network. For 5G, the mix of the three main usage scenarios (eMBB, mMTC and URLLC) needs to be considered. One key aspect to be taken into account for QoS is the prediction of traffic patterns for different services.
2) **The area spectral efficiency**. This parameter is expressed in bit/s/Hz/cell and describes how efficiently, in terms of the data rate per bandwidth per cell, the available spectrum is used. Innovations currently under standardization or research will greatly impact the achievable area spectral efficiency in 5G and its foreseeable evolution. The area spectral efficiency can be increased by enhancing the achievable single link spectral efficiency within an individual transmission reception point (TRP), for instance through higher order modulation and coding schemes, massive multiple input-multiple output (MIMO) or interference cancelation technologies, as covered in detail in Chapter 11. Also, the coordination of several TRPs present in the area, for instance in the form of advanced inter-cell interference coordination (ICIC) and coordinated multi-point (CoMP) can increase the area spectral efficiency, as detailed in Chapter 12. These latter approaches can strongly benefit from C-RAN deployments and software-defined networking (SDN), as discussed in Section 6.8.
3) **Physical deployment of network, TRPs and user distributions.** The number of network TRPs in high traffic density areas has been substantially increased in current legacy networks, with the provision of different levels of small cells, in order to maintain the QoS with the same amount of spectrum bandwidth. However, this high density deployment has a clear impact on both capital expenditures (CAPEX) and operational expenditures (OPEX) associated with the provision of the service, which could be unsustainable from an economic point of view. Realistic estimations on user density distributions and TRP deployments are needed in order to perform an adequate evaluation of the required spectrum bandwidth. Parameters like the speed of users or moving TRPs will also have an impact on the final achievable spectral efficiency of the associated links and thus the required bandwidth.

In the following sub-sections, an overview of current approaches and their applicability to 5G is given, and a statistical procedure with a technology and frequency band agnostic approach is introduced.

3.3.2 State of the Art of Spectrum Demand Analysis

Currently, the spectrum requirement analysis for new terrestrial IMT radiocommunication networks such as 5G is evolving from [8]. The parameters for the bandwidth demand analysis introduced in Section 3.3.1 are implemented in this tool in the following manner:

1) The targeted QoS is characterized by the parametrization of the requirements of twenty different service categories. The parameters associated to these service categories are the foreseen traffic models of different services based on market forecasts, defined by: session arrival rate [session/s/user], mean device bit rate [kbps] and mean session duration [s/session].
2) The area spectral efficiency is directly associated with each of the four different TRP layers considered: macro cells, micro cells, pico cells and hot spots, with different values associated to the three considered deployment scenarios (dense urban, sub-urban and rural). Therefore, each TRP is characterized by the area spectral efficiency [bits/s/Hz/cell] value used in the respective radio access technology (RAT), taking into account the estimated traffic per area.
3) The physical deployment scenarios are characterized by the service environment and by the user density [users/km^2]. Depending on the service environment, three main scenarios are considered: dense urban (with three different densities), sub-urban (with two different densities) and rural. Furthermore, each user in a scenario is associated with a probability of being in one of the following three mobility states: stationary (0-5 km/h), low mobility (5-50 km/h), and high mobility (50-250 km/h).

The results of studies on estimated spectrum requirements for terrestrial IMT (pre-5G technologies) in the year 2020, as provided in [8], are shown in Table 3-1.

Spectrum demand estimates for 5G (IMT-2020) are even higher. For example, with a technical performance-based framework for spectrum bandwidth demand analysis, a demand of about 7 GHz of additional spectrum in bands above 6 GHz is estimated for a dense urban information society scenario, and more than 14 GHz for a virtual reality office environment [6].

3.3.3 Spectrum Demand Analysis on Localized Scenarios

When planning mobile networks, mobile network operators (MNOs) aim to achieve a targeted QoS using the limited assigned bandwidth within one or several bands, with minimum deployment and operational costs. Usually, advanced radio frequency propagation tools are used, based on ray tracing techniques and thus being capable of evaluating the achievable signal-to-interference-plus-noise ratio (SINR) once TRP positions, carrier frequency and technology parameters are set.

The SINR values for randomly distributed locations in a TRP coverage area follow a Gaussian curve. Usually, the figure of merit of a new technology is evaluated as the achievable increase of the

Table 3-1. Estimated spectrum requirements for pre-5G technologies in the year 2020 [8].

User Density	Total requirement by 2020 (MHz)	pre-IMT & IMT-2000 (MHz)	IMT-Advanced (MHz)
Low	1340	440	900
High	1960	540	1420

mean value of this Gaussian curve for any specific scenario. This approach is also used in 3GPP benchmarking. For example, in [9], the performance degradation of the SINR distribution for different channel calibration errors in a 32 TRP scenario (assuming a downlink CoMP joint transmission technique) is described as variations in the Gaussian SINR values.

Based on this principle, a method for the spectrum demand analysis in a limited coverage area, with statistical model assumptions for the SINR has been developed [6]. The main parameters for spectrum bandwidth demand estimations are implemented as follows:

1) The targeted QoS is characterized by the achievable throughput for user sessions, taking into account that for each session type various levels of QoS are established, since services like video streaming may be delivered in different qualities, linked to different compressing levels. Moreover, for each type of user equipment (UE), a different probability to be connected with a different session type (data transfer, video streaming, web browsing, etc.) can be applied by a throughput requirement statistical model.

2) The area spectral efficiency is calculated for each scenario. The TRPs deployed in the scenarios (assuming a heterogenous, layered deployment with macro, micro and pico cells) are characterized - for each frequency band considered in the scenario - by the statistical distribution of the achievable SINR in their coverage area. In order for a UE to be schedulable, the UE-TRP link must present a spectral efficiency above the value established as the minimum level. The radio links to the various types of UEs considered in the scenarios (at each carrier frequency) are characterized by: a degradation coefficient (from values achievable applying Shannon theorem), the maximum values of spectral efficiency (depending on the number of antennas involved in the MIMO system), and a minimum value for SINR below which the UE is considered out of coverage.

3) The physical deployment scenarios are characterized by the user density (i.e., users/km^2) and the TRP positions. Three different levels of user densities are considered: general density over the full scenario, high density spots with increased density (campus, office areas, etc.), and ultra-dense hot spot areas. The TRPs are deployed according to these three levels, taking into account a realistic deployment of the macro layer, i.e. including random phase and distance errors from the canonical regular grid used in 3GPP evaluations, leading to realistic coverage areas provided by the Voronoi cells (see Figure 3-2) in accordance with the distance to actual positions of TRPs in the scenario.

Since the achievable QoS may vary greatly depending on the statistical models included in the evaluation tools, a Monte Carlo computational approach is required to get a reliable evaluation of the QoS, achievable with different bandwidths available in different frequency bands.

3.4 Frequency Bands for 5G

3.4.1 Bands Identified for IMT and Under Study in ITU-R

The widespread usage of smartphones and tablets is causing a continuing growth in mobile data communication. According to [10], mobile data traffic will increase with an estimated compound annual growth rate (CAGR) of 47% from 2016 to 2021. In order to cope with such a huge mobile traffic demand, spectrum regulators are working together with involved industries in order to

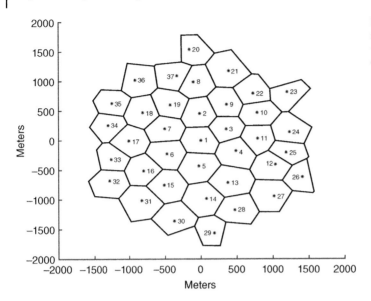

Figure 3-2. Voronoi cells in a 37 cell scenario, with 0.5 km inter-site distance (ISD) and 20% error in phase and distance.

identify bands to make sufficient spectrum available for 5G, focusing on bands that have the potential to be harmonized globally. In consideration of the diverse usage scenarios, technologies and applications enabled by 5G, access to different spectrum bands with different characteristics is required:

- spectrum at lower frequencies to enable coverage of wide areas;
- spectrum at higher frequencies with larger bandwidths to provide necessary capacity and enable higher data rates;
- spectrum at very high frequencies (above 24 GHz) and with very large bandwidths, for providing ultra-high capacity and data rates.

Concerning spectrum above 24 GHz, the World Radiocommunication Conference in 2015 (WRC-15) approved by its Resolution 238 [1] to conduct studies in the respective groups within the ITU Radiocommunication sector (ITU-R), in order to determine spectrum needs and to define sharing and compatibility conditions in the frequency ranges between 24.25 GHz and 86 GHz, as listed in the right column of Table 3-2. Most of these bands already have an allocation to the mobile service on a primary basis, except the bands 31.8-33.4 GHz, 40.5-42.5 GHz and 47-47.2 GHz, which may require such an allocation in addition.

Based on the results of the studies mentioned above, according to agenda item 1.13, the WRC-19 will consider the identification of these frequency bands for the future development of International Mobile Telecommunications (IMT), including possible additional allocations to the mobile service on a primary basis.

In order to conduct the appropriate studies, a Task Group (TG 5/1) was established in ITU-R under the Study Group 5 as being responsible for WRC-19 agenda item 1.13, while the ITU-R Working Party 5D was tasked to conduct and complete the studies with regard to spectrum needs, technical and operational characteristics including protection criteria, and deployment scenarios for

Table 3-2. Bands identified for IMT and under study in ITU-R [1].

Global identifications (in all three regions) for IMT		Regional (in one or two regions) or national identifications for IMT		Under study for IMT-2020 (RESOLUTION 238 in WRC-15)
Band	Bandwidth	Band	Bandwidth	Range
450-470 MHz	20 MHz	470-960 MHz	490 MHz	24.25-27.5 GHz
1427-1452 MHz	25 MHz	1452-1492 MHz	40 MHz	31.8-33.4 GHz
1492-1518 MHz	26 MHz	3300-3400 MHz	100 MHz	37-40.5 GHz
1710-1885 MHz	175 MHz	3600-3700 MHz	100 MHz	40.5-42.5 GHz
1885-2025 MHz	140 MHz	4800-4990 MHz	190 MHz	42.5-43.5 GHz
2110-2200 MHz	90 MHz			45.5-50.2 GHz
2300-2400 MHz	100 MHz			50.4-52.6 GHz
2500-2690 MHz	190 MHz			66-76 GHz
3400-3600 MHz	200 MHz			81-86 GHz
Σ	**966 MHz**	Σ	**920 MHz**	

the terrestrial component of IMT. These studies were completed by March 2017. TG 5/1 is also responsible for the input to the Conference Preparatory Meeting (CPM-19) concerning WRC-19 agenda item 1.13.

For the sharing and compatibility studies to be performed, the protection criteria of the radio services which have already a service allocation in the respective band or adjacent to this band have to be taken into account, and all relevant interference scenarios need to be considered. According to the TG 5/1 work plan, the results of the sharing and compatibility studies will be finalized in 2018.

3.4.2 Further Potential Frequency Bands

Beside the frequency bands above 24 GHz under study for IMT-2020 (see right column of Table 3-2), spectrum below 6 GHz is required to fulfil the requirements of all potential 5G use cases. For this purpose, bands identified for IMT (see Table 3-2), but not yet in usage by cellular mobile systems, offer the best opportunities. Examples are the 600/700 MHz bands considered in US and Europe, and the 3.3-3.8 GHz range from which parts are in focus for 5G in Europe, China, Japan and South Korea. The range 4.4-4.9 GHz, although predominantly not identified for IMT, is under consideration in Japan and China [11].

Spectrum at 28 GHz is one of the main potential candidates for first deployments of 5G above 24 GHz, as the band 27.5-28.35 GHz is put into focus for initial 5G commercialization in the US [12] and in other large markets such as Korea (26.5-29.5 GHz) and Japan (27.5-29.5 GHz)[11]. There is also some interest from the mobile industry in investigating bands in the range of 6-24 GHz [4], although corresponding proposals were not supported by WRC-15.

Applications for which spectrum bands are already harmonized [13] may also be realized with 5G technology, for example traffic safety applications within the band 5875-5925 MHz in support for the automotive sector, or wireless industrial applications within the band 5725-5875 MHz for the factories sector. Further applications with already harmonized frequency bands are for instance Public Protection and Disaster Relief (PPDR) or Programme Making and Special Events (PMSE).

3.4.3   5G Roadmaps

The European Commission signed an agreement with the 5G Infrastructure Association, representing major industry players, to establish the 5G PPP, in order to accelerate research developments in 5G technology, supported by a public funding of around €700 million through the Horizon 2020 Programme, and the same amount from the private side. Furthermore, the telecommunications industry will invest five to ten-times this amount in 5G deployments outside the partnership [14]. The EU industry is set to complement this investment to more than €3 billion. Moreover, the European Commission adopted an Action Plan [2] for a coordinated 5G deployment across all EU member states, targeting early network introduction by 2018, and moving towards commercial large scale introduction by the end of 2020. In order to support this timeline, spectrum should be made available in the 5G pioneer bands [3]: at 700 MHz, within 3.4-3.8 GHz, and at 26 GHz. Corresponding activities at national level have been initiated, e.g., in Germany [15] and in the UK [16].

In the US, the Federal Communications Commission (FCC) adopted a Report and Order [12] with new rules to enable rapid development and deployment of next generation 5G technologies and services in spectrum bands above 24 GHz. These new rules open up nearly 11 GHz for flexible, mobile and fixed wireless broadband, comprising 3.85 GHz of licensed spectrum in the bands 27.5-28.35 GHz, 37-38.6 GHz, 38.6-40 GHz, and 7 GHz of unlicensed spectrum at 64-71 GHz. In addition, a government research initiative has been launched [17], including an $85 million investment in advanced wireless testing platforms by a public-private effort, and an additional $350 million over the next seven years in academic research that can utilize these testing platforms. An auction for spectrum in the 600 MHz band in March 2017 resulted in 70 MHz for licensed use and 14 MHz for wireless microphones and unlicensed use [18], to be freed from television usage by early 2020.

In South Korea, a pilot 5G mobile service is planned for the 2018 Winter Olympics in Pyeongchang in February 2018, and the rollout of 5G commercial services is foreseen for 2019 [19]. In Japan, commercial 5G networks are expected by 2020 in time for the Summer Olympics in Tokyo, or possibly even for the rugby world cup in 2019 [11]. In China, 5G trials are scheduled in two phases, first technology trials from 2015 to 2018, and second product trials from 2018 to 2020, with 5G commercial deployment envisaged by end of 2020 [20]. Note that further details on 5G pilots and early commercial 5G deployments in the different geographic regions are provided in Chapter 17.

3.5   Spectrum Usage Aspects at High Frequencies

In this section, the propagation challenges at high frequencies are discussed first in order to give an understanding of the operational environments. Then, the capabilities of beamforming for compensating the propagation loss at high frequencies are investigated, followed by an analysis of suitable deployment scenarios in various frequency ranges. Finally, the coexistence of 5G with fixed service links and system operations under license-exempt operation at high frequencies is evaluated.

3.5.1 Propagation Challenges

The system coverage of wireless networks becomes worse at higher carrier frequencies due to the increase of propagation loss. While propagation modelling is covered in detail in Chapter 4, the key challenges related to propagation at high carrier frequencies are shortly listed here. Depending on the deployment scenario, different propagation aspects are involved, as for instance captured in [21] and also studied in [22].

The base line propagation loss, as detailed in Section 4.2.1, is derived from the frequency dependent free space loss which is distance-dependent. For high frequencies, depending on the locations of user terminals and base stations (BSs), further propagation components are to be considered in addition, leading to more severe propagation challenges.

In the scenario where user terminals are located indoors and served by outdoor macro BSs, for instance, the building penetration loss, which is typically dependent on the type of material of the outer building walls, is the main challenge, see Section 4.2.3. The incident angle of the main antenna beam to the building entry is another important component which is usually modeled as angular loss. Since macro BSs are usually placed on the rooftop of a building, the main radiation is above or around the edge of the building, causing additional loss known as diffraction loss, see Section 4.3.3. Inside the building, internal wall losses may occur. In addition to these losses related to environment and deployment, also the human body loss may need to be considered, depending on the position of the user device. When user terminals are located outdoors, e.g., in streets, the coverage is not affected by building loss, angular loss, and indoor wall loss. However, due to the placement of macro BSs on rooftops, the coverage is still affected by the diffraction loss and the body loss. In the scenario where indoor users are served by indoor BSs, free space loss, internal wall loss and body loss are affecting the coverage.

3.5.2 Beamforming and 5G Mobile Coverage

With the increase of the carrier frequency, the physical size of one antenna element can be reduced due to decreasing wavelength. Therefore, by keeping the antenna size, the number of antenna elements can be increased, resulting in higher beamforming gain. However, as the propagation loss also increases over the frequency, it is not obvious that the beamforming gain is sufficient for compensating the propagation loss.

In theory, the relation between the power gain G of an antenna, the effective antenna area A, and the wave length λ is as follows [23]:

$$G = \frac{4 * \pi * A}{\lambda^2} \tag{3-1}$$

Since the wave length is inversely proportional to the carrier frequency, the antenna gain in dBi grows over the frequency f according to the following formula when A is fixed:

$$G_{dBi} = K + 20 * \log_{10}(f) \tag{3-2}$$

with $\quad K = 10 * \log_{10} \dfrac{4 * \pi * A}{c^2} \tag{3-3}$

where c is the speed of light

For Equation (3-2), the antenna gain increases logarithmically with the carrier frequency when the antenna size is kept unchanged. However, the key frequency-dependent propagation components in the outdoor to indoor scenarios might increase more rapidly over frequency than the antenna gain. In [21], it is indicated that the building penetration loss in decibel (dB) increases linearly over frequency. For instance, the loss of concrete wall material is given by $5 + 4f$ (dB).

In reality, the antenna gain at high frequencies might be more limited for a number of reasons. First of all, the beamforming gain might not be continuously increasing over frequency due to hardware limitations or antenna design. Particularly, this would apply for hand-held devices. Therefore, there will be an upper limit on the number of antenna elements. For instance, it is currently assumed in 3GPP that each TRP is capable to have up to 256 elements in the 30 GHz range, and up to 1024 elements in the 70 GHz range [24]. However, the number of elements may be limited to 32 elements for the UE antenna. In high mobility scenarios, the use of antenna beams smaller than the angular extent of the intended coverage area requires beam steering and tracking functionalities. Even if channel state information (CSI) is available at the transmitter (Tx) and the receiver (Rx) for optimal beamforming, the optimal beam pointing to a mobile user becomes increasingly challenging due to smaller beam widths in higher frequency ranges. Furthermore, regulatory limits for transmission power and electro-magnetic field (EMF) exposure are to be taken into account.

3.5.3 Analysis of Deployment Scenarios

In this sub-section, a rough coverage analysis for carrier frequencies up to 100 GHz is performed. For the downlink coverage, three different deployment scenarios, with node types and user locations leading to different propagation characteristics, are examined: outdoor to indoor (O2I), outdoor to outdoor (O2O), and indoor to indoor (I2I).

As mentioned in the previous subsection, there is still uncertainty about the hardware capabilities which might evolve over time. Therefore, analyzing the coverage at high frequencies is a challenging task. Especially, the realizable beamforming gain and the power amplifier efficiency are not easily predictable. Thus, a required system gain is defined as the required average transmit power plus the Tx/Rx beamforming gain in order to reach a certain performance target, and the system gain is estimated over frequencies. Although this metric may not provide the coverage feasibility directly, because the realizable transmission power and the Tx/Rx beamforming gains are still needed, it gives good guidance on the coverage sensitivity at high carrier frequencies to achieve a specific target data rate. In principle, one can conclude that the higher the value of the required system gain, the more difficult it is to achieve the intended target.

In Figure 3-3, the system gain is shown as a function of the carrier frequency for different deployment scenarios. In this example analysis, a performance target of 50 Mbps was chosen by assuming a channel bandwidth of 100 MHz and a 2x2 MIMO stream at a given thermal noise power with a noise figure of 5 dB. Two different building types are assumed in the O2I scenario: a modern building type consisting of 70% infrared rejection glass windows and of 30% concrete wall, which is similar to a modern building with coated windows, and a classical building type with 30% of standard windows and 70% of concrete, which represents classical buildings in Western Europe. For the O2I and the O2O non-line-of-sight (NLoS) case, a cell edge user located 100 m away from a macro BS is considered. Users in the I2I scenario are assumed to be 10 m away from an indoor BS.

The results show that the O2I scenario is the most challenging one, and most sensitive with regard to a change of the carrier frequency. This indicates that the spectrum up to 30 GHz appears suitable

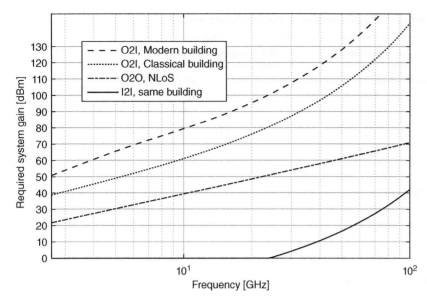

Figure 3-3. Required system gain for different deployment scenarios in relation to frequency ranges [22].

for O2I coverage. In addition, there is a very large variance on the O2I coverage feasibility due to the variety of building types and materials. Compared to the O2I case, the O2O and I2I scenarios have more relaxed requirements on the transmit power and beamforming capability so that carrier frequencies above 30 GHz could be still suitable. However, the beamforming capability is essential to compensate the propagation loss at high frequencies.

3.5.4 Coexistence of 5G Systems and Fixed Service Links

The fixed service is heavily used in some of the 5G candidate frequency bands above 6 GHz. Therefore, the coexistence of rooftop 5G macro-cell systems and fixed links operating in adjacent channels is to be investigated.

In Table 3-3, the aggregate adjacent channel interference at the fixed service receiver is shown [22], in dependence of the dish size of the fixed link receiver and the traffic load level within the 5G system. These results are based on system-level coexistence evaluations in a realistic three-dimensional (3D) dense urban city with a random height of buildings. In the simulations, the macro BSs and the fixed service link stations were placed on rooftop. A 3D ray-tracing-based propagation model was assumed for explicitly modeling diffraction and reflection. In addition, frequency dependent building penetration and wall loss were included. The antenna used in the study assumes UE specific beamforming such that the BS adjusts the beam direction towards a specific UE in a scheduling instance.

The 99th percentile of the aggregate adjacent channel interference in dBm/MHz received at the fixed service receiver is selected as the worst case. The estimated adjacent channel interference level is lower than the typical thermal noise level. In addition, an increase in the dish size of the fixed service station decreases the interference level further due to the higher directivity of the antenna,

Table 3-3. Aggregate adjacent channel interference at the fixed service receiver [22].

Fixed link receiving station dish size (m)	Load level within the 5G system	99% of adjacent channel interference (dBm/MHz)
0.3	Low	−133.7
0.3	High	−119.83
3.5	Low	−140.9
3.5	High	−127.74

allowing for a better selection of the wanted signal. The simulations were performed at a carrier frequency of 15 GHz. At higher frequencies, the potential for interference would be less, as the implementation of smaller beams creates more spatial separation in the 5G system, thus enlarging coexistence feasibilities.

3.5.5 Coexistence under License-exempt Operation

5G is designed to operate in different spectrum usage scenarios (see Section 3.2.1), including the "unlicensed user mode". Listen-before-talk (LBT) is one well-known mitigation technique to enable the coexistence of wireless systems under license-exempt operation [25]. This section investigates the usage of LBT and its impact when high gain beamforming is used at high carrier frequencies.

Figure 3-4 illustrates the downlink system level performance of street micro cells in a dense urban deployment scenario, with and without applying LBT. Two networks, each with four access points (APs) and forty UEs, are considered. The APs are wall mounted and the UEs are

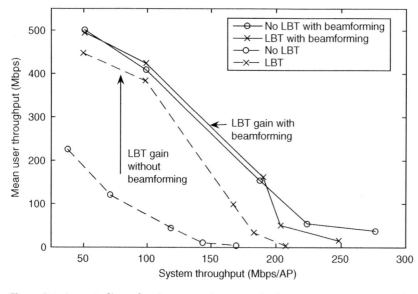

Figure 3-4. Impact of beamforming on coexistence under license-exempt operation [6].

distributed randomly outdoor in the streets. In case that beamforming is applied, it is implemented with 100 antenna elements at the BS.

It can be concluded that in general higher system load results in lower throughput per user, but generates an increase of the overall system throughput which is measured by aggregating the traffic from multiple users in the whole system. If beamforming is not applied, it can be observed that a better performance can be achieved with LBT. However, when beamforming is implemented, the impact of LBT becomes marginal. This implies that there is the potential of using high gain beamforming as a mitigation technique for inter-network spectrum sharing at high frequencies.

3.6   Spectrum Management

In order to cope with the versatile spectrum requirements of 5G, a flexible and effective spectrum management system is required. Additionally, in the advent of software-defined networking (SDN) and network function virtualization (NFV), mobile network operators are looking for new ways for further network optimization. Various levels of sharing such as radio access network (RAN) sharing with spectrum resource virtualization, infrastructure sharing, multi-operator core network (MOCN), multi-operator radio access network (MORAN), see also Section 6.7, entail the need for efficient information exchange and network management. Again, sophisticated solutions for radio resource management will be highly beneficial. In this section, evolutionary paths for the practical implementation of dynamic spectrum management are discussed, concluded by the introduction of a possible functional architecture covering both, the operator and the regulator domain.

3.6.1   Evolutions in Dynamic Spectrum Management

Accurate cooperation between MNOs and national regulatory authorities (NRAs) in terms of effective management of RAN and spectrum resources will be possible only in the presence of a stable, dedicated spectrum management system (SMS) being able to efficiently manage the utilization of spectrum resources available under diverse authorization schemes (e.g., exclusively licensed, LSA, license-exempt, see Section 3.2.1), i.e. to coordinate and control the realization of agreements, and to enforce the execution of decisions made. In consequence, such a SMS has to provide numerous functionalities to different stakeholders. On one hand, it should be fully automated, realized fully in a software manner, allowing for accurate and real-time access to shared resources. In order to achieve this goal, a set of standardized interfaces and protocols have to be defined and incorporated into the overall wireless network architecture. On the other hand, rules and policies on how the spectrum and other resources are managed might be modified in various ways, and these modifications can be implemented into the system either automatically or manually by qualified personnel. For example, new policies on spectrum sharing may be provided by the NRA in form of new regulations, or may be the result of mutual agreements between MNOs cooperating in LSA mode. Naturally, the features of a SMS have to be supplemented by effective access to the storage functionalities. The rules for spectrum management, for example those defining the way how operators share spectrum in LSA mode or how the NRA defines the spectrum access policies, have to be effectively stored and should be easily accessible for authorized users of the SMS. In consequence, the presence of dedicated databases or other forms of big data processing and management solutions have to be implemented as well.

Two key solutions for vertical spectrum sharing already exist: an architecture based on the LSA approach, and a three-tier sharing model controlled by a dedicated Spectrum Access System (SAS). The first one is implemented in Europe by a dedicated standard for the band 2.3-2.4 GHz [26], and the second one is promoted in the USA for the dynamic management in the 3.5 GHz band [27]. Both systems comprise of a central coordination entity equipped with some intelligence for proper decision making, by using dedicated protocols with a set of messages to be exchanged with the stakeholders. Moreover, the central coordination entity is able to query ancillary databases which are used for storing context information about the ambient environment. One may notice that these two solutions are not well advanced in terms of opportunities for its users, in particular as they are limited to certain frequency bands and to certain rules of spectrum sharing. Thus, in order to move a step forward towards effective dynamic spectrum management, the application of radio-environment maps (REM) are proposed in the research community.

REMs, also referred to as geolocation databases (GLDB), are considered as an advanced and dedicated tool for effective storage and management of available rich context information, which may be helpful for efficient flexible spectrum access [28]. The REM approach has been considered as a real and pragmatic solution for the problem of unreliable spectrum sensing algorithms which were developed mainly in the context of cognitive radio. REMs are applied as an entity for guaranteeing reliable access to various types of gathered information, and in consequence, for facilitating flexible spectrum access. Most frequently, the databases are considered in the context of storage of interference maps between two radio systems operating in a certain geographical area. Moreover, these structures can store information about the location of wireless transceivers, their transmit parameters and coverage area. Fast and reliable access to such accurate information will enable better spectrum management and thus lead to more efficient spectrum utilization. In a broader sense, REM databases can contain not only interference or coverage information, but may be used also for keeping typical traffic distribution of the mobile users, dedicated and unique, yet anonymized history maps associated with each user. In such a map the traffic distribution over time and space may be provided. In order to realize sophisticated spectrum management, a dedicated REM management system is needed, equipped with an entity responsible for database queries and management, and a dedicated engine for the analysis of the data on users and inference. Thus, a REM based spectrum management system seems to be highly similar to the generic concept of the SMS considered earlier.

The future SMS has to be discussed also in the context of effective coexistence between cellular mobile and Wireless Local Area Network (WLAN) deployments. Mobile data offloading from cellular to other types of networks is one solution for effective management of high user-data traffic. MNOs and hardware manufacturers are widely considering coexistence and even cooperation between technologies operating in licensed and in license-exempt spectrum, for example the License Assisted Access (LAA) scheme using carrier aggregation to combine LTE in the license-exempt 5 GHz spectrum with LTE in licensed bands. With this aggregation, where control and signaling information is transmitted in the licensed spectrum, higher user data rates can be supported, while the user experience remains seamless and reliable. The convergence of cellular and non-cellular networks is supported by technical developments in both research and standardization communities. In 3GPP-based networks, the presence of the Access Network Discovery and Selection Function (ANDSF) entity, as well as Local IP Access (LIPA) or Selected IP Traffic Offloading (SIPTO), are the exemplary steps toward this direction. In the IEEE community, the introduction of the Wi-Fi Certified Passpoint™ and the Hotspot 2.0 concept, together with advanced databases and the dedicated Access Network Query Protocol (ANQP), pave the way for tighter cooperation between cellular and non-cellular networks. Further information can be found in [29] and [30]. The discussion above leads to the following conclusions: First, there are already technologies and technical solutions available for an effective management of traffic between these two types of networks. Second, there is a strong need for a RAT-independent SMS. The latter should be treated as a solution for the effective organization of various types of spectrum resources and RATs. Again, the application of such a RAT-agnostic SMS requires the presence of a dedicated inference engine, standardized protocols and interfaces for message exchange, as well as accurate databases.

Finally, one may observe that in fact most of the contemporary wireless networks are based on IP-centric solutions. Thus, MNOs may benefit from the virtualization of their resources. Indeed, SDN together with the application of NFV opens the door for a fully software-controlled management and optimization of networks. In such a context, the data of mobile users connected to the network may be managed by means of virtualized entities. One can even claim that if the service-level agreements (SLAs) between end-users and service providers are fulfilled, the way how the data is transported, i.e., over cellular or non-cellular connection, is of secondary importance. As the virtualization techniques are now widely applied to various types of networks, one may again assume that there is a need for an advanced, flexible spectrum management system that will be able – in a wide sense – to coordinate and control the usage of spectrum resources from a common pool. A comprehensive discussion on wireless network virtualization may be found for instance in [31].

3.6.2 Functional Spectrum Management Architecture

Advanced spectrum management systems are already under consideration in different research projects [6], [32]. A possible functional architecture of a holistic SMS is illustrated in Figure 3-5, embracing the key functionalities required from a regulatory (NRA domain) as well as from an operational (MNO domain) point of view.

SMS functionalities within the NRA domain:

- **Regulatory Spectrum Coordination**: From NRA perspective, the management and coordination of spectrum requires the existence of a central coordination point which is permitted to perform any necessary action to guarantee proper execution of NRA rules. This entity has to handle also the communication with the MNO domains.
- **System Management**: As already mentioned, the SMS should be able to incorporate any modifications of the existing (applied) spectrum management policies or to include new ones. These policies may be prepared in human- and/or machine understandable form, and be provided either in an automated way, or implemented by qualified personnel, while the system management is providing functions to perform operation administration, and maintenance tasks.
- **Spectrum User Authorization**: From the NRA perspective it is evident that the SMS system should provide functionality for secure user authorization. Thus, this entity is containing MNOs licensing and registration information.
- **Data storage**: Information of interest from the perspective of NRA (such as spectrum resources, protection and usage rules, etc.) has to be stored in an efficient form. Dedicated repositories and databases shall be available in the SMS, which may create a fragment of the overall REM-based structure.
- **Monitoring functionality** (including reporting and information entry): In order to realize fair and effective management of the spectrum, but also to avoid violation of policies provided by the

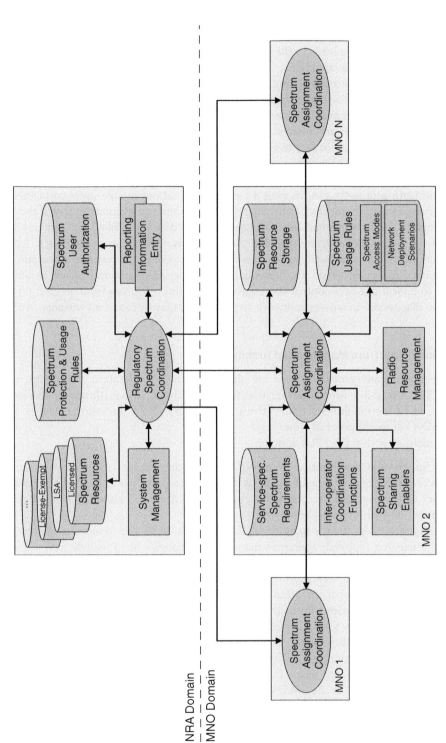

Figure 3-5. Functional architecture of a holistic spectrum management system.

stakeholders, advanced monitoring functionalities have to be available to the NRA. These modules will provide updates to the Regulatory Spectrum Coordination entity, which in turn will provide updates to the databases.

SMS functionalities within the MNO domain:

- **Spectrum Assignment Coordination**: This entity plays a central coordination role. It is responsible for spectrum assignment (radio resource management) by utilizing information from various data bases and spectrum sharing enablers if applicable. It also coordinates the mutual dependencies between operators (defined and stored in the databases), and communicates with the Regulatory Spectrum Coordination entity in the NRA domain.
- **Dedicated interfaces due to ownership issues**: There will be a dedicated client of the SMS system per each MNO due to the fact that each one needs to manage sensitive information (e.g., data related to customers, business etc.). However, in a virtualized scenario, such a spectrum management application (client) needs to communicate with other involved stakeholders to exchange necessary data for proper spectrum management. Thus, both dedicated interfaces between the involved MNOs and between MNOs and the NRA, and advanced inter-operator functionalities have to be incorporated in the system.
- **Policy analysis**: some of the policies may be generic (e.g., those provided by the NRA and defining the rules of thumb for a certain services, frequencies or geographical areas), whereas others may be more specific (such as those defined by the mutual agreements between two MNOs). As the SMS is assumed to be RAT-agnostic (i.e., various RATs may be used simultaneously), there is a risk that these policies may be somehow mutually depended, or even incoherent. In order to avoid any problems related to this aspect, the SMS should be equipped with a dedicated reasoning and inference engine being able to apply advanced inference and machine learning algorithms.
- **Access to Databases**: Information received from the NRA but also from other MNOs, or gathered by dedicated monitoring modules, have to be stored in the databases. These databases may be part of broader REM-based structure. In case of network virtualization, these databases will include for example service-specific requirements or rules related to spectrum usage. Three types of repositories are included in Figure 3-5: Service-Specific Spectrum Requirements, Spectrum Resource Storage, and Spectrum Usage Rules.

It is to be noted that the above split of functionalities does not provide any conclusion on how the SMS is implemented. In particular, although the spectrum coordinator entities are centralized in a logical way, in practice there may be a number of coordination modules, each responsible for a certain geographical area. Moreover, a hierarchical deployment scenario may be considered, where the long-term analysis (such as reasoning for potential policy updates, analysis of the traffic, application of LSA rules etc.) may be implemented in the core network, and short term decisions (such as scheduling, vertical and horizontal handover management, etc.) will be placed in the RAN.

3.7 Summary and Outlook

In 5G networks, data rates from a few kbps up to several Gbps will need to be supported. A significant amount of additional spectrum, preferably globally harmonized, needs to be made available for mobile communications to cope with the predicted extreme high mobile traffic for specific usage scenarios. Therefore, a number of frequency ranges between 24.25 GHz and 86 GHz are studied with regard to coexistence feasibilities between the services currently in operation and future 5G implementations. Furthermore, initiatives have been established in different regions and countries to foster the timely availability of 5G. For instance, the European 5G Action Plan is promoting pan-European multi-stakeholder trials and early deployment in major urban areas and along major transport paths.

For eMBB applications, a mixture of frequency bands is required: lower bands for coverage purposes and for low to medium data traffic, and higher bands with large contiguous bandwidth to deal with the expected extremely high traffic demand. For mMTC applications, mainly frequency spectrum below 6 GHz is demanded, and spectrum below 1 GHz for wide area coverage and reliable outdoor to indoor penetration. For URLLC applications, high and reliable spectrum availability is required, for which licensed spectrum is considered most appropriate. In order to ensure high QoS, good system performance, and incentives for investment in network infrastructure, exclusively licensed spectrum on a technology-neutral basis is essential. Shared spectrum and license-exempt spectrum can play a complementary role to increase capacity and user experience.

The implementation of diverse spectrum usage scenarios, using frequencies from below 1 GHz up to almost 100 GHz with possibly diverging authorization schemes, requires a flexible and effective spectrum management, both in the regulatory domain as well as within the operational system. Based on already established approaches like LSA, sophisticated architectures are under development in different research projects.

The bandwidth required for a radio communications network depends mainly on the following three parameters: (1) the targeted QoS, (2) the area spectral efficiency, and (3) the physical deployment and distribution of transmitters and receivers. Since the achievable values of QoS may vary greatly, a Monte Carlo computational approach is required, in order to obtain a reliable evaluation of the QoS achievable with different bandwidths available at each frequency band.

The move of mobile services to higher frequencies goes along with less favorable propagations conditions, in particular for outdoor-to-indoor coverage. The higher propagation loss with increasing carrier frequency might be compensated to some extent by the implementation of advanced antenna systems like beamforming, but this effect will in practice be restricted by hardware design and electro-magnetic field exposure limits. Beamforming can be considered also as a mean to ease coexistence between radiocommunication services operating in higher frequency bands.

There is a trend towards using even higher frequency bands than currently under consideration in ITU-R. For example, in the Horizon 2020 project TWEETHER [33] a wireless distribution system in the W-band (92-95 GHz) is studied, to be linked to fixed fiber networks and to mobile networks deployed in bands below 6 GHz, in order to have finally a hybrid solution providing seamless connectivity with high capacity and wide coverage. Even at 300 GHz, a range which may have a potential for ultra-fast future wireless short range services, data transmission of 12.5 Gbps could be experimentally demonstrated with a wireless link operating generated and modulated with photonic technologies [34].

References

1 ITU Radio Regulations, Edition of 2016
2 COMMUNICATION FROM THE COMMISSION TO THE EUROPEAN PARLIAMENT, THE COUNCIL, THE EUROPEAN ECONOMIC AND SOCIAL COMMITTEE AND THE COMMITTEE OF THE REGIONS, COM(2016) 588, "5G for Europe: An Action Plan", September 2016

3 RADIO SPECTRUM POLICY GROUP Opinion on spectrum related aspects for next-generation wireless systems (5G), "Strategic Roadmap Towards 5G for Europe", November 2016

4 GSMA, 5G Spectrum, Public Policy Position, November 2016

5 5G PPP METIS-II (Mobile and wireless communications Enablers for Twenty-Twenty Information Society II) project, see https://5g-ppp.eu/metis-ii/

6 5G PPP METIS-II project, Deliverable D3.2, "Enablers to secure sufficient access to adequate spectrum for 5G", June 2017

7 RECOMMENDATION ITU-R M. 2083-0, "IMT Vision – Framework and overall objectives of the future development of IMT for 2020 and beyond", September 2015

8 Report ITU-R M.2290-0, "Future spectrum requirements estimate for terrestrial IMT", December 2013

9 3GPP TSG RAN WG1 Meeting #88bis, Document R1-1705579, "OTA calibration for multi-TRP transmission", April 2017

10 Cisco, White Paper, "Cisco Visual Networking Index: Global Mobile Data Traffic Forecast Update, 2016–2021", February 2017

11 GSA Executive Report from Ericsson, Intel, Huawei, Nokia and Qualcomm, "The case for new 5G spectrum", November 2016

12 Federal Communications Commission, "REPORT AND ORDER AND FURTHER NOTICE OF PROPOSED RULEMAKING", July 2016

13 ERC REPORT 25, "THE EUROPEAN TABLE OF FREQUENCY ALLOCATIONS AND APPLICATIONS IN THE FREQUENCY RANGE 8.3 kHz to 3000 GHz (ECA TABLE)", June 2016

14 European Commission, "5G Infrastructure PPP: The next generation of communication networks will be 'Made in EU'", see: http://ec.europa.eu/research/press/2013/pdf/ppp/5g_factsheet.pdf

15 Bundesnetzagentur, "Points of Orientation for the provision of spectrum for the rollout of digital infrastructures", December 2016

16 Ofcom, "Update on 5G spectrum in the UK", February 2017

17 The White House, "Fact Sheet: Administration Announces an Advanced Wireless Research Initiative, Building on President's Legacy of Forward-Leaning Broadband Policy", July 2016, see: https://obamawhitehouse.archives.gov/the-press-office/2016/07/15/fact-sheet-administration-announces-advanced-wireless-research

18 FCC, "Broadcast Incentive Auction and Post-Auction Transition", May 2017, see: https://www.fcc.gov/about-fcc/fcc-initiatives/incentive-auctions

19 Ministry of Science and ICT, Press Release: "Ministry Highlights 5G During G20 Digital Ministerial Meeting", April 2017, see: http://english.msip.go.kr/english/msipContents/contents.do?mId=Mjc0

20 Ministry of Industry and Information Technology of the People's Republic of China, "5G Progress in China", November 2016, see: https://5g-ppp.eu/wp-content/uploads/2016/11/Opening-1_Qian-Hang.pdf

21 3GPP TR 38.900, "Study on channel model for frequency spectrum above 6 GHz", V14.2.0, December 2016

22 5G PPP METIS-II project, Deliverable D3.1, "5G spectrum scenarios, requirements and technical aspects for bands above 6 GHz", May 2016

23 C. A. Balanis, "Antenna Theory: Analysis and Design", 4th Edition, Wiley, 2016

24 3GPP TR 38.802, "Study on New Radio Access Technology Physical Layer Aspects", V1.1.0, January 2017

25 Draft ETSI EN 302 567, "Multiple-Gigabit/s radio equipment operating in the 60 GHz band; Harmonised Standard covering the essential requirements of article 3.2 of Directive 2014/53/EU", V2.0.22, December 2016

26 ETSI TS 103 235, "Reconfigurable Radio Systems (RRS); System architecture and high level procedures for operation of Licensed Shared Access (LSA) in the 2300 MHz - 2400 MHz band", V1.1.1, October 2015

27 President's Council of Advisors on Science and Technology (PCAST) Report, "Realizing the Full Potential of Government-Held Spectrum to Spur Economic Growth", July 2012

28 V. Atanasovski et al., "Constructing radio environment maps with heterogeneous spectrum sensors", 2011 IEEE International Symposium on Dynamic Spectrum Access Networks (DySPAN), May 2011

29 3GPP TR 36.889, "Study of Licensed-Assisted Access to Unlicensed spectrum", V13.0.0, June 2015

30 Dino Flore, Chairman of 3GPP TSG-RAN, "3GPP & unlicensed spectrum", IEEE 802 Interim Session, Atlanta, USA, January 2015

31 C. Liang and F. R. Yu, "Wireless Network Virtualization: A Survey, Some Research Issues and Challenges", IEEE Communications Surveys & Tutorials, Vol. 17, Issue 1, pp. 358–380, Firstquarter 2015

32 5G PPP COHERENT project, Deliverable D4.1, "Report on enhanced LSA, intra-operator spectrum-sharing and micro-area spectrum sharing", June 2016

33 H2020 TWEETHER project, see: https://tweether.eu/

34 H.-J. Song, K. Ajito, A. Wakatsuki, Y. Muramoto, N. Kukutsu, Y. Kado, T. Nagatsuma, "Terahertz wireless communication link at 300 GHz", IEEE Topical Meeting on Microwave Photonics, October 2010, see: https://www.researchgate.net/publication/224204378_Terahertz_wireless_communication_link_at_300_GHz

4

Channel Modeling

Shangbin Wu[1], Sinh L. H. Nguyen[2] and Raffaele D'Errico[3]

[1] *Samsung R&D Institute, UK*
[2] *Aalto University, Finland*
[3] *CEA-LETI, France*

4.1   Introduction

The emerging 5th generation (5G) cellular systems raise unprecedented requirements on data rates, link reliabilities, end-to-end (E2E) latencies, and the support of a portfolio of different scenarios and service types. To satisfy 5G requirements, cutting-edge technology components are proposed both in academia and in industry. First, targeted frequency bands are expanded from legacy below 6 GHz to above 6 GHz millimeter-wave (mmWave) bands, which are able to provide large frequency resources to boost throughput. Second, recently massive multiple-input multiple-output (MIMO) technology has appealed to the communication community due to its promising capability of greatly improving spectral efficiency, energy efficiency, and robustness of the system by equipping a base station (BS) with a large number of antenna elements (typically tens or even hundreds). Third, massive machine-type communications (mMTC) are emerging to provide scalable, energy-efficient and smart services for sensors, healthcare, and consumer goods. The design and evaluation of these technology components demand an accurate and efficient channel model.

The most widely used channel model family is the well-known 3rd Generation Partnership Project (3GPP) spatial channel model (SCM) and Wireless World Initiative New Radio phase II (WINNER II) family, which follows the geometry-based stochastic channel model (GSCM) approach. As shown in Figure 4-1, the SCM/WINNER II family has been continually evolving for more than a decade. Each evolution of the family introduces further key features, such as time variation modeling in the SCM extension (SCME), support of wider bandwidth and more scenarios in the WINNER II and International Mobile Telecommunications-Advanced (IMT-A) channel models, three dimensional (3D) propagation modeling in the WINNER+ channel model and 3GPP 3D-SCM [1], and early 5G support in the Mobile and Wireless Communications Enablers for the Twenty-twenty Information

5G System Design: Architectural and Functional Considerations and Long Term Research, First Edition.
Edited by Patrick Marsch, Ömer Bulakçı, Olav Queseth and Mauro Boldi.
© 2018 John Wiley & Sons Ltd. Published 2018 by John Wiley & Sons Ltd.

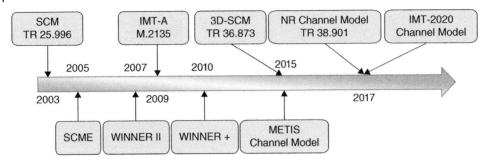

Figure 4-1. Evolution of the SCM/WINNER II channel model family.

Society (METIS) channel model [2]. Among these evolved models, the 3GPP 3D-SCM was most popular, as it was previously utilized in the standard for system evaluation. In the 3GPP 3D-SCM, like other members of the SCM/WINNER II family of channel models, the scattering environment between the BS and user equipment (UE) is abstracted by a number of effective clusters. These clusters are characterized by angular spectrum, delay spectrum, and cluster power. An important feature of 3GPP 3D-SCM is that both azimuth and elevation angles are modeled, capturing the essence of 3D channel properties. Large-scale parameters (LSPs) such as path loss parameters, shadow fading, line-of-sight (LOS) factor, delay and angular spreads are determined stochastically based on measurements. Beside channel generation, the concept of user dropping is also applied, where a drop is defined as one simulation run with constant channel parameters for a certain set of cells or sectors, BSs and UEs over a period of time.

However, 5G requirements in the 3GPP New Radio (NR) standard introduce new challenges to channel models. For example, NR is expected to support mmWave bands, massive MIMO and new use cases, which are not sufficiently supported by existing channel models. As a result, 3GPP has developed the channel model for NR [3] with a supported frequency range from 0.5 to 100 GHz. The support of large antenna arrays and up to 2 GHz bandwidth enables the NR channel model to capture essential characteristics of mmWave channels. Additionally, mmWave propagation aspects such as blockage modeling and atmosphere attenuation are considered. The NR channel model should take various types of links into account to satisfy scenarios ranging from ordinary cellular scenarios to mMTC communications and high mobility scenarios. The modeling of space-time-frequency consistency is also regarded as an important additional feature for multi-user MIMO (MU-MIMO), dual connectivity and beam tracking simulations. In the NR channel model, in addition to the extension of GSCM, a map-based approach is proposed to enable site-specific simulations. For the purpose of evaluating 5G proposals, the International Telecommunication Union (ITU) working group 5D (WP5D) proposed the IMT-2020 channel model [4] which further extended the NR channel model by considering more scenarios and including optional parameter tables and features.

In this chapter, core features of new channel models, such as path loss modeling and fast fading generation, are initially described in Section 4.2. Then, additional features such as the modeling of large antenna arrays, blockage, scattering, etc. are introduced in Section 4.3. Finally, the chapter is summarized in Section 4.4, where also potential future research directions for channel models capable of meeting the requirements of the evolving 5G system are described.

4.2   Core Features of New Channel Models

The core features in this chapter form the basic building blocks of channel coefficient generation. These include the path loss model, LOS probability modeling, outdoor-to-indoor (O2I) penetration loss modeling, and fast fading modeling.

4.2.1   Path Loss

Path loss, the electromagnetic signal power attenuation from the transmitter to the receiver, is a major measure of the propagation channel quality and is important for assessing the wireless system performance. Path loss models in different scenarios for cellular applications have been extensively studied during the past for bands below 6 GHz. In recent years, focus has been shifted to bands from 6-100 GHz for developing 5G mmWave small-cell networks. The general finding for path loss from numerous measurements in mmWave bands is that, due to high penetration and scattering loss, there is usually more excess attenuation (i.e., excluding free space path loss) on non-line-of-sight (NLOS) paths when compared to existing cellular bands. This results in a sharp difference in the path loss exponent between LOS and NLOS mmWave links.

The two classical models for path loss, i.e., the float-intercept (FI)[1] and close-in (CI) models have been used in the literature and in different study groups and projects [3] [5] [6], and have been parameterized for various environments and frequency bands from 0.5 up to 100 GHz. The path loss (in dB) in FI and CI models is a function of link distance d (in meters) and carrier frequency f_c (in GHz), and given as

$$PL^{\mathrm{FI}}\left(f_c,d\right)=10\alpha\log_{10}\left(\frac{d}{1\mathrm{m}}\right)+\beta+10\gamma\log_{10}\left(\frac{f_c}{1\mathrm{GHz}}\right)+\mathrm{N}\left(0,\sigma^{\mathrm{FI}}\right) \tag{4-1}$$

$$PL^{\mathrm{CI}}\left(f_c,d\right)=\mathrm{FSPL}\left(f_c,1\mathrm{m}\right)+10n\log_{10}\left(\frac{d}{1\mathrm{m}}\right)+\mathrm{N}\left(0,\sigma^{\mathrm{CI}}\right) \tag{4-2}$$

where σ^{FI} and σ^{CI} are the standard deviation of path loss (i.e., shadowing fading term) in FI and CI models, respectively. The parameters α, β and γ in Eq. (4-1) are the path loss decay component, FI path loss at $d = 1\,\mathrm{m}$, and the frequency dependency coefficient, respectively. The term γ is dismissed in Eq. (4-1) when a single frequency is modeled. In Eq. (4-2), FSPL(f_c, 1m) is the free-space path loss at $d = 1\,\mathrm{m}$, and n is the path loss exponent. Table 4-1 and Table 4-2 present the curve-fitting omnidirectional path loss model parameters using FI and CI models. These are obtained from numerous measurements and ray-tracing simulations in outdoor and indoor environments, respectively, reported in [5], [6] and [7]. The curve fitting process is done via minimizing the standard deviation of the measured or simulated path loss data using the above model equations, where the model parameters are the variables to be optimized. It is emphasized here that the applicable ranges of antenna height, link distance, and frequency are different in [5], [6] and [7]. In addition, different data fusion methods are used in the models for the scenarios where measured data is collected from multiple campaigns.

---

1 The term "alpha-beta-gamma" (ABG) is also used for the floating-intercept model.

Table 4-1. Parameters of path loss models for different outdoor scenarios.

Model scenario and frequency range		Model type	n or α	β	γ	σ
UMi Street Canyon in [7] Frequency range: 2-73.5 GHz	LOS	FI	2	31.4	2.1	2.9
		CI	2	–	–	2.9
	NLOS	FI	3.5	21.4	1.9	8.0
		CI	3.1	–	–	8.1
UMi Street Canyon in [5] Frequency range: 2-86 GHz	LOS	FI	1.92	32.9	2.08	2.0
	NLOS	FI	4.5	31	2.0	7.82
UMa in [5] Frequency range: 2-73.5 GHz	LOS	FI	2.8	11.4	2.3	4.1
	NLOS	CI	2.7	–	–	10

Table 4-2. Parameters of path loss models for different indoor scenarios.

Model scenario and frequency range		Model type	n or α	β	γ	σ
Indoor in [5][2] Frequency range: 2-86 GHz	LOS	FI	1.38	33.6	2.03	1.18
	NLOS	FI	3.69	15.2	2.68	8.03
InH Indoor Office in [6] Frequency range: 2-73 GHz	LOS	CI	1.73	–	–	3.02
InH Shopping Mall in [6] Frequency range: 2-73 GHz	LOS	CI	1.73	–	–	2.01

In the standardized 3GPP NR channel model [3], additional refinements are considered for urban micro (UMi) Street Canyon, indoor hotspot (InH), and urban macro (UMa) scenarios to incorporate 3D link distance, effective antenna heights, and LOS breakpoint into the path loss. Here, the same model parameters are used for the whole range of applicable frequencies, i.e. from 0.5 to 100 GHz (see Table 7.4.1-1 in [3]). ITU-R Recommendation P.1411-8 models path loss differently for ultra-high frequency (UHF), super-high frequency (SHF) up to 15 GHz, and mmWave frequency ranges, while additional losses caused by atmospheric gases and rain, and directive antennas are considered for the latter [8]. Specifically, using the CI model, the directional path loss when both transmit and receive antennas are pointing to each other in LOS condition is given by

$$PL(d) = PL_0 + 10n\log_{10}\left(\frac{d}{d_0}\right) + L_{\text{gas}} + L_{\text{rain}} \qquad (4\text{-}3)$$

---

2 Indoor scenario in [5] is a merged scenario including office, lecture room, and airport.

where PL_0, L_{gas}, L_{rain} are the path loss at the reference distance d_0, and attenuations caused by the atmospheric gases and by rain, respectively. The model is parameterized for the urban LOS scenario with different building types (high-rise and low-rise) and for transmit (receiver) antenna half-power beam widths of 30 (10) degrees in the 28 GHz band, or 15.4 (15.4) degrees for 60 GHz bands (see Table 6 in [8]).

The above FI and CI have the same set of optimized parameters including the path loss exponent and shadow fading standard deviation for the whole range of applicable frequencies. Recent works have also investigated the frequency dependency of the path loss exponent in the CI model [6], or all of the parameters in [9]. Specifically, in [9], each of $(\alpha, \beta, \gamma, \sigma)$ in the FI model or (n, σ) in the CI model are functions of carrier frequency, and given as

$$\chi = a - b \log_{10}\left(\frac{f_c}{1\text{GHz}}\right) \tag{4-4}$$

where χ represents a parameter of interest, and a and b are model parameters. The frequency-dependency of these path loss model parameters in the range of 0.8 to 60.4 GHz was revealed in [9] for the Street Canyon scenario for both LOS and NLOS conditions. Using the new frequency-dependent model parameter set, it was shown in [9] that both FI and CI models outperform the ITU-R M.2135 UMi models in term of path loss prediction accuracy.

4.2.2 LOS Probability

As different path loss model parameters are applied for LOS and NLOS links, it is important to identify whether a link is in LOS or NLOS condition. There are several different analytical models to characterize LOS probability, including the random shape theory model [10], the LOS ball model [11], and the standardized 3GPP and ITU-R models [3], [8]. In the ITU recommendation, the LOS probability for UMi and InH are calculated based on the link distance d as

$$P_{\text{LOS}}(d) = \min\left(\frac{d_1}{d}, 1\right)\left(1 - e^{-\frac{d}{\alpha}}\right) + e^{-\frac{d}{\alpha}} \tag{4-5}$$

and

$$P_{\text{LOS}}(d) = \begin{cases} 1 & \text{if} & d \leq d_1 \\ e^{-\frac{d-d_1}{\alpha}} & \text{if} & d_1 < d < d_2 \\ P_0 & \text{if} & d \geq d_2 \end{cases} \tag{4-6}$$

respectively, where d_1 is the distance up to which LOS is guaranteed, α is the decaying parameter, and d_2 is the distance at which $P_{\text{LOS}} = P_0$. In 3GPP, a similar exponential model is used, but the parameters are different for different environments. They include UMa Street Canyon, UMi Street Canyon, rural macrocell (RMa) and indoor office, where antenna heights are assumed to be at 25 m, 10 m and 3 m, respectively (see Table 7.4.2-1 in [3]).

In [12], a linear model and a more generic exponential model are considered, where the LOS probability for a link distance d is respectively given by

$$P_{\text{LOS}}(d) = \begin{cases} 1 & \text{if} \quad d \le d_1 \\ \dfrac{(d_2 - d)(1 - P_0)}{d_2 - d_1} + P_0 & \text{if} \quad d_1 < d < d_2 \\ P_0 & \text{if} \quad d \ge d_2 \end{cases} \tag{4-7}$$

and

$$P_{\text{LOS}}(d) = (1 - P_\infty) \exp\left(-\left(\frac{d}{\alpha}\right)^N\right) + P_\infty \tag{4-8}$$

where $P_\infty = P_{\text{LOS}} |_{d \to \infty}$ and N is the exponent. The models were parameterized for open square, shopping mall and indoor office scenarios using accurate point cloud models, and were demonstrated to have a very good agreement with available measurements.

Except for the results in [12], the curve-fit parameters in the above models are mainly obtained from measurements in existing cellular bands and independent of the carrier frequency. It is essential to investigate the environment-specific LOS probability in higher frequency bands, as the clearance of blockages from the first Fresnel zone of a link is smaller. Specifically, a link is obstructed-LOS or NLOS if there is at least one obstacle, having the total link distance to the BS and UE satisfying [12]

$$d_{\text{BS}-p} + d_{\text{UE}-p} < d_{\text{BS-UE}} + \frac{\lambda}{2} \tag{4-9}$$

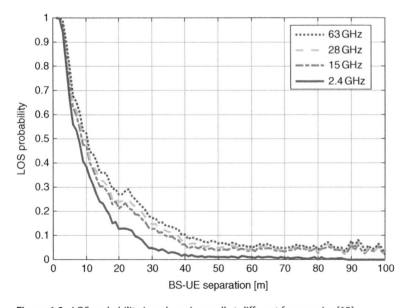

Figure 4-2. LOS probability in a shopping mall at different frequencies [12].

where $d_{\text{BS}-p}$, $d_{\text{UE}-p}$ and $d_{\text{BS}-\text{UE}}$ are the 3D distances between the BS to the obstacle, between the UE to the obstacle, and between the BS and the UE, respectively, and λ is the wavelength of the carrier frequency. It was revealed in [12] that while significantly different LOS probabilities are applied for below and above 6 GHz, the same model parameters can be used for mmWave bands, as only small variations in probability values were found in those bands.

4.2.3 O2I Penetration Loss

For O2I link analysis and coverage evaluation, the signal attenuation through building walls and objects inside the building needs to be added into the basic path loss, producing total path loss. The O2I path loss can in general be expressed as [8]

$$PL_{\text{O2I}} = PL_{\text{bpl}} + PL_{\text{bel}} + PL_{\text{in}} + N\left(0, \sigma_{\text{dB}}^2\right) \tag{4-10}$$

where PL_{bpl} is the basic path loss (formulated and modeled as in Section 4.2.1), PL_{bel} is the building entry loss through external walls, PL_{in} is the indoor path loss dependent on the depth into the building, and σ_{dB} is the standard deviation of the O2I path loss.

The building entry loss level is highly dependent on building materials and structures, incidence angle, and varies across frequencies. For a rigorous characterization, the penetration loss through a building wall or window can be calculated using the Fresnel transmission coefficients [13], given that the wall or window material and structure are accurately modeled. Using the model parameters for electrical properties of the materials given by the ITU recommendation (see Table 3 in [13]), the calculated penetration losses with a perpendicular polarization of a single-layer concrete and glass wall are shown in Figure 4-3. As can be seen from the subfigures, the loss is material-dependent, and increases with the slab width, incidence angle[3] (left subfigure) and frequency (right subfigure).

The curve-fitting model parameters for the electrical properties of popular materials in the frequency range of 1-100 GHz provided in [13] can be used for calculating site-specific building penetration loss. In reality, however, the calculation is often just an approximation, since the building material structure is much more complex, as it is not just single or homogeneous, but normally a combination of different materials. Therefore, an accurate profile of the building walls and windows is not always available, as it becomes hard for such a calculation to predict the overall penetration loss, especially for such a large environment. It is observed by measurements that compound material can have very different electrical properties, and hence cause a different transmission loss as compared to homogeneous material. For example, coated-glass windows in modern office buildings cause a signal attenuation as strong as 30 dB at mmWave frequencies, much higher than the 10 dB signal attenuation of the non-coated glass windows exhibited at the same frequencies [5].

---

3 This trend does not apply for an incidence angle close to the pseudo-Brewster angle due to the effect of a minimum reflection coefficient for the perpendicular polarization [13].

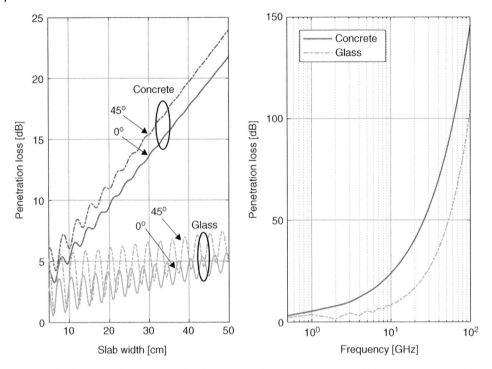

Figure 4-3. (Left) Calculated penetration loss at 2 GHz for transverse-magnetic (TM) polarization with incident angles of 0° (solid) and 45° (dashed); (Right) Calculated penetration loss for TM polarization for concrete of 15 cm width, and glass slabs with incident angle of 0°.

The experiment-based 3GPP building entry loss model takes into account the penetration losses through different building materials and the loss from the non-perpendicular incidence angle loss formulated as

$$PL_{bel} = PL_{npi} - 10\log_{10} \sum_{n=1}^{N} p_n 10^{-\frac{L_{m_n}}{10}}$$ (4-11)

where PL_{npi} is the non-perpendicular incidence loss, p_n and L_{m_n} are the proportion of and the loss through the n-th material, respectively, and N is the total number of materials. A simple linear model for L_{m_n} taking into account the frequency dependency based on curve-fitting to the measurements is provided in [1] for popular materials including concrete, wood and glass.

The 3GPP total O2I loss model in Eq. (4-11) was parameterized for UMi and UMa with different types of high-loss and low-loss buildings, for a frequency range from 0.5 up to 100 GHz [3]. It should be noted that the 3D-SCM O2I penetration loss and the NR O2I penetration loss are modeled under different modeling frameworks. The former is generated in a *link-specific* manner, i.e., each link between a BS and UE has an individual O2I penetration loss value. Conversely, the latter is *UE-specific*, i.e., each UE has an O2I penetration loss value which is shared by all links between BSs and the UE. In [5], the O2I loss model is basically the same as the corresponding model agreed in

3GPP with the additional refinements, including a lognormal frequency-dependent spread and elevation angle dependence. The model was parameterized for the UMi scenario for the frequency range of 10-60 GHz [5].

4.2.4 Fast Fading Generation

When considering a channel with N clusters, the MIMO channel impulse response between the uth receive antenna and the sth transmit antenna $h_{u,s}(t, \tau)$ at time t and delay τ can be expressed as

$$h_{u,s}(t,\tau) = \sum_{n=1}^{N} h_{u,s,n}(t)\delta(\tau - \tau_n) \qquad (4\text{-}12)$$

where $h_{u,s,n}$ is the complex coefficient between the uth receive antenna and the sth transmit antenna of the nth cluster, τ_n is the delay of the nth cluster, and $\delta(\cdot)$ is the Dirac function defined by $\delta(x)$ which equals 1 if $x = 0$ and equals 0 otherwise. The complex coefficient $h_{u,s,n}$ of the nth cluster is obtained by summing the contributions of M subpaths as

$$h_{u,s,n}(t) = \sum_{m=1}^{M} \sqrt{P_{n,m}} \mathbf{F}_{RX,u,n,m}^{T} \cdot \mathbf{\Gamma}_{n,m} \cdot \mathbf{F}_{TX,s,n,m} \cdot e^{j\varphi_{n,m}} \qquad (4\text{-}13)$$

where $P_{n,m}$ is the power, $\mathbf{F}_{RX,u,n,m}$ and $\mathbf{F}_{TX,s,n,m}$ are the antenna patterns of the receive and transmit antenna elements, $\mathbf{\Gamma}_{n,m}$ is the 2×2 polarization matrix, and $\varphi_{n,m}$ is the aggregate phase term including the phase difference due to antenna positions and the Doppler phase shift.

Prior to generating these channel coefficients, the small-scale parameters need to be computed. Cluster delays are randomly generated according to the power delay profile (PDP). Then, cluster delays are mapped to compute cluster powers. The strongest two clusters are divided into sub-clusters to support up to 100 MHz bandwidth. Next, arrival angles and departure angles are calculated based on the angular spectrum and cluster powers. With these angles, antenna patterns of the receive and transmit antenna elements can then be calculated. After generating the polarization matrix and the aggregate phase term, the final channel coefficients can be computed according to Eq. (4-12).

4.3 Additional Features of New Channel Models

To fully evaluate 5G systems, a set of additional features can be included during the channel generation procedure. In this subsection, these additional features as well as their impacts on the 5G system design will be introduced.

4.3.1 Large Bandwidths and Large Antenna Arrays

One benefit of deploying 5G networks in mmWave bands is the large bandwidth these can provide, as already discussed in Section 3.4. Meanwhile, mmWave bands also bring severe propagation loss which needs the massive array gain of large antenna arrays to compensate. Hence, support of large bandwidths and large antenna arrays in the 5G channel model is required. The key impact of large bandwidths and large antenna arrays is that they introduce finer resolutions in both the temporal and spatial domains, resulting in more resolvable subpaths in each cluster. Although intra-cluster subpaths are already supported in legacy channel models, only the strongest two clusters are so far considered to have subpaths, and an equal power distribution is assumed across these subpaths. Moreover, legacy channel models consider up to 400 paths in a channel impulse response, and the maximum number of paths is irrelevant to bandwidth or antenna array size. However, measurement campaigns in [5] in mmWave bands have shown that in order to capture 95% of the channel power, the number of paths to consider is significantly larger than those in below 6 GHz bands, when large bandwidths and large antenna arrays are applied. This is because wavelengths in mmWave are smaller, and diffusive reflected paths are more significant than in below 6 GHz bands. The increase in resolvable paths implies a richer scattering environment in large bandwidth and large antenna array scenarios.

In the 3GPP NR channel model [3], the modeling of intra-cluster delay and angular spreads was enhanced to better support large bandwidths and large antenna arrays. Unlike legacy channel models, where the number of subpaths per cluster is constant, the number of subpaths per cluster M in the NR channel model is proportional to the bandwidth, size of antenna aperture, and carrier frequency. Subpath-specific power, delay, and angles are generated according to distributions obtained via measurements. Another fundamental difference to the 3GPP 3D-SCM [1] is that intra-cluster delay and angle are independently generated. Both delays and intra-cluster angular offsets are generated by uniformly distributed random variables (r.v.s). Subpath powers are obtained via Monte Carlo sampling in the temporal-spatial spectrum. After all necessary small-scale parameters are generated, the overall channel impulse response with large bandwidth and antenna array support can be expressed as the sum of the subpaths

$$h_{u,s}\left(t,\tau\right)=\sum_{n=1}^{N}\sum_{m=1}^{M}h_{u,s,n,m}\left(t\right)\delta\left(\tau-\tau_{n,m}\right) \tag{4-14}$$

where $h_{u,s,n,m}$ is the complex coefficient between the uth receive antenna and the sth transmit antenna of the mth subpath within the nth cluster, given as

$$h_{u,s,n,m}\left(t\right)=\sqrt{P_{n,m}}\mathbf{F}_{\mathrm{RX},u,n,m}^{\mathrm{T}}\cdot\mathbf{\Gamma}_{n,m}\cdot\mathbf{F}_{\mathrm{TX},s,n,m}\cdot e^{j\varphi_{n,m}}. \tag{4-15}$$

It should be noted that intra-cluster characteristics are highly dependent on the clustering algorithm used during post-processing. Typical clustering algorithms such as the Kmeans++, the agglomerative algorithm and the time cluster and spatial lobe clustering algorithm approach were introduced in [6].

Due to spatial and temporal richness and intra-cluster characteristics in mmWave channels, the number of radio frequency (RF) chains in massive MIMO systems with hybrid beamforming should be carefully chosen to reach a balance between implementation complexity and utilization of the richness of the mmWave channel, as elaborated further in Section 11.5. Also, frame structures of different numerologies should be adaptive, i.e., the length of (shortened) transmission time intervals (TTIs) and subcarrier spacings should be designed according to the richness of the channel, as detailed in Section 11.6.

4.3.2   Spatial Consistency

The modeling of spatial consistency is an important feature in the 5G channel model. Spatial consistency describes how the channel evolves as a function of the spatial location in a continuous manner, which is important for beam tracking system design for mmWave. When two users are located near each other, their channels should experience similar characteristics, such as similar angles of arrivals and delay spreads. In legacy channel models, channel characteristics of these two users are generated independently resulting in an overestimated MU-MIMO gain. In the 5G channel model, this defect is solved by introducing correlations in the channel characteristics of these two users. The correlation of two users is calculated as

$$R(\Delta x) = \exp\left(-\frac{|\Delta x|}{d_{\text{corr}}}\right) \tag{4-16}$$

where Δx is the distance between the two users in the horizontal plane and d_{corr} is the environment-dependent correlation distance. The values of d_{corr} for parameters such as cluster-specific parameters, LOS states, and indoor states are different. Typical values of d_{corr} are between 10 and 60 meters. There are two spatial consistency modeling methods in the NR channel model: one is the spatially correlated random variable based method, and the other is the geometric stochastic approach.

In the spatially correlated random variable based method, the generation of small-scale parameters is based on two-dimensional (2D) spatially correlated random variables. As a result, clusters of co-located UEs are expected to have similar delays and angle information. In the NR channel generation procedure, random variables are either Gaussian distributed or uniformly distributed. Although the method of generating these random variables is not specified in the NR channel model, a computationally efficient method is to generate spatially correlated Gaussian random variables via the Gaussian random field method, and then map them to [0, 1] using the cumulative distribution function (CDF) of the Gaussian distribution to obtain spatially correlated and uniform random variables.

Consider a $n \times n$ matrix representing the discrete points in the area of interest, and let $\mathbf{R}$ be the correlation matrix where $[\mathbf{R}]_{ij} = \exp(-|\Delta x_{ij}| / d_{\text{corr}}) \forall i, j \leq n$ is the correlation value between the point (i, j) and point $(1, 1)$ in the area of interest. Then, compute its 2D fast Fourier transform (FFT) by $\mathbf{G} = \text{FFT2}(\mathbf{R})$, where FFT2 is the 2D FFT operator. Next, generate a zero-mean unit variance Gaussian matrix $\mathbf{V}_{\text{iid}}$ with independently and identically distributed (i.i.d.) entries. The matrix $\mathbf{V}$ storing correlated Gaussian random variables can then be calculated by

$$V = \text{real}\left\{\text{FFT2}\left\{\text{sqrt}(\mathbf{G})\right\} \circ \frac{\mathbf{H}_{\text{iid}}}{n}\right\} \tag{4-17}$$

where $\text{real}\{\cdot\}$ obtains the real part of a matrix, $\text{sqrt}\{\cdot\}$ computes the element-wise square root of a matrix, and $\circ$ is the Hadamard product which computes the element-wise product of two matrices. In this case, any two entries in $\mathbf{V}$ will have the target correlation value. According to the property of a Gaussian copula, spatially correlated uniform random variables with the same correlation matrix $\mathbf{R}$ can be obtained by mapping the spatially correlated Gaussian random variables to [0, 1] via the CDF of a standard Gaussian distribution.

The spatially correlated random variable based method provides a more accurate solution to spatial consistency modeling and is already used in LSP generation. This is extended to small-scale channel parameters. However, the spatially correlated channel parameters will be stored with respect to different positions and require large storage space.

The geometric stochastic approach assumes that the last bounce of a cluster is known, and updates the cluster delays, cluster power, and cluster angles according to the relative motion of the UE. This approach is similar to the one introduced in the SCME, which may lead to less accurate descriptions of 2D spatial consistency. For example, assume that two users are far apart at the beginning of the simulation. When these two users now move into proximity, the geometric stochastic approach may not necessarily guarantee that the correlations of their channel parameters match the target correlation values. However, a benefit of the geometric stochastic approach is that it updates small-scale channel parameters during channel generation, without requiring any extra storage space such as the spatially correlated random variable based method.

Beside the two aforementioned methods used in 3GPP, the COST 2100 channel model [14] introduced a concept known as *visibility region* to model spatial consistency. A visibility region is a circular area with a smooth power transition boundary and is randomly placed in the area of interest. Each cluster is associated to a visibility region. A UE can see this cluster if it is within its visibility region. When the UE is moving, it can transit from one visibility region to another smoothly, such that spatial consistency is maintained. The COST 2100 channel model later generalized this idea to both the transmitter and receiver sides to support dual mobility and even massive MIMO.

Better spatial consistency support in the 5G channel model allows more accurate performance evaluations of MU-MIMO and coordinated multi-point (CoMP) systems. Also, the design of beam tracking algorithms would need to estimate the variation of the channel in a continuous manner.

4.3.3 Blockage

Geometry-induced blockage is a static condition, caused by obstructing objects in the map environment. In this situation, the propagation is dominated by diffraction and diffuse scattering in those locations that are blocked. This results in additional loss, but also in different channel statistics as compared to those based on shadow fading.

The knife edge diffraction (KED) theory provides a relatively simple and robust mean for calculating the diffracted fields at surface edges. Blocking studies in literature have accounted for blockage between a transmitter and a receiver by means of the KED approach. Each blocking object is approximated by a rectangular screen as shown in Figure 4-4. The total blockage attenuation will be

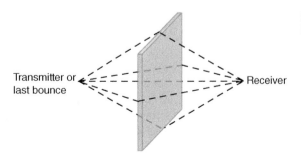

Figure 4-4. Schematic of knife edge diffraction blockage model [16].

Transmitter or last bounce

Receiver

calculated using the four links from the four edges of the screen and the link between the transmitter or the last bounce and the receiver. To model more complex shapes, a combination of multiple rectangular screens may be used. In particular, this approach has been adopted by [3]. However, a change of receiver and transmitter location will imply that as a path changes, the angle of the screen will also change. In order to avoid the shadowing variation, each screen can be continuously rotated so that it is perpendicularly oriented with respect to each path.

In [5], a substantially improved blockage model has been developed based on the KED blockage model. This model accounts for the fast fading through the summation of the complex amplitudes of the paths from the four edges of a rectangular screen. This approach has been validated with 5G radio access measurements considering blocking by a truck [15]. It was shown that summing up the complex amplitudes results in a better agreement with measurement data before and during the blocking, and avoids an underestimation of attenuation in deep shadowing zones. Also, this model has been validated towards measurements of human body shadowing.

A blockage object can significantly degrade signal strength in a certain angular area. As such, wider beams may perform better than narrow beams in case of blockage, since wider beams can capture or propagate signals in more directions. This is counter-intuitive, as it is normally encouraged to use narrow beams in 5G systems. Also, fast beam finding and tracking methods are essential in the overall system design, such that the system can flexibly sense strong reflected paths and switch beamforming directions to these paths, as elaborated further in Section 13.2.5.

4.3.4 Correlation Modeling for Multi-Frequency Simulations

In 5G NR, UEs may receive data simultaneously from different BSs operating at different frequency bands, as detailed in Section 6.5. One typical example is to receive control signaling at below 6 GHz channels while boosting data throughput through the usage of mmWave bands. These two links at different frequencies may not be fully independent as they may share similar LOS states, blockage objects, and even some LSPs. Therefore, correlation modeling for multi-frequency simulations is needed in the channel model.

In the NR channel model, a procedure is added to handle this. First, after user dropping, LOS angles for all frequencies are determined identically based on geometry. Second, if soft LOS states are considered, they are frequency-dependent. Otherwise, LOS states of all frequencies are the same. Third, LSPs are the same for all the frequencies, except for possibly frequency-dependent scaling, according to the LSP table. For example, in the 3GPP NR channel model, first- and second-order statistics of the delay and angular spreads are modeled as linear functions of the frequency (in logarithmic scale). Depending on the parameter and the type of the environment (outdoor or indoor), the frequency-dependency in the 3GPP NR channel model is strong for some LSPs, while it is rather weak or not modeled at all for others (i.e., the mean and variance values of LSPs are constant over the considered frequency range, from 0.5 to 100 GHz). The frequency-dependency of LSPs, however, is still a topic of ongoing research, and new models with additional refinements may be needed in the future. As can be seen from Figure 4-5, data from numerous recent multi-frequency channel measurement campaigns in the mmMAGIC project [5] shows smaller variations of channel delay spread with frequencies in both UMi Street Canyon and Indoor Office scenarios. To have a fair comparison and avoid biased conclusions, a set of technical requirements for comparability including equal measurement bandwidths, equal antenna patterns, equal dynamic ranges for analysis in delay and angle domains, equal angle resolutions, the same environment and same antenna locations are

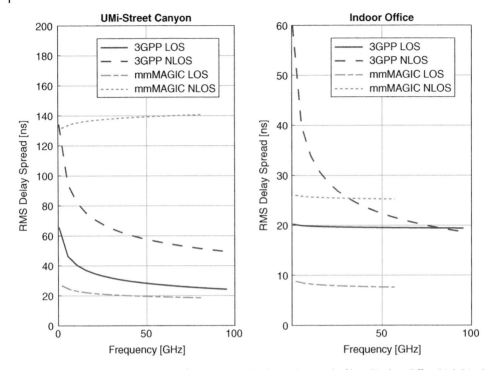

Figure 4-5. RMS delay spread versus frequency in UMi-Street Canyon (Left) and Indoor Office (right) in the 3GPP NR channel model [3] (0.5-100 GHz) and the mmMAGIC channel model [5] (2-96 GHz for outdoor and 2-60 GHz for indoor).

defined. These requirements are assured in all multi-frequency channel measurement campaigns in [5]. According to results in [5], the frequency trend of the channel delay spreads is small, and the linear relations between the channel delay spreads and the frequency are not statistically significant. This suggests that for a simpler approach, the same model parameters for LSPs can be used across considered frequency bands, from 2-86 GHz.

After necessary LSPs are obtained, channel coefficients of multi-frequency links can be generated. The accurate modeling of the root mean squared (RMS) delay spread is essential, as it can highly influence the design of the cyclic prefix of orthogonal frequency division multiplexing (OFDM) symbols. Additionally, if blockage is taken into account in the simulation, the locations of blocking objects are assumed the same across all frequencies. The procedure of multi-frequency simulation provides an evaluation framework for dual connectivity and carrier aggregation. The additional correlation introduced in the procedure helps to avoid a biased estimation of throughputs of multi-frequency links due to overestimating diversity gains of these links.

4.3.5 Ground Reflection

Although the LOS path can provide good channel quality, the superposition of the ground-reflected path and the LOS path can significantly degrade the received signal power even in LOS condition. In channel models for below 6 GHz channels, e.g., 3D-SCM, ground reflection is considered by using a two-slope LOS path loss model. The two slopes are separated by a breakpoint at a distance to the transmitter which is proportional to the carrier frequency. The LOS path loss exponent is approximately 2 before the breakpoint. Due to the ground-reflected path, the path loss exponent then increases to approximately 4 after the breakpoint. Gaussian distributed random shadow fading is added to the LOS path loss, which partially models the effect of superposition of reflected and scattered paths. However, this is not sufficient in above 6 GHz channels, where the ground-reflected path creates a deterministic fading pattern. In below 6 GHz channels, most of the area of interest lies after the breakpoint. However, as frequency increases towards mmWave spectrum, and cells become smaller and smaller, the value of the breakpoint surges, and most of the area of interest is before the breakpoint. A comparison between UMi path loss with and without explicit ground reflection modeling is illustrated in Figure 4-6. It can be observed that the path loss exponent has a hard transition from 2 to 4 at the breakpoint in the UMi path loss model without ground reflection. When ground reflection is explicitly modeled, a smooth transition of the path loss exponent and a deterministic fading pattern can be seen.

Explicit ground reflection modeling has been included as an additional feature in the NR channel model. A separate ground-reflected path is added to the original LOS channel impulse response. It should be noticed that the ground-reflected path is likely to be superimposed to the LOS path due to limited bandwidth during system evaluation using the NR channel model. The complex coefficient, delay and angles are determined geometrically.

Ground reflection has a significant impact on the overall system design. For instance, the deployment of relatively static links, such as mmWave wireless backhaul, should avoid deeply faded locations caused by the ground-reflected path. The modeling of ground reflection can help system

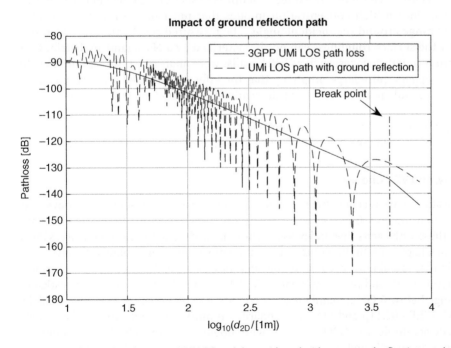

Figure 4-6. Comparison between UMi LOS path loss with and without ground reflection modeling.

designers to predict these locations and take necessary actions such as more robust link adaptation. Also, dual connectivity among different frequency bands, possibly based on carrier aggregation, can help to overcome the deep fade caused by ground reflection, see also Section 6.5.

4.3.6 Diffuse Scattering

Diffuse scattering occurs when waves impinge on a rough surface where the roughness of the surface is comparable to the wavelength. The standard deviation of surface-roughness σ_h of rough construction materials such as concrete, bricks and asphalt is of the order of 1–2 mm. The condition of roughness is usually defined by the Rayleigh criterion $\sigma_h < \lambda / 8\cos(\theta_i)$ [17], where λ is the carrier wavelength and θ_i is the incidence angle. Due to the decrease in wavelength, mmWave channels are less likely to satisfy the Rayleigh criterion, meaning that the diffuse scattering effect in mmWave channels is more significant than in below 6 GHz channels. More precisely, higher frequencies result in a smaller specular reflectivity of the surface, and each impinging ray may be backscattered into many low-amplitude rays having random propagation directions. Based on this assumption, one can hypothesize stronger dense multipath components (DMCs) at mmWave frequencies compared to lower frequencies. However, some studies have shown that the ratio of the power of the DMCs to that of the specular component is similar [18] or even lower than for lower frequencies.

The small-scale variations instigated by the propagation mechanism of diffused scattering have been investigated through the process of wall reflection measurements at a central frequency of 60 GHz and a bandwidth of 2 GHz [5]. As a metric of comparison, the notion of *power concentration*, which can be defined as the angular span corresponding to 90% of the power in the angular profile, was used. Therefore, the impact of diffused scattering is smaller with higher power concentration, since a high concentration implies that most of the received power is found within a small range around the incident angle. The measured results in [5] showed that scattering caused by the surface roughness can be characterized by a Nakagami-m distribution. When reflections from rough surfaces, such as brick walls, were analyzed, it was found that the value of this metric appears quite large, implying that the effect of diffused scattering is quite significant and should be taken into account for deterministic channel prediction tools such as ray-tracing.

Diffuse scattering may increase the richness in mmWave channels, and hence the design of beamforming techniques in 5G should take this characteristic into consideration, see Section 11.5.

4.3.7 D2D, Mobility, and V2V Channels

Vehicular communications have recently drawn a significant increase in interest in the framework of intelligent transportation systems (ITS), where vehicles are envisioned to collect data on traffic dynamics and share these with each other and possibly with the road infrastructure via wireless links, as covered in detail in Chapter 14. In this context, one of the most important points is to have accurate channel models that are able to predict the peculiarities of vehicular propagation channels. Vehicle-to-infrastructure (V2I) and vehicle-to-vehicle (V2V) channels have been widely investigated in the frequency range below 6 GHz, mainly motivated by the deployment of the 3GPP vehicle-to-everything (V2X), IEEE 802.11p and ITS-G5 standards. In mmWave bands, the V2V channels, and more generally, device-to-device (D2D) channels with mobility at both ends of the communication link, are so far poorly investigated.

A generic channel model based on a simplified or analytical channel representation could be easily extended to the D2D case. For instance, the map-based model proposed by the METIS project [2] can be used to generate a channel trace according to the transmitter and receiver positions. The authors in [19] employ a combined two-ring model and ellipse model, where the received signal is constructed as a sum of LOS, single- and double-bounced rays. In addition, this model is able to include vehicular traffic density information, and therefore mobile scatterers, to analyze their impact on channel characteristics. This has been extended to 3D in [20] using a two sphere model and an elliptic-cylinder model. The approach used in [21] to model V2V channels below 6 GHz was to define static and moving scatterers, the latter representing vehicles, and calculate the multipath components whose statistics were derived from measurements. The same approach could be extended to mmWave frequency bands, but the parametrization of the model should be corroborated by measurements. Vehicular channel angles-of-arrival at 60 GHz have been measured in [22], [23]. In [24], path loss characteristics have been derived at mmWave frequencies for LOS conditions and moving vehicles, and in the case of blockage by multiple vehicles. In [25], V2V channel measurements at 60 GHz and 1 GHz bandwidth were performed with antennas close to the bumpers. The results show that the channel impulse response is composed by a direct component, which is the sum of the LOS path and the one reflected from the ground, and delayed contributions coming from guard rails on the roads.

4.3.8 Oxygen Absorption, Time-varying Doppler Shift, Multi-Frequency Simulations, and UE Rotation

Dry air and water vapour can generate specific attenuation at frequencies up to 1000 GHz, which can be evaluated by means of a summation of the individual resonance lines from oxygen and water vapour. Significant oxygen absorption can be observed around 60 GHz [26]. Since the legacy 3D-SCM did not consider this frequency band, 3GPP added oxygen absorption modeling in the NR channel model to better support mmWave band modeling. A frequency-dependent oxygen loss can be added to each cluster in the frequency range of 53 to 67 GHz. This oxygen loss is cluster-specific and is proportional to the propagation distance of the cluster. As a result, it can impact the delay spread of channels at around 60 GHz. The impact of this gas attenuation is strongly dependent on the targeted application. While for point-to-point long range communication (e.g., backhaul) it should be included, in short range applications, like small cells and indoor hotspots, it could be neglected.

A time-varying Doppler shift is modeled to capture the phase change of a UE whose motion is not with constant velocity. The Doppler term within $[t_0, t]$ can be computed via the aggregated Doppler phase as

$$\exp\left\{ j2\pi \int_{t_0}^{t} \frac{\hat{r}_{\text{rx},n,m}^T(\tilde{t}) \cdot v(\tilde{t})}{\lambda_0} d\tilde{t} \right\} \tag{4-18}$$

where λ_0 is the carrier wavelength, $\hat{r}_{\text{rx},n,m}^T(\tilde{t})$ is the direction vector of the mth ray in the nth cluster at the receiver, and $v(\tilde{t})$ is the velocity vector of the receiver.

In mmWave bands, directive antennas are expected to be deployed at both the UE and the base station, as discussed in Section 11.5. As a result, the received signal power is sensitive to the orientation of the UE because of narrow beams. When the antenna beams are not aligned or are blocked, significant attenuation can be observed. This is modeled by rotating the bearing angle, downtilt angle and slant angle of the UE in the NR channel model.

4.3.9 Map-based Hybrid Modeling Approach

Besides the GSCM approach, another highlight of the NR channel model is the introduction of the map-based hybrid modeling (MBHM) approach. This approach aims at generating channel coefficients considering the digital map of the network layout. A ray tracing technique is used in the MBHM. Each ray-traced path (also known as deterministic cluster) from the transmitter to the receiver is tracked by modeling its interactions with the propagation environment. These interactions can be categorized into five types including LOS, reflections, diffraction, penetration and scattering. After importing the 3D digital map of the network layout and identifying the interaction type of each deterministic cluster, electric field calculations can be performed to determine its path power, angle and delay. Additionally, a configurable number of random clusters are generated following a similar procedure of the GSCM. The final channel output is then based on the combination of deterministic and random clusters. MBHM is able to provide site-specific channel characteristics, which can better support system design and network deployment.

4.4 Summary and Outlook

Bringing appealing features to the evolution of cellular systems, the upcoming 5G systems have introduced challenges to the modeling of propagation environments for 5G evaluations as well. Legacy channel models targeted below 6 GHz bands only and did not sufficiently describe 5G propagation channel characteristics, and were hence required to be extensively enhanced. The 3GPP NR channel model and IMT-2020 channel model have been developed based on the legacy 3D-SCM by additionally considering new frequency bands and new features. This includes better support for larger bandwidth and antenna arrays, more accurate modeling of space-time-frequency consistency, blockage, oxygen absorption, ground reflection, diffusive scattering, and new map-based hybrid modeling approaches.

More recently, attention has been paid to study path loss models for up to 100 GHz carrier frequency. Although classic FI and CI models are used, new path loss parameters for various scenarios such as UMa, UMi, and InH have been proposed between 6 GHz and 100 GHz. Another important environment-specific property of the channel is the LOS probability, as the clearance of blockages from the first Fresnel zone of a link is smaller in higher frequency bands. The NR channel model further introduces a O2I penetration loss model which can be characterized by simulation parameters to represent the use of metal-coated glass in buildings and the deployment scenarios.

A key characteristic in mmWave bands is the frequency-dependency of channel parameters. In the 3GPP NR and IMT-2020 channel models, frequency-dependency of delay and angular spreads is quite significant. However, recent measurements in [5] show that there is a downward trend in delay spread in indoor environments as frequency increases, while frequency-dependency is insignificant in O2I or outdoor environments. As delay spread has a large impact on the cyclic prefix length of OFDM symbols, it is important to take this into account when designing different numerologies in 5G systems. The impact of ground reflection becomes significant in mmWave bands, resulting in a deterministic fading pattern. Static links, such as those related to low-mobility users or wireless backhaul, can suffer from deep fading caused by ground reflection. An explicit modeling of ground reflection can help predict fading patterns and design necessary measures to compensate deep fading. Large bandwidths and large antenna arrays increase the richness of mmWave channels in delay and spatial domains. Intra-cluster delay and angular spreads can influence the design of hybrid beamforming techniques and decisions on TTI lengths and subcarrier spacing. Blockage effects become significant in higher frequency bands, which results in additional loss in signal power. The KED theory is used in the development of blockage modeling in the NR and IMT-2020 channel models. However, this model underestimates the loss in deep shadowing zones. Further enhancements have been proposed in the literature by accounting for the fast fading due to summing of the complex amplitude of paths from the edges of the blockage object, which provides a more accurate estimate of the signal loss in deep shadowing zones. For the evaluation of user mobility, multi-user multi-node communications and beam tracking design for 5G communication, it is important to include a spatial consistency procedure in the channel modeling. Investigations on reflections and scattering lead to the conclusion that a Nakagami-m distribution can adequately model the random fluctuations caused by surface roughness of various building materials.

The evolution of 5G will continue including more features and use cases. A number of improvements to channel modeling should be considered accordingly. For future work, a unified path loss model from 0.5 MHz to 100 GHz should be studied. Similarly, the link-specific O2I penetration loss model in 3D-SCM and the UE-specific O2I penetration loss model in the NR channel model should also be unified under the same modeling framework. The modeling of space-time-frequency consistency needs to be enhanced. Spatial consistency in the 2D plane has been considered in the NR channel model. However, in order to fully support vertical applications such as drone-to-drone communications and non-terrestrial networks, 3D spatial consistency needs to be considered as well. In 5G, dual connectivity in multiple frequency bands is a promising concept to separate control plane and user plane. The control plane, for instance, can be transmitted via below 6 GHz channels, and the user plane can be conveyed in mmWave bands for throughput boosting, as detailed in Section 6.7.2. However, there exists inconsistency between below and above 6 GHz channels in the NR channel model, hence a better frequency-consistent channel model should be developed. Furthermore, current state-of-the-art MBHMs focus mainly on cellular applications and have been validated only for below 6 GHz bands for outdoor or 28-30 GHz bands for indoor and static scenarios, and should be further extended for new vertical applications. Finally, new measurement campaigns are always encouraged to study channel characteristics such as time-variability of scattering clusters, shadowing, dynamic blockage or movement effects in dual mobility scenarios, aerial scenarios and other vertical applications.

References

1 3GPP TR 36.873, "Study on 3D channel model for LTE", V12.4.0, March 2017
2 FP7 METIS project, Deliverable D1.4, "METIS channel models", Febr. 2015
3 3GPP TR 38.901, "Study on channel model for frequencies from 0.5 to 100 GHz", V14.2.0, Sept. 2017

4 ITU-R 5D/610-E, "Updated combined proposal on IMT-2020 channel model", June 2017

5 5G PPP mmMAGIC project, Deliverable D2.2, "Measurement results and final channel models for preferred suitable frequency ranges", May 2017

6 Aalto University et al., "5G channel model for bands up to 100 GHz", Rev. 2.3, Oct. 2016

7 S. Sun et al., "Propagation path loss models for 5G urban micro- and macro-cellular scenarios", IEEE Vehicular Technology Conference (VTC Spring 2016), May 2016

8 ITU-R P1411-8, "Propagation data and prediction methods for the planning of short-range outdoor radiocommunication systems and radio local area networks in the frequency range 300 MHz to 100 GHz", July 2015

9 K. Haneda, N. Omaki, T. Imai, L. Raschkowski, M. Peter and A. Roivainen, "Frequency-agile pathloss models for urban street canyons", IEEE Transactions on Antennas and Propagation, vol. 64, no. 5, pp. 1941–1951, May 2016

10 T. Bai, R. Vaze and R. W. Heath, "Analysis of blockage effects on urban cellular networks", IEEE Transactions on Wireless Communications, vol. 13, no. 9, pp. 5070–5083, Sept. 2014

11 T. Bai and R. W. Heath, "Coverage in dense millimeter wave cellular networks", Asilomar Conference on Signals, Systems and Computers, Nov. 2013

12 J. Jarvelainen, S. L. H. Nguyen, K. Haneda, R. Naderpour, and U. T. Virk, "Evaluation of millimeter-wave line-of-sight probability with point cloud data", IEEE Wireless Communications Letters, vol. 5, no. 3, pp. 228–231, June 2016

13 ITU-R P.2040-1, "Effects of building materials and structures on radiowave propagation above about 100 MHz", July 2015

14 R. Verdone and A. Zanella, "Pervasive Mobile and Ambient Wireless Communications: COST Action 2100", Springer, 2012

15 J. Medbo et al., "Radio propagation modeling for 5G mobile and wireless communications", IEEE Communications Magazine, vol. 54, no. 6, pp. 144–151, June 2016

16 P. Okvist, N. Seifi, B. Halvarsson, A. Simonsson, M. Thurfjell, H. Asplund and J. Medbo, "15 GHz street-level blocking characteristics assessed with 5G radio access prototype", IEEE Vehicular Technology Conference (VTC Spring 2016), May 2016

17 P. Beckmann and A. Spizzichino, "The Scattering of Electromagnetic Waves from Rough Surfaces", Artech House, 1987

18 D. Dupleich et al., "Directional characterization of the 60 GHz indoor-office channel", URSI General Assembly and Scientific Symposium (URSI GASS 2014), Aug. 2014

19 X. Cheng, C. -X. Wang, D. Laurenson, S. Salous and A. Vasilakos, "An adaptive geometry-based stochastic model for non-isotropic MIMO mobile-to-mobile channels", IEEE Transactions on Wireless Communications, vol. 8, no. 9, pp. 4824–4835, Sep. 2009

20 Y. Yuan, C.-X. Wang, X. Cheng, B. Ai and D. Laurenson, "Novel 3D geometry-based stochastic models for non-isotropic MIMO vehicle-to-vehicle channels", IEEE Transactions on Wireless Communications, vol. 13, no. 1, pp. 298–309, Jan. 2014

21 J. Karedal, F. Tufvesson, N. Czink, A. Paier, C. Dumard, T. Zemen, C. F. Mecklenbräuker, and A. F. Molisch, "A geometry-based stochastic MIMO model for vehicle-to-vehicle communications", IEEE Transactions on Wireless Communications, vol. 8, no. 7, pp. 3646–3657, July 2009

22 A. Kato, K. Sato, M. Fujise and S. Kawakami, "Propagation characteristics of 60-GHz millimeter waves for ITS inter-vehicle communications", IEICE Transactions on Communications, vol. E84-B, no. 9, pp. 2530–2539, Sep. 2001

23 E. Ben-Dor, T. S. Rappaport, Y. Qiao, and S. J. Lauffenburger, "Millimeter-wave 60 GHz outdoor and vehicle AOA propagation measurements using a broadband sounder", IEEE Global Communications Conference (GLOBECOM 2011), Dec. 2011

24 A. Yamamoto, K. Ogawa, T. Horimatsu, A. Kato and M. Fujise, "Path-loss prediction models for intervehicle communication at 60 GHz", IEEE Transactions on Vehicular Technology, vol. 57, no. 1, pp. 65–78, Jan. 2008

25 M. G. Sánchez, M. P. Táboas and E. L. Cid, "Millimeter wave radio channel characterization for 5G vehicle-to-vehicle communications", Measurement, vol. 95, pp. 223–229, Jan. 2017

26 ITU-R P.676-11, "Attenuation by atmospheric gases", Sep. 2016

27 S. Wu, C. -X. Wang, el-H. M. Aggoune, M. M. Alwakeel and Y. He, "A non-stationary 3-D wideband twin-cluster model for 5G massive MIMO channels", IEEE Journal on Selected Areas in Communications, vol. 32, no. 6, pp. 1207–1218, June 2014

28 S. Payami and F. Tufvesson, "Channel measurements and analysis for very large array systems at 2.6GHz", European Conference on Antennas and Propagation (ECAP 2012), Mar. 2012

29 K. Haneda, J. Järveläinen, A. Karttunen, M. Kyro and J. Putkonen, "A statistical spatio-temporal radio channel model for large indoor environments at 60 and 70 GHz", IEEE Transactions on Antennas and Propagation, vol. 63, no. 6, pp. 2694–2704, June 2015

30 A. F. Molisch, F. Tufvesson, J. Karedal and C. F. Mecklenbrauker, "A survey on vehicle-to-vehicle propagation channels", IEEE Wireless Communications, vol. 16, no. 6, pp. 12–22, Dec. 2009

Part 2

5G System Architecture and E2E Enablers

5

E2E Architecture

Marco Gramaglia[1], Alexandros Kaloxylos[2], Panagiotis Spapis[3], Xavier Costa[4], Luis Miguel Contreras[5], Riccardo Trivisonno[3], Gerd Zimmermann[6], Antonio de la Oliva[1], Peter Rost[7] and Patrick Marsch[8]

[1] Universidad Carlos III Madrid, Spain
[2] University of Peloponnese, Greece
[3] Huawei German Research Center, Germany
[4] NEC Laboratories Europe, Germany
[5] Telefónica I + D, Spain
[6] Deutsche Telekom, Germany
[7] Nokia Bell Labs, Germany
[8] Nokia, Poland (now Deutsche Bahn, Germany)

5.1 Introduction

This chapter describes the current trends in the overall design of 5th generation (5G) network architecture. While the previous chapters were mainly focused on the description of the requirements and the problem space, from now on we will provide details on the currently envisioned 5G solutions. This is intended to be a starting point of the next chapters for the reader of this book who will be made familiar with common concepts that are the fundamental pillars of 5G research and standardization, with more details on individual parts of the 5G system in the following chapters.

More precisely, this chapter describes the high-level ideas behind the common architectural trends, with a first definition of their common capabilities and how they can be leveraged to provide the enhanced key performance indicators (KPIs) expected by 5G networks [1], as elaborated in Chapter 2.

This chapter is structured as follows. We first describe the common requirement and design principles that drive the current efforts in both standardization and research. We then describe how the end-to-end (E2E) architecture looks like, with the description of the main modules and which fundamental technical problems they will solve. The main part of this chapter is an overview of the innovative principles that are currently seen at the heart of the 5G network architecture. Finally, we describe the open issues on E2E architecture and a possible migration path.

5G System Design: Architectural and Functional Considerations and Long Term Research, First Edition.
Edited by Patrick Marsch, Ömer Bulakçı, Olav Queseth and Mauro Boldi.

5.2 Enablers and Design Principles

Future 5G networks will build on novel concepts that were not envisioned by the previous generation network architecture. The revolution provided by the introduction of software-defined networking (SDN) and network function virtualization (NFV), as detailed in Section 10.2, opens the door to a large list of possible applications. On the other hand, heterogeneous requirements, as highlighted by the major stakeholders' associations, such as Next Generation Mobile Networks (NGMN) or the 5G Public Private Partnership (5G PPP), will need to be addressed by novel architectural trends. In this section, we introduce some of the novel enablers that will be a fundamental part of future 5G networks and the associated requirements.

5.2.1 Modularization

The evolution of mobile communication systems towards 5G was intentionally aiming at achieving architecture flexibility, heterogeneous accesses and vertical business integration, leveraging on the significant advance on NFV and SDN. To enable the design of logical architectures tailored to performance and functional requirements of different use cases, the principle of architecture modularization and network function decomposition was proposed at the earliest 5G research stages. In particular, conventionally monolithic network functions, in 4th generation (4G) cellular often corresponding to physical network elements, are proposed to be split into basic modules or network functions (NFs) defined with proper granularity, both for the control plane (CP) and user plane (UP), thus allowing the definition of different logical architectures via the interconnection of different subsets of CP and UP NFs.

In the process of decomposing the NFs into basic modules, the distinction between NFs relating to the access network (AN) and core network (CN) emerged. To minimize the dependency of the 5G core on the access (and vice versa), and achieve the definition of a convergent network, providing connectivity via a multitude of accesses not only including cellular radio, a different AN/CN functional split and an interface model are necessary.

Besides flexibility, the architecture modularization provides the essentials to support network slicing, as a network slice can be defined as an independent logical network shaped by the interconnection of a subset of NFs, composing both CP and UP, and which can be independently instantiated and operated over physical or virtual infrastructure. Please see further details on network slicing in Chapter 8 and in the following section, and find a further elaboration on modularization in Section 5.4.

5.2.2 Network Slicing

In order to create tenant- or service-specific networks, NGMN has proposed the concept of network slicing [1], as detailed in Chapter 8. While legacy systems host multiple telecommunication services, such as mobile broadband, voice, SMS, etc., on the same mobile network architecture, for instance composed of Long Term Evolution (LTE) radio access and the Evolved Packet Core (EPC), future 5G networks should also support shared or dedicated logical architectures customized to the respective telco or vertical services, such as enhanced mobile broadband (eMBB), vehicular communications, ultra-reliable low-latency communications (URLLC), and massive machine-type communications (mMTC), as introduced in Section 2.2. These services need very different KPIs that are hard to be fulfilled by legacy systems, as they are characterized by monolithic network elements that have tightly coupled hardware, software, and functionality. In contrast, future architecture leverages the decoupling of software-based network functions from the underlying infrastructure resources by means of utilizing different resource abstraction technologies.

Furthermore, as explained in the previous paragraph, modularization will play a fundamental role. For instance, well-known resource sharing technologies such as multiplexing and multitasking, e.g., wavelength division multiplexing (WDM) or radio scheduling, can be advantageously complemented by softwarization techniques such as NFV and SDN. Multitasking and multiplexing allow sharing physical infrastructure that is not virtualized. NFV and SDN allow different tenants to share the same general-purpose hardware, such as commercial off-the-shelf (COTS) servers. In combination, these technologies allow building fully decoupled E2E networks on top of a common, shared infrastructure. Consequently, multiplexing will not happen on the network level anymore, but on the infrastructure level, as depicted in Figure 5-1, yielding better quality of service (QoS) or quality of experience (QoE) for the subscriber (as different slices will have tailored orchestration for a given service) as well as improved levels of network operability for the mobile service provider or mobile network operator.

In principle, a network slice is a logical network that provides specific network capabilities and network characteristics and comprises NFs, computing and networking resources to meet the performance requirements of the tenants, for instance verticals. This comprises both radio access network (RAN) and CN NFs and, depending on the degree of freedom that a tenant may have, also the management and orchestration (MANO) components. A network slice may be dedicated to a specific tenant or partially shared by several tenants that have the same performance requirements but different security or policy settings. The decoupling between the virtualized and the physical infrastructure allows for the efficient scaling in, out, up or down of the slices, hence suggesting the economic viability of this approach that can adapt the used resources on demand.

Network slices are created mostly with a business purpose: Following the 5G vertical markets paradigm [2], in which different tenants can get their own network customized for a specific purpose, an

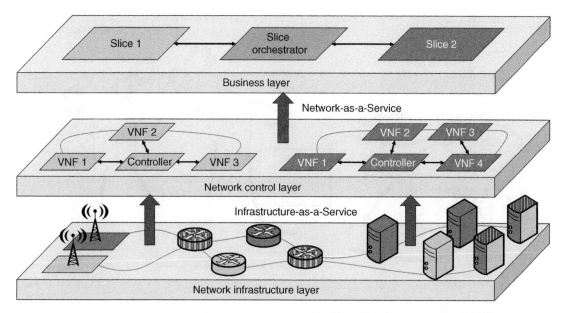

Figure 5-1. An example of a network-sliced architecture.

infrastructure provider will assign the required resources for a network slice which in turn realizes each service of a service provider portfolio (e.g., the vehicular URLLC network slice, the Factory of the Future URLLC network slice, and/or the health network mMTC network slice). The required resources are provided according to different resource commitment models, ranging from rather static reservations to on-demand provisioning.

Network slicing calls for a novel architecture capable of flexibly orchestrating and configuring all the resources, functions, and entities used by a network slice. This role is played by the network softwarization concept described in the following section. Additional information and detailed aspects about network slicing can be found in Chapter 8.

5.2.3 Network Softwarization

Future 5G networks will bring the concept of network programmability beyond what is now possible with SDN. While SDN splits routing and forwarding capabilities of a switch and reassigns the former to an SDN controller, this split between logic and agent should be performed for any NF, including the ones related to the CP. That is, the SDN principles are extended to all control and data layers as well as management functions usually deployed in mobile networks. An example of this view is depicted in Figure 5-2, in which an enhanced controller extends the SDN capabilities to different kinds of functions in the network.

The following three categories can be identified: (i) networking control functions (e.g., mobility and session management, and potentially QoS/QoE control); (ii) connectivity control functions (mainly packet forwarding or SDN-based packet forwarding); and (iii) wireless control functions (e.g., radio link adaptation and scheduling). This last category, however, may not be fully implemented using a software-defined approach, for instance due to scalability reasons.

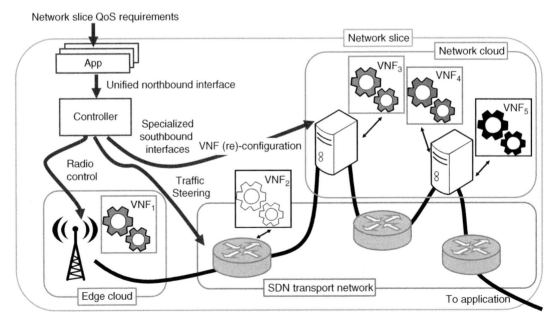

Figure 5-2. Applying softwarization in a network slice.

There are many advantages of implementing selected wireless control functions in a way such that they are not bound anymore to specialized hardware, such as an LTE enhanced Node-B (eNB), but rather become independent pieces of software. Network functionalities are then performed by a (virtualized) programmable and logically centralized controller that abstracts and, thus, homogenizes different network technologies. Such a controller will make network slices programmable by controlling the topology and functionality of the service chains as well as resource control inside the network slices. Further, this approach implies to have a unique control point for the network: By operating a small number of such controllers, network operators reduce the complexity of the network management and control.

Dense wireless networks, as envisioned in 5G, will benefit from this extension of the SDN approach. The control of mobility support schemes and adaptation to dynamic radio characteristics, such as the ones used in a multiple radio access technology (RAT) scenario, are performed by the controller that can employ especially tailored algorithms per network slice they are deployed in. Moreover, if needed, virtual NFs (VNFs) can be deployed closely to the users (e.g., in a network slice supporting vehicular URLLC) reducing their experienced latency. This feature is particularly desirable for the verticals market, as several network operators can provide their services to verticals by using a softwarized approach.

New services can, hence, be enabled by just modifying the controller functions: Services that were not initially included by an operator in its architecture design can now be introduced and implement service-specific enhancements. A good example is the operation of base station (BS) schedulers: As the controller has a global view of the network slice, it can optimize (through an application) the mid- to long-term behavior of scheduling algorithms and the resource allocation across them independently of the functional split implemented in the slice. This concept can be extended to the resource control across network slices. Controllers facilitate the optimization of network utilization: A network infrastructure provider may allocate unused resources to demanding network slices, provided that the service-level agreement (SLA) is satisfied for all the hosted network slices. In this way, more verticals may share the same infrastructure, reducing operational costs as well as avoiding a time-consuming deployment of dedicated infrastructure.

Another possible usage of such controllers is mobility management. As stated above, they extend the SDN concept to any kind of NF in the mobile network. So, a straightforward amendment of the SDN dialect, capable of directly handling user data tunnels, may be used to directly control the gateway entities of the network. However, the same idea can be used to directly control other low-level user flows, steering traffic through NFs implementing a cloud RAN. That is, one centralized, flexible application can control heterogeneous NFs through specialized interfaces. Mobility management is just an example, but verticals may directly provide their customized applications if they desire to do so, focusing on a customized function behaviour. As depicted in Figure 5-2, the controller offers a unique interface to several heterogeneous NFs ranging from wireless control to traffic steering.

5.2.4 Multi-Tenancy

With the trend of increased heterogeneity in terms of KPIs among different services sharing the same infrastructure, 5G networks must be designed embracing this from the very beginning. Nowadays, very different applications share the same communication infrastructure, but communication networks were not designed for this purpose as the primary one. Moreover, the final goal of 5G is not only to support heterogeneous services, but also reduce the costs, i.e. operational expenditures (OPEX) and capital expenditures (CAPEX).

Theoretically, this customization goal can be achieved by having several physical networks deployed, one for each service (or even one for each business entity). Isolated services can hence use their resources in an optimal way, avoiding difficult re-configuration of hardware and network entities as well as the need to accommodate possibly conflicting performance objectives. Clearly, this approach cannot be applied to real networks in a cost-efficient manner; therefore, it calls for a solution that allows for both an efficient resource sharing and multi-tenancy infrastructure utilization.

An intermediate approach to multi-tenancy, which does not involve on-the-fly reconfigurations, is already standardized for 3rd generation (3G) cellular [3] and applied by many operators who currently share physical cell sites. However, the equipment still belongs to each operator or a joint venture of involved operators, limiting hence the achievable cost reduction. 5G networks will go one step further, pushing for the active sharing of resources within the same infrastructure among different tenants, allowing for the monetization of so-called vertical sectors. In such scenarios, also non-operators may utilize network resources and functions to compose their own (virtualized) network instance.

The future 5G network E2E architecture should hence build on the network programmability and network slicing concepts described in the previous two subsections, to create completely novel deployment choices. More specifically, the future E2E architecture of 5G networks shall not only allow for (re)programming the behavior of individual NFs, but also for the adaptation of the network topology, i.e., the geographical distribution of NFs. In other words, while the option of re-configuring the behavior of network elements was already available in legacy systems (even though in a limited fashion only), the topology of such networks was static and bound to the geographical distribution of network nodes. In this way, the deployment location of a NF can a) be modified more easily, and b) be different for two instances of the very same NF, e.g., when instantiated for different network slices.

The result is an environment such as the one depicted in Figure 5-3: Different verticals may create their own network slice to serve their terminals by having a business relationship with a telecom

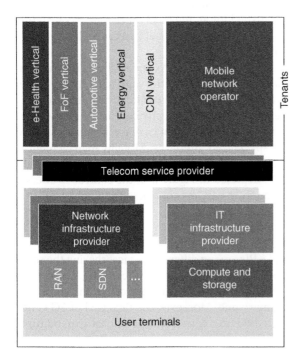

Figure 5-3. A tenant-enabled network.

service provider which, in turn, orchestrates the NFs building a network slice by acquiring the necessary resources (both network and cloud) from different infrastructure providers.

By tenant, we here consider different vertical operators, such as an automotive vertical, a Factory of the Future (FoF) vertical or a content delivery network (CDN), but it may also be a standard mobile network operator (MNO). Further, an MNO can act as a telecom service provider and infrastructure provider (e.g., when owning the required resources, such as the RAN). We remark that Figure 5-3 does not show the tenant vertical split on the user equipment (UE) basis.

5.2.5 Mobile or Multi-Access Edge Computing

A new communications and computing paradigm that has already been introduced in LTE, but which will become more essential and will require native support in the 5G system, is that of multi-access edge computing (MEC), previously known as mobile edge computing. Here, the key concept is to move applications close to the radio, for instance, physically co-located with base stations, as already depicted in Figure 5-2 for the benefit of

- reduced E2E latency for, e.g., URLLC use cases, as, for instance, the communication between a device and an application server can be kept in local proximity and need not involve a long-distance transport network or cloud environment;
- increased networking efficiency, as application data which is only needed locally can also be kept locally, and need not be unnecessarily routed over a large-scale communications infrastructure;
- increased security, because application data can be confined within areas where it is actually needed;
- providing applications access to local context and communications-related information (for instance, an application may make use of proximity information among devices).

The European Telecommunications Standards Institute (ETSI) is the body that started the standardization of MEC, and it has issued a MEC architecture and framework in March 2016 [4], which is depicted in a highly simplified form in Figure 5-4. Here, one can see MEC applications running on virtual machines (VMs) on a MEC hosting infrastructure, which would typically be the equipment of a network vendor, such as a base station. The main focus of the work of ETSI has been to standardize the MEC application platform, which aims to provide a common functional environment for MEC applications, so that (in principle) any MEC application from any vendor could run on any MEC hosting infrastructure from a different vendor. Also, the platform aims to support portability (e.g., the dynamic instantiation and execution of MEC applications in different environments based on application needs) and scalability. The MEC application platform, for instance, provides application programming interfaces (APIs) through which applications can register themselves, utilize communications services for, e.g., the communication among distributed MEC applications, or access local radio network information which could be of relevance for applications.

While the communication between MEC applications and the MEC application platform is now already standardized, the interaction of the latter and the 5G communications system is not. It has, for instance, been discussed in the 3rd Generation Partnership Project (3GPP) that the information provided from the 5G RAN or the 5G CN to the MEC hosting infrastructure and application platform should be standardized, and consequently, interfaces should be defined between related network entities and the MEC environment. However, at the time this book is written, it appears that

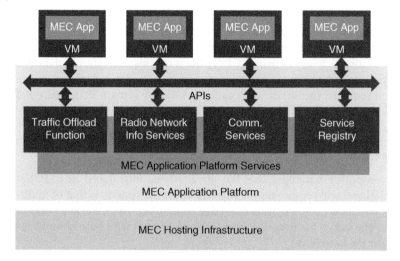

Figure 5-4. High-level overview on a typical MEC environment.

3GPP will likely not fully take up this topic before 3GPP Release 16. For the first commercial 5G products, it is, hence, left to proprietary implementation which information is passed from the 5G network to the MEC environment.

5.3 E2E Architecture Overview

The mobile network functional architecture has traditionally been divided into two main components: RAN and CN. With the current trend of network softwarization, the decomposition and the subsequent centralization of monolithic NFs, the border between these two components in 5G may not be as sharp as it was in previous generations. This certainly provides advantages when enabling the novel concepts described in the previous section, but it also entails a set of new challenges that have to be addressed by the new E2E architecture and the modules that will provide these functionalities.

Overall, there is a general agreement on how the E2E architecture will look like, as depicted in Figure 5-5. This representation, as agreed by several relevant stakeholders in [5], includes several modules that will be described throughout this section, and will support the enabling concepts, such as network slicing, as described in Section 5.2 and detailed in Chapter 8.

5.3.1 Physical Network Architecture

Future 5G networks will have a heterogeneous physical deployment, in terms of different frequency bands, different cell sizes, but also the co-existence of different RATs and so-called air interface variants (AIVs), as detailed in Section 6.4.1. In contrast with the current architecture, the introduction of network softwarization techniques will be the key to the physical deployment of different processing sites that will be used for different purposes according to the orchestrating services. We can mostly distinguish between two categories: edge clouds or central clouds.

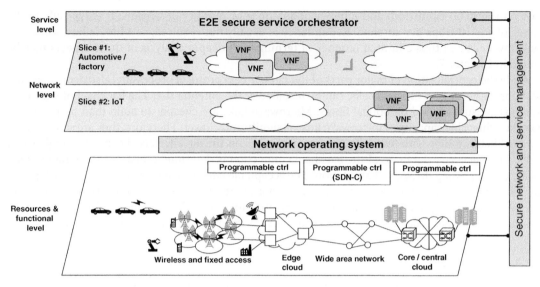

Figure 5-5. E2E architecture overview [5].

A central cloud (or centralized cloud) typically comprises multiple powerful computing and storage resources which may be several hundred kilometres apart [6]. These data centres are connected among each other by a wide area network (WAN). The WAN also connects the data centres of the central cloud to the data centres of the edge cloud. Physically, this WAN is based on optical fibers with capacities of 10 Gbps and higher, see also Section 7.3. Its topology can differ significantly according to the needs and preferences of the network operator: It may have multiple hierarchy levels, e.g., long-haul links on a high level that interconnect regional and metropolitan networks on the underlying level. On each of these hierarchy levels, star, ring, tree or chain topologies may be deployed. Redundancy must be foreseen in the WAN, because otherwise a router or link failure might affect a huge number of terminals. It should be noted that usually this WAN will not be available for exclusive use by 5G networks. Typically, it is shared between fixed and mobile services with the larger demand portion originating from fixed services.

The edge cloud is located in the vicinity of the RAN, which may reduce the delays and the jitter at the cost of reduced computational and storage resources [6]. The dominant requirements for the fronthaul connectivity between an edge cloud data center and the radio access point at the antenna site are low latency and high capacity, as detailed further in Sections 6.6.2 and 7.6.1. In the case of centralizing radio access protocol layers, latency should be less than $100 - 200\,\mu s$. Therefore, the distance between edge cloud and antenna site should not exceed a few 10 km, and a dedicated point-to-point connection should be used. For efficiency reasons, multiplexing (i.e., wave or time division multiplexing) several of these connections onto a single fiber are necessary. The suitability of Ethernet switching requires further studies as it introduces additional delay and delay jitter. The required speed of the fronthaul connection depends on the implemented fronthaul split, as detailed in Section 6.6.2, the properties of the radio signal that shall be transmitted over the air interface, and on the radio bandwidth and number of antennas or number of streams. It is typically in the range of 1 – 10 Gbps per antenna. Redundancy is usually not required, as a link failure will affect only few cells and thus a limited number of terminals. A deeper analysis of these aspects may be found in Chapter 7.

Aside the centralization of major parts of the radio access protocol processing, also dedicated base stations, either macro or small cells, can be installed at the antenna site. The backhaul of such base stations is usually based on optical fibers. Microwave links are cheaper to build than optical fibers but the achievable capacity of microwave links is significantly lower. Therefore, they are suitable only for backhauling sites with few cells and low data rates on the air interface, i.e. mostly single small cells. The necessary data rate of the backhaul connection is determined mainly by the rate of user plane traffic passing through a base station. The acceptable latency of backhaul connections depends on the requested radio functionality, e.g., when coordinated multi-point (CoMP) transmission and reception shall be applied, latencies must be significantly lower than without CoMP. Redundancy mechanisms are not required for the backhaul for the same reason as in the case of fronthaul.

5.3.2 CN/RAN Split

Related to the functional split between the core and access networks, 3GPP has identified the following split for Release 14 [7]. The considered RAN functions include legacy RAN functions such as: *transfer of user data, radio channel (de)ciphering, integrity protection and header compression, handover, inter-cell interference coordination, load balancing, connection setup and release, routing of non-access stratum (NAS) signaling, NAS messages and NAS node selection, synchronization, paging and positioning.* Additionally, there are functions that have been introduced to support the new RAN. The new features include the tight interworking (including handovers) of the new radio with the Evolved Universal Terrestrial Radio Access (E-UTRA), and its functionality to support slice selection and also session management. The latter provides the means for the CN to create, modify or release a context and related resources in the new RAN associated with a particular protocol data unit (PDU) session. Note that the new radio supports the mapping between a QoS flow and a data radio bearer and the marking of QoS flow ID, as detailed in Section 5.3.3. This new information can be used by radio resource management (RRM) functions like scheduling or even admission control and offers customized support for slices. Also, some functions like paging may be moved towards the RAN to decrease the signaling load of the CN.

The new interface between the RAN and CN is a logical point-to-point interface and will support control and user plane separation. It is required to be open and future-proof, and it will be decoupled within the possible RAN deployment variants. This interface is used for UE context management between core and access networks and the transport of NAS messages. The target is that the new interface should allow an independent evolution of the core and access networks and enable access-agnostic CN functions.

The same principles are also followed in Release 15 [8], where the new RAN nodes are called Gigabit NodeBs (gNBs) and are interworked with eNBs. These nodes are using the Next Generation (NG) interface to connect to the CN functions, such as the Access and Mobility Management Function (AMF), User Plane Function (UPF) and Session Management Function (SMF), as detailed in Section 5.4.1. gNBs have similar functions like in Release 14 with the appropriate modifications that are needed due to the introduction of the new CN functions. A graphical representation of the

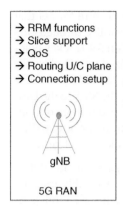
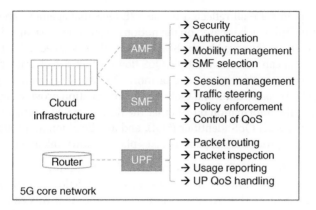

Figure 5-6. Functional split between NG-RAN and 5G core [8].

allocation of the main functions in logical nodes, and the resulting split between NG-RAN and 5G core (5GC), as defined by 3GPP, are provided in Figure 5-6.

Apart from 3GPP activities, there is also ongoing research [5] of concepts aiming to provide a high degree of architecture flexibility, e.g., a flexible assignment and integration of RAN and CN functions. Focus of future research work is to develop all options and compare them in terms of flexibility, complexity and cost involved in meeting the requirements of future use cases.

5.3.3 QoS Architecture

As mentioned before, 5G will need to simultaneously accommodate a large diversity of use case requirements, as detailed in Section 2.2. This implies that there need to be flexible and highly granular E2E means to handle and prioritize different traffic types, or even different packets belonging to the same traffic type. As an example for the latter, it is typically desired to give short packets related to Transmission Control Protocol (TCP) connection requests (and responses) a higher priority than other packets related to the same TCP traffic, in order to avoid an unnecessary delay of connection setup time.

In this respect, a key limitation in the LTE QoS architecture [9] is that the finest granularity of differentiating mobile data is on the level of radio bearers. Within one bearer, which can be established to reflect guaranteed bit-rate or non-guaranteed bit rate traffic and which is characterized by a QoS class identifier (reflecting priority, acceptable delay, and packet loss rate), all packets are treated in the same way. Also, there is a one-to-one mapping of radio bearers to EPS bearers. If now a single wireless communication link has to carry data related to very different service requirements, one would have to setup individual radio bearers to be able to sufficiently differentiate the traffic, which would be highly inefficient. Further, LTE is not able to treat packets within a PDU session differently, e.g., the aforementioned prioritization of TCP packets related to TCP session setup is not possible. Hence, there is a general consensus that 5G needs to provide a much more granular approach to QoS management, allowing for a more flexible and independent QoS handling in the CN and RAN, and in particular allowing a packet-specific traffic differentiation when needed.

The aforementioned points are all reflected in the E2E QoS management architecture for 5G that 3GPP has agreed upon in [10],which is based on the notion of QoS flows that are identified through QoS flow IDs (QFIs), and which are managed by the SMF in the CN, as detailed in Section 5.4.1. A key point is that a single PDU session can relate to multiple QoS flows, which are characterized by a QoS profile provided by the SMF to the access network, one or more QoS rule(s) provided by the SMF to the UE, and service data flow (SDF) classification and QoS-related information provided by the SMF to the UPF. The QoS profile contains the information on whether the flow is related to guaranteed bit-rate or non-guaranteed bit-rate traffic, a 5G QoS identifier (5QI), and an allocation and retention priority (ARP). Based on this, the following processing of packets takes place in downlink and uplink, respectively:

- In the **downlink**, the UPF in the CN applies the SDF classification rules to assign individual data packets to the configured QoS flows. In the access network, QoS flows can then be flexibly assigned to different data radio bearers (DRBs). It is important to note that this can be handled independently from the CN, as opposed to the LTE approach where there is a strict mapping of Evolved Packet System (EPS) bearers to radio bearers. Also, both the assignment of packets to QoS flows (e.g., based on deep packet inspection), and the assignment of flows to radio bearers, leave a lot of room for proprietary optimization. In 5G, the access network could for instance assign critical QoS flows to lower carrier frequencies, possibly applying further means for an increased reliability such as multi-connectivity, while less critical QoS flows could use mmWave frequencies;
- In the **uplink**, the UE evaluates packets against the QoS rules provided by the SMF, and assigns these accordingly to available QoS flows, which in return are mapped to available data radio bearers. Note that there is also the notion of reflective QoS handling, where the UE uses the QFI information contained in downlink packets to derive an equivalent handling of the related uplink packets.

A comparison of the E2E QoS architecture between LTE and 5G is provided in Figure 5-7. In the LTE case, on the left side, we can see the bearer-oriented approach with a strict one-to-one mapping

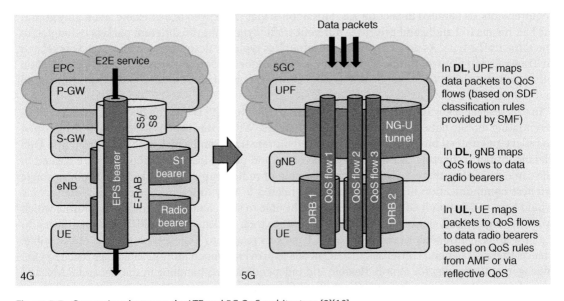

Figure 5-7. Comparison between the LTE and 5G QoS architecture [8][10].

between EPS-bearers, EPS radio access bearers (E-RAB), S1 and radio bearers. In the 5G case, on the right side, we see the flow-based approach which allows to decouple the assignment of packets to flows (controlled by the CN) from the assignment of flows to DRBs (being in the responsibility of the RAN). The particular example of three different QoS flows mapped to two different DRBs is here to be seen as an example only. Note that in the context of introducing the new QoS architecture, 3GPP has also agreed to introduce a new UP protocol layer in the RAN above Packet Data Convergence Protocol (PDCP), called Service Data Adaptation Protocol (SDAP). This is responsible to map QoS flows to radio bearers and mark the QoS flow in uplink (UL) and downlink (DL) packets, as detailed further in Section 6.4.2.

5.3.4 Spectrum Sharing Architecture Overview

Dealing with new scenarios together with the demands for high data rates and lower latencies have indicated the need for additional spectrum for cellular communications. However, since it is not trivial to allocate exclusively new spectrum bands for cellular communications, it is imperative to improve the spectrum usage through new spectrum sharing mechanisms. This implies that mobile network operators will have to cooperate and interact to cover the augmented traffic requirements. The details of the efficient spectrum sharing via coordination have been analyzed in Section 3.6. However, in summary we could say that the spectrum authorization regimes and spectrum access schemes can be divided into two categories based on the license that is provided to the users, as also detailed in Section 3.2:

1) individual authorization approaches, which comprise the exclusive, co-primary and licensed shared access (LSA) schemes, and
2) general authorization approaches (e.g., license exempt or unlicensed), which include the unlicensed shared access (i.e., Wi-Fi, Bluetooth, etc.), secondary horizontal shared access (i.e., TV white spaces) and the License Assisted Access (LAA) which enables the cellular system operation in unlicensed spectrum, as illustrated through an example mechanism in Section 12.5.1.

Figure 5-8 provides an abstract representation of the required entities for the realization of spectrum sharing [11]. In particular, the regulator provides to the entities the terms of usage and undertakes to ensure that the rights of the licensees are protected. The incumbent users and the operators inform a centralized coordination entity for the spectrum use, whereas the operators that require additional spectrum resources contact this coordination entity for reserving resources. Several schemes have been proposed in the literature for the mechanisms handling the negotiation and the identification of the most suitable spectrum according to the service that each operator targets [12][13]. More detailed descriptions of each entity can be found in Section 3.6.

5.3.5 Transport Network

5G RAN technologies will require novel fronthaul and backhaul solutions between the RAN and CN to deal with the significant traffic load increase forecasted. Furthermore, there will be a sizeable growth in the capillarity (i.e., points of attachment) of the network, since the traffic load increase in the 5G RAN will be supported by a larger number of base stations covering smaller areas (i.e., mobile network densification).

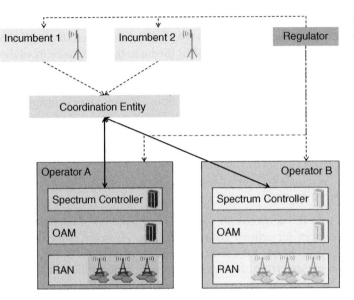

Figure 5-8. Abstract spectrum sharing architecture.

As discussed before, a promising approach to address this challenge is virtualization, which exploits the multiplexing gain of softwarized network functions on top of commoditized hardware. This has led to the cloud RAN concept where cellular base station functions are hosted in cloud computing centers. Once virtualized, base station functions can be flexibly distributed and moved across data centers, providing another degree of freedom for load balancing. Moreover, base station functions can be decomposed in many different ways, giving rise to the so-called flexible functional split, where the radio protocol split between centralized and distributed base station functions can be adjusted on a case-by-case basis, as discussed in detail in Section 6.6.

In this context, the distinction between fronthaul and backhaul transport networks blurs, as varying portions of functionality of 5G points of attachment (5GPoAs) might be moved toward the network as required for cost efficiency reasons. Thus, traditional dimensioning approaches for the transport infrastructure will no longer be applicable with 5G, and a new generation of integrated fronthaul and backhaul technologies will be needed to bring CAPEX and OPEX to a reasonable return on investment (ROI) range, as for instance analyzed in detail in Section 7.6.1. Also, for cost reasons, the heterogeneity of transport network equipment must be tackled by unifying data, control, and management planes across all technologies as much as possible, as covered in Section 7.5. A redesign of the fronthaul/backhaul network segment is a key point for 5G networks, since current transport networks cannot cope with the amount of bandwidth required for 5G. Next generation radio interfaces, using 100 MHz channels and squeezing the bit-per-MHz ratio through massive multiple-input multiple-output (MIMO) or even full-duplex radios, requires a 10-fold increase in capacity, which cannot be achieved just through the evolution of current technologies.

Current deployments of Open Base Station Architecture Initiative (OBSAI) [14] or Common Public Radio Interface (CPRI) [15] for the fronthaul traffic mainly use dark fibers to the remote radio heads (RRHs), as other technologies do not meet the stringent requirements of latency (e.g., about 100 microseconds for CPRI) and jitter (2 ppb). The usage of dark fiber might be acceptable for some markets, as the deployment of existing dark fiber is very broad. But this might not be the case for other markets, such as Europe. Ethernet, due to its dominant distribution on networks and technical equipment and the initiatives to develop it further [16], is therefore seen as the basis for combining fronthaul and backhaul technologies to provide cost-effective and well-accepted solutions. Because of such heterogeneity, the fronthaul and backhaul traffic of different gNBs or RRHs, even from different technologies, will share the same physical link. As detailed in Section 6.6, 3GPP is currently discussing various fronthaul split options that aim to alleviate the throughput, latency and jitter requirements on transport networks, for instance by applying function splits in the physical layer, such that some extent of beamforming and the conversion from frequency into time domain takes place at the radio unit as opposed to the central unit in the case of an OBSAI or CPRI based split. Similar considerations have also led to the enhanced CPRI (eCPRI) standard [17].

Considering the new challenges brought by 5G, novel architectures aimed at integrating fronthaul and backhaul in a common Crosshaul/Xhaul network are being defined [18][19]. These architectures thus aim to enable a flexible and software-defined reconfiguration of all networking elements through a unified user plane and control plane interconnecting distributed 5G radio access and CN functions, hosted on in-network cloud infrastructure.

In the following, we provide as an example a high-level overview of an integrated fronthaul/backhaul architecture available in the literature [5]. This architecture is based on three main building blocks: (i) a control infrastructure (CI, in [5] referred to as XCI) using a unified, abstract network model for control plane integration; (ii) a unified user plane encompassing innovative high-capacity transmission technologies and novel latency-deterministic switch architectures based on forwarding elements (FEs, in [5] referred to as XFEs); and (iii) a set of computing capabilities in the form of processing units (PUs, in [5] referred to as XPUs). Figure 5-9 depicts this architectural framework example.

In the example, the CI is compliant with the ETSI NFV architecture for what concerns management and orchestration. It includes additionally a set of controllers for taking control actions over networking, storage and computing resources. On top of the CI, a number of applications can be located. Of special interest for multi-domain environments is the multi-tenancy application (MTA), conceived to support per-tenant infrastructure management in multi-tenancy scenarios.

The reader is referred to Chapter 7 and [21] for further details on the challenges and solutions envisioned for transport networks in 5G.

5.3.6 Control and Orchestration

In contrast with previous approaches in 3G and 4G, with the advent of SDN, NFV and 5G, the network services are to be realized by dynamically deploying functions and programming connectivity paths among them instead of re-architecting the network. This will largely depend on the success of network-wide service management. As a whole, this concept of network-wide orchestration is associated with the replacement of the individual device configuration by a more powerful network services management mechanism able to provide network-wide services definition, configuration, deployment and monitoring. By using network-wide orchestration, services are not deployed, configured and managed in a node-by-node fashion as it is originally done in legacy systems. The increasing complexity that 5G will impose on the delivery and maintenance of services, supported by the slicing concept, force to manage the network in an integrated and coordinated way, and not as a collection of individual boxes and layers. For doing that, higher-level abstractions and automated procedures are needed to manage and configure each single component of the whole network service at once.

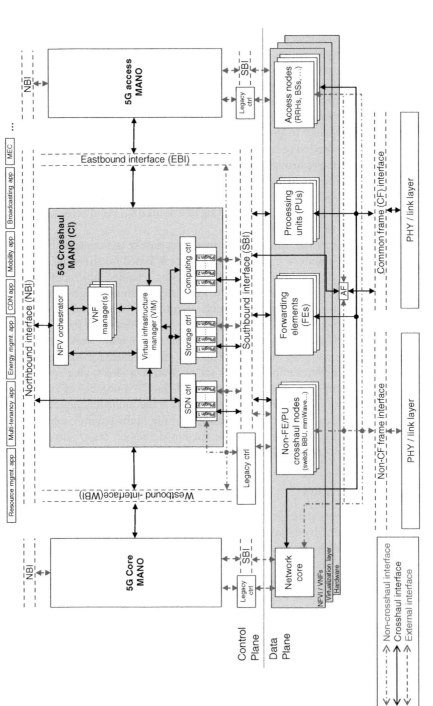

Figure 5-9. Example of a 5G transport network reference architectural framework [20].

The main advantage of this network-wide orchestration is that it gives a single point of integration, providing a centralized representation of the distributed network no matter the number of resources involved or where they could be physically placed. This provides great opportunities for accessing KPI measurements and automation, opening up possibilities for the deployment of advanced services across all network domains. In this way, the inherent complexity of deploying and managing complex and feature-rich services can be simplified by automating the underlying configuration and monitoring tasks from the orchestration module. For example, a complex distributed network service composed of multiple VNFs, physical NFs (PNFs), and network links connecting different clouds and a number of KPIs to be monitored, could be automatically deployed and managed from the orchestrator and treated as a single entity (i.e., a network service) in the network.

By the introduction of the slicing concept, 5G networks should hence consider orchestration at two different levels [22]:

- **inter-slice orchestration**, dealing with the orchestration of resources to conform the different network slices in the network, and
- **intra-slice orchestration**, referring to the orchestration of resources into the network slices.

5.4 Novel Concepts and Architectural Extensions

After having reviewed the common architectural trends in the previous section, we now highlight the specific solutions that have been devised by both research projects and standardization bodies to shape them. In this section, we provide a high-level description of some of these selected solutions. Most of these are then analyzed more in detail in specific subsequent chapters of the book, which the reader is pointed to.

5.4.1 Architecture Modularization for the Core Network

Several research projects have provided reference models for the 5G modularized architecture, and a general reference scheme has also been agreed in 3GPP [23]. The model is illustrated in Figure 5-10. The reference model formalizes the NFs as founding logical elements for different architectures and network slices, based on a separation between control and user planes and between AN and CN. Also, the model defines the complete interface model, including inter-NF interfaces, CP-UP interfaces, and

Figure 5-10. 3GPP 5G system modularized architecture.

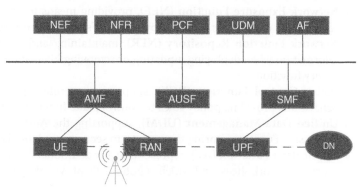

the westbound interface (WBI), which allows CP interconnection and information exchange between CN and ANs, and between CN and UEs. It should be noted that in the model, NFs relating to the access network are denoted as access functions (AFs). The model also highlights each NF/AF as composed of a set of sub-functions (SFs) by which the customization of NFs and AFs can be achieved.

The completion of the design of the 5G system architecture requires the definition and design of the set of NFs and relating interfaces, which is a high complexity engineering task, essentially for two reasons: On one hand, the problem has many variables and few constraints and hence a wide solution space. On the other hand, there are no clear and objective metrics to evaluate and compare alternative architectures. Often, novel network architectures are designed as an incremental evolution of existing ones, in the attempt of ensuring system backward compatibility and to fix specific shortcomings, but this approach does not fit the need of defining a clean-slate architecture. Some relevant modularization methodologies have been provided by the literature, e.g., [24], often results of research projects. However, at this point in time, a final choice has already been made by the relevant standardization body 3GPP, and hence it is worth referencing to the agreed modularized architecture which the first deployments of the 5G network will feature [10]. This 5G modularized architecture is illustrated in Figure 5-10, for simplicity here referring to a non-roaming scenario.

The architecture has been decomposed into a set of NFs, and orthogonal sets of NFs have been defined for CP and UP, denoted as CPFs and UPFs, respectively. CP NFs have been classified as common control network functions (CCNFs) and slice-specific control network functions (SCNFs), to distinguish between those which might be shared among multiple network slices and slice-specific NFs. Among CCNFs, a NF is dedicated to the slice selection, and is hence denoted as network slice selection function (NSSF). The following CN NFs have been so far defined. For the RAN NFs, in particular the differences to the functions in 4G, the reader is referred to Section 6.4.

- **Access and Mobility Management Function (AMF)**, including the termination of RAN CP interface (N2) and of the NAS interface (N1), NAS ciphering and integrity protection, mobility management, lawful interception, access authentication and authorization, security anchoring, and security context management;
- **Session Management Function (SMF)**, including session management, UE IP address allocation and management, UP functions selection and control, termination of interfaces towards PCF, policy enforcement and QoS, and roaming functionality. The SMF is connected to the UPF via the N4 interface;
- **Authentication Server Function (AUSF)**, providing authentication and authorization functionalities;
- **Network Exposure Function (NEF)**, providing means to collect, store and securely expose the services and capabilities provided by 3GPP NFs, e.g., to third parties or amongst NFs themselves;
- **Network Function Repository (NFR)**, maintaining and providing the deployed NF instances information when deploying, updating or removing NF instances, and supporting the service discovery function;
- **Policy Control Function (PCF)**, supporting a unified policy framework to govern network behaviours, providing policy rules to control plane function(s) to enforce them;
- **Unified Data Management (UDM)**, supporting the Authentication Credential Repository and Processing Function, storing the long-term security credentials and subscription information;
- **Application Function (AF)**, representing any additional CP function which might be required by specific network slices, and which is potentially provided by third parties;

- **User Plane Function (UPF)**, including the following functionalities: anchor point for intra-and inter-RAT mobility, external PDU session point of interconnection, packet routing and forwarding, UP QoS handling, packet inspection and policy rule enforcement, lawful interception, traffic accounting, and reporting.

Within the modularized architecture framework, several interfacing models can be envisioned. In this respect, 5G systems will bring to light additional discontinuity with respect to legacy systems. As illustrated in Figure 5-10, together with the legacy 4G point-to-point interface model (not shown in the picture, but included in [10]), an alternative southbound interface (SBI) model among CP NFs has been defined. Each CP NF exposes a SBI, by which the authorized NFs can access the services it provides. SBIs provide higher flexibility in the interaction among NFs and allow multiple alternative interconnections among those to define slice-tailored architectures. Besides, the services exposed by NFs can be used for CP procedures customization.

5.4.2 RRC States

The UE state diagram has a direct impact to the UE context placement and the UE mobility management. In LTE, the UE Radio Resource Control (RRC) state diagram has two states, RRC Idle and RRC Connected, so as to reduce the complexity of mobility and connectivity procedures arising from having a larger number of states as in High Speed Packet Access (HSPA) systems. The RRC Idle minimizes the UE power consumption, network resource usage and memory consumption, while the RRC Connected was introduced for high UE activity and network-controlled mobility. The state transition from RRC Connected to Idle and vice-versa requires a considerable amount of signaling to setup the UE's AS context in the RAN, and introduces a delay beyond the 5G CP latency requirement of 20 ms. The main characteristic of the Connected state is the RRC connection between the UE and the network and consequent allocation of logical dedicated unicast resources for the transfer of CP signaling or UP data in uplink or downlink. The UE has AS context in the RAN, and the RAN knows the cell where the UE is located. Thereafter, the RAN can transfer unicast data to or from the UE without RRC state transition related extra signaling.

However, this approach is sub-optimal for cases where the UE has limited mobility and performs transmissions of rather small data that are followed by inactivity periods. In such cases, legacy systems typically move a UE back to Idle state in inactivity periods, leading to the problem that the UE has to establish a new connection to the network and perform the large set of actions mentioned before for every state change. If, on the other hand, a UE would be kept in RRC Connected also during longer inactivity periods, this would be suboptimal from a resource and energy efficiency perspective due to the permanent communication required between the AN and the UE.

For this reason, 3GPP has agreed on a new Connected Inactive state, which allows to keep the UE in a kind of 'always ON' state from the CN perspective, hence avoiding the additional signaling and respective delays involved in switching from Idle to Connected. Additionally, this approach maintains the UE context in the RAN, thus resulting in a reduction of the frequent UE context transmission to the RAN in case of sparse transmission followed by inactivity. Once the UE is registered to the 5G network, the connection to the CN is kept alive, though the RAN anyway may release the UE and transfer it to Idle state [25]. Further, the new Connected Inactive state provides a big range of reconfigurability options related to the UE tracking in the RAN, discontinuous reception (DRX) and discontinuous transmission (DTX) cycles, camping, measurements, and initial access configuration, etc., according to the actual use case requirements. For example, in case of static devices, the UE can always remain in the Connected Inactive state requiring only RAN tracking area updates for ensuring its operation, whereas in cases of eMBB use cases more frequent UE tracking and related tracking area update (TAU) can be applied. For details on the new RRC state in 5G, the reader is referred to Section 13.3.

5.4.3 Access-agnostic 5G Core Network

As already stated, one of the goals of 5G networks is the minimization of any dependencies between core and access networks. This is desired to enable access convergence among 3GPP, non-3GPP and fixed access networks. Such a convergence requires that the core and access networks are evolved in an independent way. In other words, the CN has to be "agnostic" to the characteristics and features of the access networks. Note that in previous 3GPP releases, similar goals were pursued under the "access integration" design goal, achieving only a certain level of interworking. With the design of "access-agnostic" CNs, it is expected that a more flexible and future-proof solution will be achieved. To achieve this notion of access-agnostic core, three alternatives have been identified and examined.

The first approach, illustrated in Figure 5-11, impacts the 5G CN only. It requires the NF terminating the AN/CN interface to be able to support all access networks (i.e., 5G RAN, LTE, Wi-Fi, etc.). In practical terms, this would require the definition of a convergence sub-function of the AMF. The advantage of this approach is that only the CN would be affected, and potentially all legacy access networks could be integrated without any modification required on their side. This solution, however, would increase the system complexity, as multiple AN-CN interfaces would

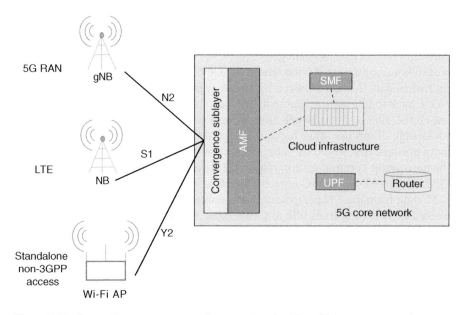

Figure 5-11. Connecting an access-agnostic core network with multiple access networks.

have to be supported (e.g., N2, S1, and Y2). Despite this complexity, this approach has finally been endorsed by 3GPP SA2 for 5G phase 1 systems (please refer to Section 17.2.1 for details of the phasing approach of 3GPP) exactly because no modifications are needed by the different access networks.

A second approach would be to define a common AN/CN interface, terminated by the AMF in the 5GC. This solution is the optimum solution from system perspective, but requires all access networks to support this single interface. Thus, the design complexity is pushed to the access, and existing interfaces (e.g., S1) would have to be replaced by the new interface. Such an approach would be less practical for already deployed and operational networks.

These two approaches would both lead to a *tight* integration of difference access networks, as actually system E2E behaviors could be consistently handled in terms of, e.g., mobility, security, QoS, etc., regardless of the access network being used.

A third approach, leading only to *loose* access integration is based on the definition of a convergence NAS layer protocol. This protocol, running at the UE and CN sides, would allow devices to roam throughout different access technologies even if mobility, security and QoS are treated differently. This approach is of lower complexity compared to the previous ones. However, it cannot be supported by legacy devices, rendering its adoption in the future complicated.

5.4.4   Roaming Support

MNOs have established since many years network capabilities for allowing the roaming of mobile users abroad. These capabilities usually involved expensive circuits and dedicated connections for handling the traffic of those users outside of the home network. The costs associated to this infrastructure setting is usually compensated by the payments of the roaming users when accessing their services.

In June 2017, the European Union has established a new roaming regulation applicable to 28 countries across Europe [26]. With this new regulation, the users from those countries will be able to use mobile services in the rest of the other European countries without extra charges.

This new situation incentivates a new manner of providing roaming services leveraging on multi-operator NFV deployments. The case is described here for LTE networks, since the architecture for roaming in 5G has not yet been completely defined.

The roaming traffic can be categorized as control plane traffic and user plane traffic. The control plane traffic includes the subscriber information and charging information and it typically has a small footprint on the exchanged data as opposed to user plane traffic. When the subscriber uses a specific service of his/her home operator or accesses external networks, the user plane traffic needs to be routed to the home network through the home packet gateway (P-GW). In roaming scenarios, the S8 interface connects serving gateway (S-GW) and P-GW nodes from different administrative domains, thus allowing the users to roam from a visited network to a home network and vice versa.

Figure 5-12 graphically represents the possible new architectures for roaming. In (a), the conventional connection of packet core entities enabling the roaming services between two mobile operators are shown. In (b), the possible new architecture is depicted, where the greyed P-GW represents the instantiation by the home operator of a P-GW network function on premises of the visited network, allowing the local breakout of services without being transported from the home network. Further details on the proposal for this architecture are described in [27].

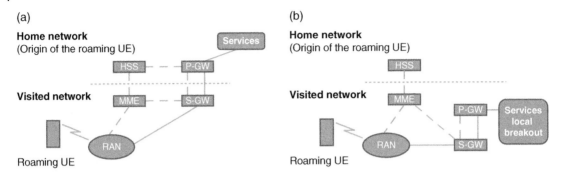

Figure 5-12. Architectural options for roaming services; (a) conventional and (b) new option.

5.4.5 Softwarized Network Control

The future 5G network will introduce a novel concept of network control by extending the software-defined routing or switching approach to all kinds of mobile NFs from both user and control plane, with a focus on wireless control functions such as scheduling or interference control. Therefore, different controllers have to be designed that apply the split between the logic of the network function and the part that can be controlled by an agent. This is a fundamental concept developed by some European research projects [22].

The role of such extended controller architecture is to abstract technology-specific or implementation-specific aspects of the network ecosystem, with interfaces towards the MANO stack and to different control applications implementing, e.g., QoE or QoS control or mobility management. A possible embodiment of this controller architecture is depicted in Figure 5-13.

An enhanced controller at the center of the architecture should have at least three kinds of interfaces: A southbound and northbound interface to and from the applications and NFs, and a third interface to the MANO. They are detailed next.

Interface A: This interface is used to enforce the conditions defined by the control applications that must be realized for a given traffic identifier on dedicated functions and resources in order to fulfil the targeted SLA. For example, via this interface the mobility management application can convey to the enhanced controller the exact network slice configuration, i.e., with the right NF type selection and the right composition and configuration of different NFs based on the selected network slice blueprint. On the other hand, this interface can convey the information regarding the current slice performance which is reported back to the corresponding control application. However, NFs may not just be assigned to a single network slice. In this case, this interface is used to enforce the conditions defined by the application that have to be realized for a given traffic type on shared resources and functions in order to fulfill the targeted SLA with respect to the relevant service policy.

Interface B: This controls and configures parameters of the dedicated or shared PNFs and VNFs which implement the NFs on the data path. This interface is hence used to configure and control these PNFs or VNFs. In case of shared NFs, these are likely to be NFs controlling scarce resources such as spectrum. The most common category of controllable NFs control SDN-compatible routers or forwarders, building the path(s) that connect the VNFs of a service chain. The interface is equivalent to the southbound interface of an SDN controller.

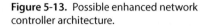

Figure 5-13. Possible enhanced network controller architecture.

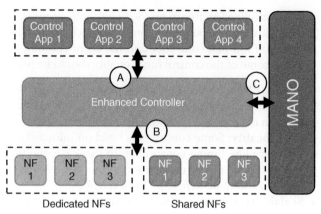

Interface C: This interface is used to convey the control application specific information derived during the translation from high-level tenant requests and established SLAs into the network slice resource provisioning, NFs logic, and lifecycle parameters. For example, with respect to the mobility management application, this interface can convey the information about the most suitable mobility management scheme and corresponding network slice template with respect to the agreed SLA and service policies. Depending on the QoS service requirements attached to the network slice, a corresponding mapping onto latency, bandwidth, computing and storage requirements, QoS thresholds to monitor, etc. can be conveyed to the QoS/QoE application via this interface.

5.4.6 Control/User Plane Split

Due to the envisioned extensive application of the software-defined networking concept to all the NFs in a mobile network, the control and user plane will necessarily be split by the controllers, as controllers split the functionality between the application *logic* (i.e., the intelligence that runs in the applications) and the *agents* running in the NFs. Therefore, the 5G control and user plane architecture pivots on the controllers.

Applications are a centralized control layer that comprise an ecosystem of applications controlling the underlying NFs (dedicated or shared), exploiting the advantages of the SDN approach:

- **Joint access and core optimization**: The transition towards a VNF based architecture is blurring the current concepts of access and CNs. The possibility of running VNFs almost everywhere in a *cloudified* network enable the possibility of jointly optimizing functions that have always been in separated boxes. This fact has important implications for the network control and orchestration;
- **One common interface to all network functions**: Having one reference point (i.e., the interface A) on the northbound interface of the controllers allows for diversity in both NF and application developers. The controller should offer the necessary abstraction to provide efficient centralized control;
- **Efficient resource sharing**: Having one centralized reference point for NFs assigned to different slices allows for their control according to the shared resources quota they are entitled to. This is a cleaner approach compared to different peer-to-peer interfaces across network functions.

The centralized approach provided by applications is, however, just one of the layers of the CP/UP architecture, as for example the one proposed in [22]. Not all the functionality can be split into a logic running on top of a controller and the agent running in the underlying NFs. That is, for different reasons, some of the control functionality should be managed in a legacy distributed way:

- **Legacy PNFs**: Legacy PNFs that should be integrated into a network slice may not support the enhanced controller approach described above. Therefore, their behavior should be integrated through the southbound interface of the controllers;
- **Data locality**: Some control NFs build on information available locally that should be processed with very low timing constraints. In this case, the performance gains obtained by a centralized approach may not be enough for certain NFs. For example, this is the case of link adaptation or Hybrid Automatic Repeat reQuest (HARQ);
- **Scalability**: For some NFs, the overhead introduced by a fully softwarized and centralized control may be too much, especially when configured for extreme situations. As an example, a centralized Medium Access Control (MAC) scheduler that controls several base stations may hardly be implementable as an application. Therefore, some of the functionality is necessarily offloaded to distributed schedulers that can operate at wire speed.

NFs are controlled through the southbound interface B described above. Also, the network functions mentioned before differ significantly from each other. Therefore, it is near at hand that interface B will require substantially different capabilities. However, this SBI might become very feature-rich and complex.

Some NFs exist for which possibilities to split control and execution parts have been discussed already in literature and for which suitable interfaces have been described:

- **Mobility management**: Aside the mobility management schemes standardized for LTE, other schemes like Proxy Mobile IP (PMIP), VertFor, OFNC and LIME have been discussed in the literature. For some of them, the OpenFlow protocol [28] would for instance be suitable to separate the controller from the router that redirects data packets when mobile terminals move;
- **Routers in the transport network**: Routers in the transport network are the "classical SDN device". Hence, for those the OpenFlow protocol would be suitable as well.

The above examples show that the properties of the southbound interfaces can vary for different NFs. It remains to be seen to which extent the requirements of different network functions on these interfaces can be homogenized. Ideally, all network functions could be served through a single common interface. If such an interface will become too feature-rich and complex, an alternative might be to define several interfaces for groups of network functions with similar requirements.

5.5 Internetworking, Migration and Network Evolution

The concepts, architectural trends and techniques presented in the previous sections are specific to 5G. However, one fundamental aspect of the research efforts in this field is the compatibility with the current architecture and the migration path from the current 4G architecture to the 5G one. In this section, we discuss about how the current RAN and core architectures can be migrated to the future architectures.

5.5.1 Interworking with Earlier 3GPP RATs

The integration of novel 5G air interfaces and the evolution of LTE that goes beyond the previous generation integration paradigms enables a high-performing multi-RAT system that is able to treat accordingly mobility between the various RATs and air interface variants (AIVs), handle UP aggregation for the various RATs and AIVs, charging, etc. For details on the definition of the notion of AIVs and further details, the reader is referred to Section 6.4.

The integration of the 5G AIVs with the previous generations requires a common interface with the 5GC for supporting E2E network slicing, new 5G services, enhanced multi-RAT integration with common CN functions (e.g., several CN functions could be designed to be independent of the access), and new CP/UP splits in the 5GC. Such interface further facilitates dual connectivity, simplifies NAS implementations (by, e.g., avoiding a dual protocol stack at the UE), UE context management, etc.

In the RAN, the various AIVs shall have their logical inter-node interface, Xn, for mobility purposes, multi-connectivity (in cases of non-collocated deployments of nodes, smarter RAN paging based schemes, interference management, traffic engineering schemes, etc.)

Furthermore, a lot of sophisticated features such as Multimedia Broadcast Multicast System (MBMS), home eNBs, self-organizing networks (SON) enhancements, IP multimedia sub-system voice over LTE (IMS VoLTE), improved access control, mobility enhancements, etc., have already been added to LTE since Release 8, such that the interworking of the previous generations' RAN with the 5GC can be straightforwardly implemented.

Figure 5-14 shows the system architecture for the interworking between 4G and 5G. The interworking architecture enables users to move between the 4G and the 5G system. Further on, the system architecture must also enable the system to serve a user with 4G and 5G simultaneously, for instance via multi-connectivity, as covered in detail in Section 6.5. As a further option, the 4G BS or eNB can also be directly connected to the 5G CN. In such a case, the EPC can be removed.

The LTE EPC and the 5GC interact to facilitate the aforementioned interworking. In particular, the Mobility Management Entity (MME) interacts directly with the 5G control plane functions (CPFs),

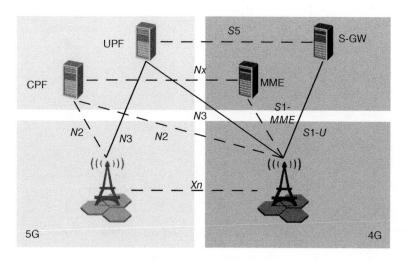

Figure 5-14. Simplified architecture for interworking between 4G and 5G [10].

such as the AMF, SMF, etc., as described in Section 5.4.1. Similarly, the S-GW should be connected with the respective user plane functions of the 5G system as shown in the figure.

In this context, it is worth to delve a bit more into the specific scenarios of 4G/5G co-deployment that 3GPP is considering [28], as illustrated in Figure 5-15. The so-called option 3 assumes that both LTE eNBs and 5G gNBs are connected to an EPC, and that LTE eNBs serve as the anchor layer for 4G/5G multi-connectivity. This option is mainly envisioned for very early 5G deployments, for instance in hotspots, as it maximally builds upon the existing installed base. Option 4 corresponds to the opposite case, where both eLTE eNBs and 5G gNBs are connected to a 5GC, and the 5G gNBs serve as the mobility anchor. Finally, option 7 corresponds to the case where a 5GC is employed, but the eLTE eNBs serve as the anchor. This option may be interesting for MNOs who still want to utilize existing LTE deployments as the layer that likely provides the better coverage for a long period to come, but want to do an early network upgrade to the 5GC. Beyond these main options, 3GPP further distinguishes whether the 4G/5G multi-connectivity is based on a bearer split at the master BS (default), whether the secondary BS has an independent UP connection to the CN (options 3a, 4a and 7a), or whether the bearer split is performed by the secondary BS (options 3x and 7x), as indicated through the three columns in Figure 5-15. The latter two bearer usage options are for instance interesting for options 3 and 7, where the default bearer split at the (e)LTE eNB likely has the disadvantage

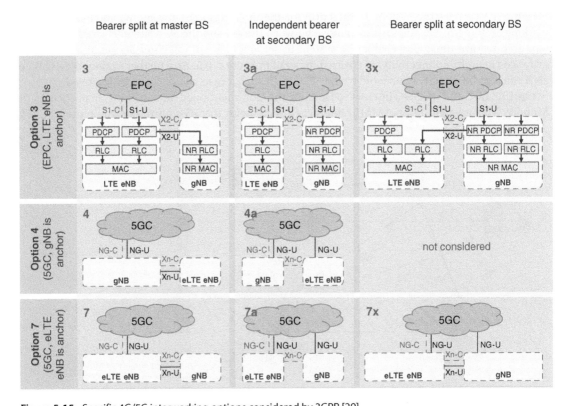

Figure 5-15. Specific 4G/5G interworking options considered by 3GPP [29].

that the LTE PDCP cannot handle the large data rates that the 5G gNB is expected to provide. Note that for brevity, the bearer split details are only shown for option 3 in the figure, as the principle is the same also for options 4 and 7.

5.5.2 Interworking with Non-3GPP Access Networks

The interworking of cellular networks with any other types of access networks has been of prime importance to 3GPP for over the past decade. 3GPP supports the integration of non-3GPP access (e.g., Wi-Fi and WiMAX) to the cellular network so as to be able to offload traffic from the cellular network. Based on the success and the penetration of these networks, the efforts to define the interworking frameworks has mainly focused on the case of Wi-Fi, where its paramount success offers new opportunities to the operators, since it allows them to take advantage of the use of unlicensed spectrum and offload traffic. Note that based on [30], the global mobile data traffic grew by 63% in 2016. Moreover, 60% of total mobile data traffic was offloaded onto the fixed network through Wi-Fi or femtocells in 2016. Thus, the need to offload part of the traffic is expected to play a significant role for 5G networks where the growth of mobile data volumes from smartphones will be a dramatic one. Although Wi-Fis are not considered to be a pervasive solution, they will provide location-specific solutions for coverage and capacity in an economically viable way. Moreover, as Wi-Fi has penetrated several application areas from sensors to vehicles' communication, it is beneficial to integrate this radio access technology with the 5G access network to improve not only the experienced performance but also the security in a seamless way.

Non-3GPP access networks may be distinguished as "trusted" and "non-trusted" depending on whether the 3GPP operator trusts or not the communication of UEs with the EPC. The interworking can be further characterized as *loose* or *tight*. In a loose integration, the non-3GPP network performance is usually not under the operator's control and is used to allow partnership of operators with wireless Internet service providers (WISPs). A loose integration may result in the end user experience of a loss of IP session continuity when there is a movement among 3GPP and non-3GPP access networks. In contrast, a tight integration suggests that the operator has performance control of the non-3GPP access, and the integration point can be a common CN infrastructure, but current efforts also suggest an integration at the radio access network level. This type of integration targets service continuity among access networks and seamless service provision [31].

As mentioned before, the interworking with non-3GPP access networks is an old story that starts with Release 6 and Universal Mobile Telecommunications System (UMTS). Initial scenarios were discussing different options and levels of loose or tighter integration of the two systems. An example of the former was the interworking wireless local area network (I-WLAN) that introduced an access and a data gateway to control the routing of the packets traveling to and from a Wi-Fi access point (AP) together with an Authentication, Authorization and Accounting (AAA) server to perform the security functions. For a tight integration, the generic access network (GAN) framework, also known as unlicensed mobile access (UMA), introduced an appropriate controller in the RAN to hide from the network the fact that the base station was essentially a Wi-Fi. This solution required for the UEs to be UMA-compatible (i.e., support the needed protocols), but could offer several benefits like the execution of seamless handoffs among the networks.

In Release 8, the EPC also provided solutions (i.e., mobility, security, policy control, and charging) for a loose interworking between these different access systems through the introduction of the evolved Packet Data Gateway (ePDG) and the respective interface to policy, security and charging components. Moreover, an extra component called Access Network Discovery and Selection Function (ANDSF) was introduced to inform end devices about the existence of Wi-Fi APs in their vicinity and the operator's policies that were essentially influencing the control of the connections to the access networks [32]. Note that Release 8 did not allow simultaneous connections via multiple access networks. Thus, the ANDSF policies were defined on a per-UE basis and not on a per-flow basis, forcing all traffic to use just a single access network. These shortcomings were addressed in Release 10 where IP Flow Mobility (IFOM) and multi-access Packet Data Network (PDN) connectivity (MAPCON) were introduced [33]. The former solution is a UE-centric solution that allows a UE to be connected to the same PDN over different access networks with the support of Dual Stack Mobile IP v6 (DSMIPv6). The latter allows to establish multiple PDN connections through different RATs, and ANDSF was extended to include routing policies based on the PDN identifier (for the MAPCON case) and/or the destination IP address and port numbers (for the IFOM case). Since many operators have deployed network instead of user mobility protocols, as for instance the General Packet Radio Service (GPRS) Tunnelling Protocol (GTP), in Release 13, 3GPP has introduced Network-Based Internet Protocol Flow Mobility (NBIFOM) [34]. This solution is based on Proxy Mobile IP (PMIP) or GTP based mobility protocols for both trusted and non-trusted 3GPP networks so as to transfer the desired support functionality from the UE to the network. As with IFOM, in this case individual IP flows are able to switch from one access network to another within the same PDN connection.

We need also to note that 3GPP has further investigated the potential interworking of Hotspot 2.0 with ANDSF to further improve the network discovery and selection of Wi-Fis over secure and trusted connections [35]. This is achieved by having the UE taking advantage of broadcast information transmitted by the Wi-Fi (e.g., numbers of users associated with an access point, load of the backhaul link, etc.).

All the aforementioned solutions improve indeed the ability of a typical 4G network to steer traffic over the most desired access network. However, it is clear that all these solutions are essentially patches to the existing architecture over well-defined interfaces and protocols (e.g., GTP). Also, improvements in the collection of information and the dissemination of policies and routing rules can enable algorithms to assist in the most optimum access selection [36]. However, for 5G networks, the possibility of introducing SDN solutions in the CN may greatly simplify the issue of traffic steering and minimize the need for complex and multiple solutions [37]. Over the past years several research proposals have analyzed the use of SDN and NFV in the CN [38]. The target is to fully separate control from data plane and also simplify routing in the CN, which also includes offloading techniques. Such solutions could be of interest to support roaming agreements between cellular operators and WISPs.

In Releases 13 and 14, 3GPP has started targeting solutions that fall under the category of a tighter integration among cellular and non-cellular access networks. More specifically, the following are supported in Release 14 [39] at the RAN level and operate only when the UE is in ECM-CONNECTED mode:

- **LTE-WLAN Aggregation (LWA)**: It capitalizes on the dual connectivity split-bearer architecture introduced in Release 12, where the aggregation of data links takes place at the PDCP layer. In this solution, the secondary link is provided by the Wi-Fi access point. The solution supports a non-collocated LWA scenario with non-ideal backhaul as well as a collocated scenario or ideal backhaul. The mechanisms essentially provide a tight resource aggregation where a data radio bearer can be either switched between LTE and Wi-Fi, or be split and provided simultaneously by the two access

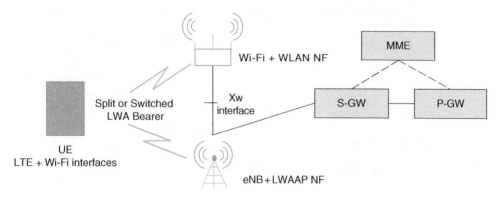

Figure 5-16. LWA radio protocol architecture for the non-collocated scenario [9].

technologies. This solution needs for the UE to be upgraded with the so-called LTE-WLAN Aggregation Adaptation protocol (LWAAP), whereas the Wi-Fi access point also needs to be upgraded with the WLAN Termination (WT) logical entity and support a new interface, called Xw, that connects it with the eNB. Figure 5-16 illustrates the LWA radio protocol architecture for the non-collocated scenario. Note that this approach is further elaborated in Section 6.5.3;

- **RAN-controlled LTE-WLAN Interworking (RCLWI):** Allows the eNB to send a steering command to the UE for WLAN offload. This solution still uses the CN-based offloading, but in this scheme the eNB is able to make the decision for traffic steering between the two access technologies, thus achieving a better performance compared to the previous solutions. This is a bearer handover solution and not a data aggregation one as LWA, and does not require the UE to be upgraded with the LWAAP protocol;
- **LTE-WLAN radio-level integration with IPsec Tunnel (LWIP):** This is based on the aggregation of resources like in LWA, but on IP level, and hence no bearer splitting is possible. The Wi-Fi link is controlled by the eNB. The solution introduces the use of IPsec tunnelling without requiring any modifications to the WLAN network, and hence supports the use of legacy WLAN networks.

Apart from the 3GPP activities, there are some research efforts to identify the possibilities to integrate 5G with Wi-Fi either in a tight or loose way. The interested reader may refer to Section 6.5.3, where the benefits of tight integration are shown based on simulation results, though the price is naturally a higher implementation complexity.

Moreover, [39] identifies the cellular RAN assistance parameters that are provided to the UE via RRC signaling. These may include signal strength and quality thresholds, WLAN channel utilization thresholds, WLAN backhaul data rate thresholds, and a list of WLAN identifiers and Offload Preference Indicators (OPI). The same document also describes the rules that apply when, apart from the aforementioned mechanisms, Access Network Discovery and Selection Function (ANDSF) rules are also available in the network. Note that CN elements like the MME may also be involved in several cases to determine the WLAN offloading permissions for the UE and the PDN connection.

The current 3GPP architectural specifications for 5G [10] also support the connectivity of the UE via non-3GPP access networks, e.g., WLAN access. Thus, in Release 15, the 5G CN will only support untrusted non-3GPP access networks. More specifically, it will access networks deployed outside the 5G-RAN. These are referred to as *standalone* non-3GPP accesses. Figure 5-17 presents the 5G

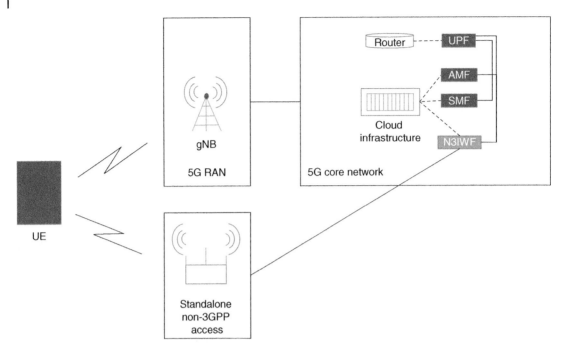

Figure 5-17. Non-roaming architecture for the 5G CN with non-3GPP access [10].

architecture where an interworking function (N3IWF) is introduced to support the access through untrusted non-3GPP access networks.

N3IWF supports the establishment of IPsec tunnels with the UE and relays the information needed to authenticate the UE and authorize its access to the 5G CN. It supports the AMF selection process and relays signaling between the UE and the AMF, and also handles signaling with the SMF. Moreover, it encapsulates packets for IPSec. A UE that accesses the 5G CN over a standalone non-3GPP access shall, after UE attachment, support NAS signaling with 5G CN control plane functions. A UE simultaneously connected to the same 5G CN of a public land mobile network (PLMN) over 3GPP access and non-3GPP access shall be served by a single AMF if the selected N3IWF is located in the same PLMN as the 3GPP access.

Finally, for the reason of completeness, we need to briefly address the case of enhanced Licensed-Assisted Access (eLAA). This solution essentially enables operators to use both licensed and unlicensed spectrum. More specifically, in Releases 13 and 14, the specifications allow to aggregate licensed with unlicensed spectrum to enable opportunistic access and improve throughput. Initially, LAA was used to improve only downlink transmission by employing carrier aggregation with the unlicensed spectrum. The eLAA is used also for uplink transmissions by taking advantage of unused or lightly used channels in the 5 GHz band. To achieve this, dynamic channel selection is performed over unused channels. This solution periodically checks for available channels and automatically signals LTE off if a Wi-Fi user is detected. It requires enhancements like channel sensing, discontinuous transmission on a carrier with limited maximum transmission duration, dynamic frequency selection, etc. These enhancements are impacting channel design, estimation and reporting, HARQ, and radio resource management.

Based on the previous analysis, it is clear that 3GPP is currently supporting a number of solutions for the integration between 3GPP and non-3GPP systems, although not all of these solutions have been widely deployed. At the same time, the decision to move forward on the specification of 5G standards in phases will not bring a complete solution in one step. The current activities have focused mainly on the latest solutions like eLWA in order to improve the performance for the operators that control Wi-Fis under their domain. On the other hand, it will still be beneficial for the operators to support roaming agreements with WISPs. Based on a recent study [40], operators have diverse data offload strategies in unlicensed spectrum. According to the study, "more than half (55%) of respondents plan to deploy Licensed Assisted Access (LAA) and prefer the notion of operating LTE in unlicensed spectrum. Wi-Fi offload solutions, such as LTE + Wi-Fi Link Aggregation (LWA) and LTE Wi-Fi Integration, (a forerunner to the LWA), were both favored by 41 percent of respondents".

In order for 5G networks to efficiently support the integration and interoperation of non-3GPP access networks, we foresee two main areas where further investigations are needed. The first one, mentioned already, is related to the potential simplifications in a number of schemes that may arise with the adoption of SDN principles in the core and transport networks. The second issue is related to the support of the different access networks by the CN in an access-agnostic way, as described in Section 5.4.3. This will require to select the most efficient option of the three alternatives and design the potential extensions needed by the 5G CN modules (e.g., AMF and SMF).

5.5.3 Network Evolution

The success of Internet data based services over the last decades has provided the fundamental basis of the global reachability of content accessed by the end-users and served by content providers, which is facilitated by the interconnection of networks. Traditional interconnection services have been merely based on a pure interchange of IP traffic, even without direct physical connection of the parties. Advertisement of IP prefixes with the usage of the Border Gateway Protocol (BGP) protocol allows the parties to know how to reach a remote entity. This peer and transit model enables global connectivity.

Nowadays, the wide deployment of computing facilities across operators' networks will bring the opportunity, enabled by multi-tenancy, of offering a subset of such infrastructure to third parties, for the deployment of network functions, or even the trading of entire resource slices.

Furthermore, the envisioned 5G ecosystem of multi-domain services provisioned in an automated on-demand manner, by using virtualization and slicing capabilities in order to accommodate services beyond the footprint of a single administrative domain, will require to overcome the administrative domain borders, extending ownership and footprint through partnership, achieving multi-provider orchestration. The opening of the network to verticals will also be an incentive for that. Evolutionary services like Virtual -Network-Functions-as-a-Service (VNFaaS) or Slice-as-a-Service (SlaaS) are the target, favored by the advent of 5G networks, with strict requirements in terms of flexibility and performance. For more details on the related change of the mobile ecosystem, the reader is referred to Section 2.6. The solutions facilitating such multi-domain environment should permit programmability, flexibility and automation, but should also allow for agile contracting, invoking and settling of services reducing significantly the time for provision.

5G networks are envisioned as a conglomerate of federated providers of both networking and compute resources, contributing to the overall resources required to fulfill a specific service level or marketplace agreement. 5G PPP advances in [5] some ideas about multi-operator orchestration, where there should be mechanisms in place for orchestrating resources and/or services using domain orchestrators belonging to multiple administrative domains. For doing that, information about resource capabilities and service capabilities (in the form of supported network functions) should be interchanged between operators. This clearly goes beyond that existing interchange of IP prefixes for ensuring reachability. Those resources and functions have to be flexibly traded and provisioned in a secure manner.

A new role seems to emerge in the form of a multi-domain orchestrator. Such an entity will be in charge of handling multi-operator information and taking orchestration decisions in such an environment. New trust models need to be considered, since the level of information required for such orchestration actions between operators will necessarily be more detailed as it is now. Also new businesses have to be developed with the intricate ecosystem of a multiplicity of provider types (infrastructure, network, transit, service, etc.) and consumers, as discussed in Section 2.6.

Key to this new ecosystem will be the grade of standardization of the resulting mechanisms, e.g., protocols and APIs. While in a single administrative domain the interactions between different elements can be to some extent tailored and customized to the need of the single provider, in a collaborative multi-operator environment it is required to define common mechanisms totally accepted by the parties. In a multi-domain environment, it is necessary to have a common understanding on the services and resources offered by each party. To do that, the same semantics and abstractions have to be handled by the different administrative domains in order to ensure consistency. Such abstractions at technical level imply the utilization of common information and data models for the resources to be configured and used. It is then needed to work in a standard way of interchanging the necessary information in this new multi-provider environment, as BGP has been for the traditional IP interconnection. That target is being developed in a number of initiatives, as described in Chapter 10.

5.6 Summary and Outlook

This chapter has reviewed the current trends in the E2E architectural design for 5G networks. In order to fulfill the heterogeneous and challenging requirements that are expected by future 5G networks, the network architecture will undoubtedly change from the one currently used by 4G networks. This comes from the recent spreading of different enablers, for instance along the lines of SDN and NFV, that finally allow for the implementation of several enhanced design requirements such as network slicing and multi-tenancy.

These new features require new elements or management techniques in the architecture that are currently being developed by both 5G research projects and standardization entities. The new elements necessarily leverage on new architectural trends, such as a C/U-plane split, a modularization of the architecture, and the definition of the necessary interfaces. Finally, the chapter has reviewed the current efforts in migrating the current architecture to the new 5G one. After this chapter has provided the big picture on the 5G E2E architecture and related novel design paradigms, the subsequent chapters will now delve into various details of the 5G architecture.

References

1 NGMN, White Paper, "5G White Paper", March 2015
2 5G PPP, White Paper, "5G empowering verticals", Feb. 2016

3 T. Frisanco, P. Tafertshofer, P. Lurin and R. Ang, "Infrastructure Sharing for Mobile Network Operators; From a Deployment and Operations View", International Conference on Information Networking (ICOIN 2008), Jan. 2008

4 ETSI Group Specification, "Mobile Edge Computing (MEC); Framework and Reference Architecture", V1.1.1, Mar. 2016

5 5G PPP Architecture Working Group, White Paper, "View on 5G Architecture", v2.0, July 2017

6 P. Mach and Z. Becvar, "Mobile Edge Computing: A Survey on Architecture and Computation Offloading", IEEE Communications Surveys and Tutorials, vol. 19, no. 3, pp. 1628–1656, Q3 2017

7 3GPP TR 38.912, "Study on New Radio access technology (Release 14)", V14.1.0, Aug. 2017

8 3GPP TS 38.300, "NR and NG-RAN Overall Description, Stage-2 (Release 15)", V15.0.0, Jan. 2018

9 3GPP TS 36.300, "Evolved Universal Terrestrial Radio Access (E-UTRA) and Evolved Universal Terrestrial Radio Access Network, Overall Description (Release 14)", V15.0.0, Jan. 2018

10 3GPP TS 23.501, "System Architecture for the 5G System (Release 15)", V15.0.0, Dec. 2017

11 K. Chatzikokolakis, P. Spapis, A. Kaloxylos and N. Alonistioti, "Toward spectrum sharing: opportunities and technical enablers", IEEE Communications Magazine, vol. 53, no. 7, pp. 26–33, July 2015

12 K. Chatzikokolakis, P. Spapis, A. Kaloxylos, G. Beinas and N. Alonistioti, "Spectrum sharing: A coordination framework enabled by fuzzy logic", International Conference on Computer, Information and Telecommunication Systems (CITS 2015), July 2015

13 B. Singh, K. Koufos and O. Tirkkonen, "Co-primary inter-operator spectrum sharing using repeated games", IEEE International Conference on Communication Systems (ICCS 2014), Nov. 2014

14 Open Base Station Architecture Initiative (OBSAI), OBSAI System Spec V2.0, see http://www.obsai.com/specs/OBSAI_System_Spec_V2.0.pdf

15 A. de la Oliva et al., "An overview of the CPRI specification and its applications to C-RAN based LTE scenarios", IEEE Communications Magazine, vol. 54, no. 2, pp. 152–159, Feb. 2016

16 IEEE 1904 Task Group, P1914.3, "Standard for Radio Over Ethernet Encapsulations and Mappings", see https://standards.ieee.org/develop/project/1914.3.html

17 Enhanced Common Public Radio Interface (eCPRI), see http://www.cpri.info/spec.html

18 5G PPP 5G-Crosshaul project, see http://5g-crosshaul.eu/

19 5G PPP 5G-Xhaul project, see http://www.5g-xhaul-project.eu/

20 5G PPP 5G-Crosshaul project, Deliverable D1.1, "5G-Crosshaul Initial System Design, Use Cases and Requirements", Nov. 2016

21 X. Costa-Perez, A. Garcia-Saavedra, X. Li, T. Deiss et al., "5G-Crosshaul: An SDN/NFV Integrated Fronthaul/Backhaul Transport Network Architecture", IEEE Wireless Communications Magazine, vol. 24, no. 1, pp. 38–45, Feb. 2017

22 5G PPP 5G NORMA project, Deliverable D5.2, "Definition and specification of connectivity and QoE/QoS management mechanisms – final report", July 2017

23 3GPP TR 23.799, "Study on Architecture for Next Generation System", V14.0.0, Dec. 2016

24 X. An et al., "Architecture modularisation for next generation mobile networks", European Conference on Networks and Communications (EuCNC 2017), June 2017

25 I. L. Da Silva, G. Mildh, M. Säily and S. Hailu, "A novel state model for 5G Radio Access Networks", IEEE International Conference on Communications (ICC 2016) Workshops, May 2016

26 REGULATION (EU) 2015/2120 OF THE EUROPEAN PARLIAMENT AND OF THE COUNCIL laying down measures concerning open internet access and amending Directive 2002/22/EC on universal service and users' rights relating to electronic communications networks and services and Regulation (EU) No 531/2012 on roaming on public mobile communications networks within the Union.

27 J. Kim, L.M. Contreras, H. Woesner, D. Fritzsche, L. Cominardi and C.J. Bernardos, "GiLAN Roaming: Roam Like at Home in a Multi-Provider NFV Environment", submitted to the IEEE Conference on Network Function Virtualization and Software Defined Networks 2017

28 OpenFlow, see http://www.openflow.org

29 3GPP TR 38.801, "Study on new radio access technology: Radio access architecture and interfaces", V14.0.0, April 2017.

30 Cisco, White Paper, "Cisco Visual Networking Index: Global Mobile Data Traffic Forecast Update, 2016–2021", see https://www.cisco.com/c/en/us/solutions/collateral/service-provider/visual-networking-index-vni/mobile-white-paper-c11-520862.html

31 4GAmericas, "Integration of Cellular and WiFi networks", Sept. 2013

32 M. Olsson, S. Sultana, S. Rommer, L. Frid and C. Mulligan, "SAE and the Evolved Packet Core. Driving the Mobile Broadband Revolution", Elsevier Academic Press, 2009

33 F. Rebecchi et al., "Data Offloading Techniques in Cellular Networks: A Survey",. IEEE Communications Surveys and Tutorials, vol. 17, no.2, pp. 580–602, Q2 2015

34 3GPP TS 23.161, "Network Based IP flow Mobility", March 2016

35 S. Barbounakis et al., "Context-aware, user-driven, network-controlled RAT selection for 5G networks", Elsevier Computer Networks, 2016

36 K. Rao et al., "Network selection in heterogeneous environment: A step toward always best connected and served", International Conference in Modern Satellite, Cable and Broadcasting Services (TELSIKS 2013), Oct. 2013

37 O. Comeras et. al., "An evolutionary approach for the evolved packet system", IEEE Communications Magazine, vol. 53, no. 7, pp. 184–191, July 2015

38 V.G. Nguyen et al., "SDN and Virtualization-Based LTE Mobile Network Architectures: A Comprehensive Survey", Wireless Personal Communications Journal, vol. 86, no. 3, pp. 1401–1438, Feb. 2016

39 3GPP TS 23.402, "Architecture enhancements for non-3GPP accesses (Release 15)", V15.2.0, Dec. 2017

40 Telecommunications Industry Association –TIA Market Report, "5G operator Survey", Jan. 2017

6

RAN Architecture

Patrick Marsch¹, Navid Nikaein², Mark Doll³, Tao Chen⁴ and Emmanouil Pateromichelakis⁵

¹ *Nokia, Poland (now Deutsche Bahn, Germany)*
² *EURECOM, France*
³ *Nokia Bell Labs, Germany*
⁴ *VTT, Finland*
⁵ *Huawei German Research Center, Germany*
With contributions from Nico Bayer, Jakob Belschner, Salah Eddine Elayoubi, Jens Gebert, Caner Kilinc, Tomasz Mach, Vinh Van Phan, Mikko Saily, Amit Sharma and Gerd Zimmermann.

6.1   Introduction

After having ventured into the 5th generation (5G) end-to-end (E2E) system architecture in Chapter 5, we will now focus on the radio access network (RAN) architecture, i.e. covering all aspects of a 5G system in the so-called access stratum (AS). Key points we will explore here are:

- How do 5G use case requirements as listed in Chapter 2 translate into RAN architecture requirements?
- What will the 5G protocol stack architecture look like, in particular: Which differences to that of the 4th generation (4G) of cellular communications will we see, to which extent will it be possible to tailor the protocol stack architecture and related network functions (NFs) to diverse service needs, and how will multi-service and multi-tenancy be reflected?
- How will the 5G RAN natively support RAN-based multi-connectivity?
- What are reasonable function splits in the 5G RAN, which logical entities will we consequently see, and how will these enable the support of diverse deployment scenarios?
- What are the key enablers for RAN programmability, as discussed in the context of the E2E system architecture in Chapter 5?

Clearly, all the points above are currently also being addressed in the 3rd Generation Partnership Project (3GPP), and by the time this book appears, many design decisions will have been taken. In this chapter, in addition to providing a brief overview of the latest 5G RAN architecture progress and related decisions in 3GPP, we hence focus on explaining 5G RAN architecture considerations from a broader perspective. For instance, we explain in detail the rationale for possible changes in the 5G RAN architecture as opposed to legacy systems, compare the pros

5G System Design: Architectural and Functional Considerations and Long Term Research, First Edition.
Edited by Patrick Marsch, Ömer Bulakçı, Olav Queseth and Mauro Boldi.
© 2018 John Wiley & Sons Ltd. Published 2018 by John Wiley & Sons Ltd.

and cons related to different architecture options, and also discuss concepts that have been proposed but not (or not yet) endorsed by 3GPP.

Please note that most parts of this chapter assume a split between core network (CN) and RAN as agreed in 3GPP and described in Section 5.3.2, though the majority of the presented concepts are independent of the exact split. For further information on possible alternative split options or a tighter coupling of CN and RAN, please also see Section 5.3.2.

The chapter is structured as follows. Section 6.2 shortly summarizes the related work on the 5G RAN architecture in 3GPP and the 5G Public Private Partnership (5G PPP). Section 6.3 then lists the 5G RAN architecture requirements, as identified in 3GPP and 5G PPP. Section 6.4 dives into details of the 5G RAN protocol stack architecture and related NFs, elaborating on the notion of NFs in the context of different air interface variants (AIVs), possible service-specific protocol stack optimizations, and NF instantiations in the context of multi-service and multi-tenancy. Section 6.5 then ventures into the 5G RAN support of multi-connectivity, either among multiple 5G radio legs, 5G and Long Term Evolution (LTE) or enhanced LTE (eLTE), or 5G and Wi-Fi. Section 6.6 goes in detail into vertical and horizontal split options in the 5G RAN, which are seen as essential to support diverse deployment scenarios, as covered in Section 6.7. Section 6.8 finally delves into aspects of RAN programmability, before the chapter is summarized in Section 6.9.

6.2   Related Work

6.2.1   3GPP

The 3GPP RAN working groups have completed the study on scenarios and requirements for Next Generation (NG) access technologies [1] and the study on New Radio (NR) access technology [2]. The next stages for the technical specifications of 5G networks in 3GPP, including the Next Generation RAN (NG-RAN), are underway. The overall descriptions of NR and the NG-RAN are captured in [3].

Based on considerations of various 5G use cases and deployment scenarios, their associated requirements and key performance indicators (KPIs), a set of requirements for the NG-RAN architecture and the migration of NG radio access technologies (RATs) has been established in 3GPP [1]. For example, the NG-RAN architecture shall support tight interworking between NG-RATs and (e)LTE, connectivity through multiple transmission points, flexible deployment and functional split options, network slicing and network function virtualization (NFV). It shall further support multi-vendor interoperability with open interfaces between RAN and CN and within the RAN, and between logical nodes and functions of NR and (e)LTE, as detailed in Section 6.3.

In general, the agreement is that the 3GPP NG-RAN consists of NR Gigabit Node-Bs (gNBs), providing the user plane (UP) and control plane (CP) protocol terminations for the radio interfaces toward the user equipment (UE), denoted as the NG-Uu interface. The gNBs may be interconnected with each other via an Xn interface. The gNBs are (ultimately) expected to be connected to the 5G CN, referred to as 5GC, via NG interfaces, more specifically to the Access and Mobility Management Function (AMF) via the NG-C or N2 interface, and to the User Plane Functions (UPF) via the NG-U or N3 interface [4]. For details on the AMF and UPF, see Section 5.3.2. Furthermore, the gNB may consist of a centralized unit (CU) and one or more distributed units (DUs) connected to the CU via Fs-C and Fs-U interfaces for CP and UP, respectively, with different function split options between

Figure 6-1. High-level architecture of the 5G-RAN [3][5] as assumed in this chapter. Note that for brevity here only gNBs and a 5G core network are depicted - for (e)LTE / NR interworking scenarios, see Section 5.5.1.

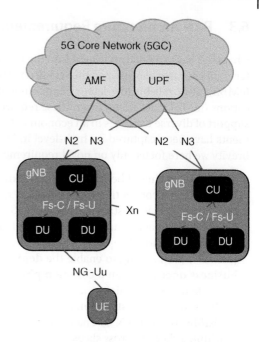

CUs and DUs [5], as discussed in detail in Section 6.6. The high-level architecture of the 3GPP NG-RAN is depicted in Figure 6-1. While the NG-RAN depicts a pure NR deployment, practical deployments will often be based on a co-existence of e(LTE) enhanced Node-Bs (eNBs) and gNBs. For details on related (e)LTE/NR interworking scenarios, see Section 5.5.1.

6.2.2 5G PPP

In 5G PPP phases 1 and 2, several projects have been looking into, or are still occupied with the 5G RAN architecture design, for instance

- **METIS-II** [6] has developed an overall 5G RAN design, focusing on the efficient integration of evolved legacy and novel AIVs, and the support of network slicing;
- **5G NORMA** [7] has developed a novel, adaptive and future-proof 5G mobile network architecture, with an emphasis on multi-tenancy and multi-service support;
- **COHERENT** [8] has developed a unified programmable control framework for coordination and flexible spectrum management in 5G heterogeneous access networks;
- **mmMAGIC** [9] has developed new RAN architecture concepts for millimeter-wave (mmWave) radio access technology, including its integration with lower frequency bands;
- **5G-MonArch** [10] aims to expand existing architecture with key innovations related to inter-slice control and cross-domain management, and a cloud-enabled protocol stack.

Furthermore, the 5G PPP Architecture Working Group facilitates the discussion among 5G PPP projects that are developing architectural concepts and components, and captures the consolidated findings across the projects in annually updated architecture White Papers [11].

6.3 RAN Architecture Requirements

Chapter 2 has provided an insight into the likely use cases that should be covered by 5G in the longer term, and in particular stressed the very diverse requirements associated to enhanced mobile broadband (eMBB), ultra-reliable low-latency communications (URLLC) and massive machine-type communications (mMTC) use cases. From these requirements as such, and from the need to enable the joint support of diverse use cases in an economically feasible way, the following RAN architecture requirements have been captured on high-level in [1] and formulated in more detail in [12]. Note that for brevity we here focus only on novel requirements specific to 5G:

- The 5G RAN should be able to scale to extremes in terms of throughput, the number of devices, the number of connections etc. To enable this, it should be able to **handle and scale CP and UP individually**, as for instance addressed in Section 6.6. Note that 3GPP explicitly also envisions that CP and UP signaling can be handled by different sites [1];
- The 5G RAN should support the **network slicing** vision from Next Generation Mobile Networks (NGMN) [13], aiming to enable the **deployment of multiple logical networks as independent business operations on a common physical infrastructure**, as detailed in Chapter 8. As a prerequisite to network slicing, it has been concluded in [14] that
  - **slices (or some abstraction thereof, such as particular groups of flows or bearers) should be visible to the RAN** to enable a treatment related to joint KPIs concerning all service flows within a slice or across slices;
  - The RAN CP and UP shall support devices with **dedicated slice selection and association mechanisms and procedures**. In addition, it should also be supported that a UE could be simultaneously connected to one or more network slice instances;
  - The RAN architecture should support **slice isolation**, e.g. by providing slice protection mechanisms so that critical fault- or security-related events within one slice, such as congestion or denial-of-service attacks, do not impact another slice;
  - The RAN architecture should support **efficient mechanisms for life-cycle management of slice instances**, i.e., to efficiently create, maintain/monitor, and finally release slice instances on the common infrastructure.
- One enabler for the system to handle the diverse service requirements stated before is that the overall network (both RAN and CN) should be **software-configurable and support software-defined networking (SDN)**. This means, for instance, that the logical and physical entities to be traversed by CP and UP packets are configurable, see Section 6.8;
- The 5G RAN architecture should support more **sophisticated mechanisms for traffic differentiation** than legacy systems in order to treat heterogeneous services differently and fulfill more stringent quality of service (QoS) requirements, see also Section 5.3.3.

Also, the following requirements have been identified, which are related explicitly to connectivity options and deployment:

- The 5G RAN architecture should **natively and efficiently support multi-connectivity**, for tight interworking among 4G and 5G, among different 5G radio variants (e.g., below and above 6 GHz carrier frequency), and among different transmission points and carriers of the same radio variant. As opposed to, e.g., the interworking between 3rd generation (3G) and 4G, all stated forms of interworking in 5G shall be possible on RAN level;

- The 5G RAN architecture should be designed such that it can **operate efficiently in a wide range of deployment scenarios**. More precisely, it should be able to maximally leverage centralized processing (e.g., in baseband hostelling scenarios), but also operate well in the case of distributed base stations (BSs) with imperfect backhaul (BH) or fronthaul (FH) infrastructure, with a soft degradation of performance as a function of BH quality. This implies the need for multiple function splits, as addressed in Section 6.6, and the usage of NFV, jointly enabling a flexible mapping of functionality to physical architecture, as detailed in Section 6.7. Note that 3GPP explicitly also envisions the support of multi-operator network sharing and open interfaces in the RAN;
- As a related point, the 5G RAN architecture should enable a native support of **self-xhauling** (i.e., self- fronthauling or backhauling), meaning that freely deployed BSs or radio frontends could autonomously establish wireless BH or FH connections using the same radio as the access, as detailed in Section 7.4. Going a step further, one may also envision that any end user device could serve as a wirelessly self-backhauled access node, which would constitute a major paradigm change to legacy systems that clearly differentiate between infrastructure nodes and devices.

6.4 Protocol Stack Architecture and Network Functions

We now investigate which implications the beforementioned RAN architecture requirements may have on the 5G protocol stack architecture and the NFs described by this.

6.4.1 Network Functions in a Multi-AIV and Multi-Service Context

One key difference between 5G and earlier cellular technology generations is that a 5G system will consist of multiple highly integrated radio variants, often referred to as *air interface variants* (AIVs) [15]. An example for an AIV could be an (e)LTE radio, which could be tightly integrated into a 5G system, but there will likely also be multiple novel radio interfaces introduced in the 5G timeframe, for instance tailored towards different carrier frequencies or cell sizes, which could also be considered as individual AIVs. For instance, it is widely understood that propagation at frequencies beyond 40 GHz carrier frequency is so different from that, e.g., below 6 GHz that at least the physical layer (PHY) and Medium Access Control (MAC) design of related radio interfaces will look quite different, as detailed in Chapter 11.

Further, the 5G RAN is expected to support diverse service types, as pointed out in Chapter 2, which naturally suggests radio functionality that is tailored to different service needs. However, from the perspective of the overall 5G RAN design, one should clearly strive to have as much communality in the NFs and overall protocol stack design as possible, despite the notion of different AIVs and different services, for the sake of an efficient standard description and efficient implementation on device and infrastructure side. In this context, it is assumed that the following terms and notions will be relevant in 5G:

- **AIV-specific** or **service-specific NFs** are functions that are designed for or tailored towards a specific AIV or service, respectively. An example for an AIV-specific NF would be a particular interference mitigation mechanism on MAC level that is designed to operate on a frame timing used for mmWave communications, and in the context of high beam directivity and link volatility;

- **AIV-agnostic** or **service-agnostic NFs** are functions that are agnostic to the protocol stack layers underneath, or to the services for which they are used. An example would be a Radio Link Control (RLC) level data segmentation function that could be applied regardless of the AIV or service;
- **AIV-overarching** or **service-overarching NFs** are functions that are specifically designed to integrate or aggregate multiple AIVs, or handle multiple services. An example could be a multi-AIV traffic steering function that dynamically maps services to different AIVs depending on instantaneous radio characteristics and QoS targets. An example for such function is provided in Chapter 12.

The aforementioned notions of NFs are illustrated in Figure 6-2. In this example, there are two AIVs (one using a carrier frequency below 6 GHz, one using mmWave frequency) and two services (eMBB and URLLC) involved. It is here assumed that a URLLC device is only served via the first AIV, while the eMBB device is served via multi-connectivity between the two AIVs. Please note that for brevity only the UP is shown, and the protocol stack at the device side is skipped. In this example, there are MAC and PHY NFs tailored towards mmWave communications. Further, there are some MAC NFs tailored towards either URLLC or eMBB. However, the RLC implementation is assumed to be the same for all AIVs and services, hence agnostic towards these. Please note, though, that there are still individual instantiations of the RLC NFs for the different AIVs and services, as discussed further in Section 6.4.4. A multi-service scheduler, as for instance detailed in Section 12.3, and multi-AIV traffic steering functionality, see Section 12.4, are here given as examples of service- and AIV-overarching NFs, as they are designed specifically to optimize operation across multiple services and AIVs, respectively. Please note that Figure 6-2 serves for illustration purposes only; in practise, the usage and granularity of service- or AIV-specific NFs may be very different and very implementation-specific, as we will see in the following sections.

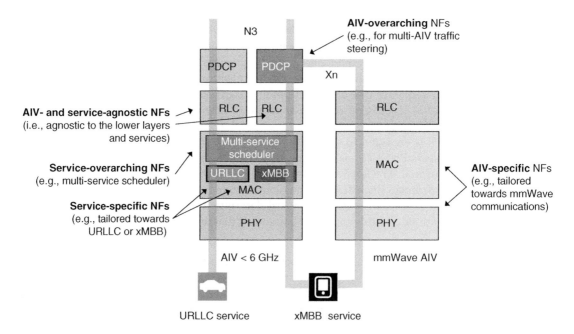

Figure 6-2. Examples for user plane NFs that are specific, agnostic or overarching w.r.t. AIVs or services.

6.4.2 Possible Changes in the 5G Protocol Stack Compared to 4G

In this subsection, we explore the different protocol stack layers, list their current functions as in LTE-A, summarize which changes 3GPP has already agreed upon for 5G, and elaborate on any further possible changes that may be introduced in 5G. Please note that the 3GPP status described here is to a large extent based on the NR UP related agreements captured in Annex A2 of [16] in March 2017. Further, we focus only on major decisions here for brevity.

6.4.2.1 Service Data Adaptation Protocol (SDAP)

3GPP has agreed to introduce the Service Data Adaptation Protocol (SDAP) [3] as a new protocol layer in NR that has the role to provide the mapping between a QoS flow and a data radio bearer, and mark the QoS flow ID in uplink (UL) and downlink (DL) packets, as described in Section 5.3.3. Typically, a single protocol entity of SDAP is configured for each individual protocol data unit (PDU) session, except for dual connectivity (DC), where two entities can be configured.

6.4.2.2 Packet Data Convergence Protocol (PDCP)

In LTE, the Packet Data Convergence Protocol (PDCP) layer is responsible for

- Header compression and decompression;
- Security aspects (i.e., ciphering and deciphering of UP and CP data, integrity protection and integrity verification of CP data);
- Maintenance of PDCP sequence numbers (SNs), detection and discarding of duplicates and timer-based discard;
- Routing and reordering of PDCP PDUs in the case of split bearers;
- Data-recovery procedure for split bearers in DC, for instance needed when part of the data transmitted over one radio leg is lost due to bad radio conditions;
- Retransmission of PDCP service data units (SDUs) in the context of handover.

For 5G, the main challenge or design goal w.r.t. PDCP lies in the reduction of duplicate functions across layers, the reduction of overhead, the decoupling of the processing related to multiple layers, and the support of enhanced reliability. Recently, 3GPP has already agreed that

- Complete PDCP PDUs, i.e. after reassembly in RLC, can be delivered out-of-order from RLC to PDCP. This allows the PDCP to already start the processing of some PDUs (for instance already starting the header processing) without having to wait for the RLC to gather and reorder the PDUs before delivery to PDCP. Note that an in-sequence delivery from RLC to PDCP, as in legacy systems, can still be configured if needed;
- Despite proposals to link the SNs in RLC and PDCP, 3GPP has agreed to keep independent SNs in RLC and PDCP for maximum segregation of the layers;
- For URLLC services, packet duplication is supported for both UP and CP in PDCP for increased reliability and handover robustness.

In general, it is considered that PDCP processing in 5G may be service-tailored, as discussed also in Section 6.4.3. For instance, header compression may be omitted if the payload is large and/or latency is crucial (e.g., for eMBB or URLLC applications), while being more pronounced for mMTC. Also, it is expected that further refinements of the PDCP and RLC handling may be beneficial in the context of multi-connectivity. For example, if retransmissions are handled in both PDCP (related to handover) and RLC (related to ARQ), and one PDCP entity is related to multiple RLC entities reflecting different radio legs, there should be means for the coordination of the re-transmissions in the different protocol entities.

6.4.2.3 Radio Link Control (RLC)

In LTE, the RLC layer can operate in acknowledged mode (AM), unacknowledged mode (UM), or transparent mode (TM), and is responsible for

- error correction through Automatic Repeat reQuest (ARQ);
- concatenation, segmentation and reassembly of RLC SDUs in order to generate RLC PDUs of appropriate size from the incoming RLC SDUs;
- re-segmentation of RLC data PDUs, if these do not fit to the actual transport blocks; and
- reordering of RLC data PDUs, duplicate detection and RLC SDU discard, RLC re-establishment, and protocol error detection.

For 5G, the main challenge or design goal w.r.t. RLC is to increase the processing efficiency, reduce overhead and duplicate functions, better segregate real-time and non-real-time functions, and enable the support of lower layers with mixed numerologies. 3GPP has already agreed that

- multiple parallel logical channels can be configured with different characteristics and priorities, e.g., involving different numerologies and transmit time interval (TTI) lengths;
- gNBs should have means to control which logical channels the UE may map to which numerology and/or TTIs with variable duration; and
- concatenation of RLC PDUs is performed in the MAC layer instead of the RLC layer. This allows the precomputation of RLC and MAC headers for faster processing and higher data rates, and the parallelization of the PHY encoding and MAC PDU construction. Further, it allows ARQ to be fully decoupled from real-time processing, and helps to avoid the duplication of concatenation-like operations in RLC and MAC.

The latter point is illustrated in Figure 6-3, where the left side shows the UP processing in LTE, i.e. where RLC PDU concatenation is performed in the RLC layer, and the right side shows the agreed 5G approach, including the introduced SDAP layer. It has been discussed to also move *segmentation*, and not only *concatenation*, from RLC to MAC layer, since also segmentation requires the knowledge on MAC transport block sizes and is hence tightly tied to MAC processing. If this would be moved to MAC - while keeping individual queues per RLC entity to avoid head-of-line blocking - all remaining functions of the RLC would be asynchronous to the radio. In this case, the interface between RLC and MAC would reflect the split between asynchronous and synchronous functions and could constitute a good CU/DU split, as elaborated in Section 6.6.2.

A further proposal for 5G is that the re-segmentation of RLC PDUs could be extended to new scenarios, e.g. involving unlicensed spectrum, where transmission may be blocked by channel acquisition, and RLC PDUs would need to be re-segmented to fit the next transmission opportunity.

6.4.2.4 Medium Access Control (MAC)

In LTE, the MAC is responsible for

- mapping between logical channels and transport channels, i.e. the multiplexing or demultiplexing of MAC SDUs belonging to one or different logical channels into/from transport blocks (TB) delivered to/from the physical layer on transport channels;

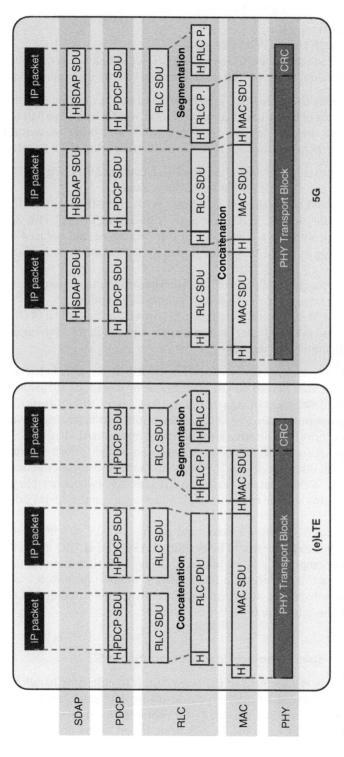

Figure 6-3. Comparison of UP processing between LTE and 5G (where the letter H indicates headers).

- scheduling information reporting;
- error correction through Hybrid ARQ (HARQ);
- initial access via the Random Access CHannel (RACH) for uplink resource request; and
- transport format selection, and priority handling among UEs via dynamic scheduling.

In 5G, key challenges are to enable MAC operation based on PHY instances with different numerology, the explicit support of beam-based communication, e.g. for mmWave bands, and the capability to meet and handle diverse QoS requirements, which will likely require a larger extent of coordination of multiple MAC entities, for instance in different cells. 3GPP has already agreed that

- a single MAC entity may support one or more numerologies and TTI lengths;
- asynchronous HARQ is preferable for NR for both UL and DL;
- MAC sub-headers should be interleaved with MAC SDUs, and MAC control elements should not be placed in the middle of MAC PDUs, but at their beginning or end. This allows to create an efficient header design with the aim of faster header processing and reduced overhead, and the option of concatenation in the MAC (although this would require the introduction of additional fields);
- UEs shall not send padding data if payload data is available and the remaining transport block size is greater than a predefined number of bytes (7 bytes in LTE); and
- it has been agreed that MAC control elements may be used for the control of UL data duplication, allowing for a faster and more granular control of data duplication than relying on Radio Resource Control (RRC) configuration for this only.

The following further considerations have been discussed: To support better service differentiation, the MAC should be able to map an UL grant to one or more logical channels with a different MAC configuration (e.g., different HARQ settings for eMBB and URLLC), instead of treating different services in the same way. Further, to enable service prioritization already during initial access, it has been proposed to introduce priority-level-specific physical RACH (PRACH) preambles that enable to increase the success probability of the initial access of high-priority services, without sacrificing the initial access of other services [17]. Note that there are also various considerations to further improve HARQ in 5G, for instance based on dynamic soft buffers, enabling to adapt HARQ behaviour to packet lengths and the extent of soft buffer available at the device side, or novel HARQ feedback concepts [18].

It is also considered for 5G to have a better and service- or AIV-specific cross-optimization of protocol stack layers. One example would be optimized ARQ and HARQ constellations, where for URLLC use cases more emphasis may be put on obtaining reliability through (improved) HARQ, while ARQ is skipped as it introduces too much latency, whereas for eMBB use cases the HARQ could be simplified, as reliability could be more efficiently gained through ARQ. Such cross-layer optimizations are also the subject of the following section.

6.4.3 Possible Service-specific Protocol Stack Optimization in 5G

As indicated before, a key requirement of the 5G system is to efficiently handle services with very diverse requirements, in particular novel services involving stringent latency requirements. In this respect, the LTE protocol stack is suboptimal, as it constitutes a fixed set of functions applied almost equally to all UP data, except for some limited flexibility in terms of the different RLC modes

Table 6-1. Latency contributions of protocol stack layers in LTE [19].

Sub-layer	Function	Overall latency contribution
PDCP	RoHC	20.0%
	De-ciphering	59.2%
	Header processing	7.8%
RLC	Reassembly	8.6%
	Re-ordering	0.4%
	Header processing	1.0%
MAC	De-mux	0.8%
	Header processing	2.2%

mentioned in Section 6.4.2.3, and Robust Header Compression (RoHC) profiles. Further, the protocol stack involves some processing inefficiency due to repetitive header processing and reorganization of data, as visible in Figure 6-3. This inefficiency is quantified in Table 6-1, which shows the contribution of the different PDCP, RLC and MAC functions to the E2E latency for an example LTE implementation [19], and which for instance emphasizes the major latency contribution of the PDCP layer and repetitive header processing.

One consideration for 5G is hence to introduce a two-layer approach [20][21], where PDCP, RLC and MAC functionalities are put into two groups for the avoidance of repetitive processing, and where the detailed functionalities are highly tailored towards or possibly even de-activated or hard-coded depending on specific service needs. For typical eMBB, URLLC and mMTC services, possible groupings and optimized functionalities are depicted in Figure 6-4.

The two-layer approach introduces a) a *Lower Layer* with traditional MAC and flexibly some RLC functions depending on service requirements, and b) an *Upper Layer* including the remainder of RLC and PDCP. With this two-layer proposal, processing and latency gains can be achieved due to less header processing, single duplication detection and re-ordering, and modular re-transmission control, where two levels of re-transmission are utilized instead of three (i.e., on MAC, RLC and PDCP level). Further, this architecture can better support multi-connectivity and RAN splits into CUs and DUs, as detailed in Section 6.6.2, where the DUs could correspond to different AIVs. Let us shortly elaborate on the key benefits for individual service types that the proposed two-layer RAN design and service-specific cross-layer optimizations could yield:

- **eMBB** services are expected to benefit strongly from the operation in higher frequencies and with larger bandwidths, which will require the usage of directional antennas to compensate for the high path attenuation, see Section 4.2.1. Directional antennas, however, may lead to challenges like single-/multi-channel directional hidden terminals, deafness, etc. Hence, the protocol design in particular for eMBB services could be optimized towards dynamic beam management, allowing for a directional MAC based on resource-to-beam allocation, beam sweeping and steering. Further, eMBB services with comparatively relaxed latency requirements can benefit from a high centralization of UP functions as well as coordinated multi-point (CoMP) transmission, in order to obtain high spectral efficiency;

	eMBB	URLLC	mMTC
5G Upper Layer **(PDCP / RLC)**	• Ciphering • RoHC • ARQ (per segment) • Packet re-ordering • Duplication detection • Segmentation	• No Ciphering • Optional RoHC • Re-transmission (PDCP level) • Semi-static service to AIV mapping	• Optional RoHC • Group-based scheduling • Semi-static service to AIV mapping
5G Lower Layer **(RLC / MAC)**	• MAC multiplexing • Dynamic beam management • Dynamic service to AIV mapping • Scheduling • Logical Channel Prioritization (LCP) • DRX / DTX, RACH	• Fixed-size LCP • Optional MAC multiplexing • Prioritized RACH, scheduling • Chase combining HARQ	• MAC multiplexing • Scheduling • Fixed size LCP • HARQ optimized for coverage • Group-based RACH

Figure 6-4. Possibly service-tailored protocol stack configurations in 5G [20][21].

- For **URLLC** services, it is expected that multi-connectivity and carrier aggregation will be important for ultra-high reliability. With the proposed layering, the upper layer will be able to connect to multiple lower layer entities, thus potentially lowering the processing delays (taking also into account the re-transmission processes). Further, due to the small and typically fixed size of the packets, functions like segmentation may not be required, fixed size Logical Channel Prioritization (LCP) may be used, and functions like RoHC and ciphering may be skipped, assuming that mission-critical services will provide application-level security mechanisms;
- For **mMTC** services, the high connection density and often low mobility suggests the usage of group-based functionalities [21], and the deactivation of mobility-related functions.

Figure 6-5 shows the possible reduction of layer 2 processing latency through the means discussed before. The benchmark is the LTE eNB UP processing latency of approximately 1 ms [22]. We show results for example eMBB, mMTC and URLLC services, where we split the latter into low mobility (e.g., factory automation) or high mobility cases (e.g., vehicular communications). Clearly, latency depends strongly on factors like TTI size and implementation details, hence the numbers are mainly to be seen as indicative of the trend. For eMBB, we see a marginal benefit from the avoidance of redundant header processing in a two-layer setup. For mMTC and low-mobility URLLC, most PDCP and RLC functions are disabled, though an aggregation layer (on top of PDCP) for group-based functionalities is assumed. URLLC shows higher latency, since it involves also packet re-ordering and assembly functions that can be skipped in the case of mMTC. For high-mobility URLLC, due to the expected frequent handovers and multi-connectivity support, the upper layer includes ciphering. Also, both URLLC cases assume no multiplexing at MAC level.

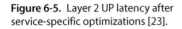

Figure 6-5. Layer 2 UP latency after service-specific optimizations [23].

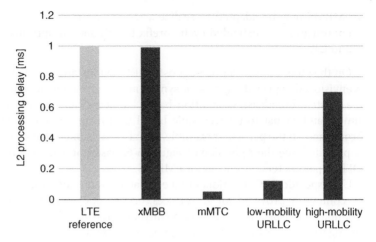

6.4.4 NF Instantiation for Multi-Service and Multi-Tenancy Support

In the past sections, we have discussed how the 5G protocol stack and NFs could be tailored towards specific service needs. We will now go one step further and elaborate on how NFs could be instantiated in a multi-service and multi-tenancy environment, in particular when multiple services and tenants are expected to utilize one common multi-service radio.

It is expected that a key to such multi-service radio will be a decomposition of NFs into service-specific and service-agnostic components [24], as already introduced in Section 6.4.1. Figure 6-6 shows an example of such decomposition in the case where a RAN supports four services eMBB, URLLC, mMTC and enhanced mobile broadband multimedia services (eMBMS). Note that the URLLC support is here explicitly split into a device-to-network (D2N) and device-to-device (D2D) part. It is further assumed in this example that all services share a common carrier and the same radio frequency (RF) circuitry, and also the same lower part of the PHY functionality, i.e. the same orthogonal frequency division multiplex (OFDM) numerology in terms of subcarrier spacing and symbol length, the same (inverse) fast Fourier transform (FFT), cyclic prefix insertion, etc. The higher part of the PHY could, however, be implemented in a service-specific manner, allowing for service-specific optimization. For example,

- eMBB may be optimized for maximum spectral efficiency, utilizing TTI durations and a demodulation reference symbol (DMRS) spacing adapted to the coherence time and bandwidth of the radio channel as well as multi-cell and multi-user capable advanced spatial multiplexing and receive processing concepts such as CoMP;
- URLLC may be optimized for lowest latency at the cost of spectral efficiency, utilizing very short TTIs and accordingly higher DMRS overhead and a radio resource management (RRM) able to pre-empt other services, as detailed in Section 12.3.3;
- The D2D part of URLLC may convey control information only (e.g., based on semi-persistent radio resource assignments) in a robust way;
- mMTC may be optimized for processing massive amounts of sporadic small packets, employing contention based access, open loop link adaptation and autonomous synchronization for maximum device energy efficiency in UL and size-optimized and robust DL control to provide extended coverage;

- eMBMS may be adapted to realize a DL single-frequency network (SFN), employing inter-cell coordination, an extended cyclic prefix length and an optimized subcarrier spacing for both CP and UP.

On the other hand, a service-agnostic RRC cell instance, as depicted on the left side of Figure 6-6, is envisioned to provide signals for synchronization and channel measurement as well as initial access (i.e., system broadcast, contention-based UL access and a basic DL channel) to the system, being simple and robust to be accessible by all UEs from low-cost machine-type to high performance broadband. In the particular example, it is also envisioned that an eMBMS-specific RRC cell instance is provided, see the right side of Figure 6-6, based on the SFN operation and the aforementioned OFDM parameters optimized for eMBMS.

User-specific RRC functionality (i.e., related to the RRC state handling, handover, etc.) may either be commonly provided for all services, or also optimized for a specific service type. In the given example, user-specific RRC functionality, denoted as "RRC user" functionality in the eMBB slice of the figure, is for instance assumed to be common to all services (except eMBMS) and to use eMBB for the transport of RRC messages.

The distinct upper PHY implementations per service also imply distinct PHY NF instances per each service. Accordingly, MAC, RLC and PDCP must also employ different NF *instances* per each service, though the same NF *type* and *implementation* may be used for each instance, e.g. with different sets of allowed options and parameter settings per service. When the system evolves over time and especially when new and not yet foreseen use cases emerge, a new NF type and corresponding implementation could be introduced for any service without impacting the others, thereby guaranteeing that the overall system design is future-proof.

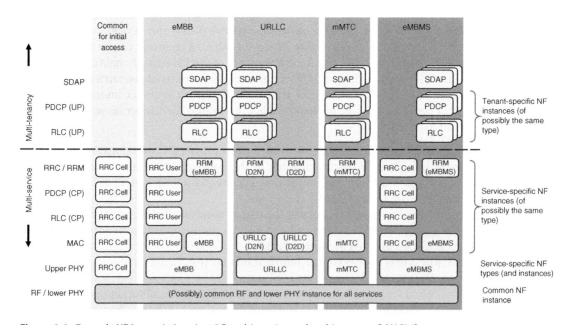

Figure 6-6. Example NF instantiations in a 5G multi-service and multi-tenancy RAN [24].

A key finding in [24], as also emphasized in Figure 6-6, is that the handling of the CP will most likely be common to all tenants (as described above and based on a limited set of defined key service types), while the UP processing could be based on tenant-specific NFs. Note that as for the CP, such tenant-specific UP NFs could be based on the same NF type and implementation, and simply be parametrized to the needs of different tenants.

It is important to stress that multi-tenancy support requires all RAN NF instances to be slice-aware. In particular, common NF instances shared by multiple slices need to maintain the one-to-one mapping of traffic to individual slices, and there must be mechanisms for slice controllers to influence how traffic of their related slice should be processed in NF instances that are common across multiple slices. Through such mechanisms, both the separation of services and separation of tenants are enabled in both common and service- or tenant-specific NFs, and network slicing effectively becomes end-to-end, spanning the whole path from data network (respectively end-user service) to the UE.

6.5   Multi-Connectivity

As stressed in Section 6.3, an essential feature of the 5G system shall be the native support of multi-connectivity between different radio legs, either related to different AIVs (i.e., different frequency bands, different RATs, etc., see Section 6.4.1) or different cells serving the same AIV. The benefits are increased throughput and/or improved reliability, for instance for 5G mmWave deployments that are subject to volatile propagation conditions, or for mission-critical use cases. We now hence elaborate on the architectural means to enable multi-connectivity, in particular on the protocol stack layers that are most suitable for the UP aggregation of different AIVs, and the way of integrating the CP functions of the involved AIVs.

6.5.1   5G/(e)LTE Multi-Connectivity

For multi-connectivity between an evolved LTE and a novel 5G AIV, there has been an early consensus that this should be realized on RAN level and avoid explicit CN/RAN signaling, which is for instance required in the case of 3G/4G multi-connectivity. Further, since the 5G PHY and MAC are agreed to be non-backward compatible to LTE-A for best possible 5G performance, 5G/(e)LTE multi-connectivity on MAC layer appears rather challenging, and is hence currently not considered by 3GPP. Instead, 3GPP is envisioning to adopt key principles of the DC concept from LTE Rel. 12 [25] for 5G/(e)LTE multi-connectivity, where the UP of LTE-A and 5G are aggregated on PDCP level. The most likely option is to have a bearer split in the master node (MN), which can be either the (e) LTE eNB or 5G gNB, and which corresponds to the option 3c in LTE DC, as depicted in Figure 6-7 (left) for the example that the (e)LTE eNB acts as the MN. In this particular example, the UP data is exchanged between MN and secondary node (SN) via the Xn or X2 interface. Please note that the figure leaves it open whether the MN is connected to a 4G or 5G CN, see Section 5.5.1, hence both related interface names are displayed.

It is further assumed for first 5G releases, as in LTE-A, that only the MN has a control plane connection to the CN. This way, a common evolved CN/RAN interface for both LTE and 5G will be used [12][14], and no extra CN/RAN signaling is needed to add or remove a secondary node.

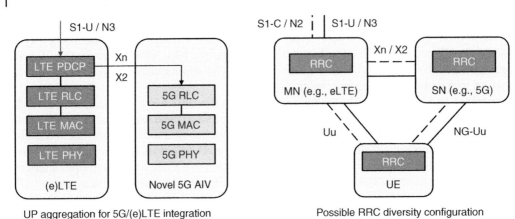

Figure 6-7. UP aggregation (left) and a possible RRC diversity option (right) for LTE-A/5G multi-connectivity [21].

Let us now focus on how RRC signaling is handled in 5G/(e)LTE multi-connectivity scenarios. For LTE DC, all RRC messages are transmitted via the MeNB. SeNB RRC messages are sent to the MeNB over the X2 interface, and the MeNB makes the final decision of whether to transmit the RRC message to the UE. This has the advantage that there is no need for coordination, since the MeNB always makes the final decision. The disadvantage is that there is no RRC diversity [26], and RRC messages from the SeNB take longer time since they are always routed via the MeNB.

For NR, the following changes have been agreed [26]:

- Master and secondary node have individual RRC instances, allowing for an independent evolution of both protocols, though a UE (still) has only one RRC state, as controlled by the MN;
- The SN can send RRC messages to the UE via the MN, as in LTE DC, but it can also send these directly to the UE, for the benefit of reduced signaling latency, as long as these do not require coordination with the MN;
- Mobility measurements (for instance 5G-specific measurements) can be sent directly to the SN;
- RRC messages originating from the MN may utilize diversity, i.e., they may be duplicated and sent from both the MN and the SN. For RRC messages originating from the SN, however, diversity is not yet specified.

It is possible that in future NR releases the RRC signaling is further evolved, for instance toward schemes involving explicit coordination of the RRC instances in the MN and SN, as proposed in [17]. By the UE and measurements, which may increase the implementation complexity on the UE side.

6.5.2   5G/5G Multi-Connectivity

Among novel 5G AIVs, more multi-connectivity options are thinkable than for 5G/(e)LTE integration, as the NFs for different novel 5G AIVs can be kept more similar also on lower layers, hence allowing to also perform bearer split and UP aggregation on lower layers. In addition to the option of aggregating the UP of different AIVs on PDCP level, as in the 5G/(e)LTE multi-connectivity case described before, the following forms of 5G/5G multi-connectivity are being discussed, as also depicted in Figure 6-8:

- **Bearer split on PDCP level, with a duplication of data packets** towards the secondary gNB. The duplicated data can then be used to enhance diversity and hence reliability whenever required by a specific service type;
- **Bearer split on RLC level,** such that the secondary gNB only implements the lower parts of the RLC functionality, for instance the NFs that are time-synchronous to the radio, while the higher parts of the RLC are centralized at the master gNB, allowing for a faster reaction of traffic steering mechanisms towards the radio legs;
- **Aggregation on MAC level,** like in LTE Rel. 10 carrier aggregation (CA). This provides the possibility of joint scheduling across the radio legs, and hence the fastest possible reaction to changing radio conditions. This level of aggregation, however, typically requires ideal BH between the AIVs, or these should be co-located, and may be complicated in the context of different numerologies involved.

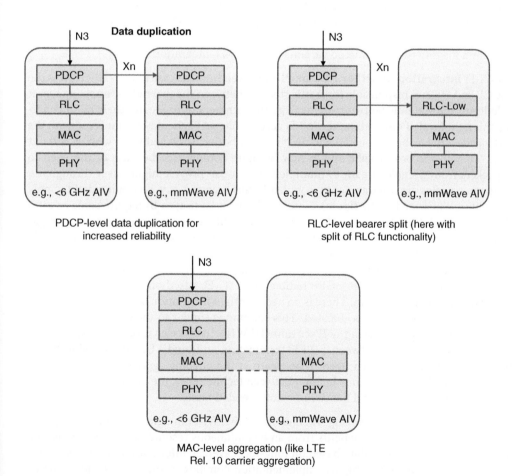

Figure 6-8. Additional bearer split and UP aggregation options for 5G/5G multi-connectivity.

It appears that the benefit of aggregation on MAC level as compared to PDCP level may in fact be quite limited, at least in terms of throughput, if a low-latency link between the master and secondary gNB is available and hence PDCP-level traffic steering can also occur on a decently fast time scale. Hence, a bearer split and UP aggregation on PDCP or RLC level may also be a preferred option for 5G/5G multi-connectivity, as it poses less requirements on the Xn interface between the gNBs and is easier to handle in the context of different air interface numerologies. Note, however, that this will still have to be investigated through detailed simulation studies.

6.5.3 5G/Wi-Fi Multi-Connectivity

The integration of 5G and Wi-Fi systems can be useful for increasing the user throughput and to offload traffic from the cellular network. This section considers the integration of Wi-Fi and 5G on spectrum below 6 GHz and compares solutions involving a tighter or looser integration on RAN level. Note that there are also other options for integrating 3GPP and Wi-Fi systems in general, such as Multi-Path Transmission Control Protocol (MP-TCP), but these are not considered in this section. Please also note that general options of integrating 5G with non-3GPP access are covered in detail in Section 5.5.2.

Figure 6-9 shows the different considered variants of 5G/Wi-Fi integration, namely:

- **Tight 5G/Wi-Fi integration on either RLC or PDCP level**. In these cases, the 3GPP eNB is the anchor for CP and UP and responsible for traffic steering between the two radios, implying that it has knowledge of the state of the two systems (e.g., the numbers of users and their radio conditions). For this tight level of integration, the Wi-Fi access point (AP) should be connected with the 3GPP eNB via ideal BH or be co-located with it. Note that both cases could work with or without bearer split. Though the bearer split option may be more complicated because of the need to schedule packets from a single stream to multiple RATs, it can provide higher user throughput, and since the two radios are anyway tightly coupled in this scenario, it is recommended to apply the bearer split (as illustrated in the figure). Note that both variants a) or b) require some adaptation of the RLC protocol on the Wi-Fi side. A MAC layer split, i.e. some form of carrier aggregation, would not be feasible, because the Institute for Electrical and Electronics Engineers (IEEE) 802.11 MAC is substantially different from the 3GPP MAC;
- **Loose 5G/Wi-Fi integration above PDCP**: In this case, higher layers decide whether to switch or split the radio bearer to transmit over 3GPP radio and/or Wi-Fi, based on longer-term averaged knowledge of the state of each system. There is no need for tight integration between 5G and Wi-Fi nodes, and they do not need to be co-located. This solution is similar to the current LTE Rel-13 LTE/Wi-Fi radio level integration using an IPsec tunnel (LWIP). In this scenario, there is an Internet Protocol Security (IPsec) tunnel between a Wi-Fi termination node and a UE connected to Wi-Fi, so the Wi-Fi network does not even need to be aware of the ongoing aggregation with a 3GPP access. Because of the lack of tight coordination between technologies, the bearer split scenario is almost impossible to implement, and hence not recommended here.

In detailed simulation studies comparing the mentioned 5G/Wi-Fi integration options [21], it is clearly visible that a tight integration benefits from an instantaneous resource allocation across the technologies, as opposed to routing data based on long-term average information on the technologies. However, the price is of course higher implementation complexity, the need for ideal BH or co-location of the radios, and an adaptation of the Wi-Fi RLC, and ultimately the difficulty of developing a solution that is subject to decisions by individual standards bodies.

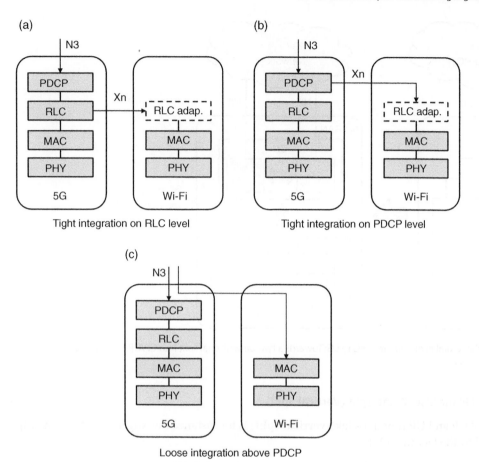

Figure 6-9. Different variants of 5G/Wi-Fi multi-connectivity [21]. Note that all 3 variants are possible with or without bearer split.

6.6   RAN Function Splits and Resulting Logical Network Entities

A likely key difference in the 5G RAN architecture to that of legacy systems will be the native support of various function splits along two dimensions: A split of CP and UP related NFs, and a split between functions which are closer to the radio, and hence tightly related to radio timing, and those that are rather asynchronous to the radio, into DUs and CUs, respectively. This is illustrated in a simplified form in Figure 6-10.

This is a paradigm shift from having few, clearly defined logical network entities as in 4G (which are in practice typically reflected in the same set of physical entities), to an architecture that allows for a more flexible mapping of NFs to physical network entities depending on deployment con-straints, use case needs, etc. Taking this to the extreme, 5G has been stated to resemble a "network of functions" rather than a "network of entities" [24].

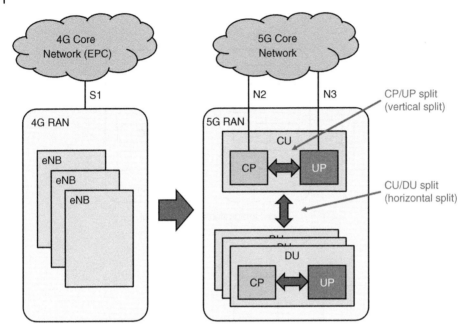

Figure 6-10. Architectural evolution from 4G to 5G: Towards a two-dimensional split into CP/UP NFs and CUs/DUs [27].

6.6.1 Control Plane/User Plane Split (Vertical Split)

A separation of CP and UP functions (aka **vertical split**) is for instance investigated in detail in [28] and motivated by the fact that [27]

- it enables the introduction of SDN principles also in the RAN [29][30];
- it allows for a separate optimization of the placement of CP and UP functionality, as detailed further in Section 6.6.3, and an independent scaling of CP and UP;
- in multi-vendor networks, a standardized interface to the CP enables a consistent control over network entities and NFs from different vendors, e.g. in terms of interference management for ultra-dense networks [20];
- due to the tight coupling of CP and UP functions in today's networks, the replacement or upgrade of a CP function often requires also the replacement of UP functions. Designing the CP and UP functions such that they are inherently less coupled and better separable might offer significant cost savings.

However, there are also major disadvantages of a CP/UP split that have to be considered:

- Standardization is required in case the interfaces between CP and UP have to be extended to introduce new features, which might slow down their introduction. Integrating additional interfaces in a proprietary manner in combination with standardized ones is not a suitable solution, as it would compromise the main benefit of a CP/UP split;
- Additional effort in terms of testing is required to guaranty the interoperability of CP and UP functions from different sources, shifting the effort from single vendors to system integrators supporting mobile network operators;

- As mentioned before, CP and UP functions are often tightly coupled, especially in the lower protocol stack layers. In particular, MAC schedulers need tight interaction with the UP since they determine the UL/DL resource usage, coding schemes, antenna mappings, precoding schemes, rank and modulation schemes to be applied by the UP. In the case of analog beamforming, e.g. for massive multiple-input multiple-output (MIMO), it is also typically assumed that the MAC scheduler selects the phase shifts or beams to be applied. All the listed interactions would require ideal BH connectivity between UP and CP.

Especially for the latter reason, a complete CP/UP split across all protocol stack layers is rather questionable. However, the following two options are possible:

- **Separated RRC**: In this case, the RRC for possibly multiple cells is handled separately, while for the complete remaining protocol stack the CP and UP are kept together. This case is very straightforward, as the related signaling between CP and UP would essentially resemble that on the X2 interface in the case of LTE Rel. 12 DC [25];
- **Separated asynchronous CP functionality**: In this case, all asynchronous CP functionality would be separated from asynchronous UP functionality, but all synchronous CP functionality requiring tight coordination with the UP would be kept together with the UP. In fact, this approach, as earlier considered in, e.g.,[21], has now been agreed in 3GPP, and the involved CP-UP interface is now specified as E1 interface [31].

6.6.2 Split into Centralized and Decentralized Units (Horizontal Split)

An architecture that allows a flexible split of RAN functions into CUs and DUs (aka **horizontal split**) is motivated by the possibility

- to obtain centralization gains, here referring to both performance gains from common multi-cell and/or multi-AIV processing, such as centralized resource management or even multi-cell joint signal processing, and gains in terms of economy of scale;
- to shift functions to different locations based on use case requirements. For instance, for latency-critical communications, it may be required to place the full protocol stack and the application closer to the edge, while for delay-tolerant applications, certain functionalities could by centralized in the cloud; and
- to adapt RAN processing to different deployments with possibly different infrastructure characteristics, such as available local processing power, different degrees of BH and FH infrastructure etc.

Various horizontal split options have been discussed, and are partially still under discussion in 3GPP and other bodies such as CPRI [32] and xRAN (see www.xRAN.org). The main options, using 3GPP terminology, are depicted in Figure 6-11. An overview on options 1-8 can be found under [5], while the variants of split option 7 have been detailed in the work item "Study on CU-DU lower layer split for New Radio" [33]. Note that there is still controversy on whether such lower layer split options should be fully standardized (i.e., "stage 3 specification"), or whether a functional level description (i.e., "stage 2 specification") is sufficient.

As indicated at the bottom of Figure 6-11, the main characteristics of the split options are the requirements posed on the related interfaces in terms of data rate and latency. Function splits that are far away from the radio, i.e. shown on the left side of the figure, have the most relaxed latency and data rate requirements, but offer the least centralization gains. For brevity, we here only shortly elaborate on the most prominent options. Regarding possible higher-layer splits,

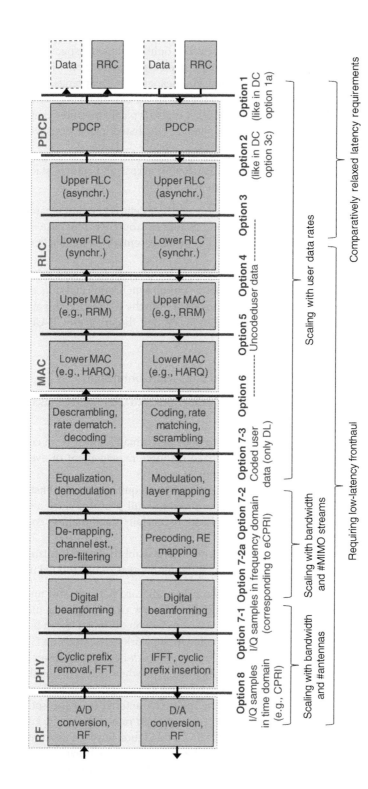

Figure 6-11. Main horizontal split options for the 5G RAN, see also [5].

- **Option 1 (RRC/PDCP split)** equals the split used in LTE Rel. 12 DC option 1a [25];
- **Option 2 (PDCP/RLC split)** corresponds to the split used for the UP in LTE Rel. 12 DC option 3c [25]. It has been chosen by 3GPP for NR Rel. 15 and is denoted as F1 interface [34]. While it may be inferior to option 3, it has the practical benefit of already being utilized in legacy systems;
- **Option 3 (intra-RLC split)** is based on a split of the RLC functionality such that the entire RLC is located in the CU, in particular all ARQ-related functionality, except for the aggregation functionality which is located in the DU. This way, the CU/DU split corresponds to the split between asynchronous (or non-real-time) and synchronous (real-time) functionality, see also Section 6.4.2.3. This option is superior to option 2 in that it offers a larger extent of centralization and pooling gains and possibly enables better flow control, though the ARQ process is more prone to interface latency;
- **Option 5 (intra-MAC split)** is in fact not pursued intensively in 3GPP, but is considered by the Small Cell Forum (SCF), as detailed in Section 6.7. In this case, high-level scheduling decisions, for instance related to inter-cell interference coordination (ICIC) or CoMP, would be performed in the CU, while time-critical MAC processing, for instance related to HARQ, would reside in the DU;
- **Option 6 (MAC/PHY split)**, also not in the main focus of 3GPP but considered by the SCF, foresees the complete MAC to be handled in the CU, with distributed PHY implementations in the DUs. Obviously, this would require sub-frame-level timing interactions between CU and DU, and FH delay would affect HARQ timing and scheduling.

A large debate has taken place in 3GPP w.r.t. split options within the PHY, where the classical Common Public Radio Interface (CPRI), based on an exchange of time-domain in-phase and quadrature phase (IQ) samples, is particularly problematic in the context of massive MIMO and mmWave communications, as interface bandwidth requirements scale with the number of antenna ports and system bandwidth. The main considered options are:

- **Option 7-1**, where the i(FFT) and cyclic prefix insertion/removal are performed at the DU, and IQ samples in frequency domain are exchanged over the interface. Compared to option 8 (CPRI) this offers the benefit that only the samples related to occupied sub-carriers need to be exchanged, instead of time domain samples reflecting the whole system bandwidth. Also, less quantization bits per symbol may be needed when quantizing in frequency domain. In the UL, some PRACH processing may also be performed at the DU;
- **Options 7-2** and **7-2a** have the additional benefit that pre-coding and digital beamforming, or parts thereof, are performed at the DU, such that the FH interface requirements scale with the number of MIMO layers, and not with the number of antenna ports as in the case of options 7-1 and 8. Options 7-2 and 7-2a differ in the extent of precoding happening at the CU or DU, or the location where channel estimation is performed in the uplink etc.;
- **Option 7-3**, considered only for the DL, further reduces bandwidth requirements on the interface, as coded user data is exchanged before modulation. As a downside, such split would likely strongly increase the complexity of the DU.

Note that RAN splits need not be the same for UL and DL, e.g., there are also considerations to use options 7-2 or 7-2a for the UL, and option 7-3 in the DL. Further note that beside the activities in 3GPP, a group of network vendors have also specified an enhanced CPRI (eCPRI) standard [32], which essentially includes all the 3GPP split options 7-x described above.

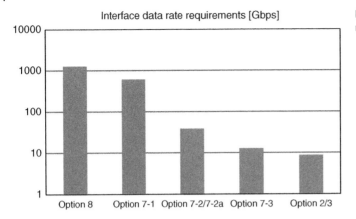

Figure 6-12. RAN split interface data rate requirements for UP traffic [15][18].

To obtain a rough feeling for the requirements associated with different UP split options, Figure 6-12 shows the interface data rate requirements for an example scenario with a system bandwidth of 400 MHz, 64 transmit antennas, 4 spatial layers, and 64 quadrature amplitude modulation (QAM) with code rate 3. For the methodology to compute the data rate needs, and all used parameters, the reader is referred to [15] and [18], respectively. For this large number of antenna ports and large system bandwidth, 3GPP split option 8 (CPRI) would require more than 1 Tbps of FH data rate. With option 7-1, this can already be reduced to about 600 Gbps, due to the possible lower bit resolution when quantizing IQ samples in frequency domain, and the fact that only occupied subcarriers have to be considered. With options 7-2 or 7-2a, the interface data rate can be further reduced to about 38 Gbps, since the quantization now happens before digital beamforming and the mapping of streams to transmit antennas. Option 7-3 again reduces the interface data rate by about a factor of 4, as now coded data bits are relayed instead of quantized IQ samples, though at the price of higher DU complexity. The higher split options, such as options 2 and 3, further reduce the data rate needs due to the forwarding of uncoded user data, and the avoidance of sending HARQ retransmissions over the FH, but the effect is rather minor.

The *latency* requirements on the interface in general become tighter the further down in the protocol stack the split is, though one can roughly classify two levels of latency requirements: For 3GPP split options 1-3, where all RAN functionality that has to operate synchronously to the radio (i.e., in real-time) is located below the split, the interface latency requirements are rather relaxed, while they are much more stringent for the other options where the RAN split cuts in between synchronous functionalities.

In consequence, even though 3GPP refers to the intra-RAN interfaces resulting from all mentioned split options as *fronthaul* (FH), the higher-layer split options 1-3 are in fact much more similar in terms of interface data rate and latency requirements to classical *backhaul* (BH) than classical FH (which in legacy systems is typically associated to lower-layer splits such as CPRI). This topic is further elaborated in the context of transport network architecture in Chapter 7.

6.6.3 Most Relevant Overall Split Constellations

In principle, many combinations of vertical and horizontal splits are thinkable, and capable to fulfil 5G requirements, but only a subset appear practically relevant. To crystallize key options, network functions in both CP and UP can be grouped into the following blocks of which logical network entities in 5G would most likely be composed:

CP Functions:

- **CP-H**: Asynchronous functionalities such as ICIC, RRC functionality, etc;
- **CP-L**: Synchronous functionality such as cell configuration, MAC scheduling and PHY layer control.

UP Functions:

- **UP-H**: QoS/slice enforcement, PDCP (possibly also the asynchronous parts of RLC);
- **UP-M**: Remainder of RLC and MAC and higher PHY layer, at least covering (de-)coding, (de-)scrambling, etc. (depending on the exact horizontal split, see Section 6.6.2);
- **UP-L**: Remainder of the PHY, at least covering (i)FFT, cyclic prefix insertion, etc.

Figure 6-13 depicts the likely most relevant overall split constellations. Constellation a) corresponds to a classical standalone BS where the complete RAN functionality is located in distributed points, also referred to as distributed RAN (D-RAN).

Constellation b) represents a large extent of centralization, in particular of all CP functions. The difference to a classical centralized RAN (C-RAN) is that the lower part of the UP PHY is located at the DU, corresponding to any of the split options 7 introduced in Section 6.6.2, and strongly reducing the data rate requirements on the FH interface. Finally, constellation c) shows a higher layer split, where only the asynchronous (i.e., non-real-time) part of the UP and the less time-critical parts of the CP functions are centralized. This constellation could be based on horizontal split options 2 (split between

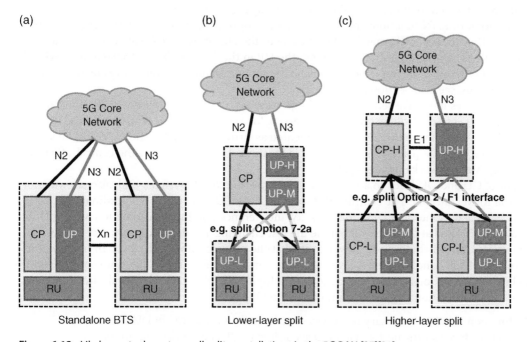

Figure 6-13. Likely most relevant overall split constellations in the 5G RAN [27][31].

PDCP and RLC) or 3 (split within RLC). All time-critical functionalities, including for instance the MAC scheduler, are here located at the DU, which allows to relax the timing requirements on the interfaces to the CU, such that these can also be based on non-ideal BH or possible wireless self-xhauling links (see Chapter 7). Since in this constellation the CU contains only asynchronous (i.e., non-real-time) functionalities, the CP and UP therein can be nicely split, allowing to exploit the benefits listed in Section 6.6.1.

3GPP has agreed to a particular embodiment of constellation c) for NR Rel. 15, where the RRC and PDCP functionalities are centralized, and RLC, MAC and PHY are decentralized, with the F1 interface in between [34], and possibly with a further split of the higher-layer functionality into CP (F1-C) and UP (F1-U), involving the mentioned E1 interface [31].

Clearly, further split constellations are thinkable, such as combinations of b) and c) with three levels of CUs or DUs, as further discussed in Section 6.7.1. Further, if a higher-layer CP and UP split is involved in case c), it is of course also possible to move either the higher-layer CP functionality to the DU (e.g., to optimize CP latency, while exploiting maximum pooling gains for the UP functionality), or move the higher-layer UP functionality to the DU (e.g., for centralized and hence better CP coordination of multiple DUs, while having latency-optimized decentralized UP processing) [31].

Regardless of whether option b) or c) is chosen, it is possible to integrate mobile or multi-access edge computing (MEC) infrastructure [29] into the CU to serve latency-critical applications. The CU approach also has a big advantage with respect to mobility handling. If a UE is moving within the range of antenna sites belonging to a single CU, mobility is handled CU-internally only, for instance through fast UP switching [20] resulting in low handover interruption time because of low latency between involved components and the avoidance of RAN/CN signaling. This is especially beneficial for ultra-dense deployments (using, e.g., mmWave bands), where the number of mobility events is typically high.

Note that independent of the split options and resulting logical entities depicted in Figure 6-13, the physical location of RAN functions may be dynamically optimized based on service requirements. For instance, the CP in the CU may be placed close to the DU or even co-located with this for the support of URLLC services. On the other hand, the UP in the CU may be centralized in a regional data centre for efficient cloud implementation and the support of multi-connectivity and other interworking scenarios, see also Section 6.7.1.

In general, it does not make sense to cover huge areas via a single CU, but one will rather implement several CUs each controlling the radio processing for a certain number of antenna sites [18][28]. Typically, the NFs running in the CU are implemented as virtual network functions (VNFs) on server platforms based on NFV principles [35]. Suitable locations for CUs are, e.g., the central offices of fixed or integrated network operators [11].

The split of CU and DU leads to a hierarchical architecture for 5G systems [36], also allowing for hierarchical RRM concepts. The central RRM located at the CU is then able to coordinate the lower layer functions across multiple DUs, for mobility, interference and load coordination among DUs. The local RRM located at the DUs will handle the fast changing and detailed parameters at low layers, for instance, for scheduling and power control at the resource block (RB) level. In other words, the central RRM will handle the high-level, less time-sensitive information and more coordination-oriented network functions for the interaction of multiple DUs. The general central RRM functions for high-level coordination include, as summarized in [37]:

- Call admission, to allow necessary resources coordinated and reserved in DUs;
- Cell selection, to select the best serving DU according to UE and service requirements;

- Load balancing between cells, for a trade-off between cell utilization and performance;
- Mobility, to improve mobility performance by centralized coordination;
- Multi-connectivity, to select serving cells and transmission mode of cells.

5G RRM architectures and particular RRM concepts are elaborated further in Chapter 12.

6.7 Deployment Scenarios and Related Physical RAN Architectures

Different deployment scenarios have been considered by different bodies, namely 3GPP, the Small Cell Forum (SCF), the Next Generation Fronthaul Interface (NGFI) forum, and NGMN, among others. 3GPP defines four main deployment scenarios as follows [5]:

- **Non-centralized deployment**, where the full protocol stack is supported by a gNB as in a macro or indoor hotspot environment (which could include public or enterprise environments), which can be connected to other gNBs and/or (e)LTE eNBs;
- **Co-sited deployment**, where the NR functionality is co-sited with (e)LTE functionality either as part of the same BS or in the form of multiple BSs at the same site;

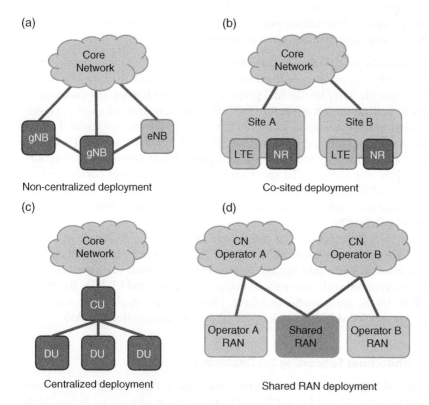

Figure 6-14. Deployment scenarios envisioned by 3GPP [5].

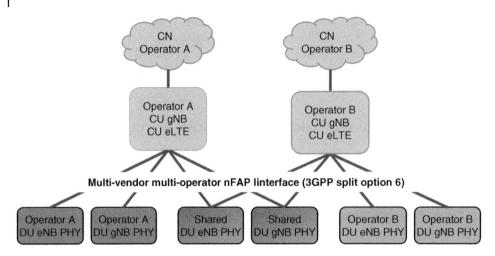

Figure 6-15. Multi-operator multi-vendor indoor deployment scenarios considered by SCF.

- **Centralized deployment**, where NR should support centralization of the upper layers of the NR radio stacks with different split options between CUs and DUs, see Section 6.6.2;
- **Shared RAN Deployment**, where NR should support RAN sharing of multiple network operators that have individual CNs but share spectrum or radio resources.

SCF adopts sharing approaches beyond multi-operator core networks (MOCNs) and mobile operator radio access networks (MORANs) by opening the internal RAN interface that allows for NFs to be shared and hence vendor specific RAN intellectual property (IPr) to be hosted in a common or separate network function virtualization infrastructure (NFVI). This new form of sharing targets multi-operator and multi-vendor deployments in an indoor small cell infrastructure enabled by the Network Functional Application Programming Interface (nFAPI) [38], see also Figure 6-15. In particular, the nFAPI interface allows to map operator-specific and (possibly) operator-shared functionality effectively to CUs and DUs, respectively. The specified nFAPI functional split corresponds to 3GPP split option 6, i.e. a MAC/PHY split as explained in Section 6.6.2, and aims to lower barriers w.r.t. shared access, in particular the multi-operator sharing of physical network functions (PNFs) and virtual network functions (VNFs) [39].

NGFI adopts a three-tier RAN architecture with a radio cloud centre (RCC) as CU entity, a radio aggregation unit (RAU) as DU entity, and remote radio units (RRUs) as further distributed entities. Three splits are considered, namely 3GPP split option 7 (intra-PHY split) with two flavours (iFFT and cyclic prefix insertion and removal with and without mapper and dedicated PRACH handling), and 3GPP split option 2 (PDCP/RLC split), see Section 6.6.2. The considered deployment scenarios are dense indoor coverage, massive MIMO, and outdoor distributed antenna systems (DAS).

6.7.1 Possible Physical Architectures Supporting the Deployment Scenarios

Figure 6-16 illustrates an evolved 3-tier RAN architecture that could support the different deployment scenarios and functional splits as considered by 3GPP, NGFI, and SCF and introduced before. Note that this architecture (with RRU, DU and CU) is an evolution of the original two-tier C-RAN (with

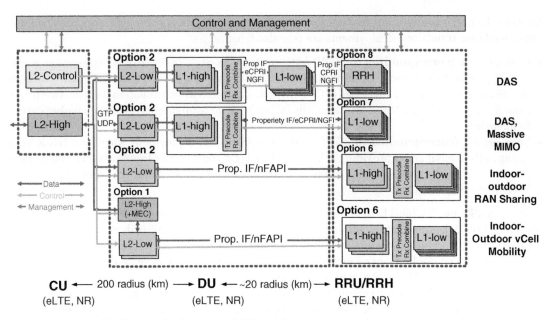

Figure 6-16. 3-tier RAN functional split in view of different deployment scenarios.

RRU and baseband unit, BBU), which provides the required flexibility for future multi-RAT deployments. Typically, a CU would cover a large area corresponding to a radius of 100-200 kilometres, whereas a DU would cover a radius of only 10 to 20 kilometres.

The exact functional splits in the 3-tier architecture are of course highly dependent on the transport network capabilities, as elaborated in Chapter 7. When using an ideal FH network (i.e., with low latency and high capacity), only the lower part of the PHY functions (L1-Low) should be placed at the RRU to maximally exploit the benefit of coordinated signal processing at the DU. DAS and Massive MIMO are two typical deployment scenarios enabled with 3GPP split options 7 and 8 and (e)CPRI/ NGFI radio-over-Ethernet (RoE) transport protocols or proprietary interfaces as implemented in OpenAirInterface [40]. Both require a common transmission precoding and reception combining operation (cell-specific and UE-specific) that must be placed either at the DU to enable the coordinated beamforming across several geographically distributed RRUs, or at a single RRU driving a massive antenna array. For a non-ideal FH network (i.e., with medium latency and capacity as in a public network), the entire PHY functions are moved to the RRU to relax the FH network and yet exploit the benefit of joint scheduling and interference coordination at DU. Indoor-outdoor RAN sharing and virtual cell mobility are two representative deployment scenarios that can be enabled with the SCF nFAPI interface or a proprietary interface corresponding to 3GPP split option 5, see Section 6.6.2. In addition, the separation of CP and UP among macro and small cell can be easily realized to seamlessly serve UP traffic via the centralized PDCP entity located either at DU or CU.

6.7.2 5G/(e)LTE and 5G Multi-AIV Co-Deployment

During the introduction of 5G, mobile network operators (MNOs) will of course aim to maximally leverage the existing (e)LTE deployments, and also need to serve legacy UEs that may be in the field for years. Hence, it is important to consider optimized co-deployments of (e)LTE and 5G. The main envisioned non-standalone deployments are (see also Chapter 5):

- **Lean on LTE (corresponding to 3GPP options 3/3a/3x or 7/7a/7x [5])**: A deployment where (e)LTE provides the basic communication layer, typically on a lower carrier frequency, acts as MeNB and provides the Primary Cell (PCell) in multi-connectivity constellations (see Section 6.5.1). In addition, one or more NR Secondary Cells (SCells) can be configured, typically on higher carrier frequencies;
- **Lean on NR (corresponding to 3GPP options 4/4a [5])**: A deployment where the NR RAT is acting as the MeNB and is providing at least the PCell (e.g., on a carrier frequency F1), and in addition an eLTE SCell is configured (e.g., on a carrier frequency F2) where both F1 and F2 are from a lower frequency band.

More details on 5G/(e)LTE tight interworking with regard to different AIVs are provided in Section 2.3 of [21]. Note that the different options listed above can work with most of the function split and physical architecture considerations described in Sections 6.6.2 and 6.7.1.

Regarding the co-deployment of multiple 5G AIVs, one strongly investigated constellation is the tight integration of mmWave AIVs with lower-frequency 5G AIVs, in order to combat the challenging propagation conditions and volatile radio links associated with higher carrier frequencies [41], as described in Chapter 4. Here, the lower- and higher-frequency AIVs would ideally share the same DUs according to Figure 6-16, such that, e.g., joint or strongly coordinated MAC scheduling across the AIVs could be performed, see also Section 6.5.2. In this case, any sudden drop of link quality on the mmWave carrier(s) could be compensated by scheduling resources to the same UE on the lower frequency AIV, to still provide a decent quality of experience (QoE). It has to be noted, though, that for most eMBB use cases with higher throughput but often comparatively relaxed latency requirements, it may also be sufficient to use DC-type multi-connectivity between a lower and higher-frequency 5G AIV (see also Section 6.5.2) and react to outages in the mmWave carrier(s) through PDCP-level flow control.

In general, a joint design and tight integration of lower- and higher-frequency 5G AIVs [41] may provide benefits going strongly beyond those of mere multi-connectivity, such as the option to utilize lower-frequency channel measurements, interference fingerprints, etc., for the decision on the fast activation and deactivation of higher-frequency AIVs. Similarly, information available on the lower frequency layer could be used to setup and dynamically update the most suitable RAN function split and CU/DU hierarchy for a given area and instantaneous UE presence (i.e., determining the best possible set of DUs that should be covered by one CU to enable fast and efficient mobility among mmWave cells for the present UEs). Finally, such lower-frequency information could be used for improved beam management and mobility among high-frequency access nodes (e.g., to determine good initial mmWave beams already before a UE is actually served by mmWave, or to determine suitable mmWave cells and related beams for handover). One has to note, though, that many of the mentioned techniques may be implemented via proprietary means, and need not necessarily be standardized.

6.8 RAN Programmability and Control

While the RAN is the most complex part of the mobile network infrastructure, it offers many opportunities to benefit from SDN principles. One reason is that strategies and technologies being adopted to improve spectral efficiency, and to scale system capacity, such as cell densification, multi-RAT, or the usage of advanced PHY techniques like CoMP, require a high level of coordination among BSs, which SDN can naturally enable since control decisions can be made and coordinated at a centralized SDN controller. As another reason, softwarization of the RAN CP not only allows for an easier evolution through programmability, but also enables a multi-service architecture supporting a wide range of use cases and novel services. One of the main design challenges of a software-defined RAN (SD-RAN) are RRM functionalities and the stringent timing constraints associated with some key RAN control operations, such as flexible functional splits, beamforming, and MAC scheduling.

A possible high-level SD-RAN architecture is shown in Figure 6-17 and realized through FlexRAN [42], a flexible and programmable platform for SD-RAN. The CP follows a hierarchical design and is composed of three entities as described below:

- **Centralized entity**: this entity controls a large network area (on the order of 1000 BSs) and is in charge of soft real-time operation on a large timescale (order of hundreds of ms);
- **Edge entity**: this entity controls a small network area (order of 100 BSs) and is in charge of time-critical operation on a small timescale (order of ms or tens of ms). It is connected to a number of RAN agents, one for each service chain;
- **RAN Agent**: This entity sits on top of the RAN service chain (i.e., one local agent per CU, DU and RRU), acts as a local controller with a limited network view, and handles the control delegated by the edge entity, or in concertation with other agents and the edge controller.

The CP/UP separation is provided by a RAN agent application programming interface (API) which acts as the southbound API following SDN principles (e.g., based on OpenFlow), with the CP protocol on one side and RAN UP on the other side. The agent abstracts the RAN-specific drivers through these APIs, and is composed of a set of control modules that interact with each layer of the protocol stack. The relationship between agent and RAN depends on the physical BS deployment and the applied functional split. For instance, in case of a legacy 3GPP Rel. 10 standalone BS, only one agent is required, whereas in case of a three-tier 3GPP RAN architecture, as discussed in Section 6.7.1,

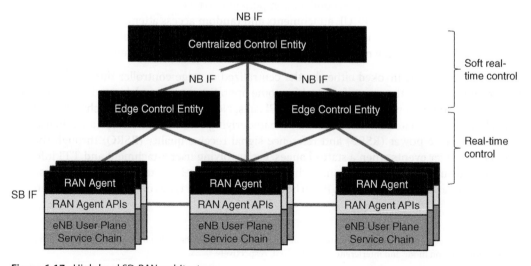

Figure 6-17. High-level SD-RAN architecture.

three agents are required. Depending on the deployment scenario, the agent-controller topology could vary from tree to mesh, and the edge entity could become part of the centralized entity. The control application can operate either on the edge entity, centralized entity, or both.

On top of the edge and centralized controller, a northbound API (labelled NB-IF) is provided, which allows RAN control applications to change the state of RAN infrastructure (BSs and UEs) based on the statistics and events gathered from the CP with the desired level of granularity. Such applications could vary from simple monitoring applications to more advanced applications that program the state and composition of the RAN (e.g., functional split and MAC scheduler).

All the access stratum protocols of LTE (RRC, PDCP, RLC, MAC) can be decomposed into two parts; the control *logic* that makes the decisions for the radio link and the control *action* that is responsible for applying those decisions. For example, the control logic of the MAC makes scheduling decisions (resource block allocation and modulation and coding scheme), while the action logic applies them to user logical channels. Similarly, part of the logic of the RRC protocol decides on UE handovers, while the actual handover operation requires RRC to perform the corresponding action. Based on this taxonomy, the separation of the RAN CP from the UP, as discussed in Section 6.6.1, can be taken a step further by also detaching the control logic from the action and consolidating all the control operations in a logically centralized controller, which in the proposed architecture is comprised of centralized and edge controllers and agents interacting via the control protocol. This allows BSs to focus on performing all the UP related action functions, such as applying scheduling decisions, performing handovers, applying DRX commands, (de)activating component carriers in carrier aggregation, etc.

To control and manage the BS UP actions, the RAN Agent API is introduced to provide a set of functions that constitute the southbound APIs. These functions allow the CP to interact with the UP in five ways, namely

- to get and set configurations like the UL/DL bandwidth of a cell;
- to request and obtain statistics like transmission queue sizes of UEs and signal-to-interference-and-noise ratio (SINR) measurements of cells;
- to issue commands to apply control decisions (e.g., calls for applying MAC scheduling decisions, performing handovers, activating secondary component carriers);
- to obtain event notifications like UE attachments and random access attempts; and
- to perform a dynamic placement of control functions to the master controller or the agent (e.g., centralized scheduling at the master controller or local scheduling at the agent-side).

These API calls can be invoked either by the centralized or edge controller through the control protocol or directly from the agent if control for some operation has been delegated to it. Table 6-2. provides a list of some exemplary RAN-specific API calls. Note that the Agent is in charge of retrieving the cell- and user-related information from the underlying eNB such as cell bandwidth, user reference signal receive power (RSRP) and reference signal receive quality (RSRQ) through the API calls, and can trigger events when a state changes, e.g. involving user attachment and TTI information (frame and subframe)[1]. In addition, such API calls may be related to the NFs, resources, UEs, etc., belonging to a particular slice [43]. It can be seen that different types of network applications can

---

1 TTI information is required to apply coordinated actions on a specific time either within one or multiple cells such as adapting the transmission mode and interference management, respectively.

Table 6-2. RAN-specific API calls [42].

API	Target	Direction	Example	Applications
Configuration (synchronous)	eNB, UE, Slice	Centralized/Edge → Agent	UL/DL cell bandwidth, reconfigure Data Radio Bearer (DRB), Measurements,	Monitoring, Reconfiguration, Self Organizing Networks (SON)
Statistic, Measurement, Metering (asynchronous)	List of eNB, UE, Slice	Agent → Centralized/Edge	Channel Quality Indicator (CQI) measurements, SINR measurements, Reference Signal Receive Power (RSRP)/Reference Signal Receive Quality (RSRQ)/ UL/DL performance	Monitoring, Optimization, SON
Commands (synchronous)	Agent	Centralized/Edge → Agent	Scheduling decisions, Admission control Handover (HO) initiation	Realtime Control, SON
Event trigger	Master	Agent → Centralized/Edge	TTI, UE attach/detach, Scheduling request, Slice created/destroyed	Monitoring, Control actions
Control delegation	Agent	Centralized/Edge → Agent	Update DL/UL scheduling, Update HO algorithm	Programmability, Multi-service

be developed ranging from monitoring for a better decision making (e.g., adaptive video optimization) to control and programmability for a better adaptability and flexibility to services (e.g., by controling resource allocation, adjusting the handover logic, changing functional splits, updating precoding matrices, or even disabling/enabling ciphering/deciphering, etc.).

6.9 Summary and Outlook

In this chapter, we have ventured in detail into the RAN architecture design for 5G, for instance covering the key changes to the RAN protocol stack that have been already agreed or discussed in 3GPP, or key considerations that have to be made in the context of the handling of novel AIVs, or for network slicing, multi-service and multi-tenancy support. We have also discussed how multi-connectivity among different 5G AIVs, or among (e)LTE and NR, will likely be realized, and have covered in detail the novel 5G design paradigm of offering various CP/UP split options and also horizontal function splits in the RAN, in order to flexibly adapt the RAN to various physical deployment scenarios. Finally, we have ventured into RAN programmability and control, which is seen as essential to meet the diverse service requirements foreseen for 5G, and to facilitate network evolution.

It appears that 3GPP has already moved fast and obtained various important agreements that pave the road towards the 5G RAN. The key challenge will likely lie in the exact implementation of the 5G system, in particular given the many (more) degrees of freedom that 3GPP foresees for 5G in comparison to legacy technology, for instance related to the usage and parametrization of network functions or a plethora of multi-connectivity and also vertical and horizontal split options in the RAN. It is expected that many details will be left to proprietary implementation and optimization, in particular how exactly network slicing, multi-service and multi-tenancy are realized in the RAN. Similarly, we will likely see proprietary RAN interfaces (e.g., related to intra-PHY split options) that are highly optimized to, e.g., specific mMIMO product offerings from network vendors, and proprietary concepts for cross-layer optimization, such as overall resource management and QoS management concepts across the layers from SDAP to PHY.

References

1 3GPP TR 38.913, "Study on scenarios and requirements for next generation access technologies", March 2017

2 3GPP TR 38.912, "Study on new radio access technology", March 2017

3 3GPP TS 38.300, "NR; Overall description; Stage-2", June 2017

4 3GPP TS 23.501, "System Architecture for the 5G System", June 2017

5 3GPP TR 38.801, "Study on new radio access technology: Radio access architecture and interfaces", May 2017

6 5G PPP METIS-II project, see https://5g-ppp.eu/metis-ii/

7 5G PPP 5G NORMA project, see https://5g-ppp.eu/5g-norma/

8 5G PPP COHERENT project, see https://5g-ppp.eu/coherent/

9 5G PPP mmMAGIC project, see https://5g-ppp.eu/mmmagic/

10 5G PPP 5G-MoNArch project, see https://5g-ppp.eu/5g-Monarch/

11 5G PPP Architecture Working Group, White Paper, "View on 5G Architecture", V2.0, July 2017, see https://5g-ppp.eu/white-papers/

12 5G PPP METIS-II project, Deliverable D2.2, "Draft Overall 5G RAN Design", June 2016

13 NGMN, White Paper, "5G White Paper", March 2015

14 5G PPP METIS-II project, White Paper, "Preliminary Views and Initial Considerations on 5G RAN Architecture and Functional Design", March 2016

15 5G PPP METIS-II project, Deliverable D4.1, "Draft air interface harmonization and user plane design", April 2016

16 3GPP TR 38.804, "Study on new radio access technology Radio interface protocol aspects", March 2017

17 5G PPP METIS-II project, Deliverable D6.1, "Draft asynchronous control functions and overall control plane design", June 2016

18 5G PPP METIS-II project, Deliverable D4.2, "Final air interface harmonization and user plane design", March 2017

19 D. Szczesny et al., "Performance analysis of LTE protocol processing on an ARM based mobile platform", International Symposium on System-on-Chip (SoC 2009), Nov. 2009

20 5G PPP METIS-II project, Deliverable D5.2, "Final Considerations on Synchronous Control Functions and Agile Resource Management for 5G", March 2017

21 5G PPP METIS-II project, Deliverable D6.2, "Final asynchronous control functions and overall control plane design", April 2017

22 3GPP TR 25.912, "Feasibility study for evolved Universal Terrestrial Radio Access (UTRA) and Universal Terrestrial Radio Access Network (UTRAN)", March 2017

23 E. Pateromichelakis et al., "Service-tailored User-Plane Design Framework and Architecture Considerations in 5G Radio Access Networks", IEEE Access, vol. 5, pp. 17089–17109, Aug. 2017

24 5G PPP 5G NORMA project, Deliverable D4.2, "RAN architecture components – final report", July 2017

25 3GPP TS 36.300, "Evolved Universal Terrestrial Radio Access (E-UTRA) and Evolved Universal Terrestrial Radio Access Network (E-UTRAN); Overall description; Stage 2", June 2017

26 3GPP TS 37.340, "Evolved Universal Terrestrial Radio Access (E-UTRA) and NR; Multi-connectivity; Stage 2 (Release 15)", V15.0.0, Dec. 2017

27 5G PPP METIS-II project, Deliverable D2.4, "Final Overall 5G RAN Design", June 2017

28 P. Arnold, N. Bayer, J. Belschner and G. Zimmermann, "5G Radio Access Network Architecture based on Flexible Functional Control/User Plane Splits", European Conference on Networks and Communications (EuCNC 2017), June 2017

29 P. Rost, A. Banchs, I. Berberana et al., "Mobile Network Architecture Evolution toward 5G", IEEE Communications Magazine, vol 54, no. 5, pp. 84-91, May 2016

30 R. Trivisonno, R. Guerzoni, I. Vaishnavi and D. Soldani, "SDN-based 5G mobile networks: architecture, functions, procedures and backward compatibility", Transactions on Emerging Telecommunications Technologies, Wiley, Dec. 2014

31 3GPP TS 38.806, "Study on separation of CP and UP for split option 2 of NR; (Release 15)", Sept. 2017

32 Industry Initiative for a Common Public Radio Interface, see www.cpri.info

33 3GPP RP-170818, "Study on CU-DU lower layer split for New Radio", March 2017

34 3GPP TS 38.475, "NG-RAN; F1 interface user plane protocol", June 2017

35 ETSI Industry Specification Group, ISG NFV, "Network Functions Virtualization"

36 3GPP R3-161120, "5G access architecture with UP/CP separation", May 2016

37 3GPP R3-171475, "Central RRM functions and gNB-DU reporting", May 2017

38 Small Cell Forum, SCF082, "nFAPI and FAPI specifications", Release 9, May 2017

39 Small Cell Forum, SCF191, "Multi-operator and neutral host small cells: drivers, architectures, planning and regulation", Release 8, Dec. 2016

40 N. Nikaein, M. Marina, S. Manickam, A. Dawson, R. Knopp and C. Bonnet, "OpenAirInterface: a flexible platform for 5G research", ACM Sigcom Computer Communication Review, vol. 44, no. 5, Oct. 2014, see www.openairinterface.org

41 5G PPP mmMAGIC project, White Paper, "Architectural enablers and concepts for mm-wave RAN integration", March 2017

42 X. Foukas, N. Nikaein, M. Kassem and K. Kontovasilis, "FlexRAN: A flexible and programmable platform for software-defined radio access networks", Conference on emerging Networking EXperiments and Technologies (CONEXT 2016), Dec. 2016

43 A. Ksentini and N. Nikaein, "Toward enforcing network slicing on RAN: Flexibility and resources abstraction", IEEE Communications Magazine, vol. 55, no. 6, June 2017

7

Transport Network Architecture

Anna Tzanakaki[1], Markos Anastasopoulos[2], Nathan Gomes[3], Philippos Assimakopoulos[3], Josep M. Fàbrega[4], Michela Svaluto Moreolo[4], Laia Nadal[4], Jesús Gutiérrez[5], Vladica Sark[5], Eckhard Grass[5], Daniel Camps-Mur[6], Antonio de la Oliva[7], Nuria Molner[8], Xavier Costa Perez[9], Josep Mangues[3], Ali Yaver[10], Paris Flegkas[11], Nikos Makris[11], Thanasis Korakis[11] and Dimitra Simeonidou[2]

[1] *National and Kapodistrian University of Athens, Greece, and University of Bristol, UK*
[2] *University of Bristol, UK*
[3] *University of Kent, UK*
[4] *CTTC/CERCA, Spain*
[5] *IHP, Germany*
[6] *i2CAT, Spain*
[7] *Universidad Carlos III de Madrid, Spain*
[8] *IMDEA Networks Institute, Spain, and Universidad Carlos III de Madrid, Spain*
[9] *NEC, Germany*
[10] *Nokia, Poland*
[11] *University of Thessaly, Greece*

7.1 Introduction

The explosive growth of the mobile Internet, and the increasing diversity of use cases that are enabled by this, force network providers to concurrently support a large variety of applications and services. This introduces the need to transform traditionally closed, static and inelastic network infrastructures into open, scalable and elastic ecosystems. At the same time, new and emerging services impose enormous capacity requirements that significantly exceed the currently installed capacity. Furthermore, tens of billions of end devices will require connectivity in a scalable and sustainable manner. To address these needs, a more efficient utilization of the radio frequency spectrum is necessary. In view of this, 5th generation (5G) mobile and wireless networks will likely be characterized by the deployment of a large number of highly densified radio access points (APs) able to operate in much higher frequency bands, such as millimeter-wave (mmWave).

One clear challenge associated with these requirements is to provide infrastructure connectivity from the APs to the core network (CN), also referred to as *transport network* connectivity, at

5G System Design: Architectural and Functional Considerations and Long Term Research, First Edition.
Edited by Patrick Marsch, Ömer Bulakçı, Olav Queseth and Mauro Boldi.
© 2018 John Wiley & Sons Ltd. Published 2018 by John Wiley & Sons Ltd.

reasonable cost. In traditional radio access networks (RANs), where baseband units (BBUs) and radio units (RUs) forming the APs are co-located, this connectivity is referred to as *backhaul* (BH). On the other hand, infrastructure connectivity within the RAN, for instance between centralized units (CUs) and distributed units (DUs) is referred to as *fronthaul* (FH). This connectivity can be based on various different RAN split options, as discussed in Section 6.6.

From now on using the terminology of CU and DU for generality, the FH network is responsible for carrying the DUs' wireless signals over the transport network, utilizing commonly digitized formats based on protocols, such as the Common Public Radio Interface (CPRI). This concept, also known as cloud RAN (C-RAN) [1], provides pooling gains due to the joint processing of signals related to many BBUs in CUs that allow an increase in spectral efficiency, for instance through coordinated scheduling or coordinated multi-point (CoMP) across many cells. The main disadvantages of C-RANs include increased transport bandwidth requirements to carry the sampled radio signals, as well as strict latency and synchronization constraints [2].

As elaborated in detail in Section 6.6.2, the 3rd Generation Partnership Project (3GPP) has already taken key steps to relax FH requirements for 5G, and has discussed several alternative interface options in the RAN that allow to adapt functional splits between CUs and DUs to use case requirements as well as transport network capabilities. A similar approach is also captured in the enhanced CPRI (eCPRI) standard [2]. The different RAN functional split points blur the distinction between traditional BH and FH. This is due to the fact that higher layer split points, such as the split between Packet Data Convergence Protocol (PDCP) and Radio Link Control (RLC) explained in Section 6.6.2, have bandwidth and latency requirements more akin to traditional BH than FH. However, it is clear that, even for such split points, the enormous increase in the volume of traffic to be aggregated will require efficient techniques for high-bit-rate transport and multiplexing. As mentioned in Section 6.2.2, for the lower RAN layer split points, latency and latency variation become an increasingly significant constraint. A more useful distinction in the requirements for the new FH may hence be between the split points defined by 3GPP as having "loose" and "tight" latency requirements [3]. The new 5G transport network will need to cater for these differing requirements of the RAN functional splits, their different bandwidth requirements and aggregation, in the form that multiple, different split point interfaces may need to be transported over the same physical infrastructure. As an example, the same transport network infrastructure may have to carry FH related to higher RAN layer splits, as well as BH.

On the other hand, it is clear that, although optical BH is a key technology solution for the realization of the C-RAN paradigm, optical fiber links are mostly available in urban environments. Their deployment is likely not fast enough to match the large number of densely deployed APs required to serve the upcoming demand in mobile data. To address this issue, low-cost mmWave links can be used for BH/FH in cases where optical fiber links are not available. Due to the very strict transport requirements and the limited availability of suitable technologies, substantial innovation in the transport network architecture and technology options is needed in support of the 5G vision, being the key focus of this chapter.

In view of this, equipment vendors are expanding their FH solutions by adopting advanced wireless technologies, for instance in sub-6 GHz and 60 GHz bands including advanced beam tracking and multiple-input multiple-output (MIMO) techniques, in conjunction with new flexible and dynamic wavelength division multiplexing (WDM) optical networks. These are also enhanced with novel control and management approaches to enable increased granularity, end-to-end (E2E) optimization and guaranteed quality of service (QoS). To further improve the efficiency of the transport network, self-backhauling may play a key role in 5G, as it allows dynamic reuse of the same radio technology and spectrum for both BH and access.

The rest of this chapter is organized as follows. Section 7.2 focuses on the description of a transport network architecture suitable for 5G in terms of both user plane (UP) and control plane (CP). An overview of the various technologies and protocols and the associated interfaces that can be used in support of both FH and BH services is provided in Section 7.3, while self-backhauling is addressed in detail in Section 7.4. Framing protocol adaption and interfacing aspects are discussed in Section 7.5. Finally, a discussion on transport network optimization and performance evaluation is presented in Section 7.6, before the chapter is summarized in Section 7.7.

7.2 Architecture Definition

7.2.1 User Plane

The 5G UP architecture considers converged optical and wireless network domains in a common 5G infrastructure supporting both transport and access. In the wireless domain, a variety of legacy and other technologies can be considered, including a dense layer of small cells that can be wirelessly backhauled through mmWave and sub-6 GHz technologies. Alternatively, small cells can be connected to a CU through optical network solutions.

In addition to BH services, the transport network needs to be able to provide operational services through the support of FH links. FH links provide connectivity services between densely distributed DUs with regional data centres hosting CUs that have very stringent delay and synchronization requirements. To maximise sharing benefits, offering improved efficiency in resource utilization and measurable benefits in terms of cost, scalability and sustainability, it is proposed to use a common network infrastructure to jointly support BH and FH functions. This can be practically supported through a suitable 5G UP architecture integrating a set of advanced wireless and wired access and transport network technologies.

In this context, a key enabler is the adoption of a high-capacity, flexible optical transport network coupled with mmWave links. This optical network may rely on hybrid approaches including passive optical networks (PONs) offering enhanced capacity through WDM and dynamic, spectrally elastic and frame-based optical network solutions to support more demanding capacity and flexibility requirements for traffic aggregation and transport as featured in [4]. A specific example for such optical networks is shown in Figure 7-1, as developed by the 5G Public Private Partnership (5G PPP) project 5G-XHaul [5]. Such transport network will be able to support the increased transport requirements of 5G environments in terms of granularity, capacity and flexibility. The mmWave technology operating in the sub-6 GHz and 60 GHz bands will provide high-bandwidth connectivity for both non-line-of-sight (NLOS) and line-of-sight (LOS) scenarios, supporting at the same time high mobility in heterogeneous environments through dynamic beamforming and programmable beam steering techniques.

Given the technology heterogeneity envisioned for 5G, a critical function of the converged wireless-wired infrastructure is the interfacing between technology domains. These domains may adopt very different protocol implementations and provide very diverse levels of overall capacity (in the wireless domain varying between Mbps up to tens of Gbps) and granularity (varying between kbps and 100 Mbps). In addition, latency remains a critical parameter to be fulfilled in such a converged transport network.

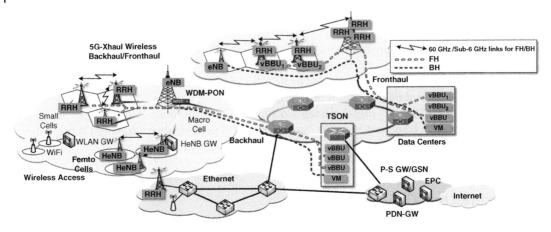

Figure 7-1. Converged heterogeneous network and compute infrastructures [5].

For the data path (physical) interconnections, the various technology domains need to support common protocols to enable seamless traffic interchange between them. A possible implementation toward this direction has been proposed in [6], being an architecture that adopts the paradigm of software-defined networking (SDN). The fundamental block of the UP architecture is the forwarding element (FE, in [6] referred to as XFE). The FE consists of a packet switch called packet forwarding element (PFE, in [6] referred to as XPFE) and a circuit switching element (CSE, in [6] referred to as XCSE).

The packet switching path is the primary path for the transport of most delay-tolerant FH and BH traffic, whereas the circuit switching path complements the packet switching path for those particular traffic profiles that are not suited for packet-based transporting (e.g., legacy CPRI or traffic with extremely low delay tolerance), or just for capacity offloading. This two-path switching architecture is able to combine bandwidth efficiency through statistical multiplexing in the packet switch, with deterministic latency ensured by the circuit switch. The modular structure of this switch, where layers may be added and removed, enables various deployment scenarios with traffic segregation at multiple levels, from dedicated wavelengths to virtual private networks (VPNs), which is particularly desirable for multi-tenancy support, as discussed in Section 5.2.4.

The DUs support a flexible functional split, with some functions of the access interface virtualized and placed at the cell site, and their complementary functions virtualized and pushed to the baseband processing nodes. The FH interface between the DUs and the CUs can be any existing or new interface, such as CPRI or future packet-based FH interfaces, see Section 6.6.2.

Some aggregation may be performed by the DUs, before interfacing the FE, in order to decrease the number of data flows, increase the bit rate on the ports, and simplify the implementation. A first example is provided by a cascade of DUs, where CPRI traffic is added and time-multiplexed at each remote radio head (RRH). In a second example, client signals at different radio carrier frequencies are multiplexed in a radio over fiber (RoF) system, and the aggregate signal is converted from analog to digital.

The DUs are connected to the FE by means of adaptation functions that perform media and protocol adaptation. The purpose of the adaptation function AF-1 is media adaptation, such as from air to fiber and translation of the radio interface (CPRI, new 5G FH packet interfaces, Ethernet used in BH links, mmWave/802.11ad frames, analog RoF, etc.) into a common frame (CF, referred to as XCF in [6]), which interfaces the packet switch. Similar to AF-1, the adaptation function AF-2 maps the radio interface into the protocol used by the circuit switch, e.g., an optical transport network (OTN) or the simpler circuit framing protocol. The packet switches communicate with each other using the common frame, which is a packet interface based on an evolution of the Ethernet Medium Access Control (MAC)-in-MAC standard, and also the interface between packet switches and the processing unit (PU, referred to as XPU in [6]), the virtualized unit in charge of hosting baseband processing and other virtual functions. The packet switch may be connected to the circuit switch through the adaptation function AF-3 that maps the common frame into the protocol used by the circuit switch. As a further advantage, this connection can be used to offload the packet switch, avoiding overload situations and, therefore, decreasing the probability of discarded packets.

7.2.2 Control Plane

Given the overall 'softwarization' principles of the 5G vision, as introduced in Section 5.2.3 and further discussed in, e.g., [6], the aspect of the transport network CP is central. In SDN, the CP is decoupled from the UP and managed by a logically centralized controller that has a holistic view of the network. This is covered in more detail in Section 10.2.1. On the other hand, network function virtualization (NFV) enables the execution of network functions on compute resources by leveraging software virtualization techniques, see Section 10.2.2. SDN is related to the control and management of the virtual resources, which, in order to reduce operational costs, need to be provisioned in an automatic way. In addition, SDN should provide the operator with the ability to easily compose and deploy new network services, which, for example, could be instantiated through different network slices, as described in Chapter 8. Virtualization and softwarization will shape the architecture of future 5G networks as recognized by the 5G PPP Architecture Working Group in [4]. In particular, the softwarization and virtualization trends will also affect the design of the future 5G transport networks. In addition to the aforementioned general requirement to support network slicing, future 5G transport networks also have to address specific requirements, such as the cost-effective transport of FH and BH traffic required to support C-RAN and distributed RAN (D-RAN) deployments.

Through joint SDN and NFV considerations, significant benefits can be achieved, including an efficient use of resources, simplification of the infrastructure management, increased scalability and sustainability as well as provisioning of orchestrated E2E services. Depending on the level of SDN and NFV integration supported, various CP approaches have been proposed [7][8][9]. Two specific examples are discussed below, which have been developed in the framework of the 5G PPP projects 5G-XHaul and 5G-Crosshaul.

The first approach [9] adopts an SDN-centered solution for the control of the UP elements. The 5G transport CP architecture is based on the following design principles:

- Full address space virtualization is offered through an overlay, implemented using encapsulation at the edge of the transport network. This means that different tenants can use overlapping layer-2 or layer-3 address spaces;
- UP scalability is achieved by isolating the forwarding tables of the transport network elements inside the 5G transport infrastructure from any tenant-related state (overlay). This is again achieved by encapsulating tenant frames at the edge of the network into transport-specific tunnels;

- Scalability of the SDN CP is achieved by introducing the concept of *areas*. An area defines a set of transport network elements that are under the control of a logically centralized SDN controller. A CP hierarchy is introduced, whereby higher-level controllers are used to coordinate the actions of area-level controllers;
- Finally, the vision of converged heterogeneous technology domains, e.g., wireless and optical segments in the transport network, is enabled by: i) the previously introduced areas, which embody a single type of transport technology (e.g., wireless mesh, optical or Ethernet), and ii) a transport adaptation function that maps the per-tenant traffic at the edge nodes to the transport specific tunnels of a given area.

To support the previous principles, three types of transport nodes are defined, which are depicted in Figure 7-2. First, edge transport nodes (ETNs) connect the tenant virtual network functions (VNFs) to the 5G transport network, maintain the corresponding per-tenant state, and encapsulate tenant traffic into transport-specific tunnels. Second, inter-area transport nodes (IATNs) support the necessary functions to connect different areas. They may be implemented using different transport technologies. Finally, regular transport nodes (TNs) support an area-specific transport technology, and provide forwarding services between the ETNs and IATNs of that area.

The bottom part of Figure 7-2 depicts the abstraction that the CP presents to each tenant, namely the transport network slice. Each tenant's goal is to connect through the transport network a set of tenant-defined VNFs, deployed in distributed compute and storage facilities. Thus, the tenant is allowed to group VNFs on virtual layer-2 segments, identified by layer-2 segment IDs (L2SIDs). Different L2SIDs can be connected through a virtual data path (vDP), which can be instantiated, for example using a software switch. Notice that the vDPs are controlled directly by the tenant, while all virtual entities, VNFs and vDPs will be physically hosted in an ETN to connect to the 5G network. For more detailed information on transport slice realization, the reader is referred to Section 8.2.2.

In the alternative system architecture [6], the CP is divided into two layers: the control infrastructure (CI, in [6] referred to as XCI) layer, and a top layer for external applications.

On top of the control infrastructure, the applications implement the service-dependent logic and the network intelligence required to optimize the utilization of the available resources, while guaranteeing the consistent coordination and orchestration of the services as well as their compliance with their QoS specification. Therefore, the control infrastructure should be able, internally, to enforce the decisions taken from the applications and, externally, to expose a suitable level of programmability and granularity at its northbound application programming interfaces (APIs) to allow the applications to specify their decisions and request the desired type of resource allocation and instantiation. On the other hand, the same CI northbound interface should expose enough information about the capabilities, status, resource availability and performance of the underlying infrastructure, as required by the applications to run their internal algorithms and take efficient decisions. Depending on the type of monitoring information, different interaction mechanisms should be supported. For example, polling of data related to resource availability is desirable for applications that need to compute paths or resource allocations for an on-demand instantiation of network services. On the other hand, applications that need to react to unexpected events (e.g., failures) or unpredictable conditions (e.g., an excessive load on a network segment) must rely on subscribe/notify mechanisms for asynchronous communications.

The control infrastructure is the 5G transport management and orchestration (MANO) platform that provides control and management functions to operate all available types of resources (i.e., networking, computing and storage). This infrastructure is based on the SDN/NFV principles and provides a unified

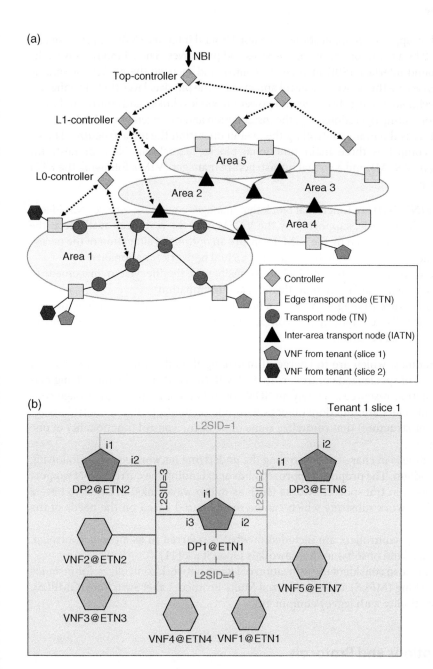

Figure 7-2. (a) 5G transport control plane architecture, (b) Slice abstraction towards tenant.

platform that can be used by upper layer applications via a northbound interface (NBI) to program and monitor the underlying UP by a common set of core services and primitives. The CI interacts with the UP entities via a southbound interface (SBI) in order to control and manage the packet forwarding behavior performed by FEs across the network, control and manage the physical layer (PHY) configuration of the different link technologies (e.g., transmission power on wireless links), and control and manage the processing units computing operations (e.g., the instantiation and management of VNFs).

The control infrastructure is the brain controlling the overall operation the infrastrucutre. The CI part dealing with NFV comprises three main functional blocks, namely an NFV orchestrator (NFV-O), VNF manager(s) (VNFM), and a virtual infrastructure manager (VIM), following the ETSI NFV architecture, see [10]:

- The **NFV orchestrator (NFV-O)** is a functional block that manages a network service (NS) lifecycle. It coordinates the VNF lifecycle, supported by the VNFM, and the resources available at the NFV infrastructure (NFVI), supported by the VIM, to ensure an optimized allocation of the necessary resources and connectivity to provide the requested virtual network functionality;
- The **VNF mangers (VNFM)** are functional blocks responsible for the lifecycle management of VNF instances (e.g., instance instantiation, modification, and termination);
- The **Virtualized Infrastructure Manager (VIM)** is a functional block that is responsible for controlling and managing the NFVI computing (via computing control), storage (via storage control) and network resources (via SDN control).

In addition to these modules, which are in charge of managing the different VNFs, the control infrastructure includes a set of specialized controllers to deal with the control of the underlying network, storage and computation resources, namely an SDN controller, computing and storage controllers. Both types of controllers are functional blocks with one or multiple actual controllers (of hierarchical or peer-to-peer structure) that centralize some or all of the control functionality of one or multiple network domains.

The SDN controller module is in charge of controlling the underlying network elements following the conventional SDN paradigm. The proposed approach aims at extending the current SDN support of multiple technologies used in transport networks (such as micro-wave links) in order to have a common SDN-controlled network substrate which can be reconfigured based on the needs of the network tenants.

The computing and storage controllers are included in what is referred to as a cloud controller. A prominent example of this kind of software framework is OpenStack [11].

The proposed architecture also considers the utilization of legacy network controllers, for instance Multi-Protocol Label Switching (MPLS) and Generalized Multi-Protocol Label Switching (GMPLS), to ensure backward-compatibility with legacy equipment.

7.3 Technology Options and Protocols

7.3.1 Wireless Technologies

- *Wireless Technologies- Sub-6, 60 GHz and MIMO*: As already discussed, the densification of the envisioned 5G RAN requires multi-hop wireless topologies for the wireless transport network, where high capacity wireless links will convey the aggregated traffic of several base stations (BSs). To cater for this traffic demand, a combination of wireless technologies has been recently considered as a viable and cost-effective approach able to provide the ideal combination of capacity and coverage, particularly in complex urban deployments.

Several technologies can be used to provide wireless transport capabilities at sub-6 GHz frequencies. Institute of Electrical and Electronics Engineers (IEEE) 802.11 technologies have been traditionally used in the industrial, scientific and medical (ISM) band, and have been based on an orthogonal frequency division multiplex (OFDM) PHY and a random-access MAC based on carrier sense multiple access/collision avoidance (CSMA/CA). However, recent IEEE 802.11 amendments, such as 802.11ac and 802.11ax, have incorporated features that make them attractive as wireless transport solutions.

Through a combination of channel bonding (up to 160 MHz) and enhanced modulation schemes, 802.11ac can deliver multi-Gigabit data rates at the PHY. Hence, transport nodes featuring multiple 802.11ac radios and directive antennas can provide sufficient performance in the chain topologies typically found in urban scenarios. In addition, 802.11ac also incorporates multi-user MIMO (MU-MIMO) capabilities (limited to the downlink), which can increase the capacity available to nodes acting as gateways to the wired network.

IEEE 802.11ax provides additional mechanisms, such as MU-MIMO in uplink and downlink, together with 1024 quadrature amplitude modulation (QAM). In addition, 802.11ax includes mechanisms for enhanced resource allocation using a combination of CSMA with orthogonal frequency division multiple access (OFDMA), allowing efficient allocation of resources between transport and access. Specifically, when a device (e.g., an AP) acquires a transmission opportunity through CSMA, then an OFDMA frame is transmitted that can provide access to other devices. The interested reader is referred to [12], [13] for an example where a time division multiple access (TDMA) overlay is used on top of IEEE 802.11 radios to allocate resources between access and BH.

- **mmWave communications in transport networks**: The industry is currently considering mmWave solutions for delivering the required high capacities in BH networks, e.g., in IEEE802.11ay [14] and FH use case scenarios, with rates in the range of 20 Gbps.

  One of the key architectural features of these systems is the availability of a large number of antenna elements forming antenna arrays at both transmitter and receiver sides, as detailed in Section 11.5.4. The antenna arrays have to provide sufficient array gain to obtain enough link margin for wide area operation. The mmWave links can deploy highly directional beams, significantly reducing the effects of interference in a multi-user environment. In a dense deployment scenario, this facilitates the configuration of meshed networks, where the nodes can be connected via steerable pencil beams. Techniques to allow dynamic and autonomic reconfiguration of the mmWave BH links, letting the CP of the network decide which nodes to connect, are key to achieve the required flexibility.

  Current mmWave systems are based on fixed-beam architectures. Switchable beam solutions are currently of interest to perform auto-alignment functions among the nodes. This is necessary to lower the deployment cost and to avoid small displacements (due to wind or accidental effects). Nowadays, beam-steering or beam-switching over a wide angle are considered to enable dynamic BH network reconfiguration.

- **Multi-PHY protocols and SDN interfaces for hybrid wireless transport**: The wireless part of the transport network architecture encompasses both mmWave and sub-6 GHz radio technologies. These technologies build up multi-hop wireless mesh islands connecting the RAN to the core. Both mmWave and sub-6 GHz nodes will be able to act as transport nodes (TNs) and APs, respectively, and are even expected to be co-located in some scenarios, the Ethernet protocol being the most suitable layer-2 transport platform.

Wireless bridges are the most natural configuration to interface these two technologies. These bridges consist of multiple radio interfaces physically attached to the same network node, which has the capacity to carry out a transparent bridging of Ethernet frames between the different radio interfaces, usually at operating system (OS) kernel level. OS-level bridging is generally outperformed, in terms of forwarding speed, by dedicated hardware, but, on the other hand, enables a seamless integration of QoS and scheduling policies.

Alternatively, the physical connection of different wireless domains can be achieved by means of wired links, provided that those links: i) support data rates in the Gbps scale to avoid adding more restrictive bottlenecks in the data path, and ii) support transparent transport of Ethernet frames. These requirements allow the use of any 1000BASE and 10GBASE (and beyond) IEEE 802.3 families.

The co-location of the nodes can reduce the complexity and communication exchange among nodes, since the available synchronization signal could be used for both interfaces, which, as a result, will mitigate contention and interference, thus improving performance. Moreover, it represents a way to reduce the amount of signaling as well as to save energy.

Assuming the adoption of the Ethernet interface by wireless transport technologies, a natural next step is to consider embedding an SDN agent in the wireless bridges. A protocol like OpenFlow (OF) can be used between the SDN agent and an external SDN controller. Adding an SDN datapath in the wireless transport segment provides a variety of benefits, such as being able to switch some devices to save energy and reconfigure the transport segment accordingly, allowing an operator to deploy per-subscriber forwarding policies, or to quickly react to external interference. The interested reader is referred to [15] for an example where General Packet Radio Service (GPRS) Tunneling Protocol (GTP) support is added to the SDN agent, and an interface between the transport SDN controller and the Long-Term Evolution (LTE) Mobility Management Entity (MME) is introduced to provide per-subscriber forwarding policies.

Several challenges, however, need to be solved in order to seamlessly incorporate SDN capabilities into the wireless BH segment. First, OF port statistics need to be expanded to convey wireless parameters to the controller, which can then be used to implement effective traffic engineering (TE) policies. Second, wireless interfaces are point-to-multi-point, in contrast to the point-to-point nature of the physical interfaces typically connected to an SDN agent; the interested reader is referred to [16] for an architecture that addresses this challenge. Third, unlike data-centered networks, in the wireless transport segment, the SDN control channel between the network elements and the controller needs to be necessarily carried in-band, which requires mechanisms to protect the SDN signaling traffic, as well as mechanisms to bootstrap the initial connection with the controller. Finally, certain transport control functions with tight delay requirements, such as fast recovery, cannot be offloaded to the controller. Therefore, in practice, the SDN wireless transport segment needs to consider a hybrid solution where some control functions are embedded in the devices, while other ones are moved to the controller. Finding the best CP split is currently an area of active research.

Finally, regarding synchronization, when a low-latency wired connection is available at the node location, standard synchronization mechanisms, such as SyncE or IEEE 1588, must be carried out through that interface. Otherwise, synchronization using mmWave links should be prioritized over sub-6 GHz when possible, due to the higher reliability and predictability of the medium access in those bands. Exceptions to these rules, however, should be contemplated in some situations. For example, this is the case when, despite the possibility of receiving synchronization from a mmWave interface, a downlink neighboring TN can only be reached by means of the sub-6 GHz interface. In such a case, the TN must either transparently forward synchronization signals, or act as a master to that downlink neighbor.

7.3.2 Optical Transport

As discussed in Section 7.2.1, RAN services can be facilitated through optical fiber technologies able to offer true broadband communications over shorter or longer distances in accordance to the concept of centralizing BBUs into CUs. In this context, two different types of optical transport technologies can be identified: the first employs PON technologies, and the second supports larger scales including also active elements.

- **Passive Optical Networks**: Nowadays, point-to-multi-point PONs are popular architectures for supporting fiber-based broadband access to a set of customers. There, an optical line terminal (OLT), which is part of a central office (CO), is connected through a tree-branched passive distribution network to the optical network units (ONUs), which are located at the customer side. Such networks have several advantages, as they are very simple, easily scalable, and do not need excessive maintenance [17]. In time division multiplexed PONs (TDM-PONs), standardized by IEEE [18] and International Telecommunication Union (ITU) Telecommunication Standardization Sector (ITU-T) [19][20], the different user signals are multiplexed in time. The latest standard envisions a joint use of TDM and WDM for increased capacity [21].

  PONs can facilitate the interconnection of remote CUs and DUs to support only FH services, as CUs can be located at a selected node of the metro network segment (for instance, an exchange node containing the CO), while the DUs could be scattered along an access tree, see Figure 7-3. The remote nodes (RNs) in the PON trees are typically implemented either employing a simple power splitter or a wavelength routing device, depending on the approach taken.

  A first approach could be to deploy a fully dedicated PON for the radio signal delivery, where communication could be achieved by means of standard PON technologies, a specifically designed hybrid TDM/WDM-PON [22], or even a flexible WDM-PON approach, as envisioned in [23]. However, as already discussed, the envisioned 5G transport is expected to jointly support BH and FH services, hence avoiding the deployment of separate networks for the two different type of services. However, this solution is challenging, as it has to fulfill the needs of the optical access network subscribers, while supporting at the same time new 5G related services. Thus, a specific wavelength plan is required to ensure full compatibility with the deployed access network. In fact, legacy access standards such as Gigabit (Ethernet) passive optical networks (GPON, GEPON) use 1490 nm wavelength for downstream, while later standards, such as 10 Gbps Ethernet passive optical network (10G-EPON, XGPON), recommend the range of 1575-1580 nm. Regarding the upstream, all the cited standards envision the use of the O-band. Consequently, the entire C-band is available for performing a wavelength overlay of channels to provide different additional services over the same access infrastructure. Interestingly, the next generation PON (NGPON2) standard supporting up to 40 Gbps also offers the option to establish virtual PtP links within the C-band, assigning different WDM channels to different metro nodes and/or services beside serving the fixed access subscribers [21].

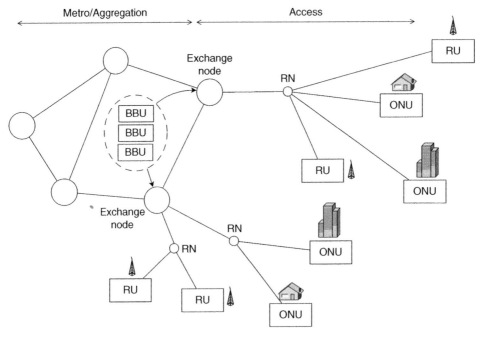

Figure 7-3. Generic network architecture. Exchange nodes contain different OLTs, which are part of a central office, in order to serve the different access trees.

Several approaches have been proposed to support wireless signal delivery over optical access networks while serving a set of fixed customers. For example, in [24], it was proposed to overlay a RAN over existing deployed TDM-PON infrastructure by using a narrowly spaced dense WDM plan with a suitable transceiver design.

- **Active Optical Networks***: The advent of elastic optical networking, enabled by the adoption of a flexible channel grid and programmable transceivers, opens the door to a truly dynamic management of optical networks [25], [26]. This is especially interesting in the context of supporting greatly varying transport services (both FH and BH) for the RAN. Furthermore, elastic networking enables to transparently set up a RAN network over the optical network. For example, a pool of BBUs concentrated in a CU can be located at a selected node of the metro network segment, while the DUs can be attached to different other nodes. To support this functionality in a cost-effective manner, it has been proposed to use programmable sliceable-bandwidth variable transponders (S-BVTs), which could be present at the 5G optical nodes in order to concurrently serve different cell sites. At the other end of the network, each cell site is equipped with a programmable BVT. The (S-)BVTs can be remotely configured by the CP for an optimal management of the network resources [26][27]. The parameters to be configured at each (S-)BVT include wavelength, spectral occupancy, and modulation format/power per flow. Therefore, the proposed (S-)BVTs deliver data flows with variable spectral occupancy and rate, according to the network and path conditions.

Among all the options for implementing the (S-)BVTs, those based on direct detection OFDM (DD-OFDM) are the most attractive for cost-effectively coping with the flexibility requirements of elastic optical networks [27]. OFDM provides advanced spectrum manipulation capabilities, including arbitrary sub-carrier suppression and bit/power loading. Due to these features, DD-OFDM transceivers can be ad-hoc configured for achieving a certain reach and/or coping with a targeted data rate adopting low-complexity optoelectronic subsystems [27][29].

Additionally, an interesting option could be to not focus either on the active or the passive optical networking (i.e., pure metro/regional or access), but to envision a transparent delivery of mobile FH and BH services in a converged elastic network [29]. There, optical channels are established across specific nodes designed to offer transparent interconnection, taking into account specific wavelength restrictions. This approach also entails some challenges, although experiments show successful connections from CUs to DUs, covering distances up to 60 km and achieving data rates beyond 50 Gbps [29]. Thus, it is another promising solution for serving the multiple endpoints employing S-BVT(s) at the 5G optical nodes.

- **Frame-based Elastic Optical Networks**: To support the varying degrees of bandwidth and latency requirements introduced by the various RAN deployments discussed above, the use of dynamic-frame-based solutions supporting a higher degree of time granularity (i.e., sub-wavelength) has been proposed. An example of this is the time shared optical network (TSON) technology [30]. These are multi-wavelength frame-based flexible systems. This type of solutions includes two different types of nodes, namely edge nodes and core nodes, incorporating different functionalities and levels of complexity. The edge nodes provide the interfaces between wireless, PON, and data center domains to the optical domain and vice versa. The ingress edge nodes are responsible for traffic aggregation and mapping, while the egress edge nodes support the reverse functionality. The edge nodes process the incoming data streams and generate optical time-slices at the ingress edge node, and also regenerate the original information from time-sliced optical frames at the egress edge node. These technologies should allow the handling of Ethernet frames, natively supporting a broad range of framing structures and communication protocols including CPRI, Open Base Station Architecture Initiative (OBSAI), and 10G Ethernet as well as protocol extensions required in support of the functional split concept under discussion in the framework of 5G, see also Section 6.6.2.

In the framework of the 5G PPP 5G-XHaul project, a hybrid WDM PON TSON optical transport solution was adopted [31]. In this context, at the optical network ingress TSON edge node, the interfaces receive traffic frames generated by fixed and mobile users, see also Figure 7-4. 1a). The incoming traffic is aggregated into optical frames, which are then assigned to suitable time-slots and wavelengths for further transmission. At the egress point, the reverse function takes place.

The optical edge node is also equipped with elastic bandwidth allocation capabilities supported through the deployment of BVTs. In addition to providing BH functionalities, this infrastructure also interconnects a number of DUs and end users with a set of general purpose servers. The use of general purpose servers enables the concept of virtual BBUs (vBBUs), allowing for an efficient sharing of compute resources. This joint functionality is facilitated by the edge nodes that comprise a hybrid subsystem of an I/Q switch and an Ethernet switch. The I/Q switch handles different functional split options as discussed in Section 6.6.2 with strict synchronization and bandwidth constraints, while the Ethernet data switch handles BH traffic and relaxed FH transport classes (i.e., split options). At the ingress part of the Ethernet switch module, the interconnection of DUs and CUs is provided, whereas at the egress part of the module, traffic is aggregated and placed in a suitable transmission queue. The elastic optical network solution adopted comprises BVTs, bandwidth-variable optical cross-connects, and fast optical switching modules. This approach enables

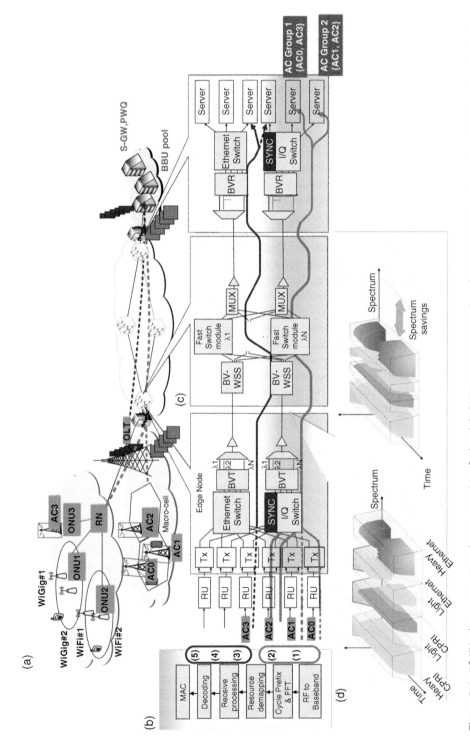

Figure 7-4. 5G-XHaul components and resources: a) Unified mobile FH/BH over converged wireless/optical data centre networks, b) Functional split of DU processing, c) Elastic allocation of TSON resources for heavy CPRI traffic and light-weight Ethernet flows.

the allocation of variable-size spectral/time slots, thus supporting services with continuous channel allocation at various bit rates (i.e., heavy and light CPRI) and services with sub-wavelength time-slot allocation (e.g., Ethernet flows). The TSON core nodes transparently switch optical frames to the appropriate output port utilizing fast optical switching with 10 ns switching speed.

7.3.3 Ethernet

Ethernet has been proposed for FH (and BH) transport [32]. It is an ubiquitous network technology, in use in corporate and public networks, and thus low-cost. Ethernet has been defined for high bit-rates up to 40 Gbps, with specifications for 100 Gbps and more being expected. Its use in public access networks provides a means towards converged fixed and mobile networks, and as Carrier Ethernet, it offers standardized Operations, Administration and Management (OAM) functions. As a packet-based technology, Ethernet can also readily take advantage of any statistical multiplexing gains offered by the new functional splits in the RAN. Its close links with and common use as the layer-2 technology in Internet Protocol (IP)-based networks has led to widespread availability of packet filtering and network monitoring tools. It is also, therefore, readily adopted into frameworks for SDN and the virtualization of network functions.

There are, however, challenges in the use of Ethernet in the mobile transport network. These are primarily related to the delay and delay variability that arises with a packet-based technology, caused mainly by output port contention and queuing. These are problems that will be more important in the context of fronthaul networks, and most important for the lower-layer split points, i.e., 3GPP split option 6 and upwards, as discussed in Section 6.6.2. One of the benefits of Ethernet, namely the ability to take advantage of statistical multiplexing gains, is in this respect a challenge, as this also means that port contention and queuing will arise.

A number of Ethernet networking techniques can be deployed to meet the delay and delay variability challenge. In the first instance, this requires a clear definition of the requirements of different traffic types. This concerns not just the particular FH interface (i.e., the exact split point), but the data types within the interface, as for instance primitives that control protocol layer interaction across the split point, global radio frame definition parameters, user data, etc. [33]. Such data can be separated for different priority queuing, as detailed further in Section 7.5.

7.4 Self-Backhauling

The concept of self-backhauling offers a flexible and cost-effective approach to address both BH and access requirements by dynamically sharing the radio resources between BH and access links as shown in Figure 7-5.

Self-backhauled BSs lack a direct connection to the BH infrastructure, but are connected via other so-called donor BSs through wireless BH links. From the perspective of a donor BS, a self-backhauled BS is served like any other user in the system, though obviously typically with larger traffic requirements than ordinary user equipments (UEs). There are two main aspects which differentiate self-backhauling from other wireless backhauling solutions:

- **Same hardware:** Both self-backhauled and donor BSs share the same type of radio interface, which also implies that the form factor of a self-backhauled BS is similar to that of a normal BS. However,

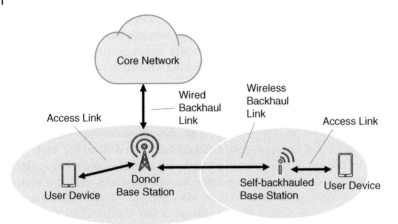

Figure 7-5. Self-backhauling, i.e., sharing of radio technology and radio resources between BH and access links.

without largely compromising the cost and/or size, self-backhauled BSs can be further optimized, e.g., by introducing multiple panels for BH and access links, while still sharing a common BB hardware and processor. In general, the only additional functionality that is required, as compared to an ordinary BS, is the UE functionality allowing a self-backhauled BS to communicate with the donor cell over the BH link;

- **Same radio access technology**: At the BH link, a self-backhauled BS uses its UE end to request connectivity to the network, and communicates with the donor BS by sharing resources with other UEs directly connected to the donor BS. At the access end, it is seen as any other BS by the end users in its coverage area.

7.4.1 Comparison with Legacy LTE Relaying

Relaying introduced in LTE-Advanced in 3GPP Release 10 has not fully capitalized on its promised potential due to several reasons. The LTE solution was mainly directed at filling coverage holes, which in practice can be circumvented using smarter deployments and a densification of macro cells. In 5G, however, self-backhauling may be essential for very dense deployments, at least until the underlying transport infrastructure reaches the same level of density, for instance in the context of mmWave access, where some deployment redundancy may be needed due to the particular signal propagation conditions at these frequencies. Moreover, since LTE relaying was designed for sub-3 GHz bands, the hardware form factor prohibits many deployments that allow for a simultaneous operation of BH and access links via different antennas without excessive self-interference. This aspect may now be overcome if BH and access in 5G use spectrum above 6 GHz with smaller antenna form factors and highly directive beamforming that reduces self-interference. Another strong limitation of the LTE relaying solution is the large number of backward compatibility issues that had to be taken into account, since relaying was added to an already matured standard. In 5G standardization, 3GPP agreed to ensure forward-compatibility for self-backhauling, meaning that self-backhauling is already considered in the overall clean-slate approach for 5G from the very beginning (even if self-backhauling itself will not be standardized yet in 3GPP Release 15), in order to maximise the offered benefits.

7.4.2   Technical Aspects of Self-Backhauling

In both LTE and 5G, any relaying or self-backhauling solution is inherently subject to losses in capacity due to half-duplex transmission constraints and the usage of additional wireless hops. In 5G, however, this can be mitigated to some extent if radio resources are utilized in a proactive and dynamic fashion at high granularity, ideally with a fully flexible usage of radio sub-frames for BH and access links depending on the traffic requirements. The actual resource split and multiplexing between BH and access on a single antenna panel can be done in time or frequency domain, depending on the general multiplexing approach used in the radio access at the level of the MAC radio scheduler. The mentioned loss in capacity can also be addressed at hardware level by deploying multiple antenna panels at the self-backhauled BS, which allow for a simultaneous operation of BH and access links while still bypassing the costs associated with a full-duplex system, as discussed in Section 16.2.4. Perceived capacity can then be theoretically doubled when BH and access links are active at the same time either transmitting or receiving (but not both at the same time), for instance facilitated through highly directive transmission at higher frequencies. Note that such a multi-panel setup does not imply that some panels are dedicated for BH and others for access, but instead all antenna panels could be used dynamically for BH or access, based on traffic needs. Since additional antenna panels contribute to the size and cost, it is also possible to deploy a self-backhauled BS with a single shared antenna panel for BH and access with the two domains multiplexed in time or frequency, as mentioned before. Since the overall benefit of self-backhauling is strongly related to the inter-site distance, it can be concluded that at the high-end of mmWave spectrum (such as 60 or 70 GHz), self-backhauling becomes more valuable as the cell sizes decrease further. For such deployments, however, multi-hop self-backhauling may become a necessity. Due to small sub-frame lengths in 5G at very high frequency, multi-hop systems (up to 4 hops) can be realized in combination with fast scheduling and packet forwarded scheduling strategies realized with configurable hardware, such as field programmable gate arrays (FPGAs).

Self-backhauling also has strong implications on the underlying RAN architecture. It is an important design goal of such a system to be as little intrusive as possible to the overall RAN design. For instance, from an architecture perspective, the impact on the UE protocol stack should be minimal. In other words, no additional design changes should be required to enable self-backhauling for a 5G user device. Similarly, from the donor cell perspective, self-backhauling shall use the same procedures to connect to the network as a normal user device to avoid any excessive implementation required at the donor end. From the perspective of the core network, a self-backhauled BS shall be managed and interfaced in a similar manner as any other 5G BS. With these high-level requirements in mind, a self-backhauling system architecture can be designed in multiple ways depending on the underlying deployment and traffic needs. For example, self-backhauled BSs can have a complete RAN protocol stack for maximum flexibility. For other deployments such as support for multi-hop and delay-critical traffic, layer-2 relays can be deployed to minimize the processing delays at each hop. To offer the full value of a 5G BS, self-backhauling shall also support advanced mobility mechanisms of underlying networks such as handovers, and a dynamic routing of traffic. Multi-connectivity is another important enabler for 5G systems, which, depending on the use-case, can offer throughput enhancements or robustness to the user traffic, see also Section 6.5. Consequently, self-backhauling is also expected to support different multi-connectivity and mobility schemes envisioned for 5G.

In conclusion, self-backhauling offers a complete set of solutions to complement a 5G system and to extend the coverage of the network overcoming infrastructure and cost constraints, and should hence be seen as an essential component in an overall transport network architecture.

7.5   Technology Integration and Interfacing

7.5.1   Framing, Protocol Adaptation, Flow Identification and Control

Assuming, as in Section 7.3, that Ethernet will form the basis of the new converged transport network, appropriate encapsulation formats will have to be used at the Ethernet level to accommodate the different traffic flows produced in the evolved FH network. These encapsulation formats will need to take into account a number of different packet types (constituting UP data, CP data and RAN processing primitives among others), each with potentially different QoS and traffic protection requirements. At the same time, each transported packet will need to be appropriately routable within the FH, while the data payload it carries will have to be properly handled by RAN nodes, for example by being addressable to a specific antenna processing element. Thus, virtual local area network identifiers (VLAN IDs) for routing and switching (also in stacked VLAN configurations), flow IDs for antenna addressing (for low PHY splits), and packet types (for split points below 3GPP split option 7, see Section 6.6.2) will likely be required in the encapsulation of FH data. Note that these are high-level encapsulation formats for which an overview is provided in Figure 7-6, together with an example of header field options within the application payload section.

Additional fields within the payload section will likely be used based on the specific implementation. For example, the IEEE 1914.3 standard for Radio Over Ethernet encapsulations and mappings has defined a number of additional control fields for CPRI encapsulation by Ethernet [34]. Similar definitions are to be expected in future standardization for new functional splits.

Ethernet allows up to eight different classes of service (CoS). A number of commercially available priority-based scheduling algorithms can be employed to reduce latency and latency variation, also referred to as packet delay variation (PDV), and these include, among others, weighted fair queueing (WFQ), weighted round robin (WRR), and strict priority (SP) [35]. These algorithms can all reduce the average PDV of a given CoS but cannot reduce the peak PDV. This can be a requirement for

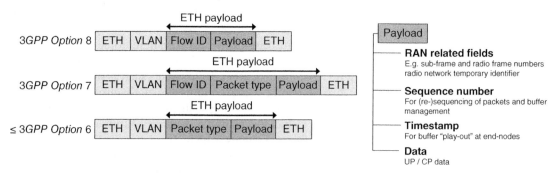

Figure 7-6. High-level Ethernet-based encapsulation formats for different functional split options and example fields within the application "payload" field.

certain traffic classes, an important example of which is the transportation of an in-line timing protocol, such as the precision-time protocol (PTP).

Ongoing standardization (IEEE P802.1CM Time-Sensitive Networking for FH) [36] is in the process of selecting and adapting time-sensitive networking (TSN) profiles for use in a bridged FH network. Currently, these are only specified for the centralized approach based on CPRI requirements. Two profiles are defined; one employing strict priority (SP) scheduling and the other pre-emption, based on IEEE 802.1Qbu frame pre-emption [37]. Pre-emption can potentially offer better performance than SP with a worst-case delay for a high-priority packet equivalent to the processing time of a 155-octet packet. Still, the advantage of pre-emption is a reduction in the E2E latency, as compared to SP for the same number of hops, thus enabling increased reach. But the improvement depends on the traffic class distribution in every aggregation point.

Time-aware scheduling based on the IEEE P802.1Qbv enhancements for scheduled traffic [38], where traffic is separated into uncontended window sections by employing port "gating", is another promising TSN profile for the FH. Simulation results have shown that such a scheduler can completely remove PDV [39][40], but its ability to do so is very dependent on the network size [39]. This method makes use of extensive buffering and can further lead to an increase in the E2E latency due to the use of guard periods (required so that low priority traffic does not overrun into the protected section) [40]. A combination of time aware-scheduling and pre-emption can be used to reduce the need for large guard periods, potentially at the cost of an increase in delay variation.

7.5.2 PBB/MPLS Framing to Carry FH/BH and its Multi-Tenancy Characteristic

To have a common transport network able to forward FH and BH payloads, a common framing format supporting both traffic flows is necessary, which is the focus of this section.

Different protocol stacks for deploying networks for LTE BH traffic have been described in [41]. Depending on the physical topology, different protocols can be used for the deployment of the BH network. In case of LTE, the BH traffic is IP-based. Depending on the functional split and its implementation, FH traffic can be IP-based, see [34]. Therefore, the frame has to support a larger variety of services for both FH and BH traffic.

Thus, the common frame has to contain enough information in the frame headers to enable switching nodes to fulfill their task as a common switching layer for both FH and BH traffic. Obviously, different functional splits (see Section 6.6.2) as well as multiple tenants (see Section 5.2.4), have to be supported. To support the migration to an integrated FH/BH 5G network, it must further be possible to interact with legacy devices. Designing the common frame, two complementary frame formats have been identified: Multi-Protocol Label Switching - Transport Profile (MPLS-TP) and Provider Backbone Bridging (PBB, IEEE 802.1ah). No clear advantages or disadvantages of PBB vs MPLS-TP as a common frame were identified, but we will focus on PBB in the following, since all the new extensions and profiles for fast forwarding of FH traffic in IEEE 802.1D bridges use it.

The PBB header, as shown in Figure 7-7, contains a new Ethernet header with MAC addresses, a Backbone VLAN Tag (B-TAG) to support VLANs, and a Backbone Service Instance Tag (I-TAG) to support further service differentiation. The outer MAC address is used to address the packet switches. The destination B-MAC address is the MAC address of the FE to which the tenant device, identified by the C-Dest address, is connected. The B-VLAN tag contains the VLAN-ID in the provider network as well as the priority code points (PCP) used to prioritize the packets appropriately. The PCP and drop eligible indicator (DEI) values of the I-Tag are redundant with respect to those in

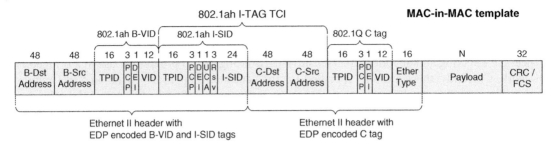

Figure 7-7. Provider backbone bridging (PBB), constituting a MAC-in-MAC format.

the B-Tag and are not used in this tagging system. The use customer address (UCA) is used to indicate whether the addresses in the inner header are actual client addresses or whether the frame is an OAM frame.

Hence, with merging of MAC-in-MAC and Ethernet features, we obtain a common frame that supports different types of traffic, such as FH and BH, and can differentiate packets of different tenants that are transported over the same FEs.

7.6 Transport Network Optimization and Performance Evaluation

7.6.1 Evaluation of Joint FH and BH Transport

To evaluate the performance of a 5G infrastructure that supports joint FH and BH services over a converged wireless optical network solution, a mathematical framework has been developed in [5] which is based on multi-objective optimization (MOP). This framework considers the user plane described in Section 7.2.1 and the details of the compute resources required to support both BH and FH services for CU processing. The proposed model takes a holistic view, jointly considering mobile FH and BH functions, to ensure an appropriate allocation of the required resources across all domains.

Specifically, the framework considers a multi-technology network infrastructure deploying a set of optical and wireless network technologies to interconnect DUs with compute resources. The DUs are uniformly distributed across the served area in a typical hexagonal cellular fashion. The CU-DU interconnection is provided through the multi-technology transport network described in the previous sections, aggregating traffic demands generated at the DUs, providing also the necessary capacity for the interconnection of compute resources.

The proposed MOP involves a two-stage optimization approach that jointly minimizes the overall power consumption of the network and compute infrastructure for the provisioning of FH and BH services. To this end, in the first step, the FH problem tries to identify the optimal placement of the CU with respect to the DUs, and the associated bandwidth requirements for their interconnection. In addition to this, recognizing the stringent delay and synchronization requirements of the existing FH protocol implementations, the concept of functional split processing is also considered. As illustrated in Figure 7-4 (b), the range of split options spans between the "traditional" distributed RAN (D-RAN) case, where all processing is performed locally at the APs, to the fully-centralized C-RAN case, where all processing is allocated to a CU. All other options allow allocating some processing functions at the DUs, while the remaining processing functions are performed remotely at the CU.

The optimal allocation of processing functions to be executed locally or remotely, i.e., the optimal "split", can be decided based on a number of factors, such as transport network characteristics, network topology and scale as well as type and volume of services that need to be supported, and is detailed in Section 6.6.2.

Once the FH problem has been modeled, the secondary objective tries to identify the optimal resources for the provisioning of BH services. In the present study, BH services are assumed to generate demands that need to be supported by specific processing resources in accordance to the content delivery network (CDN) paradigm [31]. To this end, the secondary objective of the MOP is to identify the optimal network bandwidth and processing resources for the provisioning of these services.

In addition to the service requirement mentioned above, the joint FH/BH planning problem is solved taking into account a set of constraints that guarantee efficient and stable operation of the required services. The relevant constraints included in the analysis are summarized below:

- Sufficient network bandwidth and processing capacity must be allocated to the required services;
- The physical resources that should be reserved depend on the users' mobility model assumed, the size of the wireless cells, and the traffic model adopted. In the ideal scenario, seamless handovers for a mobile user can be 100% guaranteed only if the required amount of resources is reserved for all its neighboring cells. However, to limit overprovisioning of resources, a more practical approach is to relate the reserved resources in the neighbouring cells with the handoff probability. In this study, the amount of resources that are leased in the wireless domain is assumed to be an increasing function of the handoff probability [5]. Given that both FH and BH services need to be supported by remotely located compute resources, as discussed in Section 7.1, the additional resource requirements also propagate in the transport network and the compute domain;
- Flow conservation as well as mapping, aggregation and de-aggregation of traffic between different domains is taken into consideration.

Given that both FH and BH services assumed in this use case require compute processing, there is a clear need to include the impact of the compute domain in the overall infrastructure evaluation. Therefore, the traffic associated with these services has to be mapped, not only to network resource requirements, but also to compute resource requirements. This introduces an additional constraint, associated with the conversion of network resource requests to compute resource requests. To achieve this, a mapping parameter, defined as network-to-compute, is introduced that provides the ratio of network requirements, as measured in Mbps, and computational requirements, measured in millions of instructions per second (MIPS), of a specific service demand. This parameter takes low values for cloud services requiring high network bandwidth but low processing effort (e.g., video streaming), and high values for tasks requiring intensive processing but limited bandwidth (e.g., data mining and Internet of Things). Regarding BH services, the Standard Performance Evaluation Corporation (SPEC) recently established the Cloud Subcommittee to develop benchmarks that are able to measure these parameters. Similarly, FH services require specific computing resources to support BB processing. The processing power depends on the sub-components of the BBU, calculated in Giga operations per second (GOPS). The resulting processing power depends on the configuration of the LTE system (i.e., number of antennas, bandwidth, modulation, coding, and number of resource blocks).

To address the importance of the E2E delay for specific FH or BH services, this is also considered in the analysis. In highly loaded heterogeneous networks, E2E delay can be greatly influenced by queuing delays associated with the interfaces across the infrastructure domains. Therefore, applying specific queuing policies and scheduling strategies at the interfaces across technology domains can offer significant delay benefits. Traditionally, the adoption of queuing models and scheduling strategies can be mathematically modeled applying queuing theory and modeling the different network domains as an open/closed mixed queuing network. However, such a model is not able to abide by the strict FH latency constraints. To address this issue, the queuing delays for the FH services are modeled under worst case operational conditions using network calculus theory.

The network topology assumed for the evaluation of the framework is the Bristol 5G city infrastructure, and real traffic data has been used. This infrastructure comprises TSON nodes in the optical segment, and microwave point-to-point links for backhauling the APs. TSON deploys a single fiber per link, 4 wavelengths per fiber, wavelength channels of 10 Gbps each, a minimum bandwidth granularity of 100 Mbps, and a maximum link capacity of 40 Gbps. The proposed optimization scheme focuses on the following scenarios:

- **Traditional RAN, separated BH**, where the power consumption per AP ranges between 600 W and 1200 W under idle and full load conditions, respectively, and commodity servers are used to support CDN services. In this solution, the entire RAN functionality is covered in the AP, often referred to as "all-in-one BS" solution, and hence only BH but no FH is required. For the BH services, users are assumed to generate demands that need to be processed at specific resources in accordance to the cloud computing paradigm;
- **Joint FH/BH without BBU sharing**, where remotely located specialized hardware is used for BBU processing. In this scheme, the transport network is able to support both FH and BH functionalities while BBUs are able to cover all split options ranging from traditional CPRI, i.e., 3GPP split option 8, to split option 5, which involves only the MAC processing at the CU, see also Section 6.6.2 The specific purpose hardware used for BBU processing is not re-sizable. Demands generated from BH services are processed at small scale servers as before;
- **Joint FH/BH with processing sharing**, where commodity servers are used to support both FH and BH CDN services. As before, the transport network is able to support both FH and BH services. However, compared to the previous one, this scheme allows BBUs to be instantiated as virtual functions and run on general purpose processors (GPPs). Therefore, sharing of resources and on-demand resizing of compute resources to match the FH service requirements is enabled. The main disadvantages of this approach include higher processing cost per bit associated with GPPs when compared to specific purpose hardware.

Figure 7-8 (a) shows the average load per AP generated using the dataset reported in [42]. Figure 7-8 (b) shows that when adopting the joint FH/BH approach without BBU sharing over the proposed integrated wireless-optical infrastructure and comparing it with the traditional RAN approach, significant energy savings ranging between 60-75% can be achieved. This is because the envisioned sharing of the transport network infrastructure allows a more efficient utilization of network resources and, therefore, lower power consumption. The efficiency of the system is further improved when sharing of both transport network and processing resources is allowed. However, due to overloading of transport network resources to support FH requirements, an increase of the E2E service delay in the BH is observed, as shown in Figure 7-8 (c), which however remains below 20 ms for a 100 Mbps flow request. It is interesting to note that the BH service delay calculated for the "joint FH/BH with shared processing" case is lower compared to the delay calculated for the "joint FH/BH without BBU sharing" case. This is because in the former case, lower processing times are required by the large commodity servers to execute the user cloud services.

7.6.2 Experimental Evaluation of Layer-2 Functional Splits

As already discussed, centralized RAN processing has been identified as one of the major enablers for 5G mobile network access. By moving BBUs to the cloud, multiple instances can be instantiated on the fly, serving several DUs. The goal is to satisfy the existing demand of particular geographical areas, while drastically reducing the overall capital expeditures (CAPEX) and operational expenditures (OPEX) of the mobile operators. Nevertheless, as discussed before, the realization of such

(a)

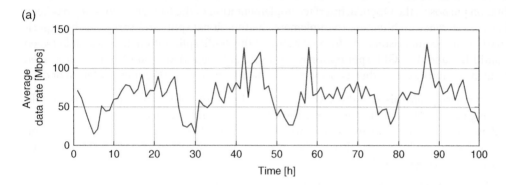

(b)

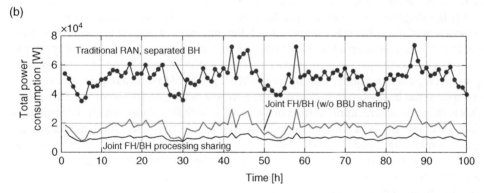

(c)

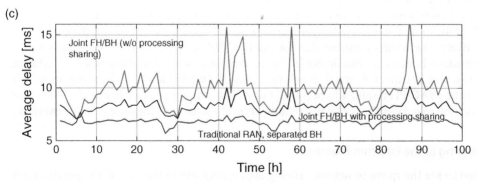

Figure 7-8. a) Average traffic per BS based on the dataset [42], b)-c) Total power consumption and total service delay over time for a traditional RAN and joint FH/BH scenarios with and without processing sharing.

cloud RANs requires high capacity on the FH interface, for transferring the data between CUs and DUs. Towards overcoming this barrier, and, thus, dropping the cost of fiber-enabled FH connections, the usage of traditional Ethernet interfaces has been proposed. Through improved split options within the PHY, such as the 3GPP split options 7-x as illustrated in Section 6.6.2, or higher-layer splits, the usage of Ethernet indeed becomes viable.

As a proof-of-concept, a detailed evaluation of layer-2 functional splits has been provided in [43], which will now be discussed in detail. The considered split options can be used for the convergence of multiple heterogeneous wireless technologies in an all-in-one unit fronthauled through Ethernet-based links.

For evaluation purposes, the OpenAirInterface implementation of the LTE protocol stack has been employed as the reference architecture to implement the functional splits and carry out the experiments. In particular, a comparison is performed between the 3GPP split option 2, i.e. a split between the PDCP and RLC, and the 3GPP split option 6, i.e. a split between MAC and PHY, both described in more detail in Section 6.6.2. Further, two approaches have been used for the transportation of data:

1) Based on stateless protocols like the User Datagram Protocol (UDP) for the lower-layer split option, being more delay sensitive regarding the scheduling of transmissions, and
2) State-full protocols like the Transmission Control Protocol (TCP) or Stream Control Transmission Protocol (SCTP) for higher layer splits, as they can operate with more slack delay requirements, if appropriate buffering of the data is employed.

For the evaluation of the chosen splits, the LTE extensions of the NITOS testbed environment have been employed [44], using the open-source OpenAirInterface platform, as mentioned before.

The implementation of the splits takes place in the following manner. Regarding the PDCP/RLC split, whenever PDCP is receiving a packet, it goes through its normal procedure before being relayed to the next layer. As soon as the packet is processed, it is sent to the DU where the RLC processes it. The resulting stream is placed in a buffer waiting for the MAC protocol to send a request. Concerning the MAC/PHY split, it has been chosen to override the part where the two layers communicate with each other. This is the point where, upon the end of the MAC scheduling algorithm, the CU instructs the DU at which subframe the data will be transmitted over the air. In the test implementation, no buffering of packets takes place inside the DU, but this is solely handled by the CU. Whenever data needs to be sent over the air, data streams are sent to the DU along with all the signaling needed to orchestrate the PHY layer, including the subframe scheduled for transmitting, the number of physical resource blocks, the modulation and coding scheme (MCS), the antennas, etc.

Figure 7-9 (a) presents the reference measurements without the splitting processes. Figure 7-9 (b) and Figure 7-10 illustrate the obtained experimental results for the MAC/PHY and the PDCP/RLC splits, respectively. The results illustrate that for the PDCP/RLC split, the transport protocols can impose performance limitations, but do not break the real-time operation of the BSs. Nevertheless, the results vary, and as expected, stateless solutions (e.g., UDP) are found to be more applicable. Moreover, for the MAC/PHY split, where the DU transmissions are solely scheduled in the cloud, real-time operation mandates the use of high bandwidth solutions, with the least possible overhead.

7.6.3 Monitoring in the Ethernet Fronthaul

5G is expected to see the move to network slicing approaches where tenants of an operator's network will be assigned a subset of the available RAN and Ethernet networking resources based on individual service-level agreements (SLAs), as described in detail in Chapter 8. From the RAN side,

(a)

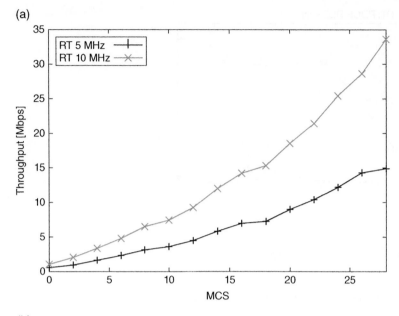

(b)

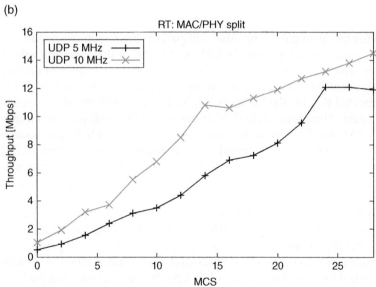

Figure 7-9. (a) Reference measurements of the platform without the splits, (b) Real-time evaluation of the MAC/PHY splits for UDP based transferring of data. RT: Real time.

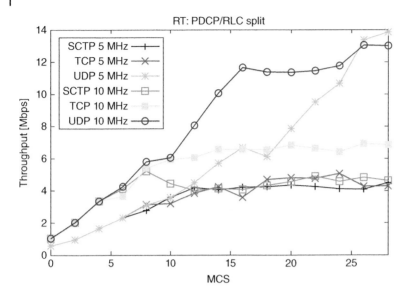

Figure 7-10. Real-time evaluation of the PDCP/RLC splits for different protocols. RT: Real time.

3GPP is in the process of defining the main parameters that will govern slice selection, slice aware-ness by the RAN entities, and resource management in RAN nodes. All of these aspects will be important for maintaining QoS, which in 5G is expected to be further optimized/targeted on user quality of experience (QoE). Slice distribution to tenants and inter-slice isolation will be coupled to NFV techniques, while it is expected that for the Ethernet FH, NFV entities will closely cooperate with SDN entities (e.g., SDN-capable Ethernet switches). Network slicing then will become a multi-dimensional concept with both RAN and FH networking processing management having to operate in unison. Adherence to SLAs will be maintained by monitoring key performance indicators (KPIs) both pertaining to the RAN and the FH networking domains. These can vary from one implementa-tion to the next, but some examples at the FH networking level include latency, PDV, packet drop rates, and throughput. KPIs can be obtained by employing deep packet filtering techniques using "smart" Ethernet small form-factor pluggable (SFP) probes [45]. An example architecture employ-ing smart probing techniques is shown in Figure 7-11(a), where multiple flows, each consisting of a number of packet types, exist over the FH network. The probes are in-line hardware probes that capture (i.e., make copies) of packets (or packet headers) that match user-defined filter definitions, and which are then sent to a central location for further processing. When not capturing packets, the probes are transparent to the FH traffic. With appropriate filter definitions, the data can be extracted at varying resolution levels (e.g., service, flow, and packet type level) and varying time resolutions. Appropriate algorithms are then used to manipulate extracted metadata from the cap-tured results, obtain different KPI updates, and use these to inform the NFV and SDN engines. Figure 7-11(b) shows an example of two monitored KPIs from an Ethernet FH section comprising a single Ethernet switch or aggregation node. The KPIs in this case include delay/latency and PDV, in the figure termed as frame-delay variation (FDV) and plotted against Ethernet frame numbers cor-responding to a specific user flow. With the move to virtualization/cloudification techniques in 5G,

(a)

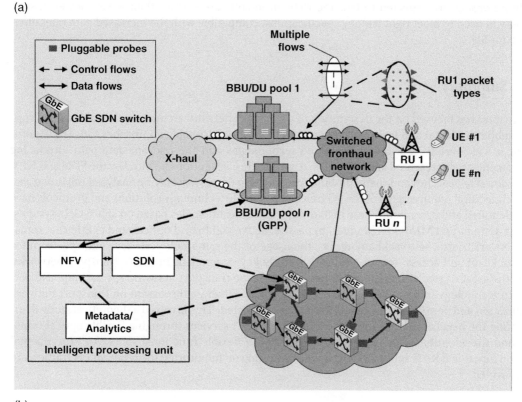

(b)

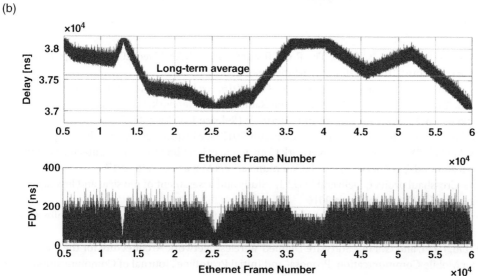

Figure 7-11. (a) Monitoring in a converged FH using in-line Ethernet "smart" probes, (b) Example of KPI extraction and monitoring (delay and frame-delay variation, FDV) in an Ethernet FH.

hardware probes are expected to become virtualized software entities themselves: however, such probes will need to meet tight latency specifications, compatible with the increased data rates envisioned for 5G.

7.7 Summary

This chapter has focused on the description of a transport network architecture suitable for 5G supporting both backhaul and fronthaul services. In this context, suitable approaches for the UP comprising a wide variety of wireless and wired technologies and the relevant data path interfacing requirements were presented and discussed. Appropriate CP solutions inspired by the SDN and NFV architectural approaches and their combination were also discussed. Having analyzed the functional and architectural requirements of the 5G transport, suitable technology solutions and protocols have been identified and described. These include: a) wireless technologies based on sub-6 GHz systems, 60 GHz systems and MIMO techniques, b) passive, active and hybrid optical, and c) Ethernet transport networks. Also, self-backhauling, i.e., the usage of the same radio interface and spectrum for both backhaul and access, was described as a likely key component in future transport networks. Given the great degree of technology heterogeneity in the transport network, special emphasis was given to technology integration and interfacing aspects. Finally, a discussion on transport network optimization and performance evaluation was also presented. This involved a theoretical study demonstrating the benefits of the support of joint FH and BH services through a flexible optical transport, and the adoption of the concept of C-RAN with flexible functional splits. An experimental study on functional RAN splits and a study concentrating on monitoring the Ethernet FH traffic were also provided.

References

1 China Mobile, White Paper, "C-RAN: The Road Towards Green RAN", Dec. 2013, see http://labs. chinamobile.com/cran/2014/06/16/c-ran-white-paper-3-0/

2 CPRI Specification V7.0, Interface Specification, Oct. 2015, see http://www.cpri.info/spec.html

3 3GPP TR 38.801, "Technical Specification Group Radio Access Network; Study on New Radio Access Technology; Radio Access Architecture and Interfaces (Release 14)", Mar. 2017

4 5G PPP, White Paper, "View on 5G Architecture", July 2017, see https://5g-ppp.eu/white-papers/

5 A. Tzanakaki et al., "Wireless-Optical Network Convergence: Enabling the 5G Architecture to Support Operational and End-User Services", IEEE Communications Magazine, Aug. 2017

6 5G PPP 5G-Crosshaul project, Deliverable D2.1, "Study and assessment of physical and link layer technologies for 5G-Crosshaul", June 2016

7 Ian F. Akyildiz, Pu Wang and Shih-Chun Lin, "SoftAir: A software defined networking architecture for 5G wireless systems", Computer Networking, Elsevier, 2015.

8 G. Sun, F. Liu, J. Lai, and G. Liu, "Software Defined Wireless Network Architecture for the Next Generation Mobile Communication: Proposal and Initial Prototype", Journal of Communications, Dec. 2014

9 H.-H. Cho, C.-F. Lai, T. K. Shih, H.-C. Chao, "Integration of SDR and SDN for 5G", IEEE Access, vol. 2, pp. 1196–1204, Sep. 2014

10 ETSI GS NFV 003, V1.2.0, Nov. 2014

11 OpenStack, see https://www.openstack.org/

12 5G PPP 5G-XHaul project, Deliverable D3.2, "Design and evaluation of scalable control plane, and of mobility aware capabilities and spatiotemporal demand prediction models", June 2017

13 E. Garcia-Villegas, D. Sesto-Castilla, S. Zehl, A. Zubow, A. Betzler and D. Camps-Mur, "SENSEFUL: an SDN-based Joint Access and Backhaul Coordination for Dense Wi-Fi Small Cells", International Wireless Communications and Mobile Computing Conference (IWCMC 2017), June 2017

14 R. Sun et al., "IEEE 802.11 TGay Use Cases", Sep. 2015

15 Betzler, D. Camps-Mur, E. Garcia-Villegas, I. Demirkol, F. Quer and J. J. Aleixendri, "SODALITE: SDN Wireless Backhauling for Dense Networks of 4G/5G Small Cells", submitted to IEEE Transactions on Mobile Computing

16 A. Hurtado-Borras, J. Pala-Sole, D. Camps-Mur and S. Sallent-Ribes, "SDN wireless backhauling for Small Cells Communications", IEEE International Conference on Communications (ICC 2015), June 2015

17 A. Girard, "FTTx PON Technology and Testing", EXFO Electrical Engineering, 2005

18 IEEE 802.3ah Standard, "Media Access Control Parameters, Physical Layers, and Management Parameters for Subscriber Access Networks", 2004

19 ITU-T Recommendation G.984.2, "Gigabit-Capable Passive Optical Networks (G-PON): Physical Media Dependent (PMD) Layer Specification", 2003

20 ITU-T Reommendation. G.987.2, "10G-capable PONs: Physical Media Dependent (PMD) Layer Specification", 2010

21 J. S. Wey et al., "Physical layer aspects of NG-PON2 standards—Part 1: Optical link design", Journal of optical Communications and Networking, vol. 8, no. 1, pp. 33–42, Jan. 2016

22 S. Liu, J. Wu, C. Koh, and V. Lau, "A 25 Gb/s/km2 urban wireless network beyond IMT-advanced", IEEE Communications Magazine, vol. 49, no. 2, pp. 122–129, Feb. 2011

23 M. H. Eiselt, C. Wagner and M. Lawin, "Remotely controllable WDM-PON technology for wireless fronthaul/backhaul application", OptoElectronics and Communications Conference (OECC 2016), International Conference on Photonics in Switching, July 2016

24 J. M. Fabrega, M. Svaluto Moreolo, M. Chochol and G. Junyent, "Decomposed Radio Access Network Over Deployed Passive Optical Networks Using Coherent Optical OFDM Transceivers", Journal of Optical Communications and Networking, vol. 5, no. 4, pp. 359–369, Apr. 2013

25 A. Napoli et al., "Next Generation Elastic Optical Networks: the Vision of the European Research Project IDEALIST", IEEE Communications Magazine, vol. 53, no. 2, Feb. 2015

26 N. Sambo et al., "Next Generation Sliceable Bandwidth Variable Transponders", IEEE Communications Magazine, vol. 53, no. 2, pp. 163–171, Mar. 2015

27 M. Svaluto Moreolo et al., "SDN-enabled Sliceable BVT Based on Multicarrier Technology for Multi-Flow Rate/Distance and Grid Adaptation", Journal of Lightwave Technology, vol. 34, no. 8, Apr. 2016

28 J. M. Fabrega et al, "Demonstration of Adaptive SDN Orchestration: A Real-Time Congestion-Aware Services Provisioning Over OFDM-Based 400G OPS and Flexi-WDM OCS", Journal of Lightwave Technology, vol. 35, no. 3, pp. 506–512, Feb. 2017

29 J. M. Fabrega, M. Svaluto Moreolo, L. Nadal, F. J. Vílchez, J. P. Fernández-Palacios and L. M. Contreras, "Mobile Front-/Back-Haul Delivery in Elastic Metro/Access Networks with Sliceable Transceivers Based on OFDM Transmission and Direct Detection", International Conference on Transparent Optical Networks (ICTON 2017), July 2017

30 B. R. Rofoee et al., "Demonstration of Service-differentiated Communications over Converged Optical Sub-Wavelength and LTE/WiFi Networks using GEANT Link", Optical Fibre Communications Conference (OFC 2015), Mar. 2015

31 A. Tzanakaki et al., "5G infrastructures supporting end-user and operational services: The 5G-XHaul architectural perspective", IEEE International Conference on Communications (ICC 2016), May 2016

32 N. J. Gomes, P. Chanclou, P. Turnbull, A. Mageec and V. Jungnickel, "Fronthaul evolution: From CPRI to Ethernet", Optical Fiber Tech., vol. 26, pp. 50–58, Dec. 2015

33 P. Assimakopoulos, G. S. Birring and N. J. Gomes, "Effects of Contention and Delay in a Switched Ethernet Evolved Fronthaul for Future Cloud-RAN Applications", European Conference on Optical Communications (ECOC 2017), Sep. 2017

34 IEEE P1914.3, "Standard for Radio Over Ethernet Encapsulations and Mappings", see http://sites.ieee.org/sagroups-1914/p1914-3/

35 M. K. Al-Hares, P. Assimakopoulos, S. Hill, and N. J. Gomes, "The Effect of Different Queuing Regimes On a Switched Ethernet Fronthaul", IEEE International Conference on Transparent Optical Networks (ICTON 2016), July 2016

36 IEEE P802.1CM, "Time-Sensitive Networking for Fronthaul", see http://www.ieee802.org/1/pages/802.1cm.html

37 IEEE 802.1Qbu, "Frame Preemption", see http://www.ieee802.org/1/pages/802.1bu.html

38 IEEE 802.1Qbv, "Enhancements for Scheduled Traffic", see http://www.ieee802.org/1/pages/802.1bv.html

39 T. Wan and P. Ashwood-Smith, "A Performance Study of CPRI over Ethernet with IEEE 802.1Qbu and 802.1Qbv Enhancements", IEEE Global Communications Conference (GLOBECOM 2015), Dec. 2015

40 M. K. Al-Hares, P. Assimakopoulos, D. Muench and N. J. Gomes, "Modeling Time Aware Shaping in an Ethernet Fronthaul", IEEE Global Communications Conference (GLOBECOM 2017), Dec. 2017

41 NGMN, White Paper, "LTE backhauling deployment scenarios", July 2011

42 X. Chen et al., "Analyzing and modeling spatio-temporal dependence of cellular traffic at city scale," IEEE International Conference on Communications (ICC 2015), May 2015

43 N. Makris, P. Basaras, T. Korakis, N. Nikaien, and L. Tassiulas, "Experimental Evaluation of Functional Splits for 5G Cloud-RANs", IEEE International Conference on Communications (ICC 2017), May 2017

44 N. Makris, C. Zarafetas, S. Kechagias, T. Korakis, I. Seskar, and L. Tassiulas, "Enabling open access to LTE network components; the NITOS testbed paradigm", IEEE Conference on Network Softwarization (NetSoft 2015), Apr. 2015

45 H2020 iCIRRUS project, Deliverable D3.3, "SLA and SON Concept for iCIRRUS", Mar. 2017, see http://www.icirrus-5gnet.eu/category/deliverables/

8

Network Slicing

Alexandros Kaloxylos[1], Christian Mannweiler[2], Gerd Zimmermann[3], Marco Di Girolamo[4],
Patrick Marsch[5], Jakob Belschner[3], Anna Tzanakaki[6], Riccardo Trivisonno[7], Ömer Bulakçı[7],
Panagiotis Spapis[7], Peter Rost[2], Paul Arnold[3] and Navid Nikaein[8]

[1] *University of Peloponnese, Greece*
[2] *Nokia Bell Labs, Germany*
[3] *Deutsche Telekom AG, Technology Innovation, Germany*
[4] *Hewlett Packard Enterprise, Italy*
[5] *Nokia, Poland (now Deutsche Bahn, Germany)*
[6] *National and Kapodistrian University of Athens, Greece and University of Bristol, UK*
[7] *Huawei German Research Center, Germany*
[8] *EURECOM, France*

8.1 Introduction

The 5th generation (5G) network is promising to upgrade not only the well-known mobile broadband services, but also enable the support of services for the so called "vertical industries" (e.g., health, transportation, factories, energy). An extensive list of 5G use cases can be found in Chapter 2. All these verticals have their own requirements and needs which may be highly divergent. Their operational requirements are translated into different key performance indicators (KPIs) such as user experienced data rate, end-to-end (E2E) latency, reliability, communication efficiency, availability, and energy consumption. These have to be satisfied in specific environments characterized by different parameters such as mobility, expected data traffic, density and types of network nodes, position accuracy, etc.

As discussed in Section 5.2.2, network slicing is introduced as one of the key enablers to support the required level of flexibility in 5G networks. Network slices are essentially multiple logical networks deployed over the same physical infrastructure. During the past years, there has been a lot of debate to reach a commonly accepted and concrete definition of network slicing. At a first glance, the notion of slicing may seem to be very similar to well-established solutions that essentially support logical networks, such as:

- Virtual local area networks (VLANs) where different hosts are logically brought under the same broadcast domain;
- Virtual private networks (VPNs) that are used to connect multiple hosts through private and secure tunnels providing logically closed groups;

5G System Design: Architectural and Functional Considerations and Long Term Research, First Edition.
Edited by Patrick Marsch, Ömer Bulakçı, Olav Queseth and Mauro Boldi.

- Dedicated mobile core networks (CNs) with standardized solutions (e.g., DECOR, eDECOR), as discussed in Section 8.2.1.

This observation would indeed be correct if the target of 5G network slicing was only to separate nodes or share resources and apply specific security and policies per service. However, these solutions do not address the need to support a number of use cases with different KPIs. This suggests that the network functions (NF) will not necessarily be the same in all slices. The formal specification of slices related to the enhanced mobile broadband (eMBB), ultra-reliable low-latency communications (URLLC) and massive machine-type communications (mMTC) families of use cases, as described in Section 2.2, is currently underway. Already, there are hints to new functions that need to be introduced (e.g., for slice selection) or others that need to be defined per use case (e.g., for session management or mobility management).

In the latest specifications, the 3rd Generation Partnership Project (3GPP) considers a network slice to be "A logical network that provides specific network capabilities and network characteristics" [1]. In [2], it is mentioned that it is a "network created by the operator customized to provide an optimized solution for a specific market scenario which demands specific requirements with end to end scope". Also, 3GPP has defined in [3] that a network slice is implemented by "slice instances", which in turn are created from a "network slice template", being a template of a logical network including the NFs and the corresponding resources. A similar definition is also provided in [4], where the use of common functions and sharing of resources among slices is possible. Thus, these definitions suggest the ability of deploying multiple logical networks possibly over the same physical infrastructure. This level of flexibility is needed to support the diverse requirements and KPIs of the 5G use cases as well as to reduce the cost for network deployment and operation. Thus, network slicing is expected to be one of the key features of 5G networks, realized by introducing solutions based on softwarization, virtualization and functional modularization.

The key services provided by network slicing and a comparison with legacy cellular systems are illustrated in Figure 8-1 [5]. On the left side, where legacy systems are depicted, one can see that the same NFs over monolithic network elements are used to support all telecommunication services

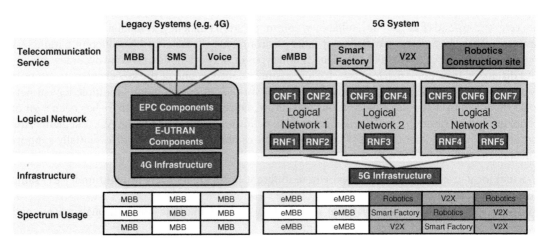

Figure 8-1. Key principles of network slicing.

(e.g., voice, SMS, web browsing, video streaming). Such homogeneous treatment of all services, even with the definition of different Quality of Service (QoS) classes, is a non-acceptable compromise for 5G services. This is because KPIs like ultra-high reliability and ultra-low delay cannot be deployed for all vertical services in a technically and economically viable way. In 5G networks, multiple verticals will be supported by dedicated or shared logical networks running on top of the 5G infrastructure. These logical networks will be a composition of core network functions (CNF) and radio network functions (RNF) and will run over the same physical network components. Note that dedicated spectrum may be allocated for each slice, or several slices may share the same spectrum, but manage to meet the service-level agreements (SLAs) with the verticals using the specialized NFs, as detailed further in Section 8.2.3.

The key enablers for the dynamic deployment of slices are considered to be a) network function virtualization (NFV), allowing the virtualization of sets of NFs and their organization into virtual blocks that may be connected together to create communication services [6] and b) software defined networking (SDN) used for separating control plane (CP) and user plane (UP) and allowing for full programmability of the network [7]. These enablers are extensively presented in Section 10.2. Moreover, contrary to the evolution of previous generations of mobile networks, it is foreseen that 5G will require not only improved networking solutions but also a sophisticated integration of massive computing and storage infrastructures into the different network domains (i.e., access, transport, and core network) to support the different use cases and services.

Although slicing of 5G networks is a rather new topic, the research area is growing rapidly. In [8] and [9], several solutions for the slicing of shared resources as well as the virtualization of NFs are discussed. This chapter contains a comprehensive summary and analysis of the latest 3GPP specifications and also findings from several 5G Public-Private Partnership (5G PPP) research projects. Section 8.2 provides detailed information for each network domain, as well as for the support of slicing across different operator administrative domains. It also discusses a realistic E2E example of network slicing. In Section 8.3, several slice operation aspects such as slice selection and isolation, context transfer, slice orchestration and management are presented. Finally, Section 8.4 summarizes the key findings and also lists the technical challenges that still remain open.

8.2 Slice Realization in the Different Network Domains

The network slicing concept refers to E2E logical networks and is meant to provide flexibility to each component of the communication system (i.e., access, core, transport). However, technological specificities of each domain require addressing different key issues for the realization of the slicing concept. For this reason, the problem of network slicing is addressed on a domain basis in the following sub-sections.

8.2.1 Realization of Slicing in the Core Network

To achieve the intended flexibility and adaptability for future 5G services, a modularization of NFs, based on detailed functional decomposition and use in dedicated slices, is a prerequisite especially in the CN [10], [11].

In 4th generation (4G) networks, there are monolithic network elements within the Evolved Packet Core (EPC), i.e., Serving Gateway (S-GW), Packet Gateway (P-GW), and Mobility Management Entity (MME), which aim to integrate hardware/software (HW/SW) implementation in physical nodes. Nevertheless, slicing approaches are initially available also in 4G. One example is a multi-operator core network (MOCN), which allows several operators to share a common radio access network (RAN), while running separated CNs with proprietary services [12]. However, 5G slicing not only enables sharing of the underlying infrastructure (core and access) by multiple logical networks, it also allows for a different configuration of these logical networks.

With a dedicated core (DECOR), initially introduced in 3GPP Release 13, operators can deploy multiple dedicated CNs (DCNs) within a single operator network [13]. A DCN may consist of one or more MMEs and one or more S-GWs/P-GWs, each element potentially featuring different characteristics and functions. The introduction of DECOR was triggered by the problem of enhancing mobile networks with architecture flexibility and enabling either resource sharing or resource isolation among specific groups of subscribers (e.g. for MTC/CIoT subscribers, subscribers belonging to specific enterprises or to separate administrative domains, etc.). The design of DECOR aimed at having no impact on legacy user equipments (UEs) and at allowing different DCNs to share the same RAN, but has some drawbacks with respect to increased initial access time and signaling efforts.

Evolved DECOR (eDECOR) [14] was introduced in 3GPP Release 14. Its aim was to support all 3GPP RANs while being backward-compatible to DECOR. DCN selection and allocation procedures, as well as slice isolation among DCNs, are improved. Also, the required CP signaling is minimized by reducing or avoiding the occurrence of redirection procedures. eDECOR introduced the concept of UE-assisted DCN selection, which is unfortunately not applicable with legacy UEs, and a network assigned DCN-ID. The DCN-ID, permanently stored at a UE, is included in Radio Resoure Control (RRC) messages piggybacked in Non-Access Stratum (NAS) signaling (e.g., Attach Request, Tracking Area Update). This allows the Network Node Selection Function (NNSF) in the RAN to directly select the proper DCN towards which the NAS signaling needs to be forwarded. eDECOR further addressed congestion control for DCN types to cope with CP NAS signaling congestion which may occur at MME serving (and hence be shared amongst) multiple DCNs. In addition, it also optimized load balancing among MMEs.

These early attempts can be considered as precursors of 5G network slicing. They even provided some customization for different use cases (e.g., mMTC). However, they were lacking the flexibility expected to be in place to support the diverse 5G use cases since their customization was always limited by the functionality of the EPC architecture. To progress towards next generation networks, the full advantage of network enablers such as SDN, NFV [6], mobile edge computing or multi-access edge computing (MEC) [15] and cloud computing technologies needs to be exploited, in combination with the introduction of the network modularization design principle as well as the service-based architecture (SBA) model. These new features target the support of multi-tenancy, the minimization of service delivery through flexible network deployment, nodes' reconfiguration and the empowerment of third parties to customize the network slices according to their needs.

Considering the 5G modularization approach, as introduced in Section 5.4.1, all relevant CN NFs should be broken down to a suitable fine-grained level. Modularization does not mean that it is always required to go down to an "atomic granularity", but to combine basic NFs in a way that they create together a functional building block with a self-contained dedicated task. This block is addressable and configurable within a network slice instance (NSI). Following the SDN principle, there will additionally be a strict separation between CP and UP NFs in the 5G CN, which is not originally the case in 4G. In 4G networks, P-GW and S-GW include both CP and UP NFs, whereas the MME entirely relates to the CP. This separation in 5G networks will enable an independent scalability of resources for CP and UP and the introduction of new NFs for both parts. It has to be noted that 3GPP is also taking care of that fact in the current standardization process by identifying and separating those NFs as well as by defining interfaces in between [16]. Based on the results of research projects and standardization organizations, orthogonal sets of CN NFs have been defined for 5G CP and UP. A more detailed description of those NFs and procedures in between is given in Chapter 5 and can be found in [1], [3] and [17]. It is still an ongoing discussion if the granularity of those NFs is sufficient to achieve the targeted flexibility for 5G, or if a finer level of separation among functions is still needed.

Slices in the 5G CN for dedicated business or service purposes are instantiated by a concatenation of selected NFs taken from available repositories based on special network slice templates (NST) [3][17]. Dependent on requirements with respect to isolation (resource, security, etc.), NSIs may consist of fully separated CP and UP NFs, but it may be also possible that some of the NFs are shared by several NSIs. Control plane network functions (CPNFs) have been classified as common control NFs (CCNFs) and slice-specific control NFs (SCNFs). Sharing of CPNFs is especially linked to Access and Mobility Management Function (AMF) instances serving UEs which are simultaneously connected to more than one NSI. In that case, the corresponding AMF instance should be common to all NSIs serving one UE. Among CCNFs, there is also one NF dedicated to slice selection, hence denoted as Network Slice Selection Function (NSSF). In Figure 8-2, an example with three NSIs is given where two of those share CCNFs. Also, the figure illustrates a slice overarching CPNF block where slice generic functions like NSSF may reside. As depicted in this figure, the 5G CN is being designed to support not only radio access networks (RANs), but any type of access networks (AN), including even fixed networks.

As NSIs have to act as separate domains, trustworthiness with respect to security aspects during the slice lifecycle is of extreme importance. Any potential cyber-attack on one NSI must have no impact on another one running on the same infrastructure or sharing common NFs [18].

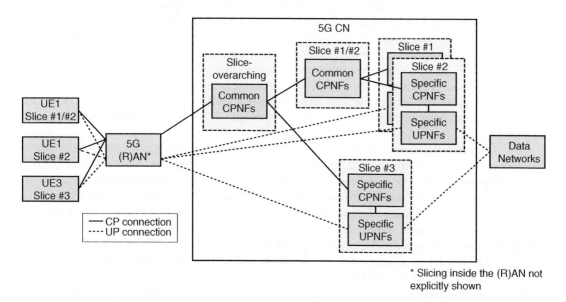

Figure 8-2. Exemplary implementation of network slices in the 5G CN with common and slice-specific NFs.

In addition to the legacy 4G point-to-point interface model, an alternative SBA model has been defined by relevant standardization organizations like 3GPP [1] for CPNFs in the 5G CN. Within that approach, each CPNF exposes a service-based interface (SBI), by which the authorized NFs can access the services it provides. SBIs provide higher flexibility in the interaction among NFs and allow multiple alternative interconnections among those to define slice-tailored architectures.

Another important aspect with respect to the differentiation against 4G is the Network Exposure Function (NEF) that provides the means to expose services and capabilities of CPNFs to third parties and application functions (AFs) such as MEC, monitoring, policy/charging or data analytics. AFs will provide information to the 5G CN, e.g., for packet flow handling (routing, QoS, etc.) and policy control. Trusted AFs may directly interact with CPNFs using application programming interfaces (APIs), while untrusted AFs have to apply the NEF framework.

Due to typically asynchronous timing behavior of NF processing in the 5G CN with respect to radio framing, CN NFs may usually be implemented as virtual NFs (VNFs) [6] in cloud infrastructures, e.g. in front- or backend data centers of operators or third parties (central clouds). In contrast to such centralized approach, some of the CN NFs as well as AFs may be located closer to the access network in so-called edge clouds, e.g. to support low latency use cases by caching, local break outs or MEC. Edge clouds may be placed, e.g., at central offices of operator networks or at larger campus locations (enterprise premises, factory halls, sport stadiums, etc.) [11][17]. This flexibility w.r.t. the placement of functions on a per-slice basis is expected to be one of the key characteristics of 5G networks.

8.2.2 Slice Support on the Transport Network

The 5G heterogeneous transport network is envisioned to rely on the convergence of a variety of technologies including wireless and optical networking, and to support a variety of services including backhaul (BH) and fronthaul (FH) services offering efficiency, scalability and management simplification, as discussed in detail in Chapter 7. In this context, transport solutions enhanced with advanced features such as slicing and virtualization will allow a pool of network and compute HW and SW resources to be shared and accessed remotely without the prerequisite of ownership. This will allow to create infrastructure slices that integrate heterogeneous technologies. These slices can transport FH services corresponding to various functional splits as well as BH services adopting novel approaches such as the notion of service chaining (SC). Network slicing and SC can be facilitated adopting and integrating architectural models such as the SDN reference architecture and the European Telecommunications Standards Institute (ETSI) NFV standard, as described in Section 10.2.

In the *highly heterogeneous 5G transport*, a critical challenge that needs to be addressed is that of "cross-domain" slicing. In this context, some fundamental incompatibilities associated with technology heterogeneity need to be addressed, such as the separation of CP and UP for some technologies (e.g., optical/wireless SDN) and close coupling between CP and UP for others. This introduces challenges in defining the relevant interfaces not only across domains, but also between the slicing systems and the orchestrators responsible for the composition and the provisioning of SCs over the transport slices. More specifically, interfacing between technology domains, including isolation of flows, flexible scheduling schemes and QoS differentiation mechanisms *across domains*, plays a key role and can be achieved by adopting flexible HW functions. These functions will be exploited to enable dynamic and on-demand sharing, partitioning and grouping of resources as required to form *independent transport network slices* with guaranteed levels of isolation and security. In this context, programmable Network Interface Controllers (NICs) that are commonly used to bridge different technology domains at the UP can have an instrumental role. These controllers have a unique ability to provide HW-level performance exploiting SW flexibility and can offer not only network processing functions (i.e., packet transactions), but also HW support for a wide variety of communication protocols and mechanisms.

5G transport slicing can be implemented through the adoption of a hierarchical architectural approach that supports management of network elements and abstracted resources by different layers [19]. Each network domain may host multiple SDN UP elements and expose its own virtualized resources through an SDN controller to the upper layer SDN controllers. A hierarchical SDN controller approach can assist in improving network performance and scalability as well as limit reliability issues. The top network controller will manage network resource abstractions exposed by the lower-level controllers that are responsible to manage the associated network elements. Orchestration of both computational and network resources can be performed by the NFV orchestrator and can be used to support multi-tenant chains, facilitating virtual infrastructure provider operational models. This will also be responsible to interact with third party operations and support systems (OSS).

8.2.3   Impact of Slicing on the Radio Access Network

Slicing support for the RAN has been vigorously researched during the past years. In this section, we capture the latest status of 3GPP standardization activities and also elaborate on the key principles. As currently reported by 3GPP [20], slicing in the RAN can be realized by Medium Access Control (MAC) scheduling and by providing different configurations for the NFs. Thus, traffic for different slices is handled by different Protocol Data Unit (PDU) sessions, for instance at MAC or Packet Data Convergence Protocol (PDCP) level. The different treatment among slices can be achieved by using specific identifiers in signaling messages to indicate specific slices. The configuration of the NFs to support the different slices is considered as implementation detail by 3GPP. The selection of the RAN part for an E2E network slice is done by assistance information, provided by the UE or the CN entities, that identifies one or more preconfigured network slices. The system supports policy enforcement between slices as per SLAs, and is able to apply the best radio resource management (RRM) policy to support the slice-specific SLA. Since the RAN can support multiple slices, resource isolation mechanisms have to be in place. These mechanisms are mainly RRM policies including scheduling schemes as well as protection mechanisms that are currently considered as implementation details in the context of the standardization activities. Nevertheless, the MAC-layer scheduling requires to be aware of slice definition and user membership in order to apply the RRM policies. Note that it is possible to fully dedicate resources (i.e., spectrum) to a specific slice and thus isolate it from other slices. The following paragraphs elaborate on these main principles.

One aspect about network slicing that is especially relevant in the RAN is the notion of **sharing the same radio resources and physical infrastructure** (e.g., processing capabilities) among multiple slices, ideally to the largest possible extent, as described in Chapters 11 and 12 (see for instance Section 12.6). Many envisioned 5G use cases are expected to be only economically viable if they can exploit significant synergies with other use cases and do not require dedicated infrastructure. However, there may be cases where some physical separation of slices is requested by involved stakeholders or even mandated by the law or some other administrative domain. For example, for highly

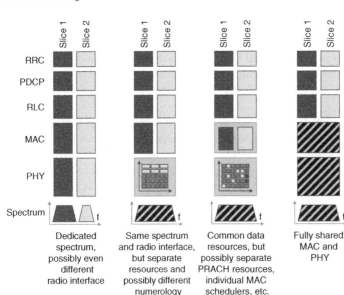

Figure 8-3. Representative RAN slicing scenarios with different level of resource sharing and isolation.

safety-critical use cases, it may be required by related regulators to keep some physical separation of the radio access for different slices.

Figure 8-3 illustrates some representative RAN slicing scenarios with different levels of resource and infrastructure reuse among slices that are thinkable, from full slice separation (left) to maximum multi-slice RAN integration (right). On the very left, we see the case where two slices use dedicated spectrum and possibly different radio interface specifications; this may be seen as the legacy case where for instance Long Term Evolution – Advanced (LTE-A) is used for an MBB slice. Then, one could have the case where slices share the same spectrum, but still use strictly separated physical resources therein, possibly involving different numerologies, being interleaved or overlaid to each other in some form. In this case, one would likely also have dedicated MAC instances and MAC schedulers for the different slices. Resembling a further extent of reuse, multiple slices could share the same spectrum, radio numerology and most of the resources, but still have some dedicated resources, such as dedicated Physical Random Access Channels (PRACH), for instance to guarantee stringent slice-specific QoS service requirements. In the case on the very right, one would have a fully shared and integrated MAC and physical layer (PHY) for both slices. Note that if the PHY is largely shared, one could further consider having individual MAC instances per slice, or a common MAC instance across multiple slices, for instance enabling multi-slice MAC scheduling. Many different flavors are possible in this respect, i.e., one may for instance assume that two slices use a same high-level MAC scheduler that allows for some flexible resource split among slices, while the slices involve some finer-granular dedicated schedulers per slice.

Figure 8-4 illustrates a slice-aware scheduler architecture with a resource visor that abstracts and shares the physical resources among slices according to the enforced RRM policies, and a slice resource manager (SRM) that allocates resources for UEs belonging to its slice according to the applied scheduling algorithm (e.g., proportional fair - PF, round robin – RR, priority-based, delay-based) [21]. It can be seen from the figure that scheduling is performed in two levels,

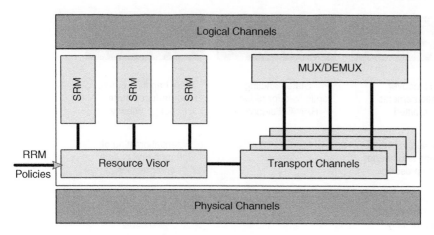

Figure 8-4. Slice-aware MAC scheduling architecture.

namely intra-slice and inter-slice, to decouple how UEs are served and how the resources are granted and mapped to the physical channels.

Another important aspect of network slicing in the RAN relates to the question of whether different services or slices use the **same or different specification and/or implementation of NFs**, and, in the former case, to which extent NFs may be **configurable** and chained to reflect service-specific or slice-specific needs. We explicitly refer here also to services and not only to slices, as even in the context of a single slice encompassing multiple services one may use different NFs or different configurations thereof to obtain service-tailored treatment. This is to some extent independent of the extent of physical resource and infrastructure sharing discussed before. For instance, two services or slices may use separated spectrum, but still reuse the same implementation and possibly configuration of, for instance, PHY network functions. On the other hand, two services/slices may use the same spectrum and radio resources, but use different implementations of some PHY functions, like for instance service- or slice-tailored encoding/decoding functions.

When it comes to actual slice implementation, the maximum reuse of resources and functions should be targeted so as to achieve the optimum use of the available resources and maximize the multiplexing gain. On the other hand, given the fact that different services with different requirements are targeted, the slices will not necessarily have the same functions or functions' configurations. Definitely, some functions need to have at least some common parts such as RRM functionalities for slices that share the same physical resources (e.g., for ensuring the slice protection), or RRC (e.g., for enabling the initial slice selection). Other functions, such as Hybrid Automatic Repeat reQuest (HARQ), random access, ciphering, etc. may be differently configured or even omitted if they are not needed.

The common understanding is that for the sake of a swift standardization process, simplified implementation and also less complexity, chip space etc., one should strive for a maximum reuse of RAN network functions among services and network slices. However, it is generally envisioned to allow the configuration of NFs such that they can be tailored to the specific needs of a given service or slice [22][23]. As an example, one could configure an RRC state machine to work very differently, depending on whether this is used in the context of an eMBB or an mMTC slice. For the former, the

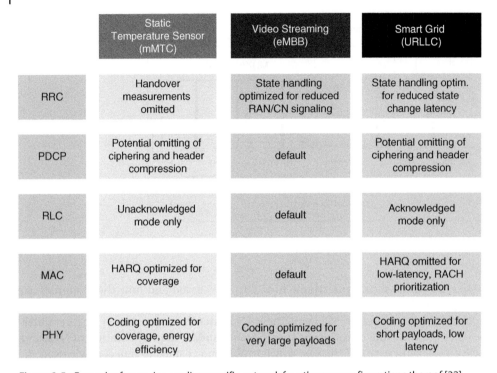

Figure 8-5. Examples for service- or slice-specific network functions or configurations thereof [22].

minimization of CN/RAN signaling could be an important objective. For the latter, the device power consumption could be the main issue of interest, as discussed further in Section 13.2. Further, it is envisioned that certain NFs, or elements of these, could be turned on or off for certain services and slices. For instance, the PDCP implementation for different services may in general be the same, except that for some services header compression is activated, while for others it is not. The stated and other examples of service- or slice-specific configuration or activation/deactivation of NFs is illustrated in Figure 8-5, and this topic will be further illustrated alongside the description of a single E2E slice example in Section 8.2.5. Note that beyond a different selection and configuration of NFs for different services and slices, some service-specific processing optimizations may be applied, as detailed in Section 6.4.2.

Please note that some slice- and service-specific processing will also inherently be facilitated by the new QoS framework that 3GPP has decided upon, as described in detail in Section 5.3.3. As a part of this, 3GPP has specified in [24] a new sublayer called Service Data Application Protocol (SDAP) to operate on top of PDCP, see also Section 6.4.2.1. The main services and functions it provides include the mapping between a QoS flow and a data radio bearer, and the marking of QoS flow identifier. The new information can be used by RRM functions like scheduling and admission control to offer customized support for slices. As will be detailed in Section 8.3.4, SDAP can also be utilized for inter-slice RRM functions in an attempt to fulfil the SLAs of all network slices. Additional information on this topic can also be found in Section 12.6.

Finally, one aspect that is likely specific to the implication of network slicing on the RAN is the introduction of **functionality that is specifically designed to facilitate the operation of network slices** with diverse and stringent requirements in a common radio infrastructure. Clearly, if one aims at a strong reuse of spectrum, radio resources and infrastructure resources such as processing capabilities, one needs other means to enable aspects such as slice protection, i.e., the guarantee that issues in one slice such as excessive load, possible malfunctioning caused, e.g., by erroneous devices or security attacks, do not have an impact on other slices. One further specific example for novel multi-slice functionality that will likely be introduced in the 5G context are means for multi-slice QoS and resource management, as outlined in Section 8.3.2 and described in more detail in Section 12.6.

8.2.4   Slice Support Across Different Administrative Domains

The slicing concept has been consistently identified so far with *single-provider* slicing. This means that a given service provider can create and deploy slices within the boundaries of its own administrative domain, using the resources at his own disposal. A key step ahead towards 5G objectives is the ability to define and provision cross-provider slicing, orchestrating resources offered by different administrations into E2E multi-provider services. This is paramount to maximizing the overall resource usage, avoiding the need of hinge overcapacity to reach the quality and performance levels demanded by 5G services.

There is no known solution to this problem yet. Resource orchestration and slice composition have until recently only been considered at intra-provider level. Main standardization working groups have recently started to take into account this scenario extension. Two relevant examples are given in [6] and [30].

From the research domain, there are also key findings for the support of cross-provider slices through the introduction of appropriate multi-provider orchestrator entities. Cross-provider slicing means that the slice can be made up by individual resources (e.g., virtual computation, storage and connectivity resources) which are partially or fully located in administrative domains different from the one spawning the slice creation and provisioning process, and owning, using and terminating the slice itself. This is also referred to as *resource slicing*. Such cross-provider slices can in turn be orchestrated with other slices, VNFs and connectivity resources to create more complex cross-provider services, also referred to as *service slicing*. The key point to make this happen is that every needed component must be searched, selected and provisioned according to a *service* model. A visiting domain entity, looking for the best resources to orchestrate and provision a given slice, must never have detailed visibility and access inside a visited domain. Instead, the visited domain must expose its available resources through a proper service catalog, where the resources themselves are presented in an abstract, mutually understood description. The orchestrating provider must be able to select the resources for its slice and access them for their requested usage, keeping a full separation from the rest of the visited domain.

Figure 8-6 shows how slicing is envisioned in a recent research project [25] tackling cross-provider orchestration, considering both resource slicing and service slicing. This architecture is based on a clear layer separation. Individual domains continue operating their own infrastructure, low-level controllers and intra-provider resource orchestrators. The I5 interface shown in the diagram is the legacy intra-domain interface between local orchestrators and infrastructure resources.

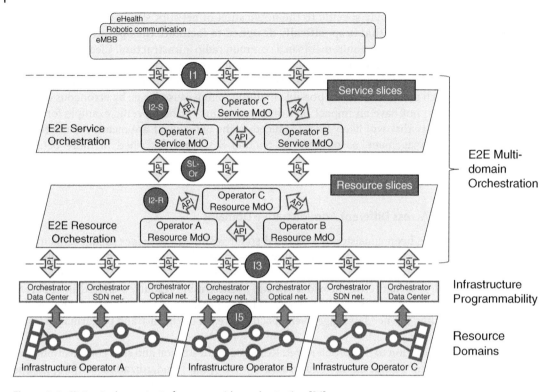

Figure 8-6. Slicing in the context of cross-provider orchestration [26].

On top of the legacy intra-provider layer, a multi-domain orchestrator (MdO) is positioned, communicating with the domain orchestrators via the interface labeled I3. Each MdO is logically decomposed into two submodules: a Resource MdO and a Service MdO, reflecting the architectural evolution currently undergoing in the ETSI NFV Management and Orchestration (MANO), as detailed in Section 10.2, and in charge of implementing resource slicing and service slicing as described above. MdO orchestrators in different domains communicate through the east-west interface generally labelled I2, used to expose the available resources from each provider, negotiate resource/service inter-provisioning, exchange information aimed at slice lifecycle management, and share a business layer (not highlighted in the picture). The I2 interface specifies a number of sub-interfaces, each one fulfilling a specific duty. Figure 8-6 highlights two such sub-interfaces:

- I2-R allows to share the resource-level details (computational resources and network topologies) available to the cross-provider ecosystem;
- I2-S facilitates service lifecycle management operations (instantiation, termination, and so forth).

The resource orchestration and service orchestration layers are interconnected through the Sl-Or interface, exposing to the Service MdOs a view of the available resources.

In the slice provisioning flow, one of the participant providers acts as *front-end provider*, and is the one who directly interacts with the user requesting the slice provision, through the interface labeled I1. Such provider is the ultimate owner of the user (customer) relationship, though of course, there must be mechanisms in place ensuring that all the providers cooperate to properly manage the slice, and business models duly apportion liabilities and responsibilities among the different providers.

The provisioning and usage of a slice built across different administrative domains requires clarifying a number of questions, among which the following should be stressed:

- **Specification**: a common data model must be defined, allowing to create slice abstraction descriptors or templates that can be automatically mapped and provisioned to external providers. This can be done by extending legacy data models (e.g., the ETSI NFV Network Service Descriptor) to incorporate the needed cross-provider add-ons;
- **Exposure**: the client MdO (i.e., the one residing at the front-end provider premises) must be able to search, in a service catalogue type repository, a directory of slice templates made available by other providers, to be selected, purchased along with a related SLA, and provisioned;
- **Slice control**: the client MdO must be given, beside UP endpoints to interconnect the slice with other service components, an additional control endpoint to internally access its assigned slice, configure its internal resources, and integrate them with others residing at different providers. This Fault Configuration, Accounting, Performance, Security (FCAPS) path typically goes through a service-specific component like the Element Management (EM) of the ETSI NFV MANO architecture;
- **Isolation**: the client MdO must be able to access the slice and its composing resources, while at the same time being fully isolated and unable to access any other resource or object inside the visited domain. Security provisions must hence be integrated in the design of the multi-provider orchestration framework;
- **Slice lifetime management** (see also Section 8.3.5): the client MdO must be able to monitor some given slice KPIs, detect possible failures and shortcomings, and trigger due actions when needed. Again, this must happen while safeguarding the privacy and non-accessibility of every off-slice resources in the visited domain. This is realized by sharing the relevant local providers' metric measures through the sub-interface I2-R, with the front-end provider collecting and combining all the partial metrics into E2E slice-level metrics. Monitoring of these latter metrics triggers recovery actions in case of need (e.g., when a given resource of the slice needs to be scaled out), enacted by the front-end provider, again through the proper I2 sub-interface conveying the due action requests to the external peer providers.

8.2.5 E2E Slicing: A Detailed Example

This section elaborates on an illustrative example for deploying multiple network slice instances in a mixed environment consisting of public, i.e. mobile network operator (MNO) owned, networks and private network infrastructure owned by a vertical enterprise. The infrastructure is used to commission an Internet of Things (IoT) network slice and two eMBB network slices.

5G Services on Factory Premises

As an exemplary deployment, a process automation use case from industrial manufacturing is considered. Traffic is composed of sensor readings, actuator control signaling, and eMBB services providing access to local as well as remote applications (e.g., augmented reality for machine maintenance). In a process automation environment, IoT devices include actuators, such as pumps, valves, etc., and sensors for capturing heterogeneous physical and logical quantities. The latter may for instance

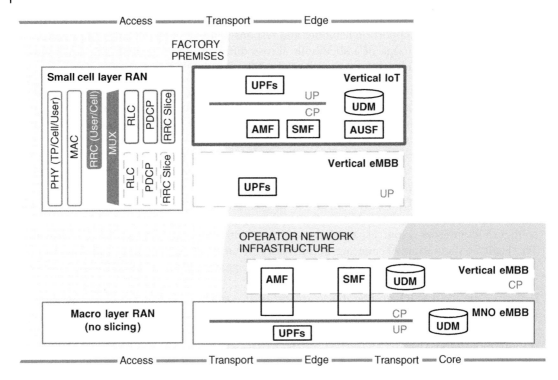

Figure 8-7. E2E network slice example for 5G services on factory premises.

include sensors for "reverse engineering" supporting maintenance processes or for critical safety applications, aiming to improve the overall operational efficiency and safety of the factory. When connecting such IoT devices, latency and bandwidth requirements can be very diverse. Such a setup requires two types of network slices, one covering machine-type communications for IoT devices and one covering eMBB traffic from smartphones, tablets and similar terminals. Figure 8-7 shows an example scenario with two E2E network slices (IoT and vertical eMBB) for the factory owner (also referred to as "vertical") as well as an eMBB network slice for the MNO. The network slices run on MNO infrastructure as well as the vertical's telecommunication infrastructure on the factory premises.

Ownership of Infrastructure, Spectrum, and Subscriber Data

In the given scenario, the vertical provides the small cell layer RAN equipment for all network slices that require coverage on the factory premises, i.e., both the IoT slice and vertical eMBB slice. Beyond the RAN, this includes the transport network and edge cloud resources, such as general purpose hardware for computing and storage. For operating the small cell layer RAN, the vertical rents dedicated spectrum resources from the MNO, such as higher frequency spectrum (e.g., above 6 GHz) with coverage strictly limited to the factory premises. For the vertical eMBB service, both small cell layer RAN and, if required, the MNO-provided macro layer RAN are utilized. RRM functions for the macro layer strictly remain under control of the MNO, which further owns a 5G-compatible network infrastructure consisting of centralized datacenters and distributed edge clouds comprising general-purpose as well as application-specific hardware. Subscriber information data including long-term security credentials are in possession of the vertical for the IoT devices. This assures full isolation of the vertical's IoT subscriber information from the MNO. For the eMBB subscribers, the MNO holds the corresponding data for own eMBB subscribers as well as vertical eMBB subscribers.

Domain-specific Network Slice Deployment and Operation incl. CP and UP Considerations

For the deployment of the individual network slices, multiple options exist. As depicted in Figure 8-7, the IoT network slice deploys all functions in the domain of the vertical, and it is only used by IoT devices that are registered in the vertical's home subscriber system (HSS) or Unified Data Management (UDM) and Authentication Server Functions (AUSF). These devices are mostly stationary and never leave the factory premises. The small cell layer RAN as well as the IoT-specific User Plane Processing Functions (UPFs) and control plane functions, in particular Access and Mobility Management Function (AMF) and Session Management Function (SMF), are operated locally under full control of the vertical. Since also the security mechanisms are strictly realized locally (i.e., access stratum security, optional "over-the-top" security), the entire network slice operates in the shielded factory environment without exposure of any data to the MNO. In contrast, the vertical eMBB slice is deployed in an inter-domain manner, see also Section 8.2.4. The CN control plane is shared with the MNO eMBB network slice and operated by the MNO outside the factory, including AMF and SMF as well as AUSF and UDM for authentication towards the core network and Non-Access Stratum (NAS) ciphering and integrity protection, respectively. Regarding the transport network, independent slices with guaranteed levels of isolation and security are used by both the vertical and the MNO, see also Section 8.2.2. In the vertical's small cell layer RAN, on the factory premises, PHY and MAC in the UP and RRC in the CP are shared by both slices. This approach limits the complexity because resource multiplexing is implemented across all network slices on MAC level, forcing each network slice to make use of the same efficient flexible RAN implementation. On the other hand, each network slice may still customize the operation through configuration and parameterization of Radio Link Control (RLC), PDCP, and RRC-Slice functions. RRM for the small cell layer realizes resource allocation according to the defined SLAs for the IoT and eMBB slices of the vertical. Further, the UP (i.e., the UPFs) is realized in a completely local manner if only access to local services is required, for instance provided by enterprise application servers in the factory's edge cloud. If a UE requests a "remote" service, such as a national voice call or Internet access, UPFs are realized on the MNO's infrastructure nodes, comparable to the regular MNO eMBB slice.

Security and Isolation

The vertical's IoT slice is completely isolated from the operator network by using own resources and hosting all functions locally. Transmitted user data and security termination points are strictly kept locally, sensitive subscriber data is maintained by the vertical, and the vertical has full control over the network.

For the eMBB service, the vertical and the MNO have a "roaming" agreement established that assures that vertical eMBB UEs (e.g., smartphones and tablets) that leave the factory premises connect to the MNO macro layer RAN to assure service continuity in the vertical's eMBB slice. In contrast, eMBB UEs subscribed directly to the MNO are continuously served by the macro layer RAN also when entering the factory premises. For vertical eMBB UEs, NAS ciphering and integrity protection is provided by the MNO, while access stratum security is terminated in RAN equipment owned by the vertical. Such a setup requires a minimum level of trust between the MNO and the vertical, since the operator owns the subscriptions including long-term credentials for vertical eMBB UEs. Therefore, the vertical can additionally employ over-the-top security (e.g., based on IPsec or TLS) to protect UP traffic from the MNO. However, this would require additional security functions to be maintained by the vertical (not shown in Figure 8-7).

8.3 Operational Aspects

After having described the E2E view for network slices, we now provide additional details on a number of slice operational aspects such as slice selection, connectivity to multiple slices, inter-slice RRM functions, and the overall management of network slices.

8.3.1 Slice Selection

Slice selection refers to the mechanisms used to identify the NSIs for a UE. The type of network slicing can affect the slice selection. In case of hard slicing, including the transmission and reception points, the initial access procedures can be very similar to legacy networks, such as LTE. Nevertheless, when the RAN supports multiple NSIs and these are sharing the same base stations, the slice selection can also influence the initial UE access procedures.

The subscription of a UE to NSI(s) can be determined via slice identifiers (IDs), such as the configured Single Network Slice Selection Assistance Information (S-NSSAI) stored in the subscription database. The S-NSSAI is used to identify a slice and thus to assist the 5G network in selecting a particular NSI [1]–[3]. It comprises a slice/service type (SST) referring to the expected slice behavior (i.e., features, services, etc.) and a slice differentiator (SD). The SD optionally allows further differentiation for selecting a dedicated NSI from potentially multiple NSIs complying with the indicated SST. The S-NSSAI can have standard values or Public Land Mobile Network (PLMN) specific values. In the latter case, S-NSSAIs are associated to the PLMN identifier of the PLMN that assigns it. The E2E slice selection consists of the selection of the CN [3] and the selection of the RAN part of the NSI [2].

Different alternatives for slice selection can be considered. Firstly, the slice IDs that a UE can be associated to can be configured a priori. In such a case, during the initial access, e.g., an RRC connection request, the slice IDs of the UE can be sent to the RAN. These slice IDs are utilized to determine the RAN policies for the associated NSIs as well as the selection of the CN part of the NSIs, such as the slice-specific CP NFs at the CN. If the slice IDs are not configured a priori, the UE subscriber information can be retrieved by the default NSI at the CN, e.g., by the CCNF, and the slice IDs of the UE will then be configured by the slice-specific CP NF over NAS messaging. Alternatively, the information about the available slices can be broadcasted in order to save signaling exchange and time during the attach procedure [27]. The penalty one has to pay for such an approach is the waste of radio resources used for broadcasting this information.

Providing the slice IDs by a UE during the initial access procedures can have the advantage of applying slice-specific RAN policies right away. This can be particularly advantageous in case of mission-critical NSIs with strict latency requirements. Yet, for non-mission-critical NSIs, determining slice IDs from the UE subscription information via the default NSI can reduce the signaling overhead during initial access procedures.

8.3.2   Connecting to Multiple Slices

Simple devices, such as sensors within the mMTC framework, will typically be associated to a single slice. For more complex devices, due to the mixed service needs, multiple slice associations can be realized. A good example for a multiple-slice UE are vehicles that will require multiple 5G services. For example, these may require both infotainment services, which are related to eMBB, and services for autonomous driving, which are mission-critical and thus related to URLLC. Another example that indicates the need for multi-slice connectivity is provided in Section 8.2.5. Some devices like tablets or smartphones may run factory-related applications that require to have access only to the local slice inside a factory over secure links, while other applications may require typical access to the Internet.

If slices are logically separated, a device will have to somehow be associated with all of them. Having UEs being associated to multiple networks may increase their complexity as well as the overall signaling considerably. When the same NFs (e.g., RRC or mobility management) are implemented for multiple slices, signaling overhead and UE complexity can be substantially reduced. However, the customization level per NSI is reduced in this case. For instance, mobility management procedures can be significantly different for an eMBB slice as compared to a URLLC slice, as discussed in Section 8.2.3. It is worth noting that according to the latest specification [20], only one signaling connection per UE is maintained even if it is associated with more than one slices.

In case a UE has simultaneous access to multiple slices, and depending on the use case requirements, it is possible to consider a context transfer from one slice NF to another. For example, the information about the current location of a vehicle in the URLLC slice may be transferred to the corresponding entity (e.g., the AMF) in the eMBB slice, thus reducing the signaling overhead that is required during a location update process. Similarly, subscription information related to a smartphone operating inside a factory can be transferred from the eMBB slice to the local URLLC slice minimizing the need for the local network operator to manage duplicated information already available in the typical operator's NFs.

Finally, the plethora of available network slices will require some adaptability from the UE side, especially for general purpose devices such as smartphones, tablets or laptops. This indicates that UEs will have to be to some extent open for programmability and reconfigurability to meet the slicing needs. Such abilities will constitute the UEs to be part of an E2E slice. However, further investigations are needed to clarify how UEs can be flexibly adapted to different slices without the need of extended operating system updates.

8.3.3   Slice Isolation

Network slicing targets the facilitation of different businesses that use the same infrastructure but possibly have diverse requirements. The deployment of multiple slices over the same infrastructure inside the network of one operator should enable the reuse of resources such as physical or software resources, which brings very big benefits compared to a hard splitting of the resources among slices. In [28], examples are provided that demonstrate that hard splitting of resources among slices can lower significantly the overall (busy hour) capacity of the network.

On the other hand, specific slice instances should be protected from the performance of other slices deployed over the same infrastructure. This feature is called *slice isolation* and relates to the ability of the network to minimize negative inter-slice effects. In the RAN, 3GPP already supports a partial protection of certain services under an extensive load of others through mechanisms such as Access Class Barring and extensions of such schemes [29], but such solutions cannot ensure the serviceability of one slice instance under excessive load of other instances.

Especially in case of slice instances that share radio resources such as common radio channels or common mobility management functions, congestion in one slice should not have a negative impact on another slice instance. A simple example of such case could be two slices that share the same Random Access CHannel (RACH) for initial access and also the same preambles. In this case, if one slice is overloaded, it will have a tremendous negative effect on the performance of the other, due to the large number of collisions.

In general, slice isolation may be achieved by:

- **Horizontal separation of resources**, based on the separation of physical resources for the different slices. This approach may lead to inefficient resource usage as explained above;
- **Efficient scheduling/coordination mechanisms**, where slice-overarching functions such as scheduling functions, QoS schemes and initial access mechanisms ensure that the service requirements of each slice are met. In any case, a soft separation of resources is required so as to prioritize certain slice instances over others.

The isolation of slice instances may also be indicated by regulations which necessitate the different treatment of certain slices. One potential example for such case could be autonomous driving slice instance(s) which may need to have their own spectrum resources. Moreover, another requirement for separating slices is security. As explained in the example of Section 8.2.5, specialized local networks such as factories or medical facilities require that even the serving operator will not have any access to local data of sensitive nature. For this reason, it is considered that even some of the core NFs will have to be re-deployed in the local network. Some context transfer may take place, for instance related to subscription information, but only from the operator network towards the local network.

8.3.4 Radio Resource Management Among Slices

With respect to the RAN, the management of the scarce radio resources is a critical issue. Thus, pooling and sharing these resources among network slices in an efficient manner is an important target. The RRM is also responsible for allocating the resources in a way that the SLAs of all network slices are fulfilled.

The basis for allocating resources in a slice-aware manner is to monitor the status of the network slices with respect to their SLAs. This could take place in a new entity of the RAN, e.g., an access controller, which has to be aware of the existing network slices and their SLAs, as well as which data stream belongs to which network slice. Corresponding information can be obtained via signaling from the CN. The enforcement of the network slice specific requirements can be realized with different levels of complexity as shown in Figure 8-8. On the right side of the figure, a lightweight implementation of multi-slice resource management is depicted. Based on the outcome of the SLA monitoring, the QoS Class Identifiers (QCIs) of the individual data streams are adjusted and traffic steering is executed. An enforcement of SLAs happens by adapting the QoS classes of individual data streams. If, for example, the SLA of a network slice guaranties a certain latency, any data stream of this slice could be mapped to a corresponding QoS class. In the case of 5G NR, this functionality can

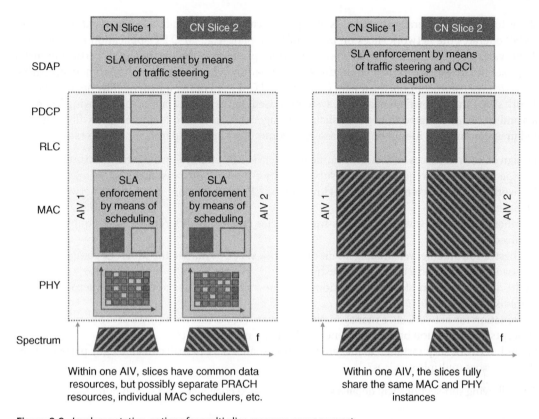

Figure 8-8. Implementation options for multi-slice resource management.

be part of the new SDAP sublayer. In this implementation, the individual AIVs can operate in a slice-agnostic way, but have to fulfil the QoS defined.

On the left side of the figure, the functionality of SDAP is enhanced by a slice-aware real-time RRM that also performs the scheduling in a slice-aware manner based on the status of the SLA monitoring. The QoS mapping or the slice-aware scheduling, respectively, is a dynamic process, which is supposed to solve conflicts between network slices in a way that all SLAs can be fulfilled. More details on multi-slice resource management can be found in Section 12.6.

8.3.5 Managing Network Slices

Current business support systems (BSS) and operating support systems (OSS) of communication service providers (CSPs) expose service management capabilities to customers. In contrast, network management and infrastructure management functions, such as the 3GPP Network Management System (NMS) and Element Management System (EMS) are typically not exposed to customers. With network slicing, CSPs will need to extend the current service management offers to a certain level of management exposures for both network and infrastructure. The extent of such exposure depends on the level of expertise of the network slice instance customer and the CSP's readiness to open internal systems to tenants. Generally, three levels of exposure can be differentiated: In the *monitoring* option, the CSP operates the network slices on behalf of the tenant (e.g., an OTT application provider) and only provides slice-specific KPIs. The *limited control* option gives the tenant (e.g., a vertical industry enterprise) the possibility to (re-) configure selected parameters of NFs associated to network slices. In the *extended control* option, the tenant (e.g., a virtual CSP) can rather independently operate its own network slices and use own management systems. Based on these constraints, the following key questions and challenges related to network slicing management have been identified and will be investigated in upcoming activities in both the research community and standardization processes:

- Design of network slices to host communication services supported by the infrastructure: How can the specified service requirements be supported by a network slice instance?
- Network slice instance management: How can FCAPS management be used for network slicing management?
- Conflict resolution: How can conflicts from policies created by different service requirements be resolved?
- Orchestration and lifecycle management of network slices: What are the different compositions of a network slice and how are they orchestrated?
- Multi-domain network slice orchestration: How can a slice be created and deployed across multiple administrative domains?
- Automation for network slice management: How can, e.g., evolved self-organizing network (SON) concepts and cognition be applied to network slicing management?
- Shared network slice instance management: How shall a network slice instance (or parts of it) be shared across multiple services?

The following sub-sections will detail a subset of the listed challenges and sketch potential solutions.

8.3.5.1 Managed Objects and Network Slice Instance Lifecycle

Next Generation Mobile Networks (NGMN) [4] has introduced the concepts of 'Network Slice Blueprint' and 'Network Slice Instance' which have largely been taken over by 3GPP as 'network slice template' and 'network slice instance', respectively [30]. Instance-specific policies and configurations are required when creating a network slice instance from network slice templates. A network slice is composed of one or multiple 'network slice subnets' which in turn contain one or multiple physical or virtualized network functions.

The lifecycle management of these NFs (both CNFs and RNFs) comprises both 3GPP domain-specific FCAPS management as well as domain-agnostic lifecycle management and orchestration. Regarding the lifecycle of a network slice instance, three distinct phases have been defined [30], as depicted in Figure 8-9: (A) commissioning phase, (B) run-time phase, and (C) decommissioning phase. A so-called Network Slice Management Function (NSMF) oversees the respective tasks of each phase, and is detailed in the following sub-section.

8.3.5.2 Network Slice Management Function (NSMF)

The NSMF [30] is responsible for managing the lifecycle of a network slice instance. Provided that the preparatory tasks such as network slice design, network slice pre-provisioning, template on-boarding and the general preparation of the network environment have been completed, the NSMF is ready to process incoming requests for communication services. It does so by selecting a network slice template that can provide the agreed service requirements including the required levels of isolation, security, and management exposure, as shown in (Figure 8-10).

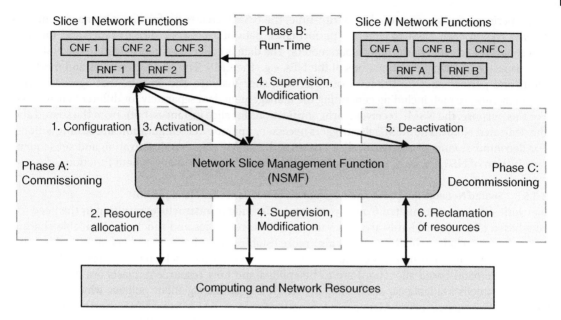

Figure 8-9. Phases of network slice lifecycle management [30].

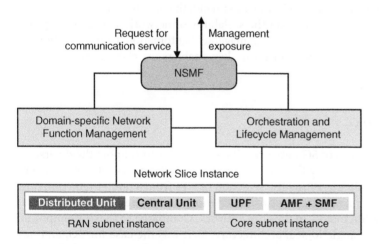

Figure 8-10. Domain-specific FCAPS management and lifecycle management of a network slice.

The NSMF commissions a network slice instance consisting of shared and/or slice-specific RAN and CN functions. Knowing which network slice subnet instances (NSSI) are associated with each network slice instance, the NSMF determines to completely reuse an existing (operating) NSI or create a new NSI. For the latter case, the NSMF instantiates and configures slice-specific VNFs and subnet instances, and re-configures already operating and shared NFs and NSSIs. Subsequently, the NSI is activated by activating all necessary CN, RAN and other functions.

For performance supervision during run-time, the NSMF creates performance management jobs for the NFs in each NSSI to generate performance data of a network slice instance and monitor thresholds for selected performance parameters. The data is collected and provided to the NSMF by the respective management functions of the NFs, e.g., the EMS. The NSMF compiles and monitors faults and performance on the level of a network slice instance and ensures that agreed service requirements are met, including generating the management data separately for different customers. For this purpose, the NSMF receives alarm notifications for slice instances, both from the shared and the dedicated NSSIs. The NSMF also triggers necessary upgrade, reconfiguration and scaling actions. For decommissioning of a network slice instance, the NSMF requests deactivation and subsequent termination of NSS instances at the respective FCAPS and lifecycle management functions.

8.3.5.3   Shared Network Functions and Automation of Network Slice Management

Network slices are instantiated from a common underlying infrastructure, resulting in the need for rule-based allocation of hardware, software, and radio resources and functions. Available sharing technologies include virtualization, but also more established techniques such as multi-tasking and multiplexing. Such sharing rules define how resource commitment schemes (e.g., static allocation, dynamic demand-based allocation) are to be applied and how resource requests are prioritized if demand exceeds available capacity. They are derived from management policies which are maintained by the resource owner. Management functions such as the EMS and NMS, together with other NFs such as radio schedulers, apply and enforce these rules in order to fulfil the SLAs associated to each network slice. On one hand, such cross-slice interdependencies considerably increase the complexity of management and orchestration tasks. On the other hand, the massive number of managed objects also significantly raises the requirement on the scalability of network management procedures. 5G networks therefore require a higher level of management automation. Here, cognition and more autonomous decision making processes constitute key enablers, as elaborated in detail in Section 10.7.

8.4   Summary and Outlook

Slicing is one of the key characteristics of 5G networks. Its main goal is to support the diverse 5G use cases and their requirements in a very flexible way, and to run them cost-efficiently over a common network infrastructure. Slicing allows the selection of NFs from a pool, their configuration, synthesis and deployment to form logical networks. Previous 3GPP releases have attempted to introduce network customization mainly in the core network. Compared to these attempts, 5G network slicing is not bound to a single logical architecture, as it was the case for 4G, and new levels of flexibility also appear in the transport and access domains. Obviously, such flexibility introduces some new architectural issues that need to be solved, such as slice selection, computation and network resources virtualization and sharing among slices, slice isolation, and support for end devices that are able to connect to multiple slices. Moreover, the 5G architecture will enable the bi-directional communication of NFs with application functions provided by the content and application service providers. This will create the ability for further tailor-cut solutions for these providers, improving multi-tenancy support, but also introducing the need to properly define the required northbound interfaces for programmable interaction with the tenant.

Although the standardization community has provided the first version of specifications for the 5G release, several issues remain open for future releases. Thus, network slicing is far from being a thoroughly studied feature, and further work on this topic is expected during the next years. In Table 8-1, the key points analyzed in the previous sections are grouped and summarized, and remaining open issues are listed.

Table 8-1. Key points and main open issues related to the support of network slicing.

Technical Area	Key Points	Main Open Issues
Core Network	• Decomposition of CN NFs • Further separation of control and user plane • Common or slice-specific NFs (which will belong to which category, how will this affect issues like security) • Service-based interfaces • Service exposure to 3rd parties	• Clear categorization is needed between common or slice-specific NFs. Impact on slice isolation • Find a balance point between complexity and flexibility that NF modularization introduces • Introduce slice policy conflict resolution and slice prioritization that may occur via the exposure of NFs to 3rd parties
Transport Network	• Convergence of highly heterogeneous transport technologies • More efficient per-slice QoS support and on demand resource allocation, adapted to dynamic workloads	• Identify potential standardization of open interfaces to support convergence in a multi-vendor environment • Identify the need for potential enforcement of policies and slice prioritization in the transport domain
Radio Access Network	• Flexibility to allow full separation of resources among slices or their sharing • NFs on different radio protocol stack layers should be configurable to support 5G use cases and allow flexible placement on physical or logical network nodes (e.g., central or distributed units)	• Identify the common functions among slices • Allow different specification and implementation of NFs without increasing the complexity of the specification • Clarify the impact of flexible NF placement on the transport domain (interfaces, latency/bandwidth requirements, etc.)
Multi-operator Slicing	• Slice support over multiple network providers through the introduction of appropriate orchestrators and their interworking • Limited exposure of available resources to other domains	• Appropriate models and interfaces need to be standardized
User Equipment	• Slice selection • Concurrent connectivity to multiple slices • Openness of UEs for programmability to slicing needs (UEs as part of the E2E slice)	• Identify how to minimize unnecessary signaling and UE complexity • Clarify how UEs can be flexibly adapted to varying needs of slicing use cases without the need of extended operating system updates
Slice Management	• Dynamically translate customer needs and instantiate slice instances • Improve network management to support a multi-slice environment (sharing of resources, functions etc.)	• Standardize a northbound interface between NMS and 3rd party application and services • Introduce schemes for conflict resolution among slice specific SON functions that operate over shared resources

References

1 3GPP TS 23.501, "System Architecture for the 5G System; Stage 2 (Release 15)", Version 15.0.0, December 2017

2 3GPP TR 38.801, "Study on New Radio Access Technology; Radio Access Architecture and Interfaces (Release 14)", Version 14.0.0, March 2017

3 3GPP TR 23.799, "Study on New Radio Access Technology; Radio Access Architecture and Interfaces (Release 14)", Version 14.0.0, December 2016

4 NGMN Alliance, "Description of Network Slicing Concept", January 2016

5 5G PPP 5G NORMA project, Deliverable D3.3, "5G NORMA network architecture – final report", Oct. 2017.

6 ETSI, Industry Specification Group (ISG) Network Functions Virtualization (NFV), see http://www.etsi.org/technologies-clusters/technologies/nfv

7 B. Blanco et al., "Technology pillars in the architecture of future 5G mobile networks: NFV, MEC and SDN", Computer Standards and Interfaces, January 2017

8 M. Richart J. Baliosian, J. Serrat and J-L Gorricho, "Resource Slicing in Virtual Wireless Networks: A Survey", IEEE Transactions of Network and Service Management, vol. 13, no. 3, Sept. 2016

9 C. Liang and F.R. Wu, "Wireless Network Virtualization: A Survey, some research issues and challenges", IEEE Communication Surveys and Tutorials, vol 17, no. 1, Q1 2015

10 NGMN Alliance, White Paper, "5G White Paper", Version 1.0, March 2015

11 5G PPP Architecture Working Group, White Paper "View on 5G Architecture", Version 1.0, July 2016

12 3GPP TS 23.251, "Network Sharing; Architecture and functional description (Release 14)", Version 14.0.0, March 2017

13 3GPP TR 23.707, "Architecture Enhancements for Dedicated Core Networks; Stage 2 (Release 13)", Version 14.0.0, Dec. 2014

14 3GPP TR 23.711, "Enhancements of Dedicated Core Networks selection mechanism (Release 14)", Version 14.0.0, Sept. 2016

15 ETSI ISG Mobile Edge Computing (MEC), see http://www.etsi.org/technologies-clusters/technologies/multi-access-edge-computing

16 3GPP TS 23.214, "Architecture enhancements for control and user plane separation of EPC nodes; Stage 2 (Release 14)", Version 14.3.0, June 2017

17 3GPP TS 23.502, "Procedures for the 5G System; Stage 2 (Release 15)", Version 15.0.0, December 2017

18 3GPP TR 33.899, "Study on the security aspects of the next generation system", Version 1.2.0, June 2017

19 Anna Tzanakaki et al, "Wireless-Optical Network Convergence: Enabling the 5G Architecture to Support Operational and End-User Services", IEEE Communications Magazine (to appear)

20 3GPP TS 38.300 "NR and NG-RAN Overall Desciption", Release 15, Version 1.2.0, May 2017

21 Adlen Ksentini and Navid Nikaein, "Toward enforcing network slicing on RAN: Flexibility and resources abstraction", IEEE Communications Magazine, vol. 55, no. 6, June 2017

22 5G PPP METIS-II project, White Paper, "Preliminary Views and Initial Considerations on 5G RAN Architecture and Functional Design", March 2016

23 P. Marsch et al., "5G radio access network architecture: design guidelines and key considerations", IEEE Communications Magazine, vol. 54, no. 11, pp. 24–32, Nov. 2016

24 3GPP TR 38.912, "Study on New Radio (NR) access technology", Release 14, Version 14.0.0, March 2017

25 5G PPP 5GEx project, White paper, "5GEx Multi-domain Service Creation - from 90 days to 90 minutes", March 2016

26 5G PPP 5GEx project, Deliverable D3.1, "Description of protocol and component design", July 2016

27 X. An, C. Zhou, R. Trivisonno, R. Guerzoni, A. Kaloxylos, D. Soldani and A. Hecker, "On end to end network slicing for 5G", Transactions on Emerging Telecommunications Technologies, Wiley, June 2016

28 I. da Silva et al., "Impact of network slicing on 5G Radio Access Networks", European Conference on Networks and Communications (EuCNC 2016), June 2016

29 3GPP TS 22.011, "Service Accessibility (Release 15)", Version 15.1.0, June 2017

30 3GPP TR 28.801, "Study on management and orchestration of network slicing for next generation networks", Release 15, Version 15.0.0, September 2017

9

Security

Carolina Canales-Valenzuela[1], Madalina Baltatu[2], Luciana Costa[2], Kai Habel[3], Volker Jungnickel[3], Geza Koczian[4], Felix Ngobigha[4], Michael C. Parker[4], Muhammad Shuaib Siddiqui[5], Eleni Trouva[6] and Stuart D. Walker[4]

[1] Ericsson, Spain
[2] Telecom Italia, Italy
[3] Heinrich Hertz Institut, Germany
[4] University of Essex, United Kingdom
[5] Fundació i2CAT, Spain
[6] National Centre for Scientific Research "Demokritos", Greece

9.1 Introduction

Future 5th generation (5G) technologies are anticipated to address next generation network's challenges and tackle the novel business requirements associated with different vertical sectors. This implies that 5G technologies will not only encompass new wired and wireless network technologies to support higher data rates, bandwidths, numbers of devices, etc., as elaborated in Chapter 2, but also need to be cohesively aligned from a technological as well as business standpoint with the different vertical sectors, for their optimized and efficient use of the network, for instance through customized network slices. Furthermore, convergence, automation and flexibility are expected to be intrinsic traits of any 5G system. The introduction of this multitude of complex new requirements and novel technologies immensely impacts the security landscape of 5G, and therefore the need to revisit its properties becomes essential.

Given the array of new technologies and vertical industries that 5G aims to support, as detailed in Section 2.2, it is obvious that different parts of the 5G ecosystem will be developed by different stakeholders. The interoperability of these different elements can be addressed via standard interfaces. However, a holistic approach to address the security of these elements, individually and more importantly when integrating them, is crucial, so that the required level of security can be guaranteed for a 5G system. The complexity of the security landscape of 5G systems increases exponentially due to the required tightly knitted alignment with vertical industries, and the heterogeneous nature of the integrating technologies (wired or wireless networks, virtualization, etc.). Thus, the security vision of

5G needs to at least comprise the security requirements associated to all of the involved technologies and vertical sectors. For example, reliability in 5G systems would go beyond availability or up-time of the network infrastructure and will also include high connectivity, virtually infinite perceived capacity and ubiquitous coverage. Furthermore, the tedious and cumbersome chores of securing legacy networks become even more complex with the inclusion of software-defined networking (SDN) and network function virtualization (NFV) technologies in 5G networks, as detailed in Section 10.2. These are just a couple of examples of how the 5G system security landscape is impacted.

The chapter is organized as follows: Section 9.2 describes the envisioned security threat landscape. Corresponding requirements for 5G security are derived in Section 9.3. Then, the main characteristics of the envisioned 5G security architecture are described in Section 9.4, before the chapter is summarized in Section 9.5.

9.2 Threat Landscape

As mentioned before, 5G enables innovative scenarios and applications making use of ultra-high speed, low-latency telecommunication networks for fixed and mobile users, and machine-to-machine communications. These scenarios, together with the introduction of the new paradigm for computing and network infrastructure which decouples the actual functionality from the underlying hardware and middleware functions, for instance through cloud computing and SDN, as detailed in Section 5.2, further reinforces the need for automated management and control of the telecommunication infrastructure. In particular, since a cloud-based paradigm that promotes that infrastructure is highly accessible and shared by multiple tenants, as for instance virtual network operators, the concept of a highly secure network gains even more relevance.

There are two different aspects that need to be taken into account in order to address the security of the upcoming 5G network: On one hand, the need to address the overall security functionalities of the network, composed of virtualized and non-virtualized functionalities, where the latter are often referred to as "traditional" or "classical" functionalities. Due to the increasingly virtualized nature of the 5G network, the effectiveness of traditional security approaches using physical network elements and devices (a.k.a. entities) is likely to diminish. When just instantiated in a cloud, the virtual network functions (VNFs) replacing any physical entities lack visibility w.r.t. changes performed over virtualized functions, service chains, and the traffic being exchanged on the virtualized network. Or putting it into different words, a holistic security approach comprising both virtualized and non-virtualized (i.e., traditional) security functions needs to be put in place.

On the other hand, there is a need to cater for an automated security management solution for the 5G network. Today we can't foresee the new and ever-changing threats that 5G networks will have to protect against, but we do have the basis to create autonomic network management solutions that shall cope with them, being fed with insights from governed real-time analytics systems on the one hand, and actuating on network resources in order to minimize or prevent the effects of the detected threats in real-time on the other hand. It is therefore of outmost importance to be able to provide robust, flexible and proactive mechanisms to detect and prevent security issues, and to be able to perform that in real-time and in an automated fashion.

Taking this into account, the infrastructure operator and the different tenants using part of the network need to maintain the overall end-to-end (E2E) security of the network with different degrees of responsibility and different focus areas, including E2E security, physical infrastructure security, and the security of new virtualized resources (being those applications or network functions).

In detail, the 5G network assets to be secured are:

i) **User information security**, referring to the protection of E2E user data, related to human users but also machine-type communications (MTC), including the transfer of E2E control plane and user plane data;
ii) **Network element security**, related to the protection of endpoint devices and network elements, here also referred to as user equipments (UEs), including the physical security for network elements and the application SW, which in a cloud-based architecture is composed of VNFs;
iii) **Transport/interface security**, referring to the protection of communication paths.

As general examples of types of attacks, depending on the attacked 5G asset, we could mention:

- **Attacks towards UEs or network elements (NEs)**, such as infection via malware or bots infecting subscribers' devices which can generate spurious or attack traffic, create signaling storms into the network, and drain device batteries;
- **Attacks towards the different network subsystems**, such as the radio access network (RAN) and core network (CN), causing resource exhaustion, terms and conditions violations such as service level agreement (SLA) violations, or attacks on the Domain Naming System (DNS), billing, and signaling infrastructure, etc.;
- **Attacks towards end-user applications**, such as server-side malware, application-level and protocol-specific distributed denial of service (DDoS) attacks, etc.

9.3 5G Security Requirements

5G networking imposes new security requirements as it is linked to new business and trust models involving new stakeholders, new actors, and new service deployment and delivery procedures for telecommunication services, as stressed in Section 2.6. In this section, as a result of the previous analysis of threats, we draw attention to selected requirements for 5G security. These do not solely affect the design of security services deployed to offer protection and privacy, but are expected to heavily impact the overall 5G architecture design as well.

Clearly, a general requirement is that security systems in 5G should be flexible enough to accommodate the expected diversity of connected devices and systems, provide the ability to monitor their real-time status and traffic, and provide protection against the main attack vectors listed in the previous section. In the following subsections, we will dive into more details on the requirements stemming from specific aspects of the 5G system.

9.3.1 Adoption of Software-defined Networking and Virtualization Technologies

Taking into consideration the dominant trend for programmable, SDN-based infrastructures, the transition to 5G is followed by the adoption of several emerging virtualization technologies such as NFV using SDN. These paradigms, although they facilitate the orchestration and management procedures which offer increased flexibility and reduce cost for the operators, require several additional security considerations and deeply impact network security.

The SDN architecture, with the decoupling of the network control and forwarding functions, the logically centralized network control elements and the exposed interfaces that enable programmability of the network elements (switches), as detailed in Section 10.2, imports new targets for attack and exploitation at control and user planes, beside the application layer. Attackers might attempt to compromise SDN elements, controllers or switches. Especially SDN controllers are attractive targets for attacks, as such attacks may allow to compromise the SDN control plane. A compromised SDN controller enables an attacker to install new flow rules and direct traffic to flows as desired across the controlled SDN switches, altering the network design. Another common security threat towards the control plane is to perform a denial of service (DoS) attack towards the SDN controller, depleting its resources. By leveraging the security vulnerabilities in well-known southbound SDN protocols such as OpenFlow (OF), Open vSwitch Database Management Protocol (OVSDB), NETCONF, etc., attackers could divert traffic flows or simply eavesdrop the already installed flow rules for reconnaissance purposes.

Regarding NFV, although it is justifiably connected to cost reduction through the replacement of specialized hardware with virtualized services, its adoption brings significant security risks. According to the European Telecommunications Standards Institute (ETSI) NFV security (SEC) working group [1], these include risks related to network virtualization (e.g., memory leakage, interrupt isolation), traditional networking threats (e.g., flooding attacks, routing security) and those new threats that result from the combination of virtualization and networking technologies. Additionally, other security risks to consider related to the use of NFV are those specific to the software used to implement a network function. To summarize, the anticipated introduction of virtualization techniques in 5G not only introduces the risks associated to one virtualized element in itself, but also to the combination and integration of different virtualized elements from different sources.

9.3.2 Security Automation and Management

Due to the complexity of the 5G infrastructure and the increasing automation and sophistication of attacks, the automation of security functionality is essential. Along these lines, network virtualization, orchestration and analytics contribute to creating a policy-controlled network automation cycle. Additionally, virtualization technologies enable the rapid provisioning of network services, along with security services, allowing for a rapid and inexpensive creation and removal of service chains and virtualized security functions on demand.

In order to properly secure the 5G network, it must be made with out-of-the-box security management features that are highly integrated with the rest of the network elements and that are flexible enough to evolve and adapt to the network changes as soon as these happen. That is, security and security management should be architectural network principles, per network design.

Even if it might not be possible to automate all security management procedures, many of them will be highly automated, and therefore the network management and orchestration components, such as the service orchestrator, the policy manager, the inventory systems, and the existing operations support systems (OSS) and business support systems (BSS) need to provide decision support for security threat detection and mitigation. Integration and co-operation of security systems with service orchestration, policy management systems and cloud and network managers can assist the identification of security incidents and accelerate the response process towards security attacks.

Additionally, it should be possible for different tenants to manage the security features of their network slice up to a certain point, though of course the security of the whole network infrastructure is rather in the responsibility of the network infrastructure provider.

9.3.3   Slice Isolation and Protection Against Side Channel Attacks in Multi-Tenant Environments

The design of future security solutions should deal with the fact that infrastructure resources are being shared between different operators and in addition between their services and slices. In an abstract view, exchanged traffic belongs to different tenants who operate their services over a common infrastructure. Tenant isolation is vital to support self-contained and independent networks required for slice creation, to ensure the quality of service offered to tenants depending on the agreed service-level agreement (SLA), and to provide data integrity and confidentiality. Given the shared infrastructure model, which is heavily based on a shared computing environment, special security measures should be provisioned to handle possible cross-tenant side-channel attacks and prevent leakage of sensitive information amongst tenants.

The multi-tenancy model imposes new requirements to the design of 5G security systems. Security systems should provide different levels of access to services, infrastructure resources and information to tenants with different management capabilities and scope. Consequently, flexible, scalable and possibly hierarchical access schemes are required. Tenants should not be allowed to access resources or information that belongs to other tenants, without excluding special cases in which resources should be shared amongst tenants. On the other hand, infrastructure providers should be able to monitor the whole infrastructure they own, without being able to access sensitive data that reveals business information of the tenants.

9.3.4   Monitoring and Analytics for Security Purposes

Resource and service monitoring information should be used to assist security decisions in future security systems. Information that could be used for attack detection includes generic metrics from the physical and virtual network devices, notifications received from security services deployed in the network, authentication and authorization incidents and others. Traditional monitoring data such as central processing unit (CPU) utilization, random access memory (RAM) utilization, network interface bitrate and packet rate measurements, can greatly assist in detecting zero-day attacks for which identification patterns and signatures have not been defined yet. Scalable analytics engines, designed to capture, store and analyze all network activities, should be deployed in central and edge clouds. These systems could be also offered in "-as-a-Service" schemes using NFV technology and be deployed on an on-demand basis.

9.4   5G Security Architecture

9.4.1   Overall Description

Previous generation networks involved one or more security and trust domains, or more precisely, at least as many domains as the number of home networks serving each of the participants involved in the communication. This paradigm is reflected in Figure 9-1.

In 5G networks, the security and trust paradigm gets increasingly complicated. Apart from having the security and trust domains associated to the participants of the communication link, i.e. *horizontal* security and trust domains (HSTD), those HSTDs are implemented by means of virtualized and

Figure 9-1. Security and trust domains in traditional networks.

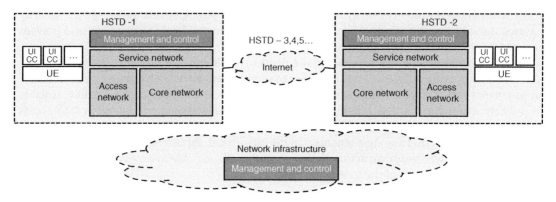

Figure 9-2. The impact of virtualization on the security and trust domains in 5G network.

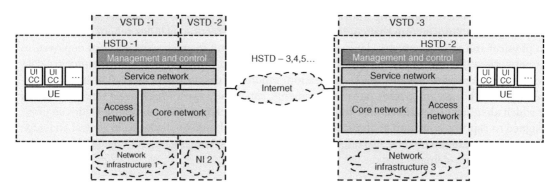

Figure 9-3. 5G security and trust domain paradigm.

non-virtualized resources which run on an infrastructure which could belong to another organization, for instance an infrastructure network provider, as reflected in Figure 9-2. For realizing a secure E2E link, obviously, joint management and control functions are needed.

We could therefore say that we have additional *vertical* security and trust domains (VSTDs) to be added to the traditional security and trust paradigm of the network. Figure 9-3 depicts the complete 5G network security and trust paradigm.

In the subsequent sections, we will focus on the security properties of the different architectural elements inside the same HSTD, taking into account that each HSTD could map to one or more

VSTDs. In particular, we here decompose each HSTD into the following elements and related security aspects: infrastructure security, physical layer security, RAN security, service-level security, and overall security management and automation.

9.4.2 Infrastructure Security

Infrastructure security (IS) can be implemented at many levels of the Open Systems Interconnect (OSI) model protocol stack in the transport network, often with a focus on the higher layers, e.g. in the form of MACSec for layer-2/data link layer, IPSec for layer-3/network layer, or transport layer security (TLS) for layer-4/transport layer. However, the layer-1/physical layer offers a complementary approach to improve communication security, particularly in the context of wireless networking, where the properties of the communications system can also be exploited, as we will see in Section 9.4.3. Thus, IS, where the transport network infrastructure inherently supports security[1] for any network functions, is becoming an important aspect of 5G security design.

Security has so far been seen as an add-on feature in the layered network design, and in particular the separation of the upper layers from the physical layer as a reliable bit pipe, and the provision of security only at those higher layers has contributed to fortifying the existing computational security techniques. Traditional solutions to mitigate the security challenges at the upper layers use various types of private and public secret keys via computation-based mechanisms, often referred to as *over-the-top* (OTT) cryptography.

While 4th generation (4G) UEs are authenticated in the mobile core, they might communicate via unsecured transport networks. The so-called Access Stratum (AS) is terminated at the enhanced Node-B (eNB) and thereby it protects control and data transport only over the air interface. In the AS, user data can be cipher-protected by the Packet Data Convergence Protocol (PDCP), where the radio control data will always be integrity-protected and can eventually be cipher-protected. The PDCP security foresees different algorithms for integrity protection, such as the Advanced Encryption Standard Counter Mode (AES-CTR), SNOW3G, and the Zu Chongzhi algorithm (ZUC). For cipher protection, it offers the same 3 schemes and also the "NULL" variant, which means that traffic is unencrypted.

In 4G, both the user plane and control plane traffic between the eNB and the Evolved Packet Core (EPC), i.e., the core network or the so-called Non-Access Stratum (NAS), is cipher- and integrity-protected. Only the encryption keys and the control traffic between eNB and Mobility Management Entity (MME) are currently protected. The NAS is realized, in practice, via a public transport network infrastructure being partly or fully shared among multiple operators and services. For UE attach procedures, the traffic is unprotected through NAS.

Now we look deeper into the methods used to further ensure secure NAS transport over the transport network infrastructure where, of course, proper protection algorithms are already available, such as TLS, IPSec and MACsec, which can be considered as potential security protocols also in 5G. These transport protocols include handshake procedures for mutual authentication and key agreement and creation.

Note that the transport network architecture is currently subject to significant changes from 4G to 5G, as stressed throughout Chapters 5, 6 and 7, one key aspect in 5G being the possibility to have

---

1 One example for IS is IPSec, where routing as an elementary network function inherently supports security.

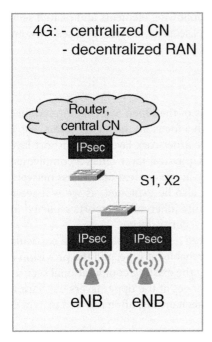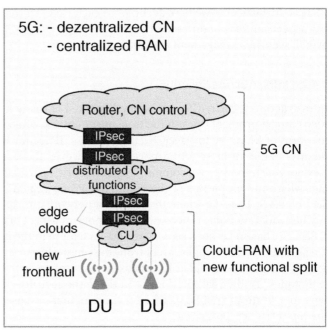

Figure 9-4. Key architectural differences between 4G and 5G, with implications on the approaches towards achieving transport infrastructure security.

more functional split options and a much more flexible assignment of network functions to physical entities, as for instance stressed in Section 6.6. This difference between 4G and 5G, and its implication on security handling, are also illustrated in Figure 9-4.

In the 4G system, depicted in the left side of Figure 9-4, the EPC is totally centralized, while the complete RAN is implemented in the form of eNBs, with an IP-based connection to the EPC. In typical 4G deployments, the complex nature of the transport network is ignored, but instead, the link between eNB and EPC is abstracted as a point-to-point (P2P) link. There is an implementation guideline to use IPSec tunnels from eNB to EPC for protecting this link, but the decision to do so is left to the mobile network operator. If not used, there is an opportunity to wiretap the user plane traffic in the transport network, for instance on unsecured microwave backhaul links.

In the 5G system [2], as shown on the right side of Figure 9-4, part of the CN functionality may be placed in edge clouds, for instance for latency-critical services, while some extent of RAN functionality may be centralized, possibly to the same edge clouds. Unlike the P2P abstraction of the 4G transport network, in 5G the complex transport network architecture, which consists of fixed core and metro networks (both typically realized as a logical ring to enhance resilience) and fixed access networks (typically realized using a logical tree), are abstracted as a hierarchical cloud network architecture with several aggregation levels at which clouds are hosted. These aggregation nodes currently evolve into data centers offering compute and storage capabilities that can be dynamically assigned to different network operators and their vertical services in 5G.

Based on a hierarchical transport network, it is possible to implement a comparatively more decentralized CN and more centralized RAN in 5G. More precisely, the 5G network will allow

CN functionality to be shifted closer to the radio, so that the future CN will be more distributed in nature. However, these distributed CN functions need to be controlled by a central CN. Starting from the top, CN clouds will host the routing into the public Internet, together with centralized CN control functions. More towards the edge, another cloud is located in which distributed CN functions can be hosted. Closer to the radio link, e.g. at the central office in the fixed access network, centralized RAN functions can be deployed in so called centralized units (CUs), whereas other RAN functionalities are implemented as distributed units (DUs) at the radio edge. From the security point of view, distributing the CN functionality and centralizing the RAN present advantages and disadvantages. On one hand, the vulnerable surface increases, while on the other hand the impact of successfully attacking one of the system sub-elements is reduced.

The functional split between centralized and decentralized RAN functions has been a matter of debate in 3GPP. One prominent higher-layer split option, the so-called 3GPP split option 2 [3], foresees the centralization of Radio Resource Control (RRC) and PDCP, while the Radio Link Control (RLC), Medium Access Control (MAC) and the physical layer (PHY) are distributed, involving a so-called F1 interface for the new fronthaul between CU and DUs. Note that this and various other RAN split options are discussed in more detail in Section 6.6.2.

The mentioned possibility in 5G to split the RAN into CUs and DUs may in fact inherently increase security for the UP traffic. In particular, if the aforementioned 3GPP split option 2 is chosen, this means that the encryption of the radio link (denoted as ciphering) that happens in the PDCP layer is placed in the CU, with the consequence that the *compound link*, i.e. from the UE over the air to the DU and then eventually via a microwave or other fixed network connection to the CU, is inherently confidentiality-protected. Because the fixed access network provides various opportunities to attack the mobile network, e.g. via unprotected microwave links in 4G, the envisioned larger degree of RAN centralization in 5G inherently closes most of the potential points of attack in the mobile network.

While the transport network can be physically deployed in various topologies (bus, star, ring), logically it is abstracted as a tree with several hierarchical aggregation nodes. The 5G network architecture foresees the availability of clouds (i.e., data centres with compute and cache capabilities) co-located with aggregation nodes where CN and RAN functions can be flexibly instantiated. If cloud resources are shared, this implies that CN and RAN network functions must be encapsulated by a secure transport infrastructure protocol, such as IPSec. Mobile network operators should aim at building secure islands inside their own cloud in each data centre, in which their own VNFs are operated. These own clouds should be isolated from the clouds of other mobie network operators or services using the same transport infrastructure. The only function that needs no further isolation is routing, which is natively safe when using IPSec. Obviously, *VNFs need security encapsulation*. Nevertheless, the cloud infrastructure is a remaining security weakness, because isolation between the tenants is virtual, and VNFs of different tenants can be physically processed in the same machine. At the low processing level, tenants are therefore not physically isolated. One way out is to only use certified cloud hardware in which interactions between the tenants can be considered to be impossible.

From a security point of view, authentication and data integrity should be inherently provided by the network infrastructure. What kind of encryption is used in CN and RAN will be a matter of future debate. Targeting an all-IP infrastructure is desired from simplified CN operation and maintenance (OAM) point of view. But the new 5G architecture forces network designers also to be aware of secure communications at layers 3, 2 and 1. Lower layer techniques may in general be simpler and have less overhead in the UP, but their OAM is more complex. A combination of most suitable techniques is needed, considering for instance physical layer security in the RAN and higher layers towards the CN.

To summarize, a combination of OTT and network-assisted E2E security techniques will be needed, both in the RAN and in the CN, to enable secure sharing of transport network infrastructures in 5G, while guaranteeing low latency and secure E2E communication.

9.4.3 Physical Layer Security

As stressed in the past chapters, 5G networking features both mobile wireless (e.g., via radio links between mobile end-users and the radio unit in the infrastructure) and fixed wireless as well as wired links (e.g., for front- and backhauling), each of which have their own security issues.

The broadcast nature of the wireless domain and the mobility of the users make them susceptible to a wide variety of security attacks such as passive (e.g., traffic analysis and eavesdropping) and active attacks (e.g., denial of service attack, resource consumption, masquerade attack, replay attack, information disclosure, and message modification [4])

Some information theory results have drawn much attention recently. Most works in this area focus on secrecy capacity, that is, the maximum rate achievable between a legitimate transmitter-receiver pair subject to the constraints on information attainable by an unauthorized receiver. Information-theoretic security is an average-information measure and it may require the knowledge of channel state information (CSI) which is not necessarily accurate. The reliability in the exchange of information between a source node "Alice" and an intended destination node "Bob", and security in terms of confidentiality and message integrity with respect to an adversary "Eve" using computational security approaches have been reported to be susceptible to attacks [5][6], and as computationally complex and intensive in a dynamic mobile environment [7][8].

From an implementation perspective, security against passive and active attacks is classified into three categories: i) channel knowledge, ii) coding and power, and iii) signal detection based techniques.

i) *Channel knowledge based techniques*

First, there is the channel characteristic which can be exploited to obtain secure keys. Knowledge about the radio channel is very specific for the position of the devices and appears random (i.e., uncorrelated) already at distances less than half a carrier wavelength apart. Sometimes the key is even reciprocal between a sender and receiver, for instance when using time-division duplex (TDD) at low mobility. The main idea is to derive a time-variant, symmetric key by using the channel as a random number generator. Radio channels exhibit a lot of randomness due to multipath propagation and time-variance, as well as mobility that introduces Doppler effects, as detailed in Chapter 4. Channel knowledge based security approaches can be classified into radio frequency (RF) fingerprinting, algebraic channel decomposition multiplexing (ACDM) pre-coding, randomization of MIMO precoding coefficients, and the introduction of artificial noise. In [9], a method has been proposed in which discriminatory channel estimation is performed by injecting artificial noise to the remaining space of the legitimate receiver's channel to degrade the estimation performance of the eavesdropper. An improvement to this approach is discussed in [10], which exploits the CSI feedback from the legitimate receiver at the beginning of each communication stage, and which is called a multi-stage training-based technique.

ii) *Coding and power based techniques*

Coding is usually considered public, as the encoder and at least one implementable decoder is described in the standard for the radio link. But if the encoder is secret, coding can be used to improve resilience against jamming and eavesdropping. The coding approach is subdivided into the use of error correction and spread spectrum coding techniques. Information protection can also be facilitated using *power* techniques. Usual schemes here also involve the employment of directional antennas and the injection of artificial noise. A directional antenna facilitates receiving data from the direction not covered by the attacking signal.

iii) Signal detection based techniques

Any E2E security approach depends on the algorithms and methods implemented at the endpoints of a data connection, i.e. at the UE, Hence, the end-user can use an own selected algorithm for cipher and integrity protection of its own E2E user connection.

To summarize, beside other techniques from physical layer security, the randomness of the radio channel allows to generate symmetric keys in specific scenarios. This new approach offers additional security, besides the traditional encryption mechanisms in the RAN, and may be interesting for some use cases, e.g., in order to make SIM cards obsolete in billions of devices anticipated in the future wireless Internet of things (IoT).

9.4.4   5G RAN Security

The general requirement for the 5G radio network is to provide at least the same security level as the previous generation network [11]. Having this in mind, the support of traditional security functions provided by the current network, like mutual authentication and key agreement between mobiles and the network, signaling data confidentiality and integrity, user data confidentiality, security visibility and configurability, are not covered in this section. It is assumed that these functionalities will be maintained and supported also in the 5G access network. However, there are also new considerations for 5G radio security design. Most notably, trust models, as discussed in Section 9.4.1, and new aspects such as potentially misbehaving entities and devices should be catered for. This section identifies the new security functionalities which should be offered by the new RAN, considering the new security requirements listed in Section 9.3.

9.4.4.1   Protection Against a Rogue 5G RAN Node

Today, there is an implicit trust relationship between the UE and the access node, which can open the door to rogue access nodes attacks, and which the 5G security architecture should overcome. In particular, the new RAN security should be designed to provide a mechanism that allows a UE to determine the legitimacy of the access node prior to engaging in any communication with it.

It has been demonstrated in [12] that current mobile networks are still vulnerable to protocol exploits, location leaks and rogue base stations. The main issue is that RRC messages that are transferred over common radio channels are transmitted without integrity and ciphering protection. The RRC protocol in LTE includes various functions needed to set up and manage over-the-air connectivity between the eNB and the UE, as covered in Section 13.3. The eNB periodically broadcasts System Information Block (SIB) messages which carry information necessary for UEs to access the network, to perform cell selection, and other information, as covered in detail in Section 13.2.2. Such broadcast messages are neither authenticated nor encrypted. Therefore, anyone owning appropriate equipment can decode them and can exploit these messages to create a targeted denial of service attack to users or to track users' movements by setting up a rogue access point. In general, a UE always scans for eNBs around it with the best signal power. Hence, in International Mobile Subscriber

Identity (IMSI) catcher types of attacks, the rogue eNB operates with a higher power than surrounding eNBs. However, there are situations where a UE, very close to a serving eNB, does not perform scanning to save power. In this case a feature called 'absolute priority based cell reselection' [13] introduced in LTE can be exploited. Based on this feature, a UE, in the Idle state, should periodically monitor and try to connect to eNBs operating at high priority frequencies. In this way, even if the UE is close to a real eNB, a rogue eNB operating on a frequency that has the highest reselection priority would force the UE to attach to it. These priorities are defined in SIB type number 4, 5, 6, and 7 messages broadcast. Using a passive attack setup, it is possible to sniff these priorities and configure a rogue eNB accordingly for malicious purposes, for example to locate the target UE. This is possible by exploiting two other types of RRC messages. A UE, when requested by the network, sends measurement reports to the eNB in RRC protocol messages. In particular, "measurement report" and "UE information-response" messages contain serving and neighboring LTE cell identifiers with their corresponding power measurements and also similar information for GSM and 3G cells. If Global Positioning System (GPS) is supported by the device, the message can also include the GPS location of the UE, and hence also that of the subscriber. Since these messages are not protected during the RRC protocol communication, an attacker can obtain these network measurements by simply decoding the radio signals and then using them to calculate a subscriber's location. These attacks might have critical impacts on the users' privacy, and are possible because of the lack of authentication of the access nodes. The mobile devices do not have the means neither to authenticate nor to validate the messages received from the access node before the authentication phase and the NAS security activation, therefore they inherently and implicitly trust all messages coming from anything that appears to be a legitimate base station.

This issue also applies when a UE is in RRC Idle mode state, as detailed in Section 13.3, and receives services such as paging, Multimedia Broadcast/Multimedia Service (MBMS), device-to-device (D2D) services etc. For example, a UE interested in the D2D service can acquire the broadcasted system information and use the radio resources configured via the system information for the D2D discovery or communication, as detailed in Chapter 14. The lack of a mechanism to verify the authenticity of this information may mislead a UE into selecting and camping on a rogue cell, which can ultimately lead to a denial of service situation, where the UE is denied access to services such as public safety warnings, incoming emergency calls, real-time application server push services, proximity services, etc.

The 5G RAN should therefore consider and provide new security features which allow a UE to determine the authenticity of the cell before camping on it and also during Idle mode. These can probably not leverage the current mobile security architecture, which is based on a symmetric key mechanism (shared key). Such a mechanism is of course suitable and best performing for UE authentication and for deriving keys for traffic protection, but it requires the involved nodes to be identified and authenticated beforehand. Instead, solutions based on a public key architecture could be taken into consideration and evaluated in 5G to counteract the problem of rogue access nodes, since they give the possibility to digitally sign the radio broadcast messages, such as Master Information Block (MIB) and SIB packets, or the sensitive information carried within, and to verify access node authenticity before a UE camps on it. The challenge lies mostly on the management side, since network-private keys are required, while mobile devices have to be provisioned with the corresponding public keys. Such keys of course have to be periodically renewed and securely re-deployed on all elements, e.g., in the case that the confidentiality of a key is compromised. This process has a non-negligible impact on the 5G security architecture and on the practical operation of the network, and hence has to be carefully evaluated.

9.4.4.2 Security Protection of the User Plane

Current 3G and 4G networks only mandate the support of encryption for the user plane data. The encryption can be enabled or not, i.e., an operator can configure the NULL cipher algorithm and in this case no encryption will be performed, based on the countries' regulations. The support for user plane integrity is not required. The reason for this choice is mainly related to the fact that the radio layer-2 utilizes Cyclic Redundancy Checks (CRCs) in the Random Access CHannel (RACH) procedure, such that there cannot be bit-errors.

However, the lack of integrity protection exposes user plane data to man-in-the-middle attacks even when encryption is enabled. It is possible in fact to change the data en route because encryption is linear (i.e., a stream cipher). In addition, an attacker can also inject rogue data into a session with the aim to either increase subscriber bills, or to waste resources carrying the data. These risks are due likely to increase with the increasing support of IoT devices in the 5G era. Even if data integrity can be provided at the transport or application layer (in addition to encryption), there may be cases in which transport or application security conflict with performance constraints, e.g. with respect to latency or battery life, and where a bearer-level integrity provides a useful compromise [14].

This requires that the 5G RAN should be able to ensure the confidentiality as well as the authenticity and integrity for the user plane. This means that ciphering and integrity protection for the UP shall at least be supported by both the access node (i.e., the gNB in 5G) and UE in so that it can be enabled by the network based on the particular usage scenario and on the required level of protection. For example, in a scenario where latency is critical, providing bearer-level protection is better than relying on transport or application security.

The support of UP security also has several security implications based on the point where it is terminated.

Considering the choices done in the previous generation network and from a mere security point of view, it is always better to have the UP protection terminated deeper in the network rather than closer to the edge of the network. One of the main objections against terminating security in the RAN is that the security endpoint would reside in an exposed location requiring additional security measures in the security endpoint and an additional UP security gateway located at the edge of the core network to ensure security protection over backhauling. This is what has happened with 4G where the termination of UP security in the eNB exposes UP traffic to interception and injection on the interfaces between the eNB and the core network (backhauling) when additional measures like IPSec are not implemented.

With the virtualization of RAN entities, more architectural options are possible for the 5G RAN. In particular the possibility in 5G to split the RAN into CUs and DUs, as described in Section 9.4.2, allows to terminate the PDCP layer, where security is performed, in the CU, being a more secure location allowing for UP traffic protection without the need for IPSec.

However, another point that requires security considerations for UP traffic protection are the gNB internal interfaces, as for instance the F1 interface connecting the CU with the DU, as explained in Section 9.4.2., for which integrity, confidentiality and anti-replay protection shall also be provided.

To minimize the susceptibility to attacks, considering that a gNB may in many cases be located in a vulnerable location, also the security procedures internal to the gNB shall be protected to avoid that an attacker may modify the gNB's settings or software configurations via local or remote access. In addition, sensitive data is stored on the gNB like keys, user data or user identifiers, which may be obtained by an attacker. As also done for eNBs in LTE networks, a gNB needs moreover to provide a secure environment, i.e., a logical entity within the gNB that provides a trustworthy environment for the execution of sensitive functions, such as the encryption or decryption of user data or boot processes, and the storage of sensitive data like long-term cryptographic secrets and configuration data. This secure environment shall ensure protection and secrecy of all sensitive information and operations from any unauthorized access or exposure.

9.4.4.3 Protection Against User Plane Denial of Service Attacks

The support of UP security, i.e., data ciphering and integrity, is mandatory in 5G on both the UE and network side. However, this does not ensure that it will be used in practice since its activation should depend on the particular use case considered. For example, there could be cases where it is not enabled because security is provided at an application layer, or because the traffic sent is small and spurious, and not particularly security-sensitive. This can be the case for some IoT use cases

The lack of protection of user plane over the air can expose the security of the network anyway. In the absence of integrity protection being provided by AS security, an attacker can launch DoS attacks on the user plane by flooding the path towards the network node with bogus packets. Though the bogus packets may be identified and filtered at the network node that can verify packets based on the security context of a device, the path towards the network node can still be flooded by bogus packets. This would lead to denial of service or at least throughput degradation caused by congestion for the devices whose traffic shares the same network links as that of the bogus packets.

To counteract these attacks, the possibility should be investigated that 5G network elements embed DoS detection and mitigation functions into the 5G RAN, e.g., via key security indicators and the adoption of dynamic resolution action according to the monitored security indicators. The DoS detection functions would include a set of measurable security indicators, as for example the detection or identification of an anomaly pattern of devices continuously streaming uplink data beyond a certain threshold, or some indicator related to the functions that monitor and detect performance and threshold alarms.

9.4.4.4 Protection Against Signaling DoS Attacks

In 5G, DoS and DDoS attacks originating from a very large number of connected devices, as envisioned for some use cases, will leverage and possibly also target the RAN. Each transaction or traffic flow among the IoT devices, other mobile devices or the Internet results in CP signaling. Unnecessary connection establishment and release signaling could potentially overburden the radio and core network and reduce the quality of service (QoS) experienced by other services.

Several attacks could be carried out by compromising a large number of devices in specific geographical locations. The foreseen increased use in 5G of low-cost machine-to-machine (M2M) devices, characterized by thin operating systems with limited patching capabilities, increases the likelihood that these devices can be tampered and used to coordinate DoS and DDoS attacks against the RAN. These attacks can occur by performing a large number of simultaneous network access attempts in specific geographical locations with the aim to cause a signaling plane overload by exhausting the local radio resources of the network.

Other DoS attacks towards the RAN can occur by coordinating a large number of compromised devices, in specific physical locations, which have already been granted access to the network, to transfer very short data followed by periods of inactivity. This forces a continuous allocation of radio and network resources to support the data transfer, and afterwards their release to allow new connections. These attacks not only deplete radio and network resources needed by new devices trying to establish new network connections, but also have an impact on the signaling plane in terms of processing and computation caused by the continuous allocation and release of the needed radio and network resources.

Current networks support an overload control mechanism that is triggered when an overload situation is detected. This mechanism is not able to distinguish between malicious devices and legitimate devices. Access is instead prevented for all new connections (malicious or not) until the overload has cleared. The effect is that legitimate devices are also not able to access the network. A more sophisticated access control mechanism is required for 5G which is able to recognize via inference data transfer patterns and network access request patterns which are not conformant and selectively targeting only those devices to be disconnected from the network. The use of analytical techniques like anomaly detection should be investigated for such analysis.

9.4.5 Service-level Security

The range of end-user services provided by 5G networks will typically rely on IP and Internet technologies, so the same considerations and mechanisms used to secure traditional web services apply, including, e.g., the use of cryptography or computation-based mechanisms using various types of private and public secret keys, etc.

9.4.6 A Control and Management Framework for Automated Security

Cloud computing, software-defined networks and the fact that security attacks are becoming more automated, reinforces the need for automated management and control of the telecommunication infrastructure. In particular, since a cloud-based paradigm promotes infrastructure that is highly accessible and shared by multiple users (e.g., telco operators), the concept of a highly secure network gains even more relevance. It is therefore of utmost importance to be able to provide robust, flexible and proactive mechanisms for detecting and preventing security issues, and to be able to perform this in real-time and in an automated fashion.

Along these lines, [15] proposes a real-time and automated security framework for the 5G telecommunication network, implementing a continuous and closed loop of real-time environment inspection, analytics, policy-based decisions and actuation/enforcement via cloud and SDN orchestration procedures, as illustrated in Figure 9-5.

Today, we can address a limited number of security threats, namely those that are currently already known to the security community, but we can't foresee the new and ever-changing threats that 5G networks will have to be protected against. However, we do have the basis to create autonomic network management solutions that should cope with them, being fed with insights from governed real-time analytics systems and actuating on network resources in order to minimize or prevent the effects of the detected threats in real-time.

In order to cater for these requirements, [15] proposes a control and management plane as depicted in Figure 9-6. It closely follows the ETSI NFV architecture [16], but introduces additional entities, depicted in grey below, and described in detail in [15].

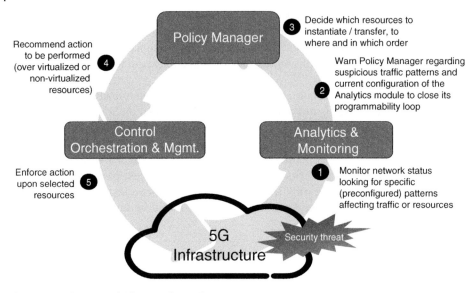

Figure 9-5. Automated 5G network security.

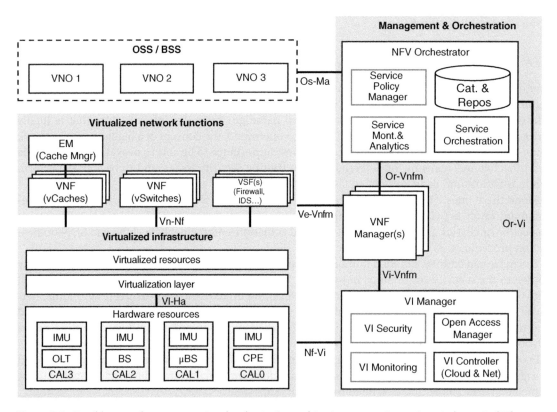

Figure 9-6. Possible control, management and orchestration architecture supporting automated security [15].

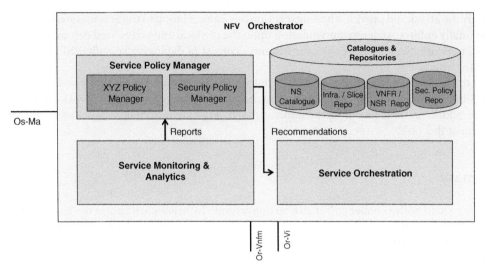

Figure 9-7. Details of NFV orchestrator.

The proposed control, management and orchestration (CMO) architecture for the 5G network consists of four groups of components: virtualized infrastructure (VI), VNFs, management and orchestration (MANO), and operations and business support systems (OSS/BSS).

Inside the MANO component, the NFV orchestrator is of special importance in order to implement the security management features of the 5G network, in particular (see also Figure 9-7):

- **Security Policy Manager (SPM):** This is in charge of making a recommendation about the best action to be taken next, taking as input events triggered by the Service Monitoring and Analytics function. The events are delivered as a result of monitoring and analyzing changes in the status of the resources. The SPM feeds recommendations into the Service Orchestration element which is in charge of enforcing these towards the network's virtualized or physical resources;
- **Service Monitoring & Analytics (SMA) component:** This is responsible for performing metrics and notifications acquisition from: i) the NFVI resources, ii) the VNFs or virtualized security functions (VSFs), and iii) the network's physical infrastructure. The NFVI resources include all physical and virtual compute, storage and network resources such as the compute resources required for the deployment VNFs. The SMA component consolidates the obtained metrics, produces events/ alarms and communicates them to the Security Policy Manager. Based on these metrics, the Security Policy Manager can derive decisions and take actions in communication with the Service Orchestration component to perform changes to the network services that are already deployed, or instantiate and deploy new services;
- **Service Orchestration (Orchestrator):** The main responsibility of the Service Orchestration component is to manage the virtualized network services (NS) lifecycle procedures, according to the recommendations provided by the Security Policy Manager.

Taking these components into account, the SMA should be able to detect an attack and notify the SPM about it. The SPM has been provisioned with the appropriate instructions (i.e., policies) in order to classify the attack and provide a best next action recommendation to the Service Orchestrator, which would finally enforce such recommendation upon the physical and virtualized resources composing the 5G network. Such action could, for instance, consist of deploying specific VSFs in order to neutralize the attack.

The goal of the previously described system is to automate the network security. However, it must be also noted that the control, management and orchestration systems are key network elements which require additional and flexible protection in itself. Policy-controlled network automation should also aim at this task.

9.5 Summary

5G networks represent both a challenge but also an opportunity from a security point of view. Among the challenges, we can mention the introduction of new actors, business models and use cases, which lead to very demanding network capabilities and security requirements. An additional challenge is the increasing number and sophistication of cyberthreats and attacks related to the new macroeconomic and geopolitical scenario.

Among the opportunities, there is the chance to design and build a network with architecturally inherent security and privacy features, donned with unprecedented capabilities for automation and self-adaptation.

References

1 ETSI GS NFV-SEC 001, "Network Functions Virtualisation; NFV Security; Problem Statement", 2014
2 5G PPP Architecture Working Group, White Paper, "View on 5G Architecture", v2.0, July 2017
3 3GPP TR 38.801, "Study on new radio access technology: Radio access architecture and interfaces", V1.1.0, March 2017
4 Y.-S. Shiu, S. Y. Chang, H.-C. Wu, S. C.-H. Huang and H.-H. Chen, "Physical layer security in wireless networks: a tutorial", IEEE Wireless Communications, vol. 18, no. 2, pp. 66–74, April 2011
5 A. Biryukov, A. Shamir and D. Wagner, "Real Time Cryptanalysis of A5/1 on a PC", International Workshop on Fast Software Encryption, Springer, 2000
6 D. Wagner, B. Schneier and J. Kelsey, "Cryptanalysis of the cellular message encryption algorithm", Annual International Cryptology Conference, Springer, 1997
7 C. Adams and S. Lloyd, "Understanding PKI: concepts, standards, and deployment considerations", Addison-Wesley Professional, 2003
8 T. Austin, "PKI: A Wiley Tech Brief", John Wiley & Sons, 2000
9 T.-H. Chang, Y.-W. P. Hong and C.-Y. Chi, "Training signal design for discriminatory channel estimation", IEEE Global Telecommunications Conference (GLOBECOM 2009), Dec. 2009.
10 I. Csiszar and J. Korner, "Broadcast channels with confidential messages", IEEE Transactions on Information Theory, vol. 24, no. 3, pp. 339–348, May 1978
11 3GPP TS 33.401, "3GPP System Architecture Evolution (SAE); Security architecture", V15.1.0, Sept. 2017
12 A. Shaik, R. Borgaonkar, N. Asokan, V. Niemi and J.-P. Seifert, "Practical attacks against privacy and availability in 4G/LTE mobile communication systems", 2015, see http://arxiv.org/abs/1510.07563

13 3GPP TS 36.133, "LTE; Evolved Universal Terrestrial Radio Access (E-UTRA); Requirements for support of radio resource management", V15.0.0, Sept. 2017

14 3GPP TR 33.863, "Study on battery efficient security for very low throughput Machine Type Communication (MTC) devices", V14.2.0, June 2017

15 5G PPP CHARISMA project, Deliverable D3.2, "Initial 5G multi-provider v-security realization: Orchestration and Management", July 2016

16 ETSI GS NFV-MAN 001, "Network Functions Virtualization", v1.1.1, Dec. 2014

10

Network Management and Orchestration

Luis M. Contreras[1], Víctor López[1], Ricard Vilalta[2], Ramon Casellas[2], Raúl Muñoz[2], Wei Jiang[3], Hans Schotten[3], Jose Alcaraz-Calero[4], Qi Wang[4], Balázs Sonkoly[5] and László Toka[5]

[1] *Telefónica Global CTO Unit, Spain*
[2] *Centre Tecnològic de Telecomunicacions de Catalunya (CTTC), Spain*
[3] *German Research Center for Artificial Intelligence (DFKI), Germany*
[4] *University of the West of Scotland, United Kingdom*
[5] *Budapest University of Technology and Economics, Hungary*

10.1 Introduction

This chapter provides an insight into network management and orchestration in the 5th generation (5G), in particular highlighting how software-defined networking (SDN) and network function virtualization (NFV) will enable increased agility, scalability, and faster time-to-market of 5G communication networks.

SDN proposes the decoupling of both the control plane (CP) and user plane (UP), which are commonly integrated nowadays in the network elements (NEs), by logically centralizing the control while leaving the NEs to forward traffic and apply policies according to instructions received from the control side. This permits the network to become programmable in a way that facilitates more flexibility than traditional networks. On the other hand, NFV enables the dynamic instantiation of network functions (NFs) on top of commodity hardware, permitting the separation of the current vertical approach. This vertical approach consists of deploying integrated functional software and hardware for a given NF. Although they have emerged as separate innovative initiatives in the industry, both SDN and NFV are complementary, with the prevalent view in the industry that 'SDN enables NFV'.

Traditional telecommunications networks have been built relying on a diversity of monolithic hardware devices designed and manufactured by distinct vendors. This approach requires complex and static planning and provisioning from the perspective of the service and the network. This static and complex approach on how the network services have been conceived and deployed over the last decades has triggered a continuous process of re-architecting the network, tailoring topologies and capacity for the design and introduction of any new service in the network.

Current telecom networks require a rapid adaptation to forthcoming 5G services and demands, and if there is not an evolution of the conventional management and operation frameworks, it would

5G System Design: Architectural and Functional Considerations and Long Term Research, First Edition.
Edited by Patrick Marsch, Ömer Bulakçı, Olav Queseth and Mauro Boldi.
© 2018 John Wiley & Sons Ltd. Published 2018 by John Wiley & Sons Ltd.

create difficulties to deploy the services fast enough. The carrier networks are usually multi-technology, multi-vendor and multi-layer, which translates into complex procedures for service delivery due to the different adaptations needed for the multiplicity of dimensions. In addition to that, the carrier networks are structured across regional, national and global infrastructures, motivating the need of managing and controlling a large number of physical NEs distributed over a multitude of locations. Furthermore, it is worth noting that the delivery of services implies the involvement of more than one single network domain (e.g., the access to contents not generated by the telecom operator), meaning that the interaction with other administrative domains is also critical.

Having networks built in the classical manner makes it tremendously difficult to cope with customized service creation and rapid delivery in very short times, as is expected to be required in 5G networks. A fundamental requirement identified by network operators' associations, such as Next Generation Mobile Networks (NGMN) [1], for 5G systems is to support flexible and configurable network architectures, adaptable to use cases that involve a wide range of service requirements. It is here where both network programmability and virtualization, leveraging on SDN and NFV, can solve (or at least mitigate) the complexity of the network management and orchestration needs for 5G.

The progressive introduction of both SDN and NFV into operational networks will introduce the necessary dynamicity, automation and multi-domain approach (with the different meanings of technology, network area or administration) to make the deployment of 5G services feasible. The target is to define management and orchestration mechanisms that allow for deploying logical architectures, consisting of virtual functions connected by virtual links, dynamically instantiated on top of programmable infrastructures. Undoubtedly, these new trends will change the telecom industry in many dimensions, including the operational, organizational and business ones [2] that should be carefully taken into account during the process of adoption of these new technologies.

The chapter is structured as follows. Section 10.2 introduces the main concepts of management and orchestration associated to SDN and NFV, with a review of the corresponding architecture frameworks. Section 10.3 profiles the main enablers for achieving the management and orchestration goals of 5G, through open and extensible interfaces, on one hand, and service and device models, on the other. Section 10.4 addresses the complexity derived from multi-domain and multi-technology scenarios. Section 10.5 describes the applicability of SDN to some of the scenarios foreseen in 5G, like the collapsed fronthaul and backhaul (known as Xhaul) and the transport networks. In Section 10.6, the main ideas of the role of NFV in 5G are stated. Section 10.7 provides insights about the autonomic network management capabilities in 5G. Finally, Section 10.8 summarizes the chapter.

10.2 Network Management and Orchestration Through SDN and NFV

The management and orchestration plane has an essential role in the assurance of an efficient utilization of the infrastructure while fulfilling performance and functional requirements of heterogeneous services. Forthcoming 5G networks will rely on a coordinated allocation of cloud (compute, storage and related connectivity) and networking resources. By resource, it can be considered any manageable element with a set of attributes (e.g., in terms of capacity, connectivity, and identifiers), which pertains to either a physical or virtual network (e.g., packet and optical), or to a data center (e.g., compute or storage).

For an effective control and orchestration of resources in both SDN and NFV environments, it is highly necessary to have proper levels of abstraction. The abstraction allows representing an entity in terms of selected characteristics, common to similar resources to be managed and controlled in the same manner, then hiding or summarizing characteristics irrelevant to the selection criteria. Through the abstraction of the resources, it is possible to generalize and to simplify the management of such resources breaking the initial barriers due to differences in the manufacturer, in particular aspects of the technology, or the physical realization of the resource itself.

The orchestration permits an automated arrangement and coordination of complex networking systems, resources and services. For such process, it is needed an inherent intelligence and implicitly autonomic control of all systems, resources and services.

In the case of NFV, orchestration is not formally defined, while, from the definition of the NFV orchestrator (NFVO), it can be assumed that this includes the coordination of the management of network service (NS) lifecycles, virtual network function (VNF) lifecycles, and NFV infrastructure (NFVI) resources, to ensure an optimized allocation of the necessary resources and connectivity. Similarly, for SDN, orchestration can be assumed to correspond to the coordination of a number of interrelated programmable resources, often distributed across a number of subordinate SDN platforms, for instance, per technology.

At the time of delivering a service, it will be needed to apply different levels of orchestration. On one hand, the resources that will be necessary to support a given service should be properly allocated and configured according to the needs of the service to be supported. This is known as *resource orchestration*. A resource orchestrator only deals with resource level abstraction and is not required to understand the service logic delivered by NFs, nor the topology that defines the relation among the NFs that are part of the service.

On the other hand, the *service orchestration* applies to the logic of the service as requested by the customer, identifying the functions needed to fulfill the customer request as well as the form in which these functions interrelate to provide the complete service. The service orchestrator will trigger the instantiation of the NFs in the underlying infrastructure in a dynamic way.

By the right combination of service and resource orchestration, the end-to-end (E2E) management and orchestration functionalities will be responsible for a flexible mapping of services to topologies of NFs, based on a dynamic allocation of resources to NFs and the reconfiguration of NFs according to changing service demands.

The next sub-sections generally introduce SDN and NFV frameworks in more detail.

10.2.1 Software-Defined Networking

While networks are based on distributed CP solutions, there is a huge interest around SDN orchestration mechanisms that enable not only the separation of UP and CP, but also the automation of the management and service deployment process. Current SDN approaches are mainly focused on single-domain and single-vendor scenarios (e.g., data centers). However, there is a need of SDN architectures for heterogeneous networks with different technologies (e.g., IP, MPLS, Ethernet, and optical), which are extended to cover multi-domain scenarios.

The SDN architecture, as defined by the Open Networking Foundation (ONF) in [3], is composed of an application layer, a control layer, and an infrastructure layer, as depicted in Figure 10-1. User or provider-controlled applications communicate with the SDN controller via an Application-Controller Plane Interface (A-CPI), also known as *northbound interface* (NBI). The controller is in charge of

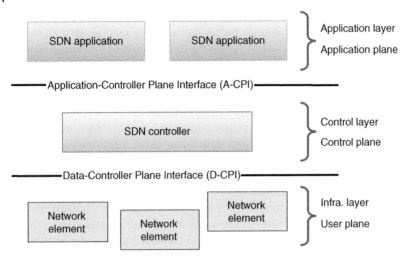

Figure 10-1. Abstract view of basic SDN components.

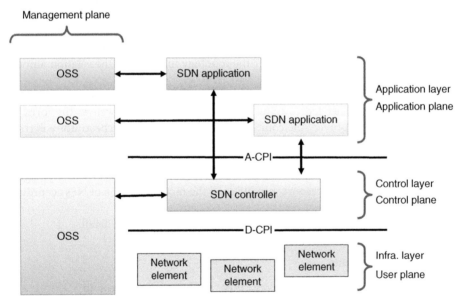

Figure 10-2. Abstract SDN architecture overview.

orchestrating the access of the applications to the physical infrastructure (i.e., the NEs), using a Data-Controller Plane Interface (D-CPI), also known as *southbound interface* (SBI).

Figure 10-2 presents a more descriptive view of a typical SDN architecture, where a management plane is also included, to carry out tasks such as registration, authentication, service discovery, equipment inventory, fault isolation, etc. In addition, Figure 10-3 shows the situation where the infrastructure owner gives away control of part of its infrastructure to a number of external entities.

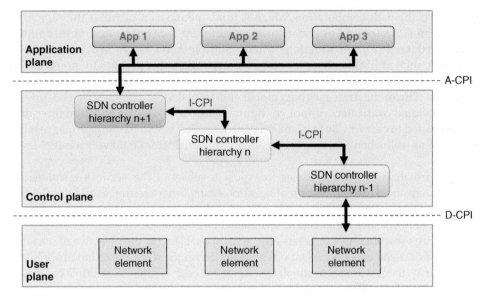

Figure 10-3. Recursive hierarchical SDN architecture.

This is relevant to scenarios where a network provider gives controlled access to equipment (or a slice of equipment through virtualization mechanisms) to some other service providers.

ONF also describes the possibilities of implementing hierarchical controllers, primarily for scalability, modularity or security reasons. Such hierarchical control structure introduces a new interface, the Intermediate-Controller Plane Interface (I-CPI), as shown in Figure 10-3. This hierarchical structure allows for recursiveness and to assure scalability, while maintaining the control of each domain in separate controllers.

In terms of functionalities, there are four main capabilities in this kind of interfaces enabling the flexible control and orchestration of different resources. Such capabilities are: (1) network topology extraction and composition, (2) connectivity service management, (3) path computation, and (4) network virtualization.

The need of network topology extraction and composition is to export the topological information with unique identifiers. Such network identifiers (such as IPv4 addresses or datapath-IDs) are required for the other functionalities. To compose the topology, it is required to export the nodes and the links in a given domain, which can be physical or virtual, as well as some parameters like the link utilization or even information about physical characteristics of the link if the operator requires the deployment of very detailed services.

The second functionality is to manage connectivity services. The operations on these services are the setup, tear down, and the modification of connections. Such services can be as basic as a point-to-point connection between two locations. Nonetheless, there are scenarios where the orchestration requires more sophistication like *(a)* exclusion or inclusion of nodes and links, *(b)* definition of the protection level, *(c)* definition of traffic-engineering (TE) parameters, like delay or bandwidth, or *(d)* definition of disjointness from another connection.

The third function is the path computation, which is fundamental as it provides the capability of properly defining an E2E service. For instance, when different controllers in a multi-domain environment are considered (e.g., in situations where multiple network segments are under a single administration, such as backhaul, metro and core networks), this permits to interact with individual controllers in each domain that are only able to share abstracted information that is local to their domain. The orchestrator with its global end-to-end view can improve end-to-end connections that individual controllers cannot configure. Without a path computation interface, the orchestrator is limited to carrying out a crank-back process that would not find proper results. This can be exploited as well when multiple technologies are considered, following a multi-layer decision approach.

Lastly, a network virtualization service allows to expose a subset of the network resources to different tenants. This advances in the direction of network slicing, where resources and capabilities of the underlying physical transport network can be offered to different users or tenants to appear as dedicated in its global network slice composition, as detailed in Chapter 8.

The ONF architecture presented here illustrates the general enablers for the objective of network programming. However, several other organizations are working on the standardization of NBIs and SBIs. In terms of maturity, there is not yet a complete solution for each model, but multiple candidate technologies for some interfaces. This is commented later on in this chapter.

10.2.2 Network Function Virtualization

European Telecommunications Standards Institute (ETSI) NFV is the most relevant standardization initiative arisen in the NF virtualization arena. It was incepted at the end of 2012 by a group of top telecommunication operators, and has rapidly grown up to incorporating other operators, network vendors, information and communications technology (ICT) vendors, and service providers. To date, the ETSI NFV industry specification group (ISG) can count on over 270 member companies. It represents a significant case of collaboration among heterogeneous and complementary kinds of expertise, in order to seek a common foundation for the multi-facet challenges related to NFV towards a solution as open and scalable as possible.

The ETSI NFV roadmap initially foresaw two major phases: The first one was completed at the end of 2014, where a number of specification documents were issued [4], covering functional specification, data models, proof-of-concept (PoC) description, etc. The second phase released a new version of the ETSI NFV specification documents. A third phase is ongoing at the time of writing, progressing the work on architectural and evolutionary aspects. The work of the ISG is further articulated into dedicated working groups (WGs). In phase 1, three WGs have been created, dealing with NFVI, Management and Orchestration (MANO), and Software Architectures (SWA). In phase 2, two additional WGs were spawned, dealing with Interfaces and Architecture (IFA) and Evolution and Ecosystem (EVE).

The currently acting specification of the ETSI NFV architecture was finalized in December 2014 [5], and its high-level picture is shown in Figure 10-4.

The ETSI NFV specification defines the functional characteristics of each module, their respective interfaces, and the underlying data model. The data model is basically made up by static and dynamic descriptors for both virtual network functions (VNFs) and network services (NSs). The latter are defined as compositions of individual VNFs, interconnected by a specified network forwarding graph, and wrapped inside a service.

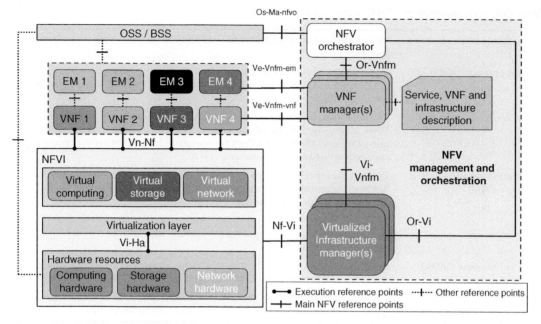

Figure 10-4. ETSI NFV architecture [5].

The ETSI NFV framework specifies the architectural characteristics common to all the VNFs. It does, though, not rule out which specific network functions can or should be virtualized, leaving this decision up to the NF provider (apart from the use cases advised for the proofs of concept).

The ETSI NFV architecture supports multi-point-of-presence (PoP) configurations, where a PoP is defined as the physical location where a NF is instantiated. A PoP can be mapped to a data center or a data center segmentation isolated from the rest of world.

A summary description of the modules in the ETSI NFV architecture is given in Table 10-1.

As it can be observed in Figure 10-4, the ETSI NFV framework assumes the existence of an outside operations support system (OSS) or business support system (BSS) layer in charge of the basic equipment and service management.

It is worth to mention that starting 2016, ETSI has launched the Open Source Mano (OSM) initiative [6]. OSM intends to develop an open-source NFV Management and Orchestration (MANO) software stack aligned with ETSI NFV specifications. This kind of open-source software initiative can facilitate the implementation of NFV architectures aligned to ETSI NFV specifications, increasing and ensuring the interoperability among NFV implementations.

10.3   Enablers of Management and Orchestration

The management and orchestration capabilities offered by SDN and NFV should be sustained by some enablers from the resource and service perspective. On one hand, there is a need for open and standard interfaces that could permit at the same time aspects like: (1) a uniform and homogeneous

Table 10-1. Components of the ETSI NFV framework.

Virtualized network function (VNF)	Virtualized instance of an NF traditionally implemented on a physical network appliance.
Element management (EM)	Component performing the typical network management functions (Fault, Configuration, Accounting, Performance and Security - FCAPS) requested by the running VNFs.
NFV infrastructure (NFVI)	Totality of hardware/software components building up the environment in which VNFs are deployed, managed and executed. Can span across several locations (physical places where NFVI-PoPs are operated). Include the network providing connectivity between such locations.
Virtualized infrastructure manager (VIM)	Provides the functionalities to control and manage the interaction of a VNF with hardware resources under its authority, as well as their virtualization. Typical examples are cloud platforms (e.g., OpenStack) and SDN controllers (e.g., OpenDaylight).
Resources	Physical resources (e.g., computing, storage, and network). Virtualization layer.
NFV orchestrator (NFVO)	Component in charge of orchestration and management of NFVI and software resources, and provisioning of network services on the NFVI.
VNF manager	Component responsible for VNF lifecycle management (e.g., instantiation, update, query, scaling, and termination). Can be 1-1 or 1-multi with VNFs.

access to the resources and services; and (2) an easy integration with supporting systems like OSS and BSS. On the other hand, a set of information and data models that could help to easily and flexible define, configure, manage and operate services and network elements in a consistent and abstract way.

10.3.1 Open and Standardized Interfaces

Through the existence of controllers allowing the programmability of the network, the operational goal is to facilitate the creation and definition of new services to be configured in the network and automatically, via OSS or directly by means of the interaction with tailored applications. The SDN controller will in this context take care of performing all the tasks needed to set up the configuration in the network (i.e., calculate the route from source to destination, check the resource availability, set up the configuration to apply in the equipment, etc.). For example, the inventory system can be better synchronized with the network, so that the provisioning can be done based on the real status of the network, avoiding any misalignment between the planning process and the deployment process.

One of the expected benefits of SDN is to speed-up the process to integrate a new vendor, or a new OSS system or application in the network. To do so, it is necessary to have standard NBI interfaces towards the OSS systems (e.g., network planning tools, inventory data bases, and configuration tools), and standard SBI interfaces towards the network element that depend only on the technology (e.g., microwave wireless transport, Metro-Ethernet/IP, or optical) and not on the vendor.

Nowadays, even for a single transport technology, the particularities per vendor implementation force a constant customization of the service constructs. This affects not only the provisioning phase, but also the operation and maintenance of the services. Activation tools (as part of current OSS and BSS) are in some cases present, being in charge of the automated configuration of network services. However, the configuration is provided by vendor-dependent interfaces, and when a service needs to be extended by configuring different network segments, the configuration process needs to be done in each network separately, and usually by means of specific or dedicated systems. For the same reason, integrating a new vendor or new equipment (or a new release of an existing vendor or equipment) is time-consuming, and needs upgrades of the interfaces and changes in the OSS tools already deployed. It delays the introduction of new technologies, de facto blocking the transformation process towards 5G with the agility and flexibility needed by the operator. All of this renders the adoption of open and extensible interfaces, for both NBI and SBI, necessary.

Currently, there is no real progress about the definition of NBIs from the orchestrator perspective that could facilitate the smooth integration referred to before with respect to OSS and BSS. All the available NBIs are platform-dependent; in consequence, there is not a common or general approach in the industry by now. However, for the SBI there is some consensus.

For the programmability and management of the network, both NETCONF and YANG, as introduced in the following, are being recognized as future-proof options.

NETCONF [7] provides a number of powerful capabilities for a uniform configuration and management of network elements. It is transport protocol independent, meaning that it does not impose restrictions for getting access towards the devices. With NETCONF, it is possible to have a separation of the configuration data from the operational data in such a way that the administrator can set some variables from features like statistics, alarms, notifications, etc. In addition to that, thanks to the support of transactional operations, it is possible to ensure the completion of configuration tasks even on a network basis. Since NETCONF supports an automated ordering of operations, the sequential actions on the network can be defined, facilitating straightforward rollback operations if needed. NETCONF is then foreseen as the manner of managing and orchestrating multi-vendor infrastructures. However, NETCONF only defines the mechanisms to access and configure the network elements, but not the configuration information to be applied.

In this sense, YANG [8], as a data modeling language, complements NETCONF by defining the way in which the information applicable to a node can be read and written. It provides well-defined abstractions of the network resources that can be configured or manipulated by a network administrator, including both devices and services. The YANG language simplifies the configuration management as it supports capabilities like the validation of the input data, and data model elements are grouped and can be used in a transaction, etc. Nowadays, there is an intensive work in the definition of general and standard YANG models especially in the Internet Engineering Task Force (IETF), but not only. Figure 10-5 presents the evolution in the number of YANG models being proposed.

Similar to NETCONF, the RESTCONF protocol [9] provides a programmatic interface for create, read, update and delete (CRUD) operations accessing data defined in YANG based on Hypertext Transfer Protocol (HTTP) transactions, allowing Web-based applications to access the configuration data, state data, data-model-specific remote procedure call (RPC) operations, and event notifications within a networking device, in a modular and extensible manner. The purpose is then similar to the one described for NETCONF.

Regarding the orchestration of services and the management of VNF lifecycles, Topology and Orchestration Specification for Cloud Applications (TOSCA) emerges as the more solid option. ETSI NFV ISG is considering it as a description language, and recently started the specification of

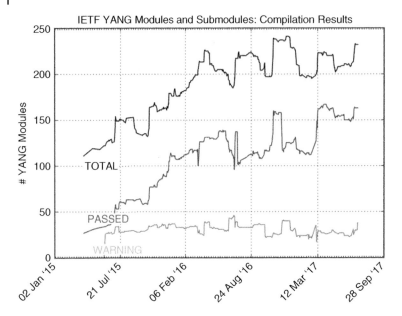

Figure 10-5. Development of YANG modules in IETF [10].

TOSCA-based descriptors [11], not yet being released at the time of writing. Nevertheless, a TOSCA template [12] is available, which is specifically designed to support describing both NS descriptors (NSDs) and VNF descriptors (VNFDs).

TOSCA is a service-oriented description language to describe a topology of cloud-based web services, their components, relationships, and the processes that manage them, all by the usage of templates. TOSCA covers the complete spectrum of service configurations, like resource requirements and VNF lifecycle management, including the definition of workflows and FCAPS management of VNFs. By this way, an orchestration engine can invoke the implementation of a given behavior when instantiating a service template.

A topology template defines the structure of a service as a set of node templates and relationships that together define the topology model as a (not necessarily connected) directed graph. Node and relationship templates specify the properties and the operations (via interfaces) available to manipulate components. The orchestrator will interpret the relationship template to derive the order in which the components of the service should be instantiated. TOSCA templates could also be used for later lifecycle management operations like scaling or software update.

From the point of view of communication method, TOSCA uses a simple REST application programming interface (API).

NETCONF/YANG and TOSCA can complement each other. Basically, the lifecycle management of the VNFs can be performed by means of TOSCA, while the VNFs can be dynamically configured at runtime by means of NETCONF/YANG. This interplay is facilitated by architectural propositions like the integrated SDN control for tenant-oriented and infrastructure-oriented actions in the framework of NFV, as described in [13]. Figure 10-6 shows the positioning of the two different levels of SDN control.

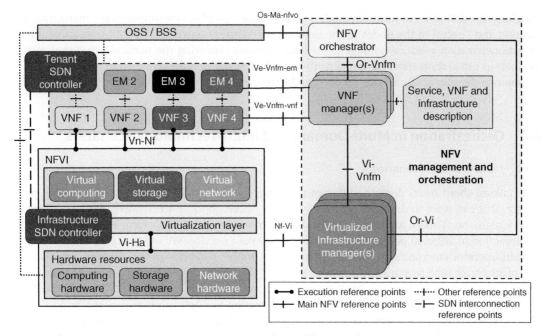

Figure 10-6. Infrastructure and tenant SDN controllers in the NFV architecture.

The SDN controller in the tenant domain can configure on-demand of NETCONF/YANG the functionality of the VNFs deployed by using TOSCA.

Furthermore, this architecture facilitates the integration of control and orchestration actions with a SDN controller at the infrastructure level for coordinating actions allowing cross-layer coordination. Both controllers manage and control their underlying resources via programmable southbound interfaces, each of them providing a different, but complementary, level of abstraction. This concept is leveraged from [14].

10.3.2 Modeling of Services and Devices

The same need for standardization as highlighted before would be also necessary for services and devices. By expressing a service to be deployed in a standard manner, it is possible to make it independent or agnostic of the actual underlying technology in which it is engineered. This provides more degrees of freedom for the decisions about how to implement a given service, and also allows for portability of such service across platforms.

Via those models, a unique entity can process all the service requests, later on triggering actions in the network for service delivery and deployment. Such an entity can be seen as a service orchestrator, which can maintain a common view across all the services deployed, instead of the legacy approach of siloed services, which renders a combined planning difficult. With such service orchestrator, dependencies can be detected in advance, allowing to improve the design by means of a coordinated usage of resources.

Similarly, the definition of common models for the same type of device simplifies the management, operation and control of the nodes in the network. A common representation of node capabilities and parametrization produce homogeneous environments removing the particularities that motivate onerous integration efforts as happens today to handle per-vendor specificities.

A generic reference about service models can be found in [15] and [16].

10.4   Orchestration in Multi-Domain and Multi-Technology Scenarios

10.4.1   Multi-Domain Scenarios

When talking about multi-domain, different meanings can be associated to the term *domain*. For instance, this can refer to different technologies, like packet, optical, microwave, etc., or different network segments. Finally, multi-domain can be understood as a multi-operator environment, with the interaction of different players for the E2E provision of a service. We use the term multi-domain for multi-operator environments and multiple administrative scenarios in this section. The importance of analyzing such scenarios was firstly raised in [17].

5G is expected to operate in highly heterogeneous environments using multiple types of access technologies, leveraging multi-layer capabilities, supporting multiple kinds of devices, and serving different types of users. The great challenge is to port these ideas to the multi-domain case, where the infrastructure (considered as network, computing, and storage resources), or even some of the necessary network functions, are provided by different players, each of them constituting a separate administrative domain.

Multi-operator orchestration requires the implementation of an E2E orchestration plane able to deal with the interaction of multiple administrative domains (i.e., different service and/or infrastructure providers) at different levels, providing both resource orchestration and service orchestration. An example would be the case of service providers offering their NFVI PoPs to host service functions of other providers, or even offering VNFs to be consumed by other service providers. However, existing interconnection approaches are insufficient to address the complexity of deploying full services across administrative domains. For instance, evolved interconnection services demanding, e.g., computing capabilities for the deployment of network building blocks as VNFs, or even inserting VNFs in the UP, cannot be satisfied with existing solutions for multi-domain environments.

An inter-provider environment imposes additional needs to be offered and served between providers like service-level agreement (SLA) negotiation and enforcement, service mapping mechanisms (in order to assign proper sliced resources to the service instance), reporting of assigned resource and service metrics, and allocation of proper control and management interfaces, to mention a few.

From the architecture perspective, an orchestration approach assuming a hierarchical top-level orchestrator playing the role of broker, with total visibility of the all providers' networks, and with the capability of orchestrating services across domains is certainly impractical, due to issues like scalability, trustiness between providers, responsibilities, etc. Instead, a peer-to-peer architecture seems to be more adequate for this kind of scenarios, as it already exists nowadays in the form of the pure interconnection for IP transit and peering.

From the point of view of SDN architecture, a primary approach to such a peer-to-peer relationship is provided by ONF in [18], which introduces an initial idea about the interaction of peer controllers,

Figure 10-7. Peer controllers in the ONF architecture.

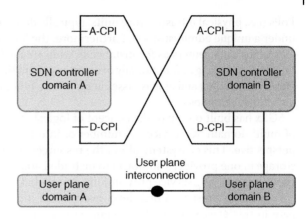

as reflected in Figure 10-7. Here, basically, each of the controllers may act as client to invoke services from the other as server, whereby A-CPI is the Application-Controller Plane Interface, and D-CPI the Data-Controller Plane Interface. The relationship among controllers is then proposed to be equivalent to a hierarchical provider/customer relationship.

For more complex orchestration scenarios, involving the provision of NFV-related services across providers, some other initiatives are in progress. To this respect, ETSI has produced a report on the description of architectural options for multi-domain [19], taking as basis for the analysis some use cases like NFVI-as-a-Service.

The Metro Ethernet Forum (MEF) Lifecycle Service Orchestration (LSO) is another initiative in the standardization arena, with a reference architecture defined in [20]. The MEF LSO architecture oversees the E2E orchestration of services where all the network domains require coordinated management and control. A shared information model for connectivity services is under definition, including the service attributes defined in MEF service specifications. Specifically, two inter-provider reference points are being proposed:

- **LSO Sonata**, which facilitates the interconnection of the BSS functions of different providers, addressing the business interactions between those providers. This includes aspects such as ordering, billing, trouble ticketing, etc.;
- **LSO Interlude**, which instead facilitates the interconnection of the OSS functions of different providers. Interlude supports control-related management interactions between two service providers and is responsible for the creation and configuration of connectivity services as permitted by service policies. It also covers notifications and queries on the operational state of services and their performance.

Co-operation between providers then takes place at the higher level, based on exchanging information, functions and control. These interfaces serve for the business-to-business and operations-to-operations relations between providers.

In addition, the 5G PPP 5G-Exchange (5GEx) project [21] has developed a multi-domain orchestration framework enabling the trading of NFs and resources in a multi-provider environment, and targeting a Slice-as-a-Service (SaaS) approach. The envisioned 5G service model is an evolution of the ETSI NFV model, proposing extensions to it. The original NFV paradigm foresees that resources used inside a service (for instance, for different VNF components) can be distributed over distinct

PoPs (i.e., physical infrastructure units, typically data centres). However, the PoPs are supposed to be under a unique administration. Furthermore, the level of control is quite limited outside the perimeter of the data centers as for instance in wide area networks (WANs). The project addressed these limitations, aiming at functionally overcoming them (i.e., enabling the integration of multiple administrative domains) and at least assessing the non-functional enablers needed to make actual business out of the technology.

5GEx has built on top of the concept of logical exchange for a global and automatic orchestration of multi-domain 5G services. A number of interfaces implement such kind of exchange for the CP perspective. This ecosystem allows the resources such as networking, connectivity, computing and storage in one provider's authority to be traded among federated providers using this exchange concept, thus enabling service provisioning on a global basis.

Figure 10-8 presents a high-level overview of the 5GEx architecture. Different providers participate in this ecosystem, each of them representing a distinct administrative domain interworking through multi-domain orchestrators (MdOs) for the provision of services in a multi-provider environment. This architecture extends the ETSI MANO NFV management and orchestration framework for facilitating the orchestration of services across multiple administrative domains. Each MdO handles the orchestration of resources and services from different providers, coordinating resource and/or service orchestration at multi-provider level, and orchestrating resources and/or services using domain orchestrators belonging to each of the multiple administrative domains.

The domain orchestrators are responsible of performing virtualization service orchestration and/or resource orchestration exploiting the abstractions exposed by the underlying resource domains that cover a variety of technologies hosting the actual resources.

There are three main interworking interfaces and APIs identified in the 5GEx architecture framework. The MdO exposes service specification APIs that allow business customers to specify their requirements for a service on business-to-customer (B2C) interface I1. The MdO interacts with other MdOs via business-to-business (B2B) interface I2 to request and orchestrate resources and services across administrative domains. Finally, the MdO interacts with domain orchestrators via interface I3 APIs to orchestrate resources and services within the same administrative domains.

Figure 10-9 presents the functional detail of the proposed architecture, showing different components identified as necessary for multi-domain service provision. In this case, all the providers are considered to contain the same components and modules, although in Figure 10-9 the complete view is only shown for the provider on the left for simplicity.

We briefly describe some of the components in the figure, particular to 5GEx.

- The **inter-provider NFVO** is the NFVO implements multi-provider service decomposition, responsible of performing the end-to-end network service orchestration. The network services operator (NSO) and resource orchestration (RO) capabilities are contained here;
- The **topology abstraction** module performs topology abstraction, elaborating the information stored in the resource repository and topology distribution modules;
- The **topology distribution** module exchanges topology information with its peer MdOs;
- The **resource repository** that keeps an abstracted view of the resources at the disposal of each one of the domains reachable by the MdO;
- The **SLA manager** is responsible for reporting on the performance of its own partial service graph (its piece of the multi-domain service);
- The **policy database** which contains policy information;

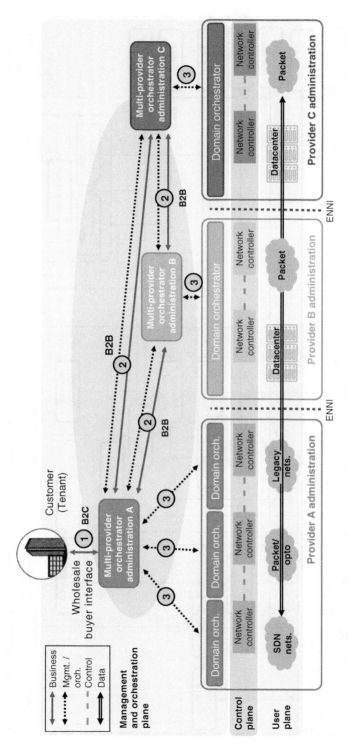

Figure 10-8. 5GEx reference architectural framework [21].

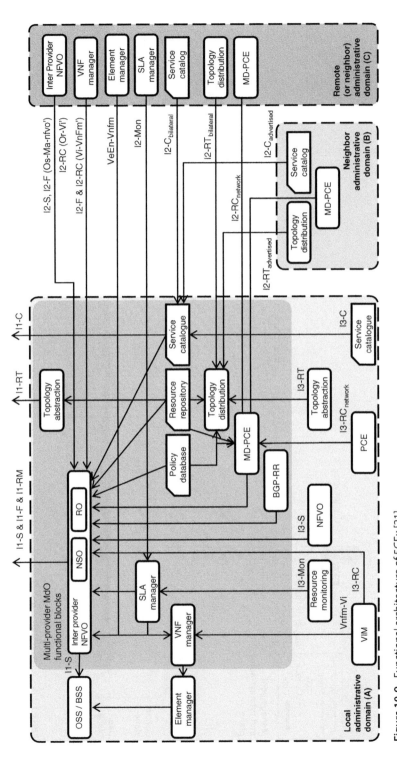

Figure 10-9. Functional architecture of 5GEx [21].

- The **resource monitoring** module dynamically instantiates monitoring probes on the resources of each technological domain involved in the implementation of a given service instance;
- The **service catalog** in charge of exposing available services to customers and to other MdOs from other providers;
- The **MD-PCE** (multi-domain path computation element) devoted to make the necessary path computations and to set up the connection between domains.

From the interfaces perspective, the functional split considered is related to service management (-S functionality), VNF lifecycle management (-F), catalogs (-C), resource topology (-RT), resource control (-RC) and monitoring (-Mon). Table 10-2 summarizes the functional needs for the mentioned interfaces as well as potential candidate solutions for their implementation. At the time of writing, the identification and specification of these interfaces is currently being defined and will be fully described in future deliverables of the project.

Figure 10-9 shows the interconnection of MdOs for three different domains. As mentioned, the left MdO is shown with full details, while the other two are skipped for simplicity. The 5GEx interfaces are presented with the corresponding functional split. The interfaces have to be considered as symmetric, since the consumer-provider role is situational in an exchange.

The left MdO is the entry point for the service request coming from the customer, through the I1 interface. Using I1-C, I1-S and I1-F, the customer (e.g., an infotainment company) will be able to request VNF instantiation and configuration, apart from expressing the way in which they are interconnected by means of a service graph.

The service will be decomposed by the NFVO of the provider A. If the service cannot be honoured by the sole use of its own resources, the NFVO will make use of resources offered by other providers in the exchange. The availability of resources from other parties is collected via I2-RT, and the availability of services offered by such parties is obtained through I2-C. Once the decision about using resources from other providers is taken, the left MdO will make use of I2-S and I2-RC for requesting and controlling the necessary resources and services. The same MdO will make use of the I3 interface for governing the own resources accordingly, in a similar manner.

In order to accomplish the negotiated SLA between the parties (i.e., both the customer and the entry provider, and the providers participating in the E2E service provision), convenient monitoring capabilities are deployed, using I1-Mon, I2-Mon and I3-Mon for the respective capabilities.

Table 10-2. Functional split of 5GEx interfaces and candidate solutions [21].

	Functional split	I1 (Customer to provider)	I2 (Inter-provider)	I3 (Intra-provider)	Candidate solutions
-S	Service management	●	●	●	TOSCA, YANG
-F	VNF lifecycle management		●	●	TOSCA
-C	Catalogs	●	●	●	Network service descriptors, TMForum
-RT	Resource topology	●	●	●	BGP-LS
-RC	Resource control		●	●	NETCONF, PCEP
-Mon	Monitoring	●	●	●	Lattice, time series data

As a reference of the different roles in the exchange, note that the provider B in Figure 10-9 (i.e., the one in the middle) participates on the E2E service only for providing UP connectivity between providers A and C.

10.4.2   Multi-Technology Scenarios

Nowadays, the automatic establishment of E2E connections is complex in a network composed of heterogeneous technological domains (that is, domains constituted by specific technologies like IP, optics, microwave, etc.). The complete process not only requires long time and high operational costs for configuration (including manual interventions), but also the adaptation to each particular technology implementation. The capability to operate and manage the network automatically and E2E is the main requirement for multi-technology scenarios. This facilitates as well the multi-vendor interworking, which is another dimension of the multi-technology issue, as already described in Section 10.2.1 in the context of the relevance of SDN. The target is to move towards a service-driven configuration management scheme that facilitates and improves the completion of configuration tasks by using global configuration procedures.

Typically, the transport network is referred to as a wide area network (WAN) in the ETSI NFV model, regardless of the complexity and diversity of the underlying infrastructure. The idea of the ETSI model is that the service orchestrator can easily interact with control capabilities that could permit the configuration and manipulations of the WAN resources to create E2E services without considering the transport domains' heterogeneity. However, this is yet far from existing capabilities and solutions.

Network operators have built their production networks based on multi-layer architectures. However, the different technologies in current transport networks are rarely jointly operated and optimized, i.e., the implications of a planning and configuration decision for different layers at the same time are typically not considered. Instead, they are usually conceived as isolated silos from a deployment and operation point of view.

This can be even more burdening across multiple domains as described before. A service deployed across domains will require actions in different networks using different technologies, inherently multiplying the intricate complexity of the E2E network provision and configuration.

A logically centralized orchestration element can have a complete and comprehensive network view independently of the technologies employed in each technological domain, and propose optimal solutions to improve the overall resource utilization. Such orchestrator, by maintaining a multi-layer view of the controlled network, can determine which resources are available to serve any connectivity request in an optimal manner, considering not only partial information (per technology domain), but the entire network resources, in a comprehensive manner. Aspects like global utilization, protection, congestion avoidance or energy saving can be optimized with such an approach. For getting the information per technology and building the multi-layer view (i.e., underlying topology, per-layer capabilities, border ports, etc.), the orchestrator could rely on lower-level controllers, for instance one per layer. In [22], an overview of the benefits obtained through a multi-layer approach is provided.

Network programmability, as enabled by SDN and already touched with relation to the radio access network (RAN) in Section 6.8, permits new ways of resource optimization by implementing sophisticated traffic engineering algorithms that go beyond the capabilities of contemporary distributed shortest path routing. Multi-layer coordination can help to rationalize the usage of technologically diverse resources. This new way of planning and operating networks requires a comprehensive view of the network resources and planning tools capable for handling this multi-layer problem.

10.5   Software-Defined Networking for 5G

5G will impose the need of a flexible network to support the diverse requirements of the distinct services and customers (i.e., verticals) on top of the providers' networks. This section introduces two particular scenarios for fronthaul, backhaul and core transport networks as examples of network segments out of the RAN also impacted by the advent of 5G. Note that SDN approaches for the RAN are covered in detail in Section 6.8.

10.5.1   Xhaul Software-Defined Networking

10.5.1.1   Introduction

The integration of fronthaul and backhaul technologies, also known as *Xhaul* and covered in more detail in Section 7.6.1, will enable the use of heterogeneous transport and technological platforms, leveraging novel and traditional technologies to increase the capacity or coverage of the future 5G networks.

The design of the Xhaul segment is driven by the detailed extracted requirements obtained from practical use cases with a clear economical target. A large number of use cases are proposed in literature, as covered in Chapter 2.

From the SDN perspective, the diversity and heterogeneity of the relevant technologies involved in the Xhaul segment means that using a single controller may not be applicable. This might be due to the need for controlling heterogeneous emerging technologies such as millimeter-wave (mmWave), while controlling a photonic mesh network. Thus, a hierarchical approach is typically proposed in order to tackle with this technological heterogeneity [23][24].

10.5.1.2   Possible Hierarchical SDN Controller Approaches for Xhaul

A possible solution to manage and control such diversity of heterogeneous technologies is to focus on a deployment model in which a SDN controller is deployed for a given technology domain (considering it as a child controller), while the whole system is orchestrated by a parent controller, relying on the main concept of network abstraction [25].

The proposed SDN architecture by ONF foresees the introduction of different levels of hierarchy, allowing for network resource abstraction and control. A level is understood as a stratum of hierarchical SDN abstraction. In the past, the need of hierarchical SDN orchestration has been justified by two purposes: a) Scaling and modularity: each successively higher level has the potential for greater abstraction and broader scope (e.g., RAN, transport, and data center network abstraction); and b) Security: each level may exist in a different trust domain, where the level interface might be used as a standard reference point for inter-domain security enforcement. The benefits of hierarchical SDN orchestration become clear in the scope of the described Xhaul with technology heterogeneity.

The Applications-Based Network Operations (ABNO) framework has been standardized by the IETF, based on standard protocols and components to efficiently provide a solution to the network orchestration of different CP technologies. An ABNO-based network orchestrator has been validated for E2E multi-layer and multi-domain provisioning across heterogeneous control domains employing dynamic domain abstraction based on virtual node aggregation [26].

Figure 10-10 shows the proposed hierarchical architecture for a future Xhaul network. It takes into account the different network segments and network technologies which are expected to be present.

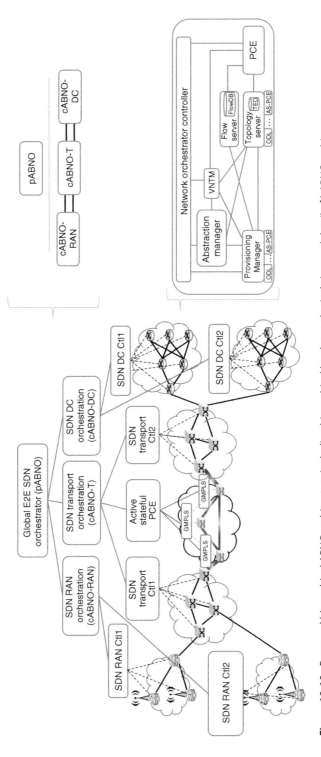

Figure 10-10. Proposed hierarchical ABNO architecture including hierarchical levels topological view and detail of hABNO architecture.

In the RAN segment, we observe several SDN-enabled controllers for wireless networks, which tackle their complexities. In a transport network, the aggregation segments and core network are taken into account. SDN-enabled Multiprotocol Label Switching - Transport Profile (MPLS-TP) can be used in the aggregation network, while a core network might use an optical SDN controller, such as an active stateful path computation element (AS-PCE) on top of an optical network. Finally, several SDN-enabled controllers are responsible for intra-data center networks, which typically run at layer-2.

Within the hierarchy, an SDN orchestrator may consider itself as the direct control entity of an information model instance that represents a suitably abstracted underlying network. It follows that, with the exception of network domain SDN controllers (which are directly related to NEs), a given SDN orchestrator might provide an abstracted network view and be present at any hierarchy level and act as parent or child SDN orchestrator. At any level of the recursive hierarchy, a resource is understood to be subject to only one controlling entity.

In the proposed architecture, several child ABNOs (cABNO) are proposed. Each cABNO is responsible for a single network segment. A recursive hierarchy could be based on technological, SDN controller type, geographical/administrative domains or network segment basis (each corresponding to a certain hierarchical level). Further, parent ABNO (pABNO) are introduced, responsible for the provisioning of E2E connections through different network segments.

For both the pABNO and the cABNO, the internal system architecture is similar, based on a set of components that are displayed in Figure 10-10 and detailed in [26]. The network orchestration controller is the component responsible for handling the workflow of all the processes involved (e.g., the provisioning of E2E connectivity services). It also exposes a NBI to offer its services to applications. For the cABNO, the NBI of the network orchestrator controller is extended to offer a REST-based interface for topology recovery and connection provisioning based on the Control Orchestration Protocol [27], which has evolved in ONF T-API and IETF TE models.

Figure 10-10 also provides the different topological views at different hierarchical levels (top hierarchical level for the pABNO, and lower hierarchical level for the different segments). The provided topological views correspond with the proposed experimental validation, where a pABNO and cABNO-T and cABNO-DC are deployed. The cABNO-T is responsible for SDN orchestration of two SDN aggregation domains and an SDN core network domain. The cABNO-DC is responsible for two intra-DC network domains.

The hierarchical SDN approach benefits single operator scenarios, where multi-layer, multi-vendor, and multi-technology SDN controllers are needed. For multi-operator scenarios, where centralized elements may be impractical, a peering model as presented in Section 10.4.1 may be the preferred option [21].

10.5.1.3 Integration with NFV Architecture

The wide adoption of NFV requires virtual computing and storage resources deployed throughout the network. Traditionally, virtual computing and storage resources have been deployed in large data centers (DCs) in the core network. Core DCs offer high computational capacity with moderate response time, meeting the requirements of centralized services with low-delay demands. However, it is also required to offer edge computing (i.e., micro-DCs and small-DCs) in different sites of the mobile network (e.g., at base stations, cell aggregation points, radio network controllers, or central offices) leveraging on low latency and high bandwidth. For example, ETSI is defining the multi-access edge computing (MEC), see Section 5.2.5, to offer applications such as video analytics, location services, mission-critical applications, augmented reality, optimized local content distribution, and data caching.

Typically, a single NFVI domain for the mobile Xhaul network is considered. The NFVI is distributed and interconnected by the Xhaul network. The VIM is commonly implemented using a cloud controller based on, e.g., OpenStack. It interfaces with the NFV reference implementations (i.e., OPNFV and OSM) using the OpenStack API. OpenStack enables to segregate the resources into availability zones for different tenants and to instantiate the creation, migration or deletion of virtual machines (VMs) and containers (CTs), related to compute services, storage of disk images (image services), and the management of the VM/CT's network interfaces and network connectivity (networking services). For example, the OpenStack compute service (named Nova) manages pools of compute nodes with many choices available for hypervisor technology (e.g., KVM, VMWare, Xen) or container technology. The OpenStack networking service (named Neutron) manages networks and IP addresses, providing flat networks or VLANs to separate traffic between hosts. Further, the Neutron service enables to configure a virtual switch such as an Open vSwitch (OVS) within a compute node (e.g., creation of new ports connecting new VMs/CTs, configuration of forwarding rules) through an SDN controller. It would allow to have a single VIM acting as global orchestrator of compute, storage and network resources. However, the current definition of the Neutron plugin does not support all the specific functionalities that would be required to control transport switches (packet or optical) external to the data center. To overcome this limitation, the ETSI NFV MANO framework has also defined the WAN infrastructure manager (WIM), as a particular VIM. In this scenario, the VIM (i.e., OpenStack cloud controller) is responsible for controlling and managing the NFVI-PoP's resources (i.e., DCs resources), whilst the WIM is used to establish connectivity between NFVI-PoP's. The WIM can be performed by a single SDN controller (e.g., OpenDaylight, ONOS, Ryu), or by an SDN orchestrator in a multi-layer (wireless, packet, optical) network with dedicated SDN controllers per technology, as explained in the previous section and further described in [28].

Additionally, each DC can be managed independently through its own cloud controller acting as a VIM. Moreover, a single cloud controller directly controlling thousands of compute nodes spread in multiple DCs does not scale. Thus, it is required to deploy a cloud orchestrator enabling to deploy federated cloud services for multiple tenants across distributed DC infrastructures. The considered cloud orchestrator may act as a parent VIM and interface with the NFVO, within a hierarchical VIM architecture. However, the cloud orchestrator should support the OpenStack API, since it has become the de-facto interface between the VIM and the reference NFVO implementations. There are two OpenStack projects aiming at developing a hierarchical OpenStack architecture. These would enable to develop a cloud orchestrator based on OpenStack (e.g., Trio2o and Tricircle) and use the OpenStack API as both the southbound interface (SBI) with the OpenStack controllers as well as the northbound interface (NBI) with the NFVO implementations. Alternatively, the NFVO should perform the orchestration of the NFV infrastructure resources (i.e., DC resources) across the multiple VIMs by directly interfacing with the multiple VIMs, instead of the cloud orchestrator.

10.5.1.4 Supporting Network Slicing over the Xhaul Infrastructure

Network slicing has emerged as a key requirement for 5G networks, although the concept itself is still not (yet) fully developed. Macroscopically and from a high-level perspective, the word slicing is understood to involve the partitioning of a single, shared infrastructure into multiple logical networks (*slices*), along with the capability of instantiating them on demand, in order to support functions that constitute operational and user-oriented services. In this setting, important characteristics of slicing are that it not only involves network resources but also computing and storage, and that such slices are expected to be customized and optimized for a service (set) or vertical industry making use of such slice [29]. Network slicing is covered in detail in Chapter 8.

In this section, we focus on the specifics related to network management and SDN/NFV control aspects of network management. Research, development and standardization work is consequently needed, not only to define information and data models for a network slice, but also mechanisms to dynamically manage such constructs, providing multiple, highly flexible, and dedicated E2E slices (considering virtual network, cloud and function resources), while enabling different models of control, management and orchestration systems, covering all stages of slice life-cycle management. This includes the ability to deploy slices on top of the underlying infrastructure, including, where appropriate, the ability to partition network elements. The existing mechanisms to carry out this resource partitioning are multiple, and there is no formal or standard mechanism to do so.

As mentioned in Chapter 8, and from the point of view of business models, network slicing allows, e.g., mobile network operators (MNOs) to open their physical transport network infrastructure to the concurrent deployment of multiple logical and self-contained slices. In this line, slices can be created and operated by the 5G network operator or enable new business models, such as "Slice-as-a-Service" (SlaaS). As a basic, canonical example, the ETSI NFV framework, conceived around the idea and deployment model where dedicated network appliances (such as routers and firewalls) are replaced with software (i.e., guests) running on hosts, can be the basis for a slicing framework, at least for a well-scoped definition of slices. From a functional architecture perspective, the ETSI NFV framework needs to be extended to support slicing natively, by means of, e.g., a slice manager (Xhaul slice control and orchestration system) or entity that performs the book-keeping of slices and maps them to slice owners and associates them to dedicated, per-slice control and management planes.

Part of the function of such control and orchestration system is thus to ensure access rights, assign resource quotas and provide efficient means for the resource partitioning and isolation. Those functions are nonetheless assumed to be part of the network slicing lifecycle management. Support of multi-tenancy has a strong impact on the SDN and MANO functions and components. For example, at the SDN controller level, multi-tenancy requirements are related to the delivery of uniform, abstract and UP-independent views of its own logical elements, while hiding the visibility of other coexisting virtual networks, including the logical partitioning of physical resources to allocate logical and isolated network elements and the configuration of traffic forwarding compliant with per-tenant traffic separation, isolation and differentiation. At the VIM and VNF MANO level, similar considerations on virtual resource allocation and isolation are extended to computing elements and a suitable modeling of the tenant and its capabilities [30].

Related to the Slice-as-a-Service, it is commonly accepted that the tenants may need to have certain control of their sliced virtual infrastructure and resources. It is part of the actual service control model to define the degree of control over the slice [30].

In a first model, the control that each tenant (i.e., owner or operator of the allocated network slice) exerts over the allocated infrastructure is limited, scoped to a set of defined operations. For example, the tenant can retrieve, e.g., a limited or aggregated view of the virtual infrastructure topology and resource state and perform some operations, using a limited set of interfaces, allowing a limited form of control, and different from controlling or operating a physical infrastructure. For example, the actual configuration and monitoring of individual flows at the nodes may not be allowed, and only high-level operations and definitions of policies may be possible.

Alternatively, each allocated slice can be operated as a physical one, that is, each tenant is free to deploy its choice of the infrastructure operating system and control. A virtual network operator

(VNO) is able to manage and optimize the resource usage of its own virtual resources. This means that each tenant can manage its own virtual resources, implemented by deploying a per-tenant controller or per-tenant management. This approach results in a control hierarchy and recursive models, requiring adapted protocols that can be reused across the controllers' NBIs and SBIs.

10.5.2 Core Transport Networks

The evolution towards fully operational 5G networks imposes a number of challenges that are usually perceived as impacting only the access networks, although this is not actually the case. Network functions, as integral parts of the services offered to the end-users, have to be composed in a flexible manner to satisfy variable and stringent demands, including not only dynamic instantiation but also deployment and activation. In addition to that, and as a complement of it, the whole network should be programmable to accomplish such expected flexibility, allowing for interconnecting the network functions across several NFVI-PoPs and scaling the connections according to the traffic demand. The versatile consumption of resources and the distinct nature of the functions running on them can produce very variable traffic patterns on the networks, changing both the overlay service topology and the corresponding traffic demand. The location of the services is not tightly bound to a small number of nodes any more, but to distributed resources that are topologically and temporally changing. The network utilization then becomes time-varying and less predictable. In order to adapt the network to the emergence of 5G services, the provision of capacity on demand through automatic elastic connectivity services in a scalable and cost-efficient way is required. The backbone or core transport networks then become a key component for E2E 5G systems.

The transformation objectives of the core transport networks have been traditionally focused towards more affordable and cost-effective technologies, being able to cope with the huge increase in traffic experienced in the latest years, at a reduced cost per bit. 5G networks, however, present innovative requirements to be faced by the transport networks, like the need to accommodate a large number of simultaneous devices, provide transport and service resources in a flexible and dynamic manner, and reduce the provisioning time to make such flexibility functional. Specifically, 5G transport networks will have to support high traffic volumes and ultra-low latency services. The variety in service requirements and the necessity to create network slices on demand will also require an unprecedented flexibility in the transport networks, which will need to dynamically create connections between sites, network functions or even users, providing resource sharing and isolation. Key aspects on the concept of network slicing are [31]: (i) resource manageability and control, (ii) virtualization through abstraction of the underlying resources, (iii) orchestration of disparate systems, and (iv) isolation of the offered compound assets in the form of slice. 5G transport shares all those goals. Moreover, the flexibility required by 5G transport, such as the dynamic creation and reconfiguration of network slices, makes some of the requirements even more stringent.

The programmability of the transport networks will be performed through open, extensible APIs and standard interfaces that permit agile E2E service creation in a rapid and reliable way. The goal is to evolve towards E2E automated, dynamically reconfigurable and vendor-agnostic solutions based on service and device abstractions, with standard APIs able to interoperate with each other, and facilitating a smooth integration with the OSS and BSS deployed by network operators.

From a complementary angle, transport networks will also have a very relevant role in the optimization of RAN resources by enabling flexible fronthaul and backhaul systems, maximizing the benefits provided by distributed and virtualized RAN environments, tailored to the needs of a variety of vertical customers. The support of different functional splits in the radio part, the packetized transport of such signals, and the dynamic location of the processing units will render a full programmability and dynamicity in the transport part necessary.

Network management and orchestration mechanisms at transport level are needed in order to create the programmable environment required for 5G networks. The purpose is to integrate this programmable transport infrastructure with the overall 5G orchestration system, creating, managing and operating slices for different customers.

10.6 Network Function Virtualization in 5G Environments

Virtualization is the technique which significantly reshaped the IT and the networking ecosystem in recent years. On one hand, cloud computing and related services such as Infrastructure-as-a-Service (IaaS), Platform-as-a-Service (PaaS), and Software-as-a-Service (SaaS) are the results of a successful story (and ongoing stories) from the IT field. On the other hand, networking is in the middle of a momentous revolution and important transition. The appearance of virtualization techniques for networks fundamentally redefines how telecommunications enterprises will soon operate. In the visions of 5G, the often-heard service-level keywords are cost-effectiveness and improved service offering with fast creation, fast reconfiguration and a larger geographical reach of customers. This paradigm shift is technologically triggered by NFV, i.e., the implementation of telco functions on virtual machines that can be run on general purpose computers instead of running them on expensive dedicated hardware as in the traditional way; and also by SDN, i.e., configuring network appliances with easily manageable, often centrally run controller software applications. Combined with the already mature cloud technologies, 5G services can be best implemented in service function chains (SFCs) in which basic functions are run separately, possibly in remote data centers, while network control ensures the connectivity among those, and of course among the end users, by steering traffic based on, e.g., network service headers (NSHs).

In order to enable carrier-grade network services and dynamic SFCs with strict QoS requirements, a novel UP is needed that supports high performance operations (comparable to traditional hardware-based solutions), controllable bandwidth, and delay characteristics between physical or logical ports and interfaces. Therefore, the flexible and fine-granular programming of the general purpose forwarding and processing elements is crucial. SDN is the key enabler of CP softwarization and targets a programmable UP split from the control part. Besides the activities addressing carrier-grade SDN CP platforms, such as OpenDaylight or Open Network Operating System (ONOS), significant efforts have been focused on UP solutions. For example, Intel's Data Plane Development Kit (DPDK) is a software toolkit which enables enhanced packet processing performance and high throughput on general purpose commodity servers. It is supported by the de-facto standard of software switches, i.e., Open vSwitch (OVS).

Many tools are already available for network service providers and network operators. There are open-source solutions for the orchestration of IT resources, e.g., OpenStack as a fully-fledged cloud operating system, and the building blocks, e.g., OVS and DPDK, to make the underlying networking UP programmable and efficient. However, as virtual machines (VMs) and containers (CTs) use the same hardware resources (CPU, memory) as the components responsible for networking, a low-level resource orchestrator is also needed (besides resource orchestrators running at higher abstraction and aggregation levels), which is capable of jointly handling the requests, and of calculating, configuring and enforcing appropriate resource allocation.

In this envisioned SFC-based 5G ecosystem, multiple novel types of actors appear, as also discussed in Section 2.6: infrastructure providers that offer compute and/or network resources for service deployment, application developers who sell the code and/or the management service of VNFs from which the SFC can be built, and the customers that are, at the end of the day, the application providers to end users. The first type of actors are mostly the traditional Telcos and Internet service providers (ISPs), while the second and third types are often merged today in the form of over-the-top (OTT) solution providers.

Future 5G services, such as coordinated remote driving, remote surgery or other Tactile Internet related applications with round-trip latency requirements on the order of few ms, pose extreme requirements on the network, and call for the joint control of IT and network resources. Moreover, typical network services, realized by SFCs, span not only over multiple domains, but over multiple operators as well, as cost-effectiveness by resource sharing is envisioned, and a wide geographical reach of customers in the 5G ecosystem. As one of the most important use cases, the Factory of the Future will make an intensive usage of 5G technologies for supporting the digitization in the way conceived by the idea of Industry 4.0. A high number of connected devices, collaborative robots, augmented reality, and the integration of manufacturing, supply chain and logistics, altogether open an opportunity window to operators for monetizing the provision of virtualized infrastructures and capabilities.

The multi-provider orchestration and management of network services involves many aspects, from the resource discovery and business negotiations between operators, to the computation and monitoring of assured quality network connections among their domains, and the efficient embedding of services into the available resource set. Novel features and technical enablers are necessary for NFVO in a flexible multi-provider setup. A multi-provider NFVO handles abstract sets of compute and network resources and provisions the necessary subset to the customer in order to deploy its service within. In addition to that, it provides an integrated view of infrastructure resources to the customer, also encapsulating managed VNF capability, and ensures that the demanded service requirements are fulfilled.

With well-defined interfaces and orchestration-management mechanisms, operators can act not only as NFVI providers, but also as integrators of VNF-as-a-Service (VNFaaS) offerings from third parties. As such, operators can also act as virtualization platform providers that open interfaces for third party components, such as VNF managers (VNFMs).

10.7 Autonomic Network Management in 5G

10.7.1 Motivation

To meet the diverse and stringent KPI requirements specified in ITU-R IMT-2020, the 5G system will necessarily become more complex [32], which can be mainly characterized by the following technical features: 1) a heterogeneous network consisting of marco cells, small cells, relays, and device-to-device (D2D) links; 2) new spectrum paradigms, e.g., dynamic spectrum access, licensed-assisted access, and higher frequency at mmWave bands, as elaborated in Chapter 3; 3) cutting-edge air-interface technologies, such as massive antenna arrays and advanced multi-carrier transmission, as detailed in Chapter 11; and 4) a novel E2E architecture for flexible and quick service provision in a cost- and energy-efficient manner, as introduced in Chapter 5.

The system's complexity imposes a high pressure on today's manual and semi-automatic network management that is already costly, vulnerable, and time-consuming. However, mobile networks' troubleshooting (related to systems failures, cyber-attacks, and performance degradations, etc.) still cannot avoid manually reconfiguring software, repairing hardware or installing new equipment. A mobile operator has to keep an operational group with a large number of network administrators, leading to a high operational expenditure (OPEX) that is currently three times that of capital expenditure (CAPEX) and keeps rising [33]. Additionally, troubleshooting cannot be performed without an interruption of the network operation, which deteriorates the end user's quality-of-experience (QoE) [34]. Without the introduction of new management paradigms, such large-scale and heterogeneous 5G networks simply become unmanageable and cannot maintain service availability.

Recently, the research community has started to explore artificial intelligence [35] in order to minimize human's intervention in managing networks to lower the OPEX and improve the system's performance. IETF has initiated a research group called Intelligence-Defined Networks to specifically study the application of machine learning technologies in networking. Moreover, the 5G PPP projects SELFNET [36] and CogNet [37] have focused on designing and implementing intelligent management for 5G mobile networks. For example, the SELFNET project has been set up to design, prototypically implement, and evaluate an autonomic management framework for 5G mobile networks. Taking advantage of new technologies, in particular SDN [38], NFV [39], self-organized networks (SON) [40], multi-access edge computing (MEC) and artificial intelligence, the framework proposed by the SELFNET project can provide the capabilities of self-healing against network failures, self-protection against distributed cyber-attacks, and self-optimization to improve network performance and end users' QoE [41]. Although the current SON techniques have a self-managing function, it is limited to static network resources. It does not fully suit 5G scenarios, such as network slicing [42] and multi-tenancy [43], where dynamic resource utilization and agile service provision are enabled by SDN and NFV technologies. Currently, existing SON can only reactively respond to detected network events, while the intelligent framework is capable of proactively performing preventive actions for predicted problems. The automatic processing in SON is usually limited to simple approaches like triggering, and some operations are still carried out manually. In addition, the self-x management mainly focuses on the RAN. An extension beyond the RAN segment to provide a self-organizing function over the E2E network is required. By reactively and more importantly proactively detecting and diagnosing differently network problems, which are currently manually addressed by network administrators, the SELFNET framework could assist network operators to simplify management and maintenance tasks, which in turn can significantly lower OPEX, improve user experience, and shorten time-to-market of new services.

In this section, a reference architecture for the autonomic management framework [36] will be introduced, including the functional blocks, their capabilities and interactions; the autonomic control loop starting from the SDN/NFV sensor and terminating at the actuators will be provided, as well as a brief exemplary loop so as to illustrate how the autonomic system may mitigate a network problem. Furthermore, several classical artificial intelligence algorithms that can be applied to implement the network intelligence are briefly shown.

10.7.2 Architecture of Autonomic Management

In addition to the software-defined and virtualized network infrastructure [44], the autonomic management framework mainly consists of: 1) SDN/NFV sensors that can collect the network metrics; 2) monitoring modules that can derive the symptoms from the collected metrics; 3) network intelligence that is in charge of diagnosing network problems and making tactical decisions; and 4) SDN/NFV actuators and an orchestrator that perform corrective and preventive actions. As shown in Figure 10-11, the potential architecture for autonomic management can be split into several layers, which are explained as follows:

- **Infrastructure layer**: All NFs managed autonomously by the framework rely on physical and virtualized resources in this layer. It encompasses physical and virtualization sublayers. The former provides an access to physical resources (networking, computing, storage, etc.), while the latter instantiates virtual infrastructures on top of the physical sublayer. It represents the NFVI as defined by the ETSI NFV terminology;
- **Data network layer**: This implies an architectural evolution towards the SDN paradigm by decoupling the CP from the UP. In this framework, the data layer represents a simple data-forwarding, which can be either a non-virtualized or virtualized NF;

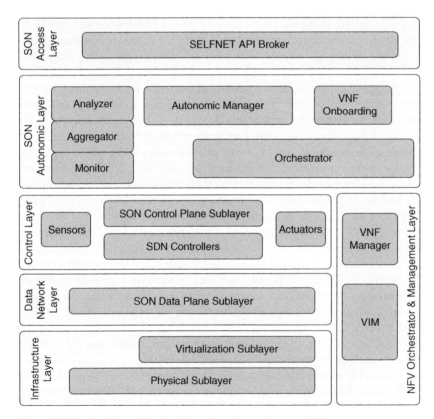

Figure 10-11. Possible architecture for autonomic management [36].

- **SON control layer**: This layer includes two internal sublayers: SDN controllers and SON CP sublayer. SDN/NFV sensors and actuators, which are capable of collecting data from the entire system and enforcing actions, respectively, are also contained. The SON control layer and data network layer have associated CPs and UPs of the network that are decoupled in the SDN paradigm;

- **SON autonomic layer**: To realize the network intelligence, this layer consists of three modules, i.e., monitor, aggregator and analyzer, autonomic manager, and orchestrator. The monitor and analyzer extract metrics related to network behavior, aggregate the collected metrics into health of network (HoN) metrics, and use these to infer the network status. The autonomic manager is in charge of diagnosing the root cause of any existing or potential network problems, and deciding which countermeasure should be conducted. Following the tactical decisions from the autonomic manager, the orchestrator coordinates the physical and virtualized resources, and manages the SDN/NFV actuators to execute the decided actions;

- **NFV orchestration and management layer**: This layer is responsible for orchestrating and managing VNFs via the VNF manager, as well as virtualized resources through VIM. It conforms to the NFV MANO specified by ETSI [5];

- **SON access layer**: This is the external interface that is exposed by the framework. Despite the fact that internal components may have specific interfaces for the particular scope of their functions, these components contribute to a general SON API, managed by the SELFNET API broker that exposes all aspects of the autonomic framework to external systems, such as BSS or OSS and administration graphical user interfaces (GUIs). The latter enable network administrators to interact with and configure the SELFNET framework and also observe the complete status of the network.

10.7.3 Autonomic Control Loop

One of the main challenging aspects of the autonomic management is the implementation of network intelligence. Apart from the underlying software-defined and virtualized network infrastructure, a closed control loop referred to as autonomic control loop, starting from the sensors and terminating at the actuators, is needed to control the processing flow. When the monitor detects or predicts a network problem, an autonomic control loop is initiated. The autonomic manager diagnoses the cause of the problem, decides on a tactic, and plans an action. Once the orchestrator receives an action request from the action enforcer (AE), it coordinates the physical and virtualized resources to enforce this action.

The autonomic manager can be regarded as the brain of the autonomic management framework and plays a vital role in the provision of network intelligence. Taking advantage of cutting-edge techniques in the field of artificial intelligence, it provides the capabilities of self-healing, self-protection and self-optimization by means of reactively and proactively dealing with detected and predicted network problems. As illustrated in Figure 10-12, the autonomic manager consists of the following functional blocks:

- A **diagnoser** is in charge of diagnosing the root cause of network problems. The monitor can derive a symptom for each detected or predicted network problem from the collected sensor data. The diagnoser processes the reported symptom to make clear its reason, and notifies the decision-maker;

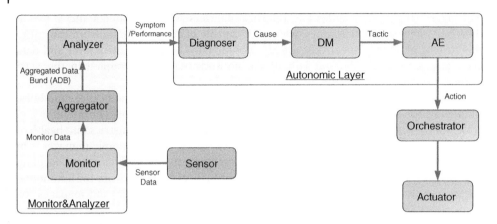

Figure 10-12. Autonomic management control loop.

- A **decision-maker (DM)** can decide a set of corrective or preventive tactics to deal with the network problems based on incoming diagnostic information. A tactic is a high-level description of a countermeasure, which needs to be transferred into an implementable action;
- An **action enforcer (AE)** is responsible for providing a consistent and coherent set of scheduled actions to be enforced in the network infrastructure. For this purpose, this module recognizes and validates these tactics by applying conflict detection and resolution in order to provide implementable actions to be enforced.

Within this control loop, the metrics collected by the sensors are processed by the monitor module first. Subsequent modules extract the required information from the previous module and provide the next-level results to the next module. The information model associated with the autonomic control loop is explained as follows:

- **Sensor data**: A range of differentiated data sources can be expected to be identified in the upcoming 5G infrastructure. All monitoring information retrieved from physical devices, UP, SDN controller, SDN/NFV sensors, VIM etc., are uniformly referred to as sensor data. The monitor is the corresponding module that is in charge of collecting sensor data from underlying infrastructures;
- **Monitor data**: The monitor regularly collects the sensor data and reports the necessary information to the aggregator. Some of the data is periodically collected, which stands for either normal or abnormal network behaviors;
- **Aggregated data bundle (ADB)**: The monitor data related to a network problem may be retrieved from a set of sensors, rather than a single one. For example, in the case of a distributed denial of service (DDoS) attack, the source and destinations are distributed. The raw information contained in monitor data should be processed to produce aggregated and correlated information, which is called *aggregated data bundle*;
- **Symptom**: A high-level health-of-network metric that may be derived from a set of correlated alarms, events, KPIs, etc., that can be evaluated to indicate the characteristics of an existing or emerging network problem, together with the additional contextual information such as metadata, is defined as a symptom;

- **Performance**: The report of achieved performance by an executed action is two-fold: i) if an action degrades performance rather than solving a problem, a roll-back mechanism will be triggered to recover the network status to the initial point before the action was performed; ii) the achieved performance, which can be regarded as the benefit or reward of action. If a large extent of operational data can be recorded, the network intelligence can be trained based on machine learning techniques;
- **Cause**: It is a description of what the reason of a network problem is or why a network problem happens or will happen. Once the diagnoser receives a symptom, it diagnoses the cause of this symptom;
- **Tactic**: After the cause of a network problem is clarified, a countermeasure that can be applied to tackle this problem needs to be decided by the decision-maker. A tactic is a high-level description of a countermeasure, which is required to be transferred into an implementable action;
- **Action**: This is an implementable version of a countermeasure with a description of how to enforce this, taking into account available physical and virtualized resources. The action provided by the AE contains more implementation details, e.g., the actuator's type, the target deployment location, and configuration information.

To close this section, let us use the following example to show the autonomic control loop and illustrate its main mechanisms. The storyline is depicted as follows: A summer concert is taking place in the city centre, where a large number of spectators gather in a small area. Some of the spectators start to share real-time videos in their social media. When the number of video users increases, especially if some of them transfer videos in ultra-high definition, the network suffers from traffic congestion and the perceived QoE deteriorates. The monitoring modules first detect this network's anomaly by means of collecting, aggregating, and analysing the sensor data. A symptom called video QoE decreasing is reported to the diagnoser. After the diagnosis, it is found that the cause of the video QoE decreasing is the increased number of video users in the zone. Then, the possible tactics, for instance, load-balancing, video coding optimizing, and admission control, are determined by the decision-maker. The AE transfers these tactics into implementable actions and notifies the orchestrator. Taking into account available resources, the action of load balancing is finally selected and executed by the orchestrator. An actuator acting as a load balancer is instantiated, configured and deployed in the local network surrounding this concert. Afterwards, the congested network is successfully recovered and the perceived QoE of end users is improved.

10.7.4 Enabling Algorithms

We will give a brief introduction about enabling intelligence algorithms. The motivation is to provide a view for the readers how to apply artificial intelligence to implement the network intelligence. Hence, only several classical algorithms are given. For further artificial intelligence technologies, such as neural networks [45], reinforcement learning [46], transfer learning [47], and deep learning [48], the reader is referred to the stated references.

10.7.4.1 Feature Selection
In practice, a large number of features (i.e., network metrics) can be extracted from the 5G infrastructure. Each feature generally needs to be periodically recorded, resulting in a huge volume of data. When the management system tackles a specific problem, e.g., traffic congestion, it is

inefficient (if not infeasible) to process all data. That is because generally only a relatively small subset of all-available features is informative, while others are either irrelevant or redundant. As a data-driven approach, the network intelligence should be built on relevant features, while discarding others, so that irrelevant and redundant features do not degrade the performance on both training speed and predictive accuracy.

Feature selection (FS) is hence one of the most important intelligence techniques and an indispensable component in machine learning and data mining. It can reduce the dimensionality of data by selecting only a subset of features to build the learning machine. A number of classical FS algorithms, such as Relief-F [49] and Fisher [50], can be directly applied to calculate the relevance of the collected features.

10.7.4.2 Classification

In the terminology of machine learning, classification is an instance of supervised learning. It is applied to identify which class a new observation belongs to on the basis of a training dataset. An example would be assigning an incoming email into SPAM or non-SPAM classes in terms of the observed features of the email (e.g., source IP address, text length, and title content). The following is a brief introduction of classification algorithms that can be used in the network intelligence:

- A **decision tree (DT)** [51] is a classical supervised learning method used for classifying. Decision rules are inferred from a training dataset and a tree-shaped diagram is built. Each node of the decision tree relies on a feature to separate the data, and each branch represents a possible decision. DT is simple, interpretable and fast, whereas it is hard to apply in a complex and non-linear case;
- A **discriminant analysis** is a classification method which assumes that different classes generate data based on different Gaussian distributions. A linear discriminant (LD) analysis [52] is to find a linear combination of features that maximize the ratio of inter-class variance to the intra-class variance in any particular dataset so as to guarantee maximal separation;
- A **support vector machine** (SVM) [53] utilizes a so-called hyperplane to separate all data points of one class from another. The number of features does not affect the computational complexity of SVM, so that it can perform well in the case of high-dimensional and continuous features. However, it is a binary classifier, and a multi-class problem can be solved only by transferring this into multiple binary problems;
- Another algorithm called **k Nearest Neighbor** (kNN) is applied for data classification following the hypothesis that close proximity in terms of inter-data distance has a similarity. The class of an unclassified observation can be decided by observing the classes of its nearest neighbors. It is among the simplest algorithms with a good predictive accuracy. But it needs high memory usage, is vulnerable to noisy data and is not easy to interpret.

10.8 Summary

This chapter has shown that the management and orchestration plane is instrumental to enable the efficient utilization of the infrastructure, while meeting the performance and functional requirements of heterogeneous services. The requirements for the forthcoming 5G networks trigger the work on a complex ecosystem where compute, storage and connectivity must be coordinated in real-time.

SDN decouples the control and user planes of the NEs to enable a central network control that can make smart decisions, while the NEs are focused on the forwarding and application of policies. Such separation enables the network to become more flexibly programmable than current networks. The programmability of SDN is required by the NFV paradigm. NFV facilitates the dynamic instantiation of VNFs on top of commodity hardware, which lets the operator separate the NFs from the hardware. Autonomics will evolve the networking technologies with the necessary support for handling its heterogeneous complexity and provide the necessary service availability and resiliency. These technologies are key enablers of the new management and orchestration technologies.

In the case of multi-operator orchestration scenarios, it is essential not only to define, but also to implement an E2E orchestration plane able to deal with the interaction of multiple administrative domains. The use of open and standard interfaces as well as the modeling of services and devices are the only way to have an ecosystem to facilitate the deployment of new paradigms in network operators. Similarly, it is the use case of multi-technology, where the scenario is a real network with legacy systems that are providing services to the end-customers.

References

1 NGMN, White Paper, "5G White Paper", Feb. 2015
2 L.M. Contreras, P. Doolan, H. Lønsethagen and D.R. López, "Operation, organization and business challenges for network operators in the context of SDN and NFV", in Elsevier Computer Networks, Vol. 92, pp. 211–217, 2015
3 ONF SDN Architecture, Issue 1, June 2014, see www.opennetworking.org
4 ETSI NFV, see http://www.etsi.org/technologies-clusters/technologies/nfv
5 ETSI GS NFV MAN 001, "Networks Functions Virtualization (NFV); Management and Orchestration", V1.1.1, Dec. 2014
6 ETSI, White Paper, "Open Source MANO", Release 2, April 2017
7 R. Enns, M. Bjorklund, J. Schoenwaelder and A. Bierman, "Network Configuration Protocol (NETCONF)", RFC 6241, IETF, June 2011
8 M. Bjorklund, "YANG - A Data Modeling Language for the Network Configuration Protocol (NETCONF)", RFC 6020, IETF, Oct. 2010
9 A. Bierman, M. Bjorklund and K. Watsen, "RESTCONF Protocol", RFC 8040, Jan. 2017
10 http://claise.be/IETFYANGPageCompilation.png
11 Draft ETSI GS NFV-SOL 001, "Network Functions Virtualisation (NFV) Release 2; Protocols and Data Models; NFV descriptors based on TOSCA specification", V0.4.0, Dec. 2017
12 OASIS, "TOSCA Simple Profile for Network Functions Virtualization (NFV)", v1.0, March 2016
13 ETSI GS NFV-EVE 005, "Network Functions Virtualisation (NFV); Ecosystem; Report on SDN Usage in NFV Architectural Framework", v1.1.1, Dec. 2015
14 L.M. Contreras, C.J. Bernardos, D.R. López, M. Boucadair and P. Iovanna, "Cooperating Layered Architecture for SDN", draft-irtf-sdnrg-layered-sdn-01 (work in progress), Oct. 2016
15 Q. Wu, W. Liu and A. Farrel, "Service Models Explained", May 2017
16 D. Bogdanovich, B. Claise and C. Moberg, "YANG Module Classification", May 2017
17 ONF Report TR-534, "Framework and Architecture for the Application of SDN to Carrier Networks", July 2016

18 ONF Report TR-521, "SDN Architecture – Issue 1.1", Jan. 2016

19 Draft ETSI GR NFV IFA 028, "Network Function Virtualisation (NFV); Management and Orchestration; Architecture options to support the offering of NFV MANO services across multiple administrative domains", V0.9.0, Dec. 2017

20 MEF, Service Operations Specification MEF 55, "Lifecycle Service Orchestration (LSF) Reference Architecture and Framework", March 2016

21 5G PPP 5G-Exchange project, see http://www.5gex.eu

22 V. Lopez, D. Konidis, D. Siracusa, C. Rozic, I. Tomkos and J.P. Fernandez-Palacios, "On the Benefits of Multilayer Optimization and Application Awareness", Journal of Lightwave Technology, vol. 35, no. 6, March 2017

23 5G PPP 5G-Xhaul project, see http://www.5g-xhaul-project.eu/

24 5G PPP 5G-Crosshaul project, see http://5g-crosshaul.eu/

25 M. Fiorani et al., "Abstraction models for optical 5G transport network", Journal of Optical Communications and Networking, vol. 8, no. 9, pp. 656–665, Sep. 2016

26 R. Vilalta et al., "Hierarchical SDN Orchestration for Multi-technology Multi-domain Networks with Hierarchical ABNO", European Conference on Optical Communication (ECOC 2015), Dec. 2015

27 A. Mayoral et al., "The Need of a Transport API in 5G for Global Orchestration of Cloud and Networks through a Virtualised Infrastructure Manager and Planner", JOCN Special Issue OFC 2016, 2016

28 R. Casellas, R. Muñoz, R. Vilalta and R. Martínez, "Orchestration of IT/Cloud and Networks: From Inter-DC Interconnection to SDN/NFV 5G Services", Optical Network Design and Modeling (ONDM 2016), May 2016

29 NGMN, White Paper, "Description of Network Slicing Concept", v1.0, Jan. 2016

30 Xi Li, et al., "5G-Crosshaul Network Slicing Enabling Multi-Tenancy in Mobile Transport Networks", IEEE Communications Magazine, vol. 55, no. 8, pp. 128–137, Aug. 2017

31 J. Ordonez-Lucena, P. Ameigeiras, D. Lopez, J.J. Ramos-Munoz, J. Lorca and J. Folgueira, "Network Slicing for 5G with SDN/NFV: Concepts, Architectures and Challenges", IEEE Communications Magazine, vol. 55, no. 5, pp. 80–87, May 2017

32 J. G. Andrews et al., "What will 5G be?", IEEE Journal on Selected Areas in Communications, vol. 32, no. 6, pp. 1065–1082, June 2014

33 Aviat Networks, "Top ten pain points of operating networks", 2011

34 B. Bangerter et al., "Networks and devices for the 5G era", IEEE Communications Magazine, vol. 52, no. 2, pp. 90–96, Feb. 2014

35 A. He et al., "A survey of artificial intelligence for cognitive radios", IEEE Transactions on Vehicular Technology, vol. 59, no. 4, pp. 1578–1592, May 2010

36 5G PPP SELFNET project, see https://selfnet-5g.eu/

37 5G PPP CogNet project. see http://www.cognet.5g-ppp.eu/

38 B. A. A. Nunes et al., "A survey of software-defined networking: Past, present, and future of programmable networks", IEEE Communication Surveys, vol. 16, no. 3, pp. 1617–1634, 2014

39 R. Mijumbi et al., "Network function virtualization: State-of-the-art and research challenges", IEEE Communication Surveys, vol. 18, no. 1, pp. 236–262, 2016

40 S. Dixit et al., "On the design of self-organized cellular wireless networks", IEEE Communications Magazine, vol. 43, no. 7, pp. 86–93, Jul. 2005

41 J. P. Santos et al., "SELFNET framework self-healing capabilities for 5G mobile networks", Transactions on Emerging Telecommunications Technology, Wiley, vol. 27, no. 9, pp. 1225–1232, Sep. 2016

42 X. Zhou et al., "Network slicing as a service: enabling enterprises' own software-defined cellular networks", IEEE Communications Magazine, vol. 54, no. 7, pp. 146–153, July 2016

43 K. Samdanis et al., "From network sharing to multi-tenancy: The 5G network slice broker", IEEE Communications Magazine, vol. 54, no. 7, pp. 32–39, July 2016

44 P. Neves et al., "The SELFNET approach for autonomic management in an NFV/SDN networking paradigm", International Journal of Distributed Sensor Networks, vol. 16, no. 2, pp. 1–17, Feb. 2016

45 G. P. Zhang, "Neural networks for classification: a survey", IEEE Transactions on Systems, Man, and Cybernetics, Part C (Applications and Reviews), vol. 30, no. 4, pp. 451–462, Nov. 2000

46 L. Buoniu, R. Babuka and B. D. Schutter, "A Comprehensive Survey of Multiagent Reinforcement Learning", IEEE Transactions on Systems, Man, and Cybernetics, Part C (Applications and Reviews), vol. 38, no. 2, pp. 156–172, March 2008

47 M. E. Taylor and P. Stone, "Transfer learning for reinforcement learning domains: A survey", Journal of Machine Learning Research, pp. 1633–1685, July 2009

48 Z. Fadlullah et al., "State-of-the-Art Deep Learning: Evolving Machine Intelligence Toward Tomorrow's Intelligent Network Traffic Control Systems", IEEE Communications Surveys & Tutorials, no. 99, May 2017

49 I. Kononenko et al., "Estimating attributes: analysis and extensions of RELIEF", European Conference on Machine Learning, April 1994

50 Q. Gu, Z. Li and J. Han, "Generalized Fisher score for feature selection", Conference on Uncertainty in Artificial Intelligence, July 2011

51 S. K. Murthy, "Automatic construction of decision trees from data: A multi-disciplinary survey", Journal on Data Mining and Knowledge Discovery, vol. 2, no. 4, pp. 345–389, Dec. 1998

52 Y. Guo, T. Hastie and R. Tibshirani, "Regularized discriminant analysis and its application in microarrays", Biostatistics, vol. 1, no. 1, pp. 1–18, 2005

53 C. J. Burges, "A tutorial on support vector machines for pattern recognition", Journal on Data Mining and Knowledge Discovery, vol. 2, no. 2, pp. 121–167, Dec. 1998

Part 3

5G Functional Design

11

Antenna, PHY and MAC Design

Frank Schaich[1], Catherine Douillard[2], Charbel Abdel Nour[2], Malte Schellmann[3], Tommy Svensson[4], Hao Lin[5], Honglei Miao[6], Hua Wang[7], Jian Luo[3], Milos Tesanovic[8], Nuno Pratas[9], Sandra Roger[10] and Thorsten Wild[1]

[1] *Nokia Bell Labs, Germany*
[2] *IMT Atlantique Bretagne-Pays de la Loire, France*
[3] *Huawei German Research Center, Germany*
[4] *Chalmers University of Technology, Sweden*
[5] *Orange, France*
[6] *Intel, Germany*
[7] *Keysight Technologies, Denmark*
[8] *Samsung Electronics R&D Institute, UK*
[9] *Aalborg University, Denmark*
[10] *Universitat Politècnica de València, Spain*
With contributions from Rana Ahmed Salem, Mario Castaneda, Xitao Gong and Dinh Thuy Phan Huy.

11.1 Introduction

The 5th generation (5G) air interface (AI) constitutes the complete radio access network (RAN) protocol stack, i.e., the physical layer (PHY), Media Access Control (MAC), Radio Link Control (RLC), Packet Data Convergence Protocol (PDCP), Radio Resource Control (RRC) and Service Data Adaptation Protocol (SDAP), and all related functionalities describing the interaction between infrastructure and device. Furthermore, it covers all services, bands, cell types, etc., expected to characterize the overall 5G system. This chapter describes the lower part of the protocol stack, namely, PHY/ MAC related technologies, and highly related aspects, such as antenna design. Before heading to the detailed elaborations in subsequent sections, we start with establishing basic design criteria and assumptions. While we keep this chapter more open than the current status of discussions in 3GPP, we still relate to 3GPP where possible and reasonable.

Earlier generations of wireless mobile communications under the framework of the 3rd Generation Partnership Project (3GPP) have exclusively used transmission frequencies below 6 GHz. While this frequency range is still of high value for 5G and thus on the agenda of 3GPP, 5G will go beyond this and deploy transmission points radiating with higher frequencies up to 100 GHz. Details on the usage

5G System Design: Architectural and Functional Considerations and Long Term Research, First Edition.
Edited by Patrick Marsch, Ömer Bulakçı, Olav Queseth and Mauro Boldi.
© 2018 John Wiley & Sons Ltd. Published 2018 by John Wiley & Sons Ltd.

of the different spectral regions can be found in Section 3.4. In particular, the exploitation of millimeter-wave (mmWave) frequencies, in the context of cellular systems typically associated with the wider range of 6-100 GHz, puts a number of challenges on the AI design, requiring special and dedicated mechanisms both at PHY and MAC. 5G will be required to support both frequency division (FDD) and time division duplexing (TDD) and will rely on heterogeneous deployment layouts being built upon a macro layer for providing ubiquitous coverage applying a frequency reuse of one and a small cell capacity layer (probably at mmWave frequencies in many of the deployments) for boosting the throughput in areas of high demand. The following sections will elaborate on various technical parts of the AI having those characteristics in mind and carve out respective particularities.

The support of multiple antennas both at the base station (BS) and at the device will be a fundamental corner stone of 5G. As with 4G, 5G will make use of this for enhancing both the throughput per area via spatial reuse and the coverage - both due to the virtue of beamforming gains and the use of diversity mechanisms, and thanks to coordinative means between adjacent cells exploiting the spatial selectivity of beamformed signals. Beamforming can be implemented in digital domain and/or in analog domain. Especially at mmWave frequencies, due to the large bandwidth, the large number of antennas and the lower efficiency of electronics, analog beamforming technologies need to be exploited to allow practical implementation. Currently, hybrid beamforming techniques that combine the merits of analog and digital beamforming have been developed and are widely considered as a design assumption for mmWave communications. Section 11.5 will provide detailed insights into the overall concept and will elaborate on the available options.

While each transition from one generation to the next (i.e., from 2G to 3G and from 3G to 4G) has introduced fundamentally different signal formats and mechanisms to multiplex users[1], the move from 4G to 5G is expected to be less radical when it comes to this choice. According to agreements at 3GPP RAN meetings[2], the early incarnation of 5G, referred to as New Radio (NR) phase 1, see also Section 17.2, will still rely on Cyclic-Prefix Orthogonal Frequency Division Multiplex (CP-OFDM) as 4G does, multiplexing users in time, frequency, and space. As with 4G, 5G will potentially allow for discrete Fourier transform (DFT) precoding to achieve more favorable peak to average power ratio (PAPR) conditions. Optionally, filtering or windowing functionalities can be used to further enhance specific characteristics of the signal. In Section 11.3, a more in-depth analysis of the waveform candidates and means for multiplexing user transmissions are provided. Beside candidates being in line with the current NR draft from 3GPP, we present further promising enhancements to provide the reader with a more comprehensive long-term view. This is especially relevant as later releases of 5G might still allow for the further introduction of those.

The most potent mechanisms for increasing the reliability of a single wireless transmission link are Forward Error Correction (FEC) and Hybrid Automatic Repeat reQuest (HARQ). 4G applies for the former convolutional Turbo codes (CTC) to protect the data channels, and a combination of tail-biting

---

1 2G: Frequency-multiplexed channels applying Time Division Multiple Access (TDMA) with Gaussian Minimum Shift Keying (GMSK), multiplexing users in time domain; 3G: Wideband Code Division Multiple Access (WCDMA), multiplexing users in code domain; 4G: Orthogonal Frequency Division Multiple Access (OFDMA) in the downlink and Single Carrier Frequency Domain Multiple Access (SC-FDMA), also known as Discrete Fourier Transform (DFT)-spread OFDM or DFT-s-OFDM, in the uplink, both making use of cyclic symbol extensions and multiplexing users in frequency and time domains.

2 In the framework of the study item 'Study on New Radio Access Technology' [1] followed by the work item 'Work Item on New Radio Access Technology' [2].

convolutional codes (TBCC) and repetition coding for the control channels. 3GPP has agreed towards the use of Low Density Parity Check (LDPC) codes for data channels and Polar codes (PC) for control channels. In the downlink (DL), 4G applies asynchronous adaptive HARQ, while the uplink (UL) is synchronous. Both variants rely on single-bit feedbacks. 5G will rely on asynchronous variants in both directions. Additionally, recent studies indicate the benefit of making the retransmissions adaptive and allowing for more sophisticated feedbacks. In Section 11.4, more details are provided on the related options that can be drawn from both FEC and HARQ, and the respective interdependencies.

The wireless channel is a shared medium, and the available spectral resources consequently have to be allocated to the respective transmission requests satisfying each user, while keeping the overall system spectrally efficient, as covered in detail in Chapter 12. 4G supports both individually scheduled access on a per transmit time interval (TTI) basis (fixed to 1 ms, at least for the earlier incarnations of 4G) and semi-persistent scheduling (SPS). Some of the new use cases being foreseen to be served by 5G and their respective requirements are implying different access types to potentially be more efficient (e.g., contention-based access and pre-emptive scheduling). In fact, 3GPP has recently agreed that NR shall support grant-free, SPS-like, Physical Uplink Shared CHannel (PUSCH) transmissions, which can be used to reduce the scheduling latency [3]. Additionally, for serving use cases with very stringent timing requirements, a more fine-grained resource allocation needs to be applied, i.e., beside the basic scheduling periodicity of 1 ms, 5G requires to support in parallel a faster scheduling process based on the time basis of, e.g., 1/4 or even 1/8 ms. In fact, 3GPP has now agreed that the time interval between scheduling request (SR) resources configured for a user equipment (UE) can be smaller than a slot (which is the basic scheduling unit). What this in practice means is an agreement by 3GPP to support shorter periodicities for transmitting SRs. Section 11.6 covers these points in more detail both for the design of control channels and data channels. Furthermore, when considering high-density deployments of mmWave small cells, reusing the same spectrum and AI for backhaul and access becomes a good option to relax backhaul and deployment cost, as detailed in Section 7.4. For this purpose, joint scheduling of backhaul and access resources across multiple cells becomes essential to allow the system to operate efficiently.

The remainder of this chapter is structured as follows. Section 11.2 covers relevant criteria for the PHY and MAC design, including considerations of harmonization, for instance between different radio access technologies (RATs). Then, Section 11.3 delves into details on waveforms, numerology and modulation schemes, followed by Section 11.4 on coding approaches and HARQ. Section 11.5 ventures in detail into antenna design, analog, digital and hybrid beamforming, Section 11.6 covers novel PHY and MAC design paradigms and specific solutions for serving and multiplexing the main service types envisioned for 5G, before Section 11.7 summarizes the chapter.

11.2 PHY and MAC Design Criteria and Harmonization

The early incarnations of 4G have focused on efficiently delivering mobile Internet services to devices, such as smart phones and tablets. For this kind of traffic, the most relevant performance indicator to improve is the throughput both per user and per area. To satisfy its customers, operators need to deliver sufficiently high and most importantly consistent data rates where and whenever needed. Hence, the main concern in designing 4G has been to maximize spectral efficiency and spatial reuse.

Especially areas not having a dominant connection to a single BS, i.e., at the cell edge, had to be treated carefully. These areas are interference-limited instead of being noise-limited and thus require special attention, e.g., by applying dedicated mechanisms, such as (further enhanced) inter-cell-interference coordination (Fe)ICIC mechanisms to coordinate the transmissions of BSs in the vicinity. For details on these mechanisms and potential improvements for 5G, the interested reader is referred to Section 12.5. 4G has additionally provided a flat network architecture and a self-organizing approach for the handover process, which has reduced latency significantly. Also, high reliability in 4G is ensured through lower-layer techniques, such as HARQ and FEC, and higher-layer techniques, such as the design of RLC with its acknowledged mode (AM) type of transfer. New and emerging use cases, such as ultra-reliable low-latency communications (URLLC), vehicular-to-anything (V2X) or industrial automation, as well as the need to simultaneously support multiple use cases, imply that the AI design needs to be substantially revisited for the 5G era.

More specifically, 5G NR is anticipated to support a much more diverse set of use cases, as outlined in Section 2.2, and with respective requirements detailed in Section 2.3. Obviously, while throughput is still of very high relevance, 5G is required to support a much wider range of requirements being related to various aspects, such as low latency, high reliability, and energy and cost efficiency at the device.

Beside the wide range of use cases to be supported, further aspects requiring special attention are the wide range of deployment types (e.g., dense urban vs. rural, pure macro-cellular vs. heterogeneous networks, and train lines) and link characteristics (e.g., a wide range of Doppler and delay spreads) as well as the wide range of spectrum below and above 6 GHz. Regarding the spectrum, the most probable mmWave frequencies would be 26 GHz, 28 GHz, 32 GHz and 40 GHz, as detailed in Section 3.4. Different frequency bands imply different bandwidths, propagation conditions and/or even regulations. Accordingly, the 5G NR should be scalable and reconfigurable to be able to properly support the different properties of these.

In a nutshell, 5G in general and the PHY and MAC layers in particular should be designed having the following criteria in mind:

- **To have a high degree of flexibility and versatility** for supporting the broad class of services with their associated broad class of key performance indicators (KPIs) and to enable efficient multi-service support (i.e., meeting the high heterogeneity of requirements) while dealing with high heterogeneity of deployment types, operating frequencies and link characteristics;
- **To be highly scalable** to efficiently support a large number of devices and a wide range of antenna system designs, for instance including different hybrid beamforming architectures, different bandwidths and carrier frequency configurations;
- **To allow for satisfactory service quality** where- and whenever needed, both related to consistent service quality (e.g., by introducing special means to improve cell edge performance, such as interference mitigation techniques) and related to the provision of capacity peaks in respective areas (e.g., by the introduction of high capacity links in crowded areas);
- **To be highly efficient** to support the requirements on energy consumption and resource utilization and to enable high spectral, energy, and cost efficiency in general;
- **To be highly robust** to hardware impairments to allow reduction of hardware costs and to enable cost-efficient operation in mmWave frequencies, where such impairments (e.g., phase noise) are in general more severe than in lower frequencies;
- **To be future-proof/forward compatible** to support easy integration of new services, functionalities and new frequencies without the need of redesigning the AI;

- **To enable tight interworking and synergies** between different RATs, such as 4G and 5G, or different 5G AI variants (AIVs), such as variants for below and above 6 GHz. Here, different levels of integration should be possible, ranging from RAN-level integration up to loose higher layer co-existence or even core network (CN) interconnection. In the following, we treat this aspect related to the lower layers of the protocol stack. Further details on the RAN-level integration of multiple RATs or AIVs can be found in Sections 6.5 and 12.4.

11.3   Waveform Design

One fundamental component of the AI is the underlying waveform, which needs to be designed to properly match to various conditions that can be expected during operation of the wireless communication system. For the most general categorization, there are two different types of waveform designs, namely single-carrier and multi-carrier. For single-carrier waveforms, a single symbol – constituted of a modulated data symbol and an appropriate pulse shape – spans the entire bandwidth B available for transmission, and symbols are transmitted consecutively at a rate of B without inserting any guard symbols or zeros in between. Transmitting these signals via delay-spread channels causes the symbols at the receiver to overlap. To compensate for this, channel estimation and calculation of the corresponding coefficients for a finite impulse response (FIR) or infinite impulse response (IIR) equalization filter is required. For channel estimation, preamble signals need to be frequently transmitted to enable capturing the channel's time variance. Based on these, the channel coefficients and appropriate equalization filter coefficients can be calculated at the receiver. The length of the preamble as well as the complexity for filter calculation and filter operations scales with the number of channel delay taps, which is one of the reasons why single-carrier waveforms are preferred for the application in channels with low delay spread. The other reason is that single-carrier waveforms cannot access the channel in a frequency-selective manner, but instead imply an averaging of the channel quality over the transmission bandwidth. Thus, deep fades that may occur in highly frequency-selective channels may severely degrade the system performance.

Multi-carrier waveforms, on the other hand, divide the available transmission bandwidth into a number of N subcarriers of equal bandwidth and thus allow transmitting independent symbols on these subcarriers in parallel. Choosing N narrowband signals instead of a single wideband signal extends the transmitted symbols in time and thus reduces vulnerabilities to delay spreads from multi-path propagation. The subcarrier bandwidth is usually chosen much smaller than the channel coherence bandwidth, so that each subcarrier signal effectively experiences a flat channel. This significantly simplifies the channel equalization process, as each subcarrier signal can be equalized by a single complex multiplication. Moreover, channel estimation is also simplified, since so-called *scattered pilot grids* can be used, where only a few of the total N subcarriers are selected to carry pilot symbols for channel estimation. From this, it becomes evident that the application of multi-carrier signals is favourable in particular for channels with high frequency selectivity, i.e., exhibiting a large delay spread. If coded transmission is applied, the effect of deep fades can further be well alleviated thanks to a frequency-selective channel access, yielding a much better performance than corresponding single-carrier systems. There is a price to pay, though, and that is related to the fluctuation of the transmit signal amplitude leading to a higher peak-to-average power ratio (PAPR) compared to single-carrier signaling, which scales with the number of subcarriers N, challenging the requirements on the power amplifier (PA).

The most prominent representative of a multi-carrier waveform is OFDM. Here, the subcarrier signals utilize the maximum bandwidth of B/N, thus attaining high spectral efficiency. Though the spectra of the subcarrier signals overlap in frequency domain, an orthogonal design of those spectra ensures an interference-free reconstruction at the receiver. The number of subcarriers for practical implementation is typically chosen as a power of 2, which allows using computationally efficient fast Fourier transform (FFT) algorithms for generating transmit signals and analyzing received signals, respectively. The spectrum of each subcarrier signal has a *sinc*-shape, which translates to the rectangular pulse in time domain. Typically, a guard interval is used for the transmission between two successive OFDM symbols in time domain, which should be larger than the channel's delay spread to protect the OFDM symbols from any inter-symbol interference. This, however, creates an additional overhead, which degrades the spectral efficiency. To keep this overhead small, the required size of the guard interval is usually decisive for the subcarrier bandwidth and thus the number of subcarriers N. Most of the OFDM systems operated today fill this guard interval with a cyclic extension of the OFDM symbol, yielding the so-called cyclic prefix of a CP-OFDM signal. However, also other measures are possible, such as zero padding yielding a zero prefix (ZP-OFDM), or using a unique word (UW-OFDM) as a prefix, which can then also beneficially be used for channel estimation purposes (see Section 11.3.1.2 and [4]). Pure ZP-OFDM is usually not favoured to be applied in practice, since it degrades the PAPR and challenges the PA due to the sudden drops of the signal to zero.

As described above, OFDM schemes as used in today's systems have indeed favourable properties for application in practice; however, they also exhibit some drawbacks. In particular, they require tight synchronization in time and frequency to maintain the signal orthogonality, and they are vulnerable to Doppler distortions in highly mobile channels. Moreover, they rely on a fixed configuration of the so-called numerology, constituted by the number of subcarriers N and the size of the guard interval, which is usually chosen as a best fit for supporting all the channel conditions that are expected during system operation. Adaptations or adjustment of the numerology during operation are not yet foreseen and are not well supported by conventional OFDM anyway, in particular if different numerologies should be supported simultaneously within the available bandwidth B to provide the system more flexibility to respond to the particular requirements of new services and use cases. The main reason for all these deficiencies is the *sinc*-shape of the subcarrier spectra, which has a poor localization of the signal power in the frequency domain due to its high side lobes. For improving the spectral containment of the subcarrier signals, OFDM can be extended by filtering components, which suppress the side lobes of the subcarrier signals. This can be done either by windowing of the time domain signal, which translates to filtering each subcarrier signal in frequency domain, or by filtering in time domain with a filter spanning a set of subcarrier signals – a so-called *sub-band*. In both cases, the filters can be designed to exhibit a steep power roll-off at the edges of a sub-band of a desired size, thus minimizing the power leaking into the adjacent band. Depending on the design constraints, successively transmitted filtered or windowed OFDM symbols may overlap, which may be accounted for at the receiver to avoid interference to arise. This becomes necessary, though, only for a larger filter length going significantly beyond the length of the prefix. While for a fixed filter length, filtered OFDM signals can attain a steeper power slope at the edge of a sub-band, windowed OFDM signals provide additional robustness against frequency errors and Doppler distortions – thanks to the fact that the signal spectra of the individual subcarriers have been uniformly modified. If properly designed, filtering and windowing do not change the orthogonality of the OFDM system, and, hence, all algorithms developed for OFDM can be reused without any alteration.

A special case of a windowed OFDM system with overlapping symbols is Filter Bank Multi-Carrier (FBMC). With FBMC, no guard interval is required, and thus the maximum spectral efficiency can be achieved. However, due to the restrictions of the Balian-Low theorem [5], which states that it is not possible to attain maximum spectrum efficiency with a well-localized pulse power in time and frequency while maintaining complex orthogonality, either the power localization needs to be compromised or the signal orthogonality needs to be relaxed. The latter can be realized with Offset Quadrature Amplitude Modulation (OQAM), where the FBMC symbols carry real-valued data on the subcarriers only, and successive symbols are transmitted at double the symbol rate $2/T = 2B/N$. A complex modulation pattern ensures that the real-valued data of symbols overlapping within a period T overlay in different dimensions of the complex signal space, so that they can be easily reconstructed at the receiver. Hence, orthogonality in OQAM-FBMC exists only in the real field and no longer in the complex field as in OFDM. As a consequence, several schemes designed for OFDM cannot be directly transferred to be used with OQAM-FBMC, but require some redesign of selected signal processing procedures. Though OQAM-FBMC received a lot of attention in research during the past years, this latter fact hampered this waveform to get commonly accepted as a mature candidate for 5G. More details on 5G waveform design can be found in [6].

In recent 3GPP discussions on the waveform to be used for NR, where the focus has been set on enhanced mobile broadband (eMBB) and URLLC services, it has been agreed that the waveform underlying NR should be based on CP-OFDM [3]. It may be extended by filtering components like filtering and windowing, but this filtering option should be transparent to the receiver. In practice, this means that the schemes should work properly even for the case that the receiver applies a simple CP-OFDM receiver without any further filtering. This transparency requirement can be supported by all extended OFDM schemes where the length of the filter tails does not go significantly beyond the length of the guard interval. Depending on the particular filter design, evaluations have shown that the tails are allowed to overlap with preceding and succeeding symbols by up to a maximum of 50% of the symbol period N/B for supporting the transparency requirement. However, it should be noted that some performance degradation will always be observed in this case: since CP-OFDM-based processing at the receiver translates to a mismatched filtering if any other filter has been used at the transmitter, the signal-to-noise ratio (SNR) at the receiver cannot be maximized, translating to an effective performance loss. The relative performance loss generally increases with the length of the filter tail. Furthermore, longer filter tails may lead to increased vulnerability to longer channel delay spreads.

To alleviate the problem of the high PAPR of multi-carrier OFDM signals, a single-carrier-like waveform based on OFDM has been introduced as DFT-spread OFDM (DFT-s-OFDM). Here, the modulated data symbols are generated in time domain with a bandwidth covering a sub-band, and this signal is then DFT-transformed and shifted to the desired frequency position. Applying the inverse FFT (IFFT) covering the entire transmission bandwidth B and adding the guard interval then creates the single-carrier-like OFDM signal. At the receiver, the signal is equalized as in OFDM and then transformed via an inverse DFT (IDFT) to obtain the data symbols. For channel estimation, full preamble (or mid-amble) signals need to be used now, meaning that a pilot signal fills the entire sub-band. This way, the simple and efficient OFDM-based processing can be maintained, while the tight envelope of single-carrier signals yielding small PAPR can be adopted. Note, however, that deep fades in the sub-band may again decrease the system performance significantly due to the implicit averaging of the channel quality over the sub-band – a feature inherent to any single-carrier transmission.

The DFT-s-OFDM scheme may also be combined with filtering and windowing, similar to its pure OFDM counterpart. Moreover, an advanced scheme with an inherent windowing operation has been proposed as zero tail (ZT) DFT-s-OFDM [7]. Here, a tail of zeros is used at the beginning and the end of the OFDM-like symbol, whose length can be adjusted to improve the resilience against inter-symbol interference in delay-spread channels.

11.3.1 Advanced Features and Design Aspects of Multi-Carrier Waveforms

In the following, we provide details on advanced features of multi-carrier waveforms and elaborate on various aspects to be accounted for in the waveform design.

11.3.1.1 Dynamic Numerology Switching

OFDM-based multi-carrier waveforms as introduced above provide three degrees of freedom for their overall design, which can be set according to the particular requirements of a desired service or use case: the numerology, consisting of the subcarrier spacing B/N and the length of the guard interval, and the filter used to attain spectral containment, which may be a time domain window or an FIR filter, or a combination of both. The spectral containment of the signal power yielded by the filtering allows partitioning the system bandwidth into separate isolated sub-bands, wherein the numerology can then be configured individually, thus supporting simultaneously different numerology configurations in the same band. This is referred to as frequency domain numerology multiplexing. Numerologies may further be changed over time following a predefined time grid, then yielding a time domain multiplexing. The time/frequency grid defining potential switching borders between different numerologies covers the total set of resources in time and frequency, and enables structuring it into so-called *tiles*, representing subsets of resources dedicated to a particular service with its unique numerology configuration [6]. Among service data, the tiles may also carry control information required for this service, yielding a self-contained structure.

The tile structure currently being discussed in 3GPP NR is based upon the length of a time slot as defined in Long-Term Evolution (LTE), constituted of 7 OFDM symbols and a constant overhead of roughly 7% for the guard interval (prefix). The subcarrier spacing (or symbol duration, respectively) is allowed to be scaled by a factor equal to an integer power of 2, i.e., 2^n with $n \in Z$. Choosing these factors, while keeping the overhead for the prefix constant, yields a nested structure of the TTI time grid, i.e., consecutively transmitted frames of different TTI size will be aligned in time at regular intervals, determined by the frame with the longest TTI. This feature allows for simple and frequent switching of TTI configurations over time without creating idle times, thus facilitating the desired flexibility in the frame design to respond to different services' demands. Note that, besides improving latency, as detailed in Section 12.3.2, shorter symbol lengths also enhance the signal quality under high mobility, as inter-carrier interference is decreased and channel estimation quality is increased thanks to a denser pilot grid in time domain [6]. However, when choosing short symbol durations with a constant overhead of 7% for the prefix, it may become too short in scenarios with large delay spread, giving rise to undesired inter-symbol interference. In this case, an enhanced prefix is ready to be chosen. If the number of OFDM symbols in the TTI is reduced from 7 to 6, some additional room for the enhanced prefix is gained, yielding an overhead of 25%. This way, we can allow for using a larger prefix within a TTI without violating the TTI time grid, enabling to change the prefix length on a TTI level.

11.3.1.2   Advanced Prefix Design

As the prefix in CP-OFDM is usually discarded at the receiver, there is high motivation to make better use of this signal overhead, which triggered endeavours on advanced prefix design. One approach is to replace the prefix with a known sequence, leading to UW-OFDM. This principle is also called *known symbol padding* (KSP) OFDM in the literature. With the unique word (UW) replacing the prefix, a periodically appearing known sequence is available, which can be used as training signal that comes without any additional cost in signaling overhead. Such training can be used for various purposes, for example, to enhance channel estimation, phase noise tracking, Doppler tracking, synchronization, and to monitor the received signal power - for instance to detect blockage of the link. Furthermore, the UW can be designed to achieve lower PAPR. In [4], it has been shown that by exploiting UW for phase noise estimation, considerable gain can be achieved compared to CP-OFDM when considering low-complexity schemes.

When replacing the prefix in CP-OFDM with a UW, the circular convolution between channel impulse response and the transmitted signal is impaired. Therefore, specific demodulation schemes are needed at the receiver. Three methods have been described in [4], showing that with proper demodulation schemes and moderate complexity increase, no performance loss is observed. Further, full compatibility with CP-OFDM can be achieved regarding multiple access, multiple-input multiple-output (MIMO), and pilot usage. Practical schemes have been proposed in [4], which exploit UW to enhance channel estimation and phase noise tracking.

11.3.1.3   Mitigating Hardware Impairments

One important challenge in the design of the multi-carrier waveform is the fact that the hardware used for implementing the transceiver functionalities typically exhibits various imperfections. This gets more pronounced with higher carrier frequencies. Therefore, the waveform design and evaluation should take into account these hardware impairments. Two most typical hardware impairments are oscillator phase noise and nonlinear characteristics of the PA. The following paragraphs will provide more details on this.

Phase Noise

Free-running oscillator and phase-locked loop (PLL) based oscillators are the most common implementations assumed in the literature [8]. The phase noise of the PLL-based oscillator consists of three main noise sources: the reference oscillator, the phase-frequency detector along with the loop filter, and the voltage controlled oscillator (VCO). Each of these noise sources includes both white noise (thermal noise) and colored noise (flicker noise). The detailed modeling of phase noise can be found in Table 4-2 of [9]. The detrimental effect of phase noise increases as a function of carrier frequency. Phase noise will cause common phase errors (CPEs) and inter-carrier interference (ICI), resulting in an increased error vector magnitude (EVM) of the desired signal. CPE refers to a common phase rotation of all sub-carriers, which can be compensated quite easily in frequency domain. The actual phase rotation requiring compensation is estimated with the help of pilot subcarriers. ICI may be modeled as additive noise (not always Gaussian) and is usually hard to be compensated. It requires denser pilots for phase noise and channel tracking, and estimation and compensation can be computationally intensive. The most straightforward method to mitigate the effect of phase noise is the use of a larger subcarrier spacing, though this may increase the vulnerability to frequency-selective channels if the overhead of the prefix is kept constant, as discussed earlier in Section 11.3.1.1.

Non-linear Characteristics of the PA

When digitally modulated signals go through a PA having non-linear characteristics, spectral regrowth appears, which in turn causes adjacent channel interference. The power series model or the polynomial model is widely used in the literature for the modeling of memoryless nonlinear PAs [10], which is given by:

$$y(t) = \sum_{k=0}^{K} c_{2k+1} \left| x(t) \right|^{2k} x(t) \tag{1}$$

where K is the non-linear order, $y(t)$ is the output signal, $x(t)$ is the input signal, and c_{2k+1} is the $(2k+1)$th complex-valued polynomial coefficient. The coefficients c_{2k+1} can be calculated by using least squares estimation (LSE). More recently, there has been growing interest in modeling nonlinear PAs with memory effects, for instance based on a memory-polynomial model or Volterra series [4].

In order to have high PA efficiency, it is required that the input signals have low PAPR. One may also employ linearization techniques, e.g., pre-distortion, to compensate for PA nonlinearity. However, such technique comes with major baseband complexity and may not be effective for large bandwidth signals and/or hybrid beamforming architectures. It remains an open question to which extent such techniques can be effective for large bandwidth signals, as for instance envisioned in the context of mmWave, and how much additional complexity is required.

11.3.1.4 PAPR Reduction Techniques

One of the main drawbacks of OFDM as compared with single-carrier waveforms is the high PAPR, which requires the PA to operate linearly in a very wide range. As explained in the introduction of Section 11.3, DFT-s-OFDM is a means to create a single-carrier-like signal based on OFDM, though its performance suffers from deep fades, which may occur in particular if the transmit signal covers a broader bandwidth. Hence, PAPR reduction techniques for conventional OFDM are worthwhile to be explored. It is noteworthy that a low PAPR is not only important in the UL, but also in the DL, in particular at very high frequencies (e.g., mmWave bands) due to the need for low-cost BSs. Various simple and effective PAPR reduction techniques for OFDM (e.g., amplitude clipping, exponential companding, and constrained clipping) have been proposed in the literature, see [4] for further details and corresponding references.

Figure 11-1 (top) shows the complementary cumulative distribution function (CCDF) of the PAPR for DFT-s-OFDM as well as for OFDM with and without various PAPR reduction schemes. Through applying appropriate PAPR reduction techniques, it is shown that OFDM can achieve a similar PAPR performance like DFT-s-OFDM.

Figure 11-1 (bottom) shows the EVM performance for DFT-s-OFDM and OFDM signals with and without the different PAPR reduction schemes at different power back-off settings. Here, the EVM is measured at the receiver, where the signal distortions are arising from the noise, from distortion being introduced by the PAPR reduction techniques, and from the nonlinear characteristics of the PA. With high power back-off, the signal is less distorted by the nonlinear PA; thus, the EVM performance of DFT-s-OFDM and OFDM is almost the same, while OFDM with the PAPR reduction schemes shows degraded EVM performance due to the distortion introduced by those schemes. With low power back-off, the nonlinear effects of the PA are introducing additional signal distortions, where the severity of the distortion depends on the PAPR of the input

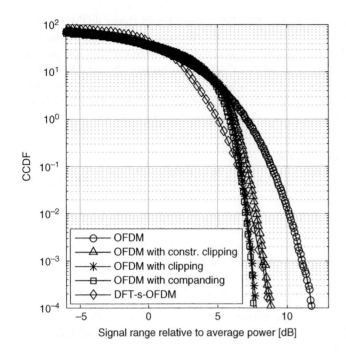

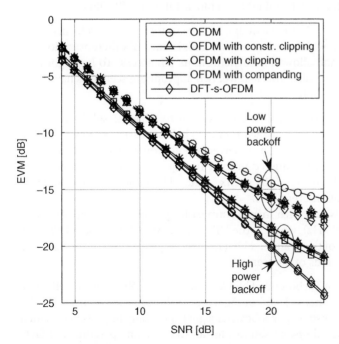

Figure 11-1. CCDF of the PAPR for DFT-s-OFDM and OFDM signals with/without PAPR reduction schemes (top) and corresponding EVM performance (bottom).

signal. As DFT-s-OFDM and OFDM with PAPR reduction have lower PAPR than OFDM, OFDM performs the worst among all schemes, while OFDM with PAPR reduction schemes achieves similar EVM performance as compared to DFT-s-OFDM. It is assumed here that the low power back-off is insufficient to provide the required "headroom" to accommodate the OFDM signal, but sufficient to accommodate the OFDM signal with PAPR reduction. Furthermore, it is shown that the degradation of EVM in OFDM with PAPR reduction is more pronounced at high SNRs than for low SNRs, where the additive channel noise is the dominant adverse factor.

11.3.2   Comparison of Waveform Candidates for 5G

Several windowed and filtered multi-carrier schemes have been investigated in recent years, targeting on finding the most suitable candidates for 5G. These have been evaluated and compared with respect to their capability to address the particular requirements of the future 5G system as well as to their behaviour under given hardware impairments, which get particularly pronounced in the context of higher frequencies, as discussed in the previous section. In this section, a summary of the most important findings from this evaluation is presented. A detailed description of all multi-carrier schemes as well as further details on their evaluation and comparison can be found in [4][6]. The following multi-carrier waveform candidates have been in the focus of those investigations:

- **Conventional (non-filtered)**: CP-OFDM; or single-carrier-like variant DFT-s-OFDM;
- **Subcarrier-wise filtered**: Windowed (W)-OFDM, pulse-shaped (P)-OFDM, UW-OFDM, flexibly configured (FC)-OFDM, and OQAM-FBMC;
- **Sub-band-wise filtered**: Universal-filtered (UF)-OFDM and block-filtered (BF)-OFDM.

As detailed at the beginning of Section 11.3, one favorable feature of the novel waveforms is the spectral containment of the signal power, which facilitates easy coexistence of different radio configurations in the same frequency band and allows for asynchronous UL access. To highlight how well the different waveform candidates can realize this feature, two appropriate scenarios have been selected for evaluation and direct comparison [6]:

- **Scenario 1 (asynchronous UL access)**: Here, 3 UEs are assigned to adjacent sub-bands, each spanning 48 subcarriers, corresponding to a total of 720 kHz for 15 kHz subcarrier spacing. The UE in the center sub-band is the one being evaluated. The receiver is synchronous with this UE, while the two UEs in the adjacent sub-bands are misaligned in their timing with a constant time offset relative to the center UE, going beyond the size of the prefix. A guard band of configurable size is used between the adjacent sub-bands of different UEs.
- **Scenario 2 (UL synchronous transmission with mixed numerology)**: Here, 2 UEs are assigned to adjacent sub-bands of equal size (720 kHz), whereas the UE located aside the evaluated UE uses double the subcarrier spacing (30 kHz) of the evaluated UE (15 kHz). A guard band of configurable size is used between the adjacent sub-bands of different UEs.

Performance comparisons in terms of the Turbo-coded block error rate (BLER) versus the effective SNR are shown in Figure 11-2 (top) for scenario 1 and in Figure 11-2 (bottom) for scenario 2. The effective SNR reflects the total average power spent per subcarrier signal, including the power contained in the prefix overhead, if applicable. Standard system settings for an LTE system operating at 4 GHz with 10 MHz bandwidth have been applied, if not stated otherwise. In Figure 11-2 (top), 16-QAM modulation has been used, and the guard band between the frequency sub-bands of different users

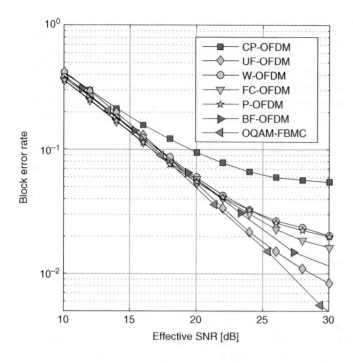

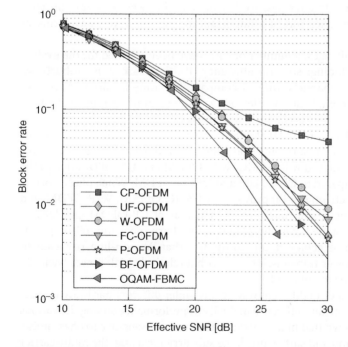

Figure 11-2. Performance comparison of the waveform candidates for asynchronous UL access (scenario 1, top) and mixed numerology coexistence (scenario 2, bottom).

is set to 30 kHz. The adjacent users are misaligned in their timing by a number of samples amounting to 1/8 of the FFT block size. From the figure, it can be observed that all evaluated waveforms attain much better BLER performance than conventional CP-OFDM. The best performance is attained by the OQAM-FBMC scheme, which is not fully OFDM-compatible, though. The sub-band filtered schemes get closest to this superior performance, while the windowed schemes follow at some distance. It is worth mentioning that with increasing the guard band size, the performance of all modified OFDM schemes converges towards that of OQAM-FBMC, while the performance of W-OFDM is kept at some distance.

In Figure 11-2 (bottom), 64-QAM modulation has been used, and the guard band between the frequency sub-bands of different users is set to 60 kHz. We observe that OQAM-FBMC again attains the best performance, followed at some distance by the modified OFDM schemes, which do not differ too much from each other, but all significantly outperform conventional CP-OFDM. Again, W-OFDM performance is kept at some distance. For further details on these waveform comparisons, the interested reader is referred to [6].

Effects of Hardware Impairments in the Context of mmWave Transmission

Non-ideal properties of the hardware implementing mmWave transceiver components cause impairments, as addressed in Section 11.3.1.3. Further impairments include in-phase and quadrature phase (I/Q) imbalance, sampling jitter and sampling frequency offset, carrier frequency offset, etc. These imperfections are present in every hardware implementation, but their impact in the mmWave frequency range is larger than in sub-6 GHz bands, because these hardware components are operated closer to the overall physical limits and therefore closer to the limit of their capabilities. In particular, the PA has lower efficiency at mmWave frequencies, so that increased power consumption for a given transmit power target is expected. Therefore, it is important to have low-PAPR waveforms, as stressed earlier in Section 11.3.1.4.

Based on those mmWave-related challenges, a number of KPIs have been selected for the evaluation of the waveform candidates, including: spectral efficiency, PAPR, phase noise robustness, robustness to frequency/time selective channels, MIMO compatibility, time localization, out-of-band emissions (with and without PA), complexity and flexibility. A summary of prominent results is presented here, while detailed evaluation results can be found in [4].

Figure 11-3 shows the power spectral density (PSD) of different waveforms with and without hardware impairments. Without any hardware impairments (top figure), it is shown that indeed very low out-of-band (OOB) emissions can be achieved with FBMC, W/P-OFDM and UF-OFDM due to the filtering/windowing operations, as compared to CP-OFDM and DFT-s-OFDM. When phase noise is included (bottom figure), the sharp spectrum roll-off provided by FBMC-OQAM, W/P-OFDM, and UF-OFDM is significantly reduced, but is still much lower than that of CP-OFDM and DFT-s-OFDM. When a nonlinear PA is further added, it is observed that the sharp spectrum roll-off promised by these waveforms is unlikely to be achieved, due to the fact that PA non-linearity leads to spectral regrowth. However, for low power transmission, i.e., with a relatively high power back-off, OOB advantages over OFDM can still be maintained.

Figure 11-4 shows the EVM performance of different waveforms under different hardware impairments. It is observed that there is no significant difference in the EVM performance among the various candidate waveforms. It is generally known that multi-carrier waveforms are sensitive to phase noise. However, with phase noise compensation and sufficiently large subcarrier spacing, the multi-carrier

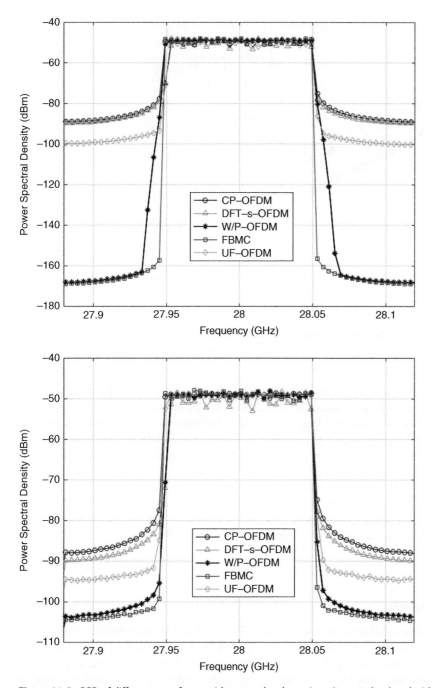

Figure 11-3. PSD of different waveforms without any hardware impairments (top) and with phase noise (bottom).

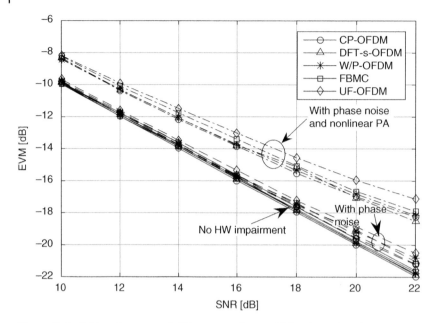

Figure 11-4. EVM performance of different waveforms with hardware impairments.

waveforms can be made robust against phase noise effects. Besides, scattered pilot-based channel estimation was used, which also compensates for the CPE caused by phase noise.

The waveform comparisons presented before have revealed that the waveforms extended by filtering or windowing provide significant gains compared to conventional CP-OFDM in scenarios reflecting novel applications envisaged for 5G. Looking at their performance under hardware impairments, it is observed that the studied waveforms have similar EVM performance as conventional CP-OFDM. The new waveforms exhibit improved OOB performance as compared to CP-OFDM, but the improvement gets smaller when hardware impairments, such as phase noise and nonlinear PAs, are taken into account. For nonlinear PAs with high power transmission, i.e. small power back-off, the OOB advantage finally vanishes, and similar OOB emissions are observed for all waveforms.

11.3.3 Co-existence Aspects

The 5G waveform should be capable to coexist with the CP-OFDM waveform, which is motivated by the following two reasons: First, during the early deployment of 5G systems, the existing 4G bands will not be re-farmed for 5G usage immediately. However, to allow for a gradual and effective penetration of the 5G system, some of the 4G bands, especially under-used UL carriers, could be shared with the 5G RAT. This kind of sharing could be semi-static or dynamic. Thus, with 4G and 5G RATs sharing the same band, the respective waveforms used in 4G and 5G, respectively, need to co-exist. Second, 5G is envisioned to support various kinds of services beyond conventional (e)MBB traffic, as outlined in Section 2.2. At the moment of writing this section, 3GPP NR has decided to adopt a CP-OFDM based waveform for eMBB services [3], while other waveforms are not precluded for other services, such as mMTC or V2X. This suggests that even within the 5G system, possible other waveforms need to coexist with CP-OFDM. For facilitating this coexistence, two options can be considered:

1) The co-existing 5G waveform should be designed as orthogonal (or quasi-orthogonal) as possible with respect to the 4G waveform to effectively minimize the mutual interference. Naturally, this will limit the degrees of freedom in the overall waveform design. Some possible solutions are: The 5G waveform can be based on subcarrier-filtered waveforms on top of CP-OFDM, with the window length being limited to the length of the prefix, or, alternatively, the 5G waveform can be based on sub-band filtering on top of CP-OFDM, where the filter length has similar restrictions.

2) Another option is to release the requirement of the 5G waveform to be fully orthogonal to CP-OFDM. In this case, the inter-system interference needs to be controlled by applying a frequency guard band, combined with a power back-off at the edge of the sub-band used by any 5G service. This will allow for more ambitious 5G waveform designs. The inter-sub-band interference from the "4G sub-band" to the "5G sub-band" can be handled by the 5G receiver thanks to spectral confinement techniques, whereas the interference from 5G to 4G is mitigated by proper guard band dimensioning and by using a power back-off at the edge of the sub-band used by 5G.

Another aspect constraining 4G-5G co-existence resides in the basic frame design of LTE. The Physical Downlink Control CHannel (PDCCH) occupies one up to three successive OFDM symbols across the entire band. Furthermore, LTE applies cell-specific reference symbols (CRS), which are scattered across the bandwidth and which are not allowed to be muted. To care for this, the 5G signals would have to be muted at the respective positions in the time-frequency grid. The introduction of mini-slots in 5G NR supports to solve this issue. In the UL, the respective frame design decisions in 4G are less restrictive, as both the control channel and the reference symbol placements are frequency-localized.

11.3.4 General Framework for Multi-Carrier Waveform Generation

Multi-waveform harmonization was not critical for the design of LTE, because the main target use case has been MBB, traditionally well-served by the use of CP-OFDM. However, in 5G, the use cases as well as the respective requirements are much more diverse, motivating the use of different waveform alternatives with different features, as already discussed in previous sections. In particular, one of the major motivations for waveform harmonization is to enable the support of different waveforms for different services with minimized implementation effort. This objective can be attained by a modular structure based on the reuse of hardware components.

In [11], a general framework based on the mathematical tool known as Gabor systems [5] was introduced, which allows to represent different multi-carrier waveforms under the same system model by selecting the appropriate prototype filter, subcarrier spacing and symbol spacing in time.

In fact, the system model is useful to represent conventional CP-OFDM, W-OFDM, P-OFDM and ZT-DFTs-OFDM. Furthermore, the framework is also valid to represent waveforms of the FBMC family, such as FBMC-QAM and FBMC-OQAM. As a result, this general framework properly facilitates a harmonized hardware implementation capable to generate multiple waveforms.

Harmonized Implementation Concept for Multiple Waveforms

The block diagram of a harmonized transmitter capable to implement the generic multi-carrier waveform following the general framework is presented in Figure 11-5. The diagram corresponds to an implementation based on poly-phase filtering, carried out in time domain through a poly-phase network (PPN) [12]. As further elaborated next, by selectively enabling or disabling particular blocks, the harmonized implementation is able to generate each of the mentioned waveform variants.

Note that the blocks necessary to generate the CP-OFDM signal are shown in white, whereas the extra blocks required for the generation of any of the other waveforms are highlighted in grey. The special blocks included in the harmonized implementation, which require specific configuration for some waveforms, are detailed next:

- **DFT spreading**: This block is intended to perform the spreading operations necessary for the generation of ZT-DFT-s-OFDM;
- **OQAM pre-processing**: This set of blocks contains the necessary preparative multiplexing steps required for FBMC-OQAM, that is, complex-to-real number conversion of QAM complex symbols, up-sampling, and time staggering;
- **PPN**: The task of this block is to perform the convolution of the discrete signals with a filter implemented through a PPN.

The harmonized block diagram illustrates the usefulness of the proposed implementation to provide flexible adaptation to a particular communication scenario, i.e. to allow for a dynamic waveform selection, and, at the same time, to reduce implementation costs. Regarding the OFDM variants, all of them will leave aside the OQAM pre-processing and will include the prefix addition, except for ZT-DFT-s-OFDM, which leaves also the prefix aside. Only ZT-DFT-s-OFDM makes use of the DFT spreading block, though. With respect to the PPN block, its inclusion will actually depend on the specific variant: Plain CP-OFDM will not require this block, whereas W-OFDM and P-OFDM will need it for windowing. Concerning the FBMC transmitters, they must enable all the blocks in Figure 11-5 except those for involving the prefix and DFT spreading. The operations in charge of OQAM generation are only necessary for the transmission of FBMC-OQAM. Further

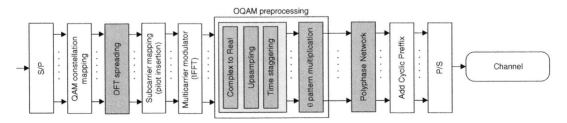

Figure 11-5. Harmonized transmitter for multi-carrier waveform generation.

implementation aspects and complexity evaluations of other waveforms and of the harmonized transceiver are covered in detail in Section 16.3.3, where the typical savings provided by the harmonized transmitter compared to the independent implementation of all constituting waveforms are shown to be in the range between 60–75%.

11.4 Coding Approaches and HARQ

11.4.1 Coding Requirements

An unprecedented variety of new applications and services are foreseen to be introduced in the future 5G communications systems, as detailed in Section 2.2. This results in challenges and constraints for the envisioned usage scenarios, such as very high user data rates for eMBB services, stringent reliability and latency constraints for URLLC, or the transmission of short packet messages with sporadic traffic for mMTC. Therefore, a special emphasis has to be placed on the design of FEC solutions able to efficiently support the underlying constraints. In this regard, three main KPIs provided in 3GPP NR [13] can be clearly identified as of high importance for FEC choice and design:

- For eMBB, the target for peak data rate should be 20 Gbps for DL and 10 Gbps for UL;
- For URLLC, the target for user plane latency should be 0.5 ms for UL, and 0.5 ms for DL;
- For URLLC, the target for reliability should be 10^{-5} of packet error rate (PER) with keeping the time constraint of 1 ms. This reliability performance shall be supported together with user experienced data rate on the order of 300 Mbps.

Unfortunately, the FEC coding and modulation components of LTE and LTE-Advanced (LTE-A) are not optimal in this respect, as they were not designed to meet such requirements. Actually, the following weak points have been identified [14] for the Turbo codes (TCs) employed in LTE:

A known issue related to TCs resides in their poor performance at low error rates when transmitting data with coding rates higher than 1/3. This is due to the so-called *error floor*, which can be observed when a TC is punctured with the rate matching mechanism. A detrimental resulting effect is the frequent resort to HARQ retransmissions. Consequently, the LTE FEC code cannot simultaneously meet the reliability and latency constraints of URLLC usage services. Moreover, the error floor issue of TCs is also not compatible with the requested increased user data rate for eMBB services, since high coding rates are then required.

Using convolutional component codes, conventional TCs have not been originally designed for encoding short blocks. So, the LTE/LTE-A FEC code does not provide capacity-approaching performance for the transmission of short data packets. In particular, it is called for using tail bits for trellis termination. On one hand, this results in a non-negligible bandwidth efficiency reduction for short blocks. On the other hand, this type of trellis termination introduces low-weight truncated codewords and does not ensure the same protection for all data bits, since tail bits are not encoded twice (i.e., Turbo encoded) as the rest of the data. Therefore, the LTE TCs need to be improved to be able to efficiently cope with the sporadic traffic of short messages, as typical for mMTC services.

The target peak rate of 20 Gbps for eMBB represents a major challenge for any family of FEC codes. This is particularly true when taking into account the required flexibility, on the order of what is specified in LTE in terms of supported packet sizes and coding rates. Regarding TCs, due to the recursive nature of the underlying convolutional codes, their decoding structure is serial by nature, and they are known to have difficulties in achieving extremely high data rates at a reasonable implementation cost.

Accordingly, coding solutions able to answer favorably to the identified constraints of the different usage scenarios of 5G have been identified and evaluated in the FEC selection process of 3GPP.

11.4.2 Coding Candidates

Defined around 10 years ago, the LTE TC is somewhat dated, delivering performance far from what can be achieved from best FEC code designs nowadays. For example, an enhanced TC family was designed to target the requirements of the different scenarios of 5G [15][16]. On the other hand, taking the envisioned peak data rates into account, LDPC codes [17] may represent a better choice as a coding solution. Furthermore, Polar codes (PCs) [18] concatenated with an outer error detecting code, such as a cyclic redundancy check (CRC) code, have recently emerged as strong coding candidates for short block sizes. In addition to these capacity-approaching coding solutions, convolutional codes and block codes, such as Reed Muller or Bose–Chaudhuri–Hocquenghem codes (BCH) can be of interest for the particular case of extremely short packet sizes, for instance less than 40 bits.

In the rest of this section, a special focus is put on describing the latest advances regarding the three main families of codes represented by TCs, LDPC codes, and concatenated PCs.

11.4.2.1 Enhanced TCs

The enhanced TC (eTC) family [15][16] was designed to address the drawbacks of the existing TC solution in LTE, when targeting the requirements of the different scenarios of 5G. The encoder structure is a parallel concatenation [19] of two 8-state recursive systematic convolutional encoders, as shown in Figure 11-6. Each component code is a modified version of the LTE TC component code with an additional parity symbol W, resulting in a TC with a mother coding rate equal to $R = 1/5$. The generator polynomials for C1 and C2 are $(1,(1+D+D^3)/(1+D^2+D^3)$ and $(1+D+D^2+D^3)/(1+D^2+D^3))$, respectively.

Tail-biting, also called circular encoding, is introduced. It ensures that, when encoding a message of length K, the initial and the final states are identical for each component encoder C1 and C2. Tail-biting is the best-known termination method for TCs, since it avoids the transmission of tail bits. Thus, there is no rate loss and the spectral efficiency of the transmission is not reduced. Moreover, with tail-biting, all the information bits are protected in the same way by the TC, and the circular property prevents the occurrence of low-weight truncated codewords. Therefore, tail-biting termination helps to lower the error floor.

Rate adaptation is performed via the application of a periodic puncturing pattern of length Q. The information block size K is assumed to be a multiple of Q. Typical values for Q are 4, 8, 16 or 32, but others values are possible, provided that Q is a divisor of K. The selection of the puncturing patterns is performed on the basis of a joint analysis of the Hamming distance spectrum of the punctured component convolutional code and of the mutual information exchange between the two component encoders [15]. Incremental puncturing patterns have been designed, enabling inherent HARQ support via incremental redundancy.

The interleaver has an important impact on the performance of TCs in the low error rate region. For implementation-friendly designs, algebraic interleaving is adopted in most standards. Amongst

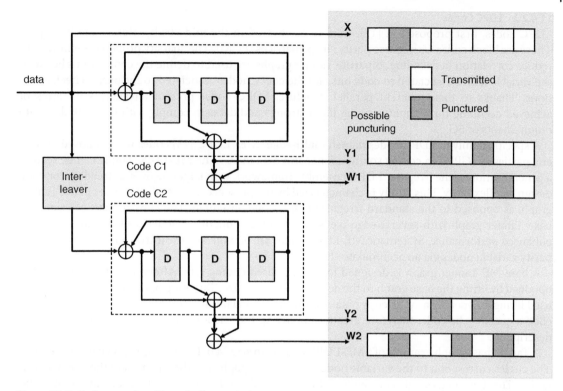

Figure 11-6. Turbo encoder with periodic puncturing.

the different existing models, it was shown in [20] that the *almost regular permutation* encompasses most of the popular algebraic interleavers, including the quadratic permutation polynomial (QPP) model of the LTE TC. Therefore, almost regular permutation interleavers can achieve at least the same minimum Hamming distances as QPP interleavers and were adopted for the proposed eTC. The corresponding interleaving function is given by the following equation:

$$\Pi(i) = (Pi + S(i \mod Q)) \mod K \tag{2}$$

where i denotes the address of the data symbol after interleaving, and $\Pi(i)$ represents its corresponding address before interleaving. P is a positive integer relatively prime to K. S is a vector containing Q integer values. The values of parameters P and $S(i)$, $i = 0 \cdots Q-1$, are chosen to support the different block sizes and coding rates. Their selection procedure follows the steps described in [16] to generate a so-called *protograph-based* interleaver design.

The eTCs are decoded using an iterative process that exchanges probabilistic extrinsic information between two component convolutional decoders, each applying a variant of the Bahl, Cocke, Jelinek and Raviv (BCJR) algorithm in the logarithmic domain, commonly named *scaled Max-Log MAP (maximum a posteriori) algorithm* [21]. After several decoding iterations, for instance 6 to 8, the final binary decision is provided.

11.4.2.2 LDPC Codes

LDPC codes, first proposed in [17], were re-discovered in the mid-90s by McKay and Neal [22]. These are block codes with sparse parity check matrices (PCM). The sparsity property is enforced to reduce correlation in decoding. Bipartite Tanner graphs are used to define the connections between the variable nodes associated to code bits and the check nodes associated to the parity-check equations. Thanks to their inherent parallel structure, LDPC codes present advantages in terms of achieved decoding throughput, making them strong candidates to comply with the peak data rate requirements of 5G.

A specific family of LDPC codes named multi-edge (ME)-LDPC [23] codes was proposed by several parties at 3GPP as a coding solution for eMBB data in 5G. The ME-LDPC family can be seen as a generalization of the irregular LDPC ensemble framework with larger degrees of freedom for code design and flexibility. This family is characterized by the presence of several edge types in the Tanner graph, as opposed to the standard irregular LDPC code ensemble. The construction consists of a base Tanner graph with several edge types and includes punctured nodes, called *state nodes*, for enhanced performance. Systematic ME-LDPC codes are obtained by the introduction of degree-two parity variable nodes via an accumulate chain.

A base ME Tanner graph is designed for each desired coding rate. Afterwards, the final PCM is obtained by lifting the base graph to the desired codeword size. Targeting hardware-friendly designs, used lifting is generally a cyclic copy obtained through circulant matrices. This makes the ME-LDPC code a quasi-cyclic code, therefore greatly simplifying the encoding operation as well as the code description.

A Tanner base graph example of a ME-LDPC code with size 24 before lifting is shown in Figure 11-7. The circles correspond to the variable nodes of the base graph, and the squares are the parity-check nodes. The T-shaped dongles on the top of each variable node represent the transmitted bits. The variable node with no dongle represents a punctured state node, intentionally introduced to improve performance. The degree-two parity node accumulate chain is shown on the right with dashed edges. For a codeword size N, the lifting size Z satisfies $N = 24 \times Z$. To obtain a cyclic lifting, each edge has to be associated with an integer from the cyclic group modulo Z.

The design of the base Tanner graph for each target coding rate has an important impact on the overall performance of the code. Parameters such as the girth, i.e., the minimum cycle length in the Tanner graph, also play an important role in resulting performance. ME-LDPC codes are generally designed using the density evolution technique [24].

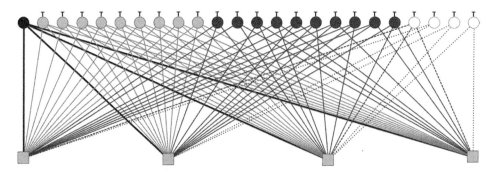

Figure 11-7. Tanner graph example of an ME-LDPC code with a base graph size of 24 nodes.

HARQ support via incremental redundancy (IR) can be implemented by designing the highest-rate code and then extending this to lower rates with the addition of extra parity bits. However, due to the incremental redundancy constraint, the base graphs of the higher-rate codes have to be sub-graphs of the base graphs of the lower-rate codes. This generally results in a low-rate base graph different from the one obtained directly by density evolution, with poorer performance.

The decoding of ME-LDPC codes requires iterative processing and exchange of extrinsic probabilistic information between variable nodes and check nodes, based on a low-complexity variant of the belief propagation principle in the logarithmic domain, called *scaled* or *offset min-sum algorithm* [25]. After several iterations, for instance 10 to 25 layered iterations, the final binary decision is provided. Since the lift size Z is generally larger than the number of columns in the base graph, a decoding hardware capable of processing Z liftings in one clock cycle would allow a high decoding throughput, which is particularly appealing for 5G. Note that the complexity of LDPC decoders is also discussed in further detail in Section 16.3.4.

11.4.2.3 Polar Codes

A PC can be viewed as a recursive concatenation of a base short block code designed to transform the encountered transmission channel into a set of virtual channels with variable levels of reliability. The first description introducing the idea of channel polarization and the framework of PC design was explicitly provided in [18].

The capacity C of a binary input symmetric discrete memoryless transmission channel T satisfies $0 \le C(T) \le 1$. The problem of designing an error correcting code for the two extreme values is simple to solve. The target of polarization is to transform N channel uses of T into N virtual channels with capacity 0 or 1, as shown in Figure 11-8. Polarization is performed via the application of XOR operations as described in [18]. It can be achieved for quasi-infinite packet sizes. However, for finite-length packets, a non-uniform reliability distribution is obtained for the different virtual channels.

The simplest application of this principle advocates the design of a base matrix respecting the polarization constraints that is recursively used until polarization is achieved over the codeword length N.

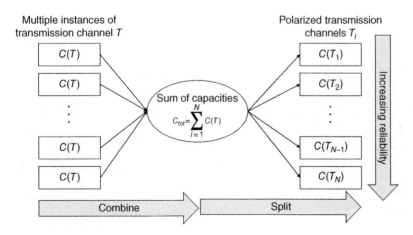

Figure 11-8. Channel polarization.

The right lower dashed rectangle in Figure 11-9 represents the encoder of a size-4 Polar codeword. The upper dashed rectangle represents the code extension to a codeword size of 8 bits. Rate compatibility is achieved by transmitting a subset of bits u_i enjoying the highest reliability levels after polarization. Bits that are not transmitted, called frozen bits, correspond to the least reliable channels after polarization.

The corresponding code shows capacity-achieving performance for quasi-infinite packet sizes with a successive interference decoder [18]. However, for most packet sizes used in practical applications, PCs decoded with this type of decoder show poor performance, far from the best achieved by TC and LDPC codes.

At a later stage, a list-based decoder was proposed in [26]. It classifies codewords in increasing reliability order. Then, a concatenation with an outer error detection code, typically based on a CRC, was introduced to eliminate the least reliable codewords from the list [27]. The resulting concatenated structure is able to bridge the existing performance gap with TC and LDPC codes. However, there still exist practical drawbacks in terms of implementation efficiency and parallelization.

During the standardization process of 5G, a novel concatenated structure, called parity-check PCs, was proposed [3]. This structure divides the codeword into independent sub-codewords

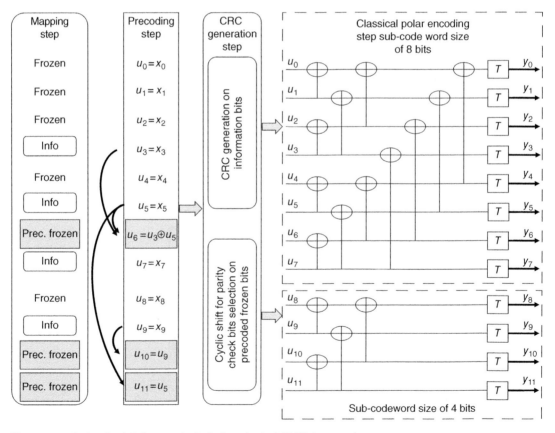

Figure 11-9. Parity-check Polar encoder including classical CRC Polar encoder.

(potentially of different sizes) linked solely by a precoding step. This latter consists of an XOR or a repetition operation that generates precoded frozen bits (u_6, u_{10} and u_{11}), as shown in Figure 11-9. The precoding step is followed by the generation of cyclic parity-check bits, separately on the pre-coded frozen bits and on the information bits, used for error detection. Two families of parity-check bits are generated from precoded frozen bits: self-parity-check bits computed from bits of the same sub-codeword (such as, u_6 and u_{10}), and cross-parity-check bits computed from bits of adjacent sub-codewords (such as, u_{11}). Finally, classical Polar encoding is performed.

For each sub-codeword, the corresponding decoder unrolls and sorts its list of candidate code-words in increasing reliability order. Then, adjacent sub-codewords exchange information on cross-parity-check bits. These bits can now be decoded as frozen bits by each individual list decoder, pruning the list of candidates of the corresponding sub-codeword. The resulting decoder, called *chained-list* decoder enjoys lower latency and can achieve higher throughputs since sub-codewords can be decoded in parallel, making it particularly appealing for the peak data rates of 5G. Moreover, a well-constructed PC with a proper parity-check function over some well-chosen frozen bits increases its minimum Hamming distance, thereby improving the error correction performance. In fact, appending CRC bits as in [27] can be seen as a special case of the parity-check PC.

HARQ support is achieved through shortening via incremental frozen bits selection. Shortening is applied on the sub-codewords, such that the resulting total size is strictly larger than the size of the codeword for the lowest supported coding rate.

11.4.3   General Summary and Comparison

Strong and weak points of each of the three different families of codes considered for 5G are identified in this section. They are based on observed tendencies regarding performance, implementation complexity, flexibility and maturity, and hence characterizing these families.

It is extremely difficult to draw conclusions from sole performance comparisons, since they should be performed at equal complexity. Moreover, computational complexity is far from being an accurate representative of final hardware complexity and power efficiency. It is indeed widely acknowledged that memory resources and memory accesses have a large impact on chip area and power consumption, respectively. Therefore, fair comparisons require the availability of hardware designs for all three coding families, and comparable assumptions related to quantization, number of decoding iterations, achieved throughput, technology, etc. Nevertheless, since performance results comparing these families have shown that the observed gaps for large block sizes (i.e., larger than $K = 1024$ bits and code rates ranging from $R = 1/3$ to $R = 8/9$) were within the limits (≤ 0.3 dB) related to the assumptions made for reduced complexity hardware designs, it can be concluded that all three families of codes offer a satisfactory level of performance for these cases. A similar conclusion was reached during the selection process of 3GPP for 5G [28].

LDPC codes are considered as widely implemented in commercial hardware supporting several Gbps throughput with attractive area and energy efficiency, but with a flexibility support which is far below the requirements for the eMBB services of 5G. Actually, the area efficiency of LDPC decoders reduces when decreasing the coding rate, and their complexity rises when the flexibility is increased. Moreover, despite the ability of achieving large parallel decoding degrees, some of this parallelism may not be exploited for all code block lengths and code rates, resulting in a penalizing impact on energy and area efficiency. In addition, for the shortest block sizes of some 5G scenarios, penalizing short cycles cannot be avoided. This leads to poor performance of this family of codes for short block sizes. On another note, as previously mentioned, IR-HARQ support by ME-LDPC codes entails performance penalties for low rate ME-LDPC codes, compared to the best-known low-rate ME-LDPC codes and to TCs. This is due to the design constraints forcing high-rate base graphs to be embedded into low-rate base graphs. To conclude, hardware implementation with attractive area and energy efficiency is considered challenging when simultaneously targeting the peak data rate and flexibility requirements of eMBB services of 5G.

PCs are considered implementable, although there are currently no commercial implementations available, and, in relation to the eMBB service of 5G, there are some concerns linked to the maturity and the availability of decoding hardware. In addition, most existing work in the literature is related to successive interference cancellation decoders and not to list-based decoders that are required to enable the excellent performance of this family of codes. The implementation complexity of list-based decoding increases with the list size, especially for large block sizes. Moreover, the area efficiency reduces for shorter block lengths and lower coding rates. A list-4 decoder is largely agreed as implementable for all codeword sizes. However, in practice, most existing simulations results considered list-8 decoders that could be argued implementable only for short block sizes. Besides, IR-HARQ support for PCs entails performance penalties because the best positions of frozen bits are not necessarily incremental when lowering the coding rate. To conclude, decoding hardware can now achieve acceptable latency, performance and flexibility for PCs, but there are still some concerns about the feasible area efficiency and energy efficiency, and about the maturity of the technology.

TCs are widely implemented in commercial hardware, supporting IR-HARQ and the flexibility constraint required for 5G, but not at the high data rates or low latencies needed for the eMBB usage scenarios. In fact, TCs meet the flexibility requirements of 5G with the most attractive area and energy efficiency except at very high throughputs. With TCs, for a given code structure, the area and energy efficiency is constant when varying the coding rate, via puncturing and HARQ. Another advantage resides in the fact that the decoding complexity increases linearly with the information block size for a given mother code rate.

Due to the peak data rate requirements of 5G, in addition to a stand-alone TC solution, a combination of TC (for flexibility) and LDPC (for high throughputs) codes was proposed [29]. It considers designing a Turbo decoder capable of decoding both LTE and, at least, lower information block sizes ($K \leq 6144$ bits) of the eMBB scenario of 5G. For the high throughput case of 5G, a LDPC code can be designed with a limited flexibility or equivalently for a few combinations of code rates ($R > 1/2$) and block sizes ($K > 6144$ *bits*). This proposal has the benefit of combining the advantages of each family (TC and LDPC) of codes without bearing the burden of their drawbacks. Indeed, it was shown in [29] that this combination of codes could answer favorably all the requirements of 5G, especially in terms of complexity.

As a general conclusion, we can state that each family of codes presents its challenges when trying to satisfy simultaneously all the requirements of 5G. Therefore, it is quite difficult to clearly identify an all-around favourite without performing a joint thorough analysis of performance, complexity and latency taking into account real implementations. Due to timing constraints, the framework for such a comparison was not agreed in 3GPP, and individual technical contributions were used as a basis for the selection process. Finally, regardless of potential technical drawbacks, a compromise was found that led to the adoption of LDPC codes for eMBB data channels and PCs for control channels [3].

The choice of the coding solution for mMTC and possibly URLLC scenarios remains an open issue. While simulation conditions for these two scenarios are quite different from those of eMBB, they are partly related to the comparisons performed for short block sizes. In fact, block sizes lower than 1024 bits were considered for coding rates from $R = 2/3$ down to $R = 1/12$. Error rates of 10^{-4} to 10^{-5} of PER are targeted.

Taking into account the performance results provided in [30] comparing PCs and eTCs, we can clearly identify these two families of codes as strong candidates for URLLC and mMTC from the performance point of view, with a slight edge for eTCs showing improved performance for these targeted low rates [31]. LDPC codes cumulate two main drawbacks: the first lies in the large performance penalty (more than 1.0 dB for short block sizes and low rates) in some cases, the second is the fact that decoding complexity increases by orders of magnitude when decreasing the coding rate, compared to the two other families of codes.

From the complexity point of view, TCs present an advantage for such low rates as $R = 1/12$ since the decoding complexity of this family of codes scales linearly with the information block size K and not with the codeword size N. This is not the case for both Polar and LDPC decoders. To identify the best technical choice, in-depth complexity comparisons, going beyond simple computational complexity, would have to be carried out at comparable performance. The framework for such comparisons should be clearly set and agreed between the parties proposing these coding solutions.

Finally, the selection process for 5G coding solutions has launched a wave of new proposals for the three families of codes. Current studies are focusing on improving the decoding efficiency of Turbo decoders when targeting high throughput scenarios. A large number of studies are also focusing on the design and implementation efficiency of PCs and related decoders. Therefore, significant improvements are being made, and a thorough investigation taking into consideration performance and hardware complexity between these two strong candidates represented by TCs and PCs should be performed before a final selection.

11.4.4 Hybrid Automatic Repeat reQuest (HARQ)

HARQ is a tool being applied in many communication systems to both increase the reliability of a single transmission and the overall spectral efficiency. The cost to pay is a higher device complexity (e.g., a buffer is needed), processing effort, and latency. With a system applying HARQ, any data transmission is to be acknowledged by the respective recipient, i.e., in the DL by the device, and in the UL by the BS. Based on a given decision criterion (e.g., a CRC check), the integrity of a received packet is checked. If this check is not passed, a respective feedback message "negative acknowledged" (NACK) is sent to request a retransmission, or otherwise the successful reception is acknowledged (ACK). In the former case, the original message or another redundancy version is subsequently transmitted. The receiver combines both the original message and the retransmission(s) to eventually detect the original message. So, HARQ is a tool to increase the reliability by making use of time and (depending on the variant) frequency diversity, energy, and (again depending on the variant) coding gain. Obviously, there is a multitude of design choices possible for implementing this mechanism. In the following we introduce these, highlight the design choices of 4G, and present potential improvements for 5G NR.

The core functionality of HARQ is to enable the system to potentially retransmit a given packet if the reception has not been successful so far. For the actual implementation various aspects have to be decided for:

How to Check if a Transmission has been Successfully Received?
The receiver needs to be able to check the integrity of each received packet. The most reliable and complex variant is to base this decision on the CRC. The CRC check relies on the received bits after

FEC and hard decision. Less complex but also less reliable is to base this decision on the statistical characteristics of the soft-symbols before the demapper or of the soft-bits before or after the decoder. The potential costs to pay are reduced reliabilities (if the check wrongly indicates a successful reception) or reduced system throughput (if the check wrongly indicates a non-successful reception). The potential merits are less processing burden and reduced processing time, which can contribute to reducing the overall latency.

How to Design the Feedback Carrying the Outcome of this Check (Single-bit vs. Multi-bit Feedback)?

Data transmission both in 4G and in 5G NR is based on so-called transport blocks. A transport block comprises all bits being transmitted within a given transmission opportunity. For efficiency reasons, a transport block may be segmented into code blocks, each being individually encoded. The variant with the least overhead is to spend a single-bit feedback referring to the whole transport block. If only parts of the transport block are corrupted (e.g., due to localized interference), resources are wasted, as the complete transport block has to be unnecessarily retransmitted. With allowing single code blocks to be (non-)acknowledged, this can be avoided. The cost to pay is the increased overhead (i.e., one bit per code block instead of a single bit for the complete transport block).

The Timing between the (re-)Transmissions and the Related Feedback (Synchronous vs. Asynchronous HARQ) and the Overall Number of (re-)Transmissions Possible

HARQ can be implemented both in a synchronous and in an asynchronous manner. The former has a fixed timing between the (re-)transmissions and the related ACK/NACK feedbacks. The 4G UL applies this variant with 4 ms, or 4 TTIs, being the time span between (re-)transmissions in the UL and ACK/NACK feedback in the DL, and allows for up to 4 retransmissions. In contrast to this, 4G applies asynchronous HARQ for DL transmissions. In this case, the BS has more degrees of freedom related to setting up the timing, as the retransmissions are treated alike scheduled transmissions. The latter, however, implies more overhead, as the BS has to communicate its decisions. The advantage of this is a higher flexibility. As outlined several times, 5G NR is foreseen to require a very high degree of flexibility to be able to serve the various use cases, and, for instance, meet the stringent latency requirements for URLLC services. Consequently, non-elastic mechanisms, such as synchronous HARQ, should be avoided in 5G, as already agreed in 3GPP [3].

How to Configure the Retransmission both w.r.t. Resources to be used and the Transmission Parameters (Adaptive vs. Non-adaptive HARQ)?

Again two variants are possible. With non-adaptive HARQ, the retransmission requires to use the very same resources (naturally, in a respectively later sub-frame) and configuration (e.g., selected modulation and coding scheme). As above, the merit of this variant is a lower overhead, since the configuration is implicitly known at the cost of fewer degrees of freedoms. In the 4G UL, non-adaptive HARQ is applied, while the 4G DL applies adaptive HARQ. 5G NR is foreseen to use in both directions the adaptive variant to avoid troublesome restrictions.

In a nutshell, compared to 4G, 5G NR requires to move the implementation of HARQ towards a higher flexibility at the cost of higher processing effort and overhead. In this section, we have treated all available options in a rather abstract manner. For further details, the interested reader is for instance referred to [6].

11.5 Antenna Design, Analog, Digital and Hybrid Beamforming

As highlighted in the previous sections, there is a need to more densely reuse the spatial domain in wireless communications. This needs to be done both at each network node and by densifying the network. More spatial efficiency at the node level can be obtained by using many antennas at the transmitter and/or receiver. This approach is in general referred to as MIMO in the literature. Depending on the channel properties and scenario, MIMO systems can be configured for spatial (transmit and or receive) diversity, beamforming, multiplexing, and spatial multiple access. The maximum diversity gain that can be achieved in a MIMO system equals the number of independent paths between antenna pairs, and the maximum number of spatial streams that can be supported equals the minimum of the number of transmit and receive antenna elements in the system. However, note that full diversity and spatial multiplexing cannot be obtained simultaneously [32].

In the diversity mode, sufficiently separated antenna elements are used to make the link more robust (i.e., lower the outage) by transmitting and/or combining different redundancy versions of the signal by taking advantage of more-or-less independently fading propagation paths between the transmit and receive antenna elements. Receive diversity is sometimes called single-input-multiple-output (SIMO), and transmit diversity is sometimes called multiple-input-single-output (MISO) in the literature, when the transmitter/receiver has only one transmit/receive chain, respectively.

In the beamforming mode, the complex baseband weights of the transmitted and/or the received signals at each antenna element are chosen to adjust and to shape the transmit and/or receive beams in order to increase the signal-to-interference-plus-noise ratio (SINR) of the link, and to avoid harmful interference on other links. Transmit beamforming can be done either in a user-agnostic manner or in a per-user channel-aware manner. The latter case is normally denoted as precoding in the literature (sometimes also the per-user power allocation is included in the notion of precoding), and requires knowledge of the users' channels at the transmitter.

In spatial multiplexing, multiple signals are sent as independent streams to obtain throughput gains whenever there is sufficient multipath propagation in the environment, and provided that channel state information (CSI) at the transmitter side can be obtained. The multi-user case of spatial multiplexing is called spatial multiple access, in which the streams can be aimed for different users.

As highlighted above, the more transmit and receive antennas can be used, the higher the potential for spatial diversity, beamforming, spatial multiplexing and spatial multiple access systems is, provided the spatial channels are sufficiently uncorrelated. In particular, with spatial multiplexing and multiple access, the capacity increases and the required transmit power decreases with the number of antennas. For this reason, intensive research has been carried out on so-called massive or large MIMO [33], addressing both theoretical and practical challenges. From the theoretical side, asymptotic results show that in rich multipath fading environments so-called *channel hardening* [34] appears, essentially eliminating the fading, and with suitable precoding orthogonal spatial channels for the users can be obtained. However, to practically implement large MIMO systems, there are challenges related to, in particular, channel state acquisition, hardware impairments and signal processing complexity. In FDD systems, CSI needs to be obtained by reference or pilot symbols in UL and DL, whereas, with transceiver chains in TDD that are calibrated to compensate for the different Rx/Tx hardware (HW) properties, it is possible to take advantage of the reciprocity of the propagation channel, and thus probing signals need to be sent in only one direction. For this reason, TDD is seen as the most practical possibility to implement large MIMO systems. A drawback with any half duplexing system, however, is the so-called half-duplex loss, since the transceivers take turn in transmitting and receiving. This loss would be avoided if full duplex (FD) at the same frequency would be possible to implement, cf. Section 16.2.4. With FD, there is the potential to double the spectral efficiency and increase the resource allocation flexibility. Perhaps even more important for some 5G use cases, FD has the potential to reduce the access delay by a factor of two in the system, since all time slots can be made available for transmission. Still, reference symbol design for large MIMO systems is a challenge in order to obtain sufficient CSI knowledge at the transmitter side with a reasonable overhead. In particular, intensive research has been carried out on the so-called pilot contamination problem [35], since orthogonal reference symbols would impose too much overhead on the system.

More spatial efficiency at the network level can be obtained by a dense deployment of infrastructure nodes. This is especially important at mmWave frequencies, due to challenging propagation conditions, such as a higher path loss, penetration losses, and shadowing, as detailed in Chapter 4. In addition, the hardware capabilities are worse with weaker output power and noisier oscillators, causing phase noise to be a design issue, as detailed in Section 11.3.1.3. On the other hand, the small wavelength at higher frequencies can be exploited to pack a large number of antennas in a small area, allowing for large beamforming gains at reasonable form factor, which allow to overcome the weaker output power. At the receiver, sufficiently large antenna area could be achieved creating a larger effective aperture, thus enabling large enough antenna array gain, and also enabling receive beamforming to be implemented for additional directivity gains. It should be noted that mmWave channels are typically wideband and spatially confined causing sparse multipath, and, hence, the massive MIMO focus for mmWave frequencies is rather on beamforming, while spatial multiplexing is less relevant for these frequencies, unless the antennas are spatially distributed.

In the following, an overview of the multi-antenna support in 3GPP NR is given, followed by a more detailed discussion on the so-called hybrid-beamforming architecture, which is a promising approach for large antenna arrays in particular at mmWave carrier frequencies. Then, a short discussion is devoted to an alternative MIMO architecture that has received increased attention to address the complexity, energy consumption, and cost: digital beamforming with finite precision digital to analog converters (DACs). Finally, although currently not supported by NR, the potential for so-called *massive multiple-input massive multiple-output* (MMIMMO) is shown to be promising to boost the spectral efficiency in specific scenarios. It might be feasible from a complexity point of view using a novel spatial multiplexing scheme.

11.5.1 Multi-Antenna Scheme Overview of 3GPP NR

In the 3GPP LTE standard, there has been support of various MIMO schemes from the start, i.e., since Release 8. In Release 13, the eNodeB can be configured in ten different so-called transmission modes (TMs), implementing transmit diversity, beamforming and spatial multiplexing. The TM can be selected per UE based on the channel properties, UE and eNodeB capabilities. The CSI feedback can contain the so-called rank indicator (RI) describing the rank of the MIMO channel, i.e., the number of sufficiently spatially separable channels, the precoding matrix indicator (PMI) describing the preferred precoder from a set of pre-defined codebooks, and a channel quality indicator (CQI) used for adaptive transmission and multi-user scheduling. Up to four spatial layers are supported, but only two codewords can be transmitted simultaneously to a given UE. Due to the potential of dense spatial reuse, in 3GPP NR, a beam-oriented approach is adopted, and there is inherent support of distributed cooperative transmission and reception schemes, as well as more advanced MIMO schemes. Below, the main novelties of NR are summarized.

11.5.1.1 Beam Management

In 3GPP NR [1], beam management is defined as a set of layer-1 (L1) and layer-2 (L2)[3] procedures to acquire and maintain a set of transmit-receive points (TRPs) and/or UE beams that can be used for DL and UL transmission and reception. Specifically, at least the following aspects are addressed by beam management procedures:

- **Beam determination**: For TRP(s) or UEs to select their own Tx/Rx beam(s);
- **Beam measurement**: For TRP(s) or UEs to measure characteristics of received beamformed signals;
- **Beam reporting**: For UEs to report information of beamformed signal(s) based on beam measurements;
- **Beam sweeping**: Operation of covering a spatial area, with beams transmitted and/or received during a time interval in a predetermined way.

According to [1], the following DL L1/L2 beam management procedures are supported within one or multiple TRPs:

- **DL beam alignment**: Utilized to enable UE measurements on different TRP Tx beams to support selection of TRP Tx beams and UE Rx beam(s);
- **DL beam refinement**: Utilized to enable UE measurement on different TRP Tx beams to possibly change inter- or intra-TRP Tx beam(s);
- **UE receive beam refinement**: Utilized to enable UE measurements on the same TRP Tx beam to change the UE Rx beam in the case that a UE uses beamforming.

At least network-triggered aperiodic beam reporting is supported under the above three beam management related operations. UE measurements based on reference signals (RSs) for beam management (at least CSI-RS) are composed of K beams, and UEs report measurement results for N selected Tx beams, where N is not necessarily a fixed number. Note that the procedure based on RS for mobility purposes, such as synchronization signal blocks, is not precluded. Reporting information at least includes measurement quantities for N beam(s) and information indicating N DL Tx beam(s), if $N < K$.

NR also supports the following beam reporting considering L groups, where $L \geq 1$ and each group refers to an Rx beam set or a UE antenna group. For each group l, the UE reports at least the following information:

- Information indicating group at least for some cases;
- Measurement quantities for $N \cdot l$ beam (s);
- Information indicating $N \cdot l$ DL Tx beam(s), when applicable.

---

3 L1 refers to the PHY sublayer, while L2 refers to MAC/RLC/PDCP sublayers.

NR supports that the UE can trigger a mechanism to recover from beam failure. A beam failure event occurs when the quality of beam pair link(s) of an associated control channel falls low enough (i.e., involving comparison with a threshold and time-out of an associated timer). The mechanism to recover from beam failure is triggered when beam failure occurs.

11.5.1.2 MIMO Schemes

For NR, the number of codewords per Physical Downlink Shared Channel (PDSCH) assignment per UE is 1 codeword for 1 to 4-layer transmission and 2 codewords for 5 to 8-layer transmission.

DL demodulation reference signal (DMRS) based spatial multiplexing is supported. At least 8 orthogonal DL DMRS ports are supported for single user MIMO (SU-MIMO), and a maximum of 12 orthogonal DL DMRS ports are supported for multi-user MIMO (MU-MIMO). At least the following DMRS based DL MIMO transmissions are supported for data in NR:

- **Scheme 1**: Closed-loop transmission where data and DMRS are transmitted with the same precoding matrix;
- **Scheme 2**: Open loop and semi-open loop transmissions, where data and DMRS may or may not be restricted to be transmitted with the same precoding matrix.

For the DL data, at least a precoding resource block group (PRG) size for physical resource block (PRB) bundling equal to a specified value is supported. A configurable PRG size is also supported for data DMRS. DL transmission scheme(s) achieving diversity gain at least for some control information transmission are supported.

11.5.1.3 CSI Measurement and Reporting

For NR, DL CSI measurements with up to 32 antenna ports are supported. At least for CSI acquisition, NR supports CSI-RS and SRS. NR further supports aperiodic, semi-persistent, and periodic CSI reporting. The periodic CSI reporting can be configured by a higher layer, above PHY. Higher-layer configuration includes at least reporting periodicity and timing offset. By semi-persistent CSI reporting, configuration of CSI reporting can be activated or de-activated.

CSI reporting with two types of spatial information feedback is supported.

Type I CSI feedback is the normal CSI feedback scheme. As in 3GPP LTE, it consists of codebook based PMI feedback with normal spatial resolution. The PMI codebook has at least two stages, where the first stage comprises of beam groups and vectors. Type I feedback supports at least the following (DL) CSI reporting parameters:

- Resource selection indicator (i.e., reference signal sequence or beam);
- RI;
- PMI;
- Channel quality feedback.

There is support for multi-panel scenarios by having a co-phasing factor across antenna panels.

Type II CSI feedback is an enhanced CSI feedback scheme, enabling explicit feedback and/or codebook-based feedback with higher spatial resolution. At least one scheme must be supported from the following Category 1, 2, and/or 3 for Type II CSI:

- **Category 1**: Precoder feedback based on a linear combination of dual-stage codebooks. Specifically, stage one consists of a set of L orthogonal beams taken from a set of 2-dimensional DFT beams, and the beam selection is wideband. The L beams with common stage one precoder are combined in stage two, which supports subband reporting of phase quantization of beam combining coefficients;

- **Category 2**: Covariance matrix feedback. The feedback of the channel covariance matrix is long-term and wideband. A quantized/compressed version of the covariance matrix is reported by the UE. Specifically, the quantization/compression is based on a set of M orthogonal basis vectors, where M is the number of supported simultaneous beam pair links, and the maximum value of M may depend at least on UE capability. The reporting can include indicators of the M basis vectors along with a set of coefficients;

- **Category 3**: Hybrid CSI feedback. A type II Category 1 or 2 CSI codebooks can be used in conjunction with 3GPP LTE Class-B CSI feedback using beamformed CSI-RS to reduce the RS overhead and improve coverage. The LTE Class B CSI feedback can be based on either Type I or Type II CSI codebook.

11.5.2   Hybrid Beamforming

Due to the large number of antennas in the transmit and receive arrays required to enable mmWave communication, equipping each antenna with a separate radio frequency (RF) transceiver chain along with a high-resolution converter, as done with smaller arrays at lower frequencies, would result in a high complexity, cost, and power consumption. This is mainly due to the implementation of RF components at mmWave frequencies, as well as the expected large bandwidths, which impose the requirement on the DACs at the transmitter and analog to digital converters (ADCs) at the receiver to operate at a high sampling rate. Thus, equipping one converter per antenna in a large antenna array translates inevitably into a high power consumption and cost. For this reason, analog beamforming with a single RF chain has been adopted in early standards, such as in IEEE 802.11ad. However, since this architecture offers limited signal processing capability, a hybrid beamforming architecture [36] using a reduced number of RF chains, and subsequently converters, has attracted substantial attention as a promising solution in particular for mmWave scenarios, and is depicted in Figure 11-10.

With hybrid beamforming, the number of RF chains N^{RF} is smaller than the number of antennas in the array, e.g., $N_{tx}{}^{RF} \leq N_{tx}$ at the transmitter. By splitting the beamforming operation between the analog RF domain and the digital baseband, this architecture provides a reduction of complexity and power consumption [36], at the expense, however, of reduced degrees of freedom for the baseband digital processing, and a consequently reduced possible number of streams $N_s \leq N_{tx}{}^{RF}$. Compared to

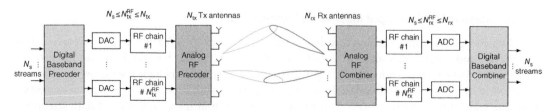

Figure 11-10. Hybrid beamforming with analogue RF beamforming.

a fully digital system, the hybrid beamforming architecture poses different challenges for the CSI acquisition and beamforming design due to the constraints on the analog processing and the need of directional transmission at mmWaves.

For example, the analog processing is frequency flat, which implies that the analog beamforming matrix is fixed for all subcarriers in a multi-carrier system, whereas the digital beamformer can be adapted for each subcarrier. For the wideband hybrid beamforming design, however, the fact that the spatial characteristics of the channel are frequency-invariant can be exploited. Similarly, for hybrid beamforming in a multi-user scenario, the design of the analog beamforming needs to consider that it is common for all users. Furthermore, the analog processing can be implemented via a network of phase shifters, RF switches or with a lens antenna array, and can be implemented at different stages including RF, intermediate frequency and baseband. In case of phase shifters, the entries of the analog beamformer are constrained to have unit modulus. An open question is, however, how the limited number of RF chains should be connected. A partially connected architecture has recently been proposed [37], where the output of each RF chain is connected only to a subset of the transmit antennas. This approach reduces the required number of phase shifters as well as the losses, thereby facilitating the implementation of hybrid beamforming at the expense of reduced design flexibility. If each RF chain (or set of RF chains) is connected to a distinct set of antennas, the architecture is based on subarrays, where each subarray is basically connected to its own transceiver. The partially connected architecture is applicable at both the transmitter and receiver side, and the performance can be rather close to a fully connected hybrid beamforming architecture. For further discussions about such a hybrid beamforming architecture, see also Section 2.3 in [38].

11.5.3 Digital Beamforming with Finite DACs

Hybrid beamforming is currently the most promising approach to tackle the power consumption and complexity bottleneck mmWave transceivers are facing. An alternative approach to address these aspects that is currently attracting increased attention is digital beamforming with low resolution DACs [39], as also discussed in detail in Section 16.2.3.2. Reducing the precision of the converters enables to reduce the power consumption, which scales roughly exponentially with the number of resolution bits. This enables to have a large antenna array with many active elements at a reduced power consumption and cost. Despite the increased signal processing capabilities compared to analog beamforming, the nonlinearity introduced by the quantization leads to limited capacity at high SNR, and imposes certain challenges on the channel estimation and data detection [39]. Still, investigations show promising performance even with 1-bit DACs in multi-user MIMO DLs, as shown in Section 2.4 in [38].

11.5.4 Massive Multiple-Input Massive Multiple-Output

The hybrid beamforming architecture has good potential to be used for access in particular in mmWave bands, but it can also be used for wireless relaying and backhauling, see for instance Section 7.4. However, in rather static high-throughput wireless backhaul scenarios, massive and symmetric MIMO might be feasible. In such a case, arrays with hundreds of antenna elements at both the transmitter and the receiver sides can be used to multiplex hundreds of data streams in the spatial domain, as illustrated in Figure 11-11. In theory, such MMIMMO could deliver spectral

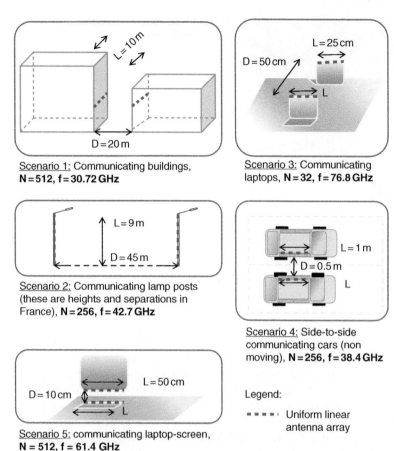

Scenario 1: Communicating buildings, **N = 512, f = 30.72 GHz**

Scenario 2: Communicating lamp posts (these are heights and separations in France), **N = 256, f = 42.7 GHz**

Scenario 3: Communicating laptops, **N = 32, f = 76.8 GHz**

Scenario 4: Side-to-side communicating cars (non moving), **N = 256, f = 38.4 GHz**

Legend:

▪ ▪ ▪ ▪ ▪ Uniform linear antenna array

Scenario 5: communicating laptop-screen, **N = 512, f = 61.4 GHz**

Figure 11-11. Practical examples of deployment scenarios of MMIMMO systems, see Section 3.4 in [9].

efficiencies of hundreds of bps/Hz, and therefore, provide multi-Gbps throughput, which are essential for the backhaul of future wireless communication systems.

So far, MMIMMO has been regarded as infeasible due to the complexity; thus, it has yet not been in focus for 3GPP NR. However, in [40] it has been recently shown that hundreds of data streams can be spatially multiplexed through a short range and line-of-sight MMIMMO propagation channel thanks to a new low-complexity spatial multiplexing scheme called block DFT based spatial multiplexing with maximum ratio transmission (B-DFT-SM-MRT). The block-based approach is beneficial to control that the spatial subchannels have similar properties, and maximum ratio transmission (MRT) is used to mitigate the effect of scattering and to deal with cases where the uniform linear arrays are not perfectly parallel. Its performance in real and existing environments was assessed using accurate ray-tracing tools and antenna models. In the best simulated scenario, depicted in Figure 11-12, 1.6 kbps/Hz of spectral efficiency is attained, corresponding to 80% of singular value decomposition (SVD) performance, with a transmitter and a receiver that are 200 and 10,000 times less complex, respectively.

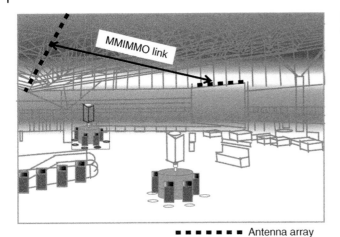

Figure 11-12. Helsinki airport simulated deployment scenario.

■ ■ ■ ■ ■ ■ Antenna array

11.6   PHY/MAC Design for Multi-Service Support

As outlined in the introduction, 5G NR is to obey to various design criteria for making the system ready for the foreseen ecosystem of devices and services benefitting from having wireless access. This requires means to configure single connections according to the respective needs of the connected device or service. The overall system should be able to do so concurrently for any given combination of devices and services requesting access at a given point in time, enabling efficient *multi-service support*. Furthermore, its basic functionalities, such as access protocols and reference symbols, need to be able to scale with reasonable effort and resource consumption, following the respective network conditions, e.g., the number of devices requesting access, or the number of antenna ports or beams that are available or active. Beyond Release 15, 5G should be able to add new functionalities and use cases with low efforts and without requiring major redesign of the initial versions, commonly referred to as the notion of *forward compatibility*. Finally, its design should allow for tight interworking (or even coexistence) with other access technologies. Naturally, while obeying to all those design criteria, all design choices have to keep track of resource, energy and cost efficiency, and allow the system to be robust against, e.g., harsh channel conditions.

While any component of the communication system needs to care for the points given above, the design of the radio frames and related functionalities is pivotal for the overall communication system in general and the AI in particular in achieving the aforementioned targets. Before heading to specific parts of the frame design and the status of the discussions in 3GPP, some fundamental aspects are outlined.

11.6.1   Fundamental Frame Design Considerations

5G requires to support a reasonable set of options with respect to supported bandwidth to cover all relevant deployment scenarios and spectral bands being available, while keeping the overall number reasonably low. For each of these options, the respective sampling rate, number of subcarriers covering the bandwidth and the supported subcarrier spacing should be integer multiples or fractions of a given baseline, to keep system complexity and testing efforts at a reasonable level. Optimally, this baseline is aligned with 4G (i.e., 15 kHz) to both allow for efficient multi-RAT implementations and to ease inter-working and multi-connectivity among 4G and 5G at low and high bands. [3] is summarizing the status of the discussions in 3GPP following similar lines as given above: For below 6 GHz, the smallest band-width to be supported is 5 or 10 MHz, while in regions above 6 GHz and up to 52.6 GHz, the smallest bandwidth is 40 or 80 MHz. To avoid excessively large FFTs, the subcarrier scales with increasing carrier frequency. The baseline is 15 kHz, and further options are 15kHz*2^n (with n being an integer).

For schemes dealing with concurrent transmissions of multiple cells (e.g., inter-cell interference coordination), it is advantageous to have their transmissions to be time-aligned on symbol level. For energy reasons, the amount of always-on signal components should be minimized, and the actual repetition rate (e.g., for synchronization signals in DL) should be configurable. The on-demand principle should be applied as far as possible to increase energy efficiency especially in low-load scenarios. As an example, system information blocks (SIBs) shall be transmitted on-demand only, while master information blocks (MIBs) would always be transmitted.

Special care has to be taken for the new requirements being brought up by the new use cases. Hence, the overall frame design requires to obey to the following rules:

- Low-end devices, as typically envisioned for mMTC services, require to be able to detect the DL signal in a sub-sampled manner for energy efficiency reasons. This calls for applying narrow-band implementations in general (e.g., DL control channel) and eventually the introduction of complementary narrow-band signals (e.g., DL synchronization signals);
- To achieve very low latencies for URLLC services, very short TTIs and very quick scheduling processes have to be enabled. To avoid inefficient implementations, these options need to be introduced in a complimentary manner;
- Efficient support of various antenna configurations by implementing scalable RS designs is needed to increase the spatial reuse of spectral resources and for improving coverage, e.g., for eMBB services.

The frame requires various control channels (both for carrying the relevant global system configurations and for maintaining/configuring the various device connections and transmissions) and data channels (potentially different formats for different use cases). Additionally, one needs to account for DL synchronization symbols, UL sounding, UL random access opportunities, and RSs for various means, such as beam selection and channel estimation. Both TDD and FDD configurations have to be accounted for, and different areas of the spectrum require at least partially varying treatments. Finally, to allow for energy-efficient devices, pipelined processing should be enabled. Figure 11-13 depicts the basic sub-frame configurations being foreseen taking those points into account.

In the figure, options (c), (d), and (e) depict bidirectional sub-frames containing both DL and UL transmissions. They are the basic building blocks to enable efficient TDD, while (a) and (b) are exclusively carrying either DL or UL transmissions. Control (including DMRS that are not depicted here) is separated from the data block for enabling pipelined processing at the device and is placed at the borders, i.e., *frontloaded* in the case of the DL, and at the end of the sub-frame for the UL.

Having laid out the fundamental design paradigms to follow, the next sections provide some further details to the single building blocks of the signaling frames of 5G NR. First, the required frame elements for enabling initial access are introduced, followed by a discussion of the design of the control channels and the different data channels variants that are foreseen. The section is then completed by covering further access variants beyond unicast, namely, device-to-device (D2D) and broadcast/multicast.

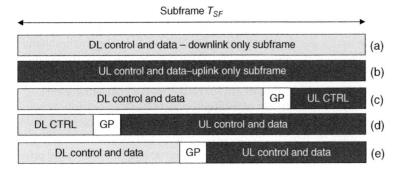

Figure 11-13. Sub-frame design variants. GP: Guard Period.

11.6.2 Initial Access

This section is closely related to Section 13.2. Before being able to transmit or receive data, a device needs to perform various steps when powered up. It has to:

1) identify close-by cells or transmission points (cell discovery);
2) align its transmission parameters (time and frequency) to the reference;
3) read the system configuration; and
4) perform the registration process, i.e., perform the random access procedure and message exchange for setting up the device configuration, e.g., related to authentication and encryption.

The BS has to regularly transmit synchronization signals for the device to be able to perform steps 1 and 2. Similarly as with 4G, 5G DL frames will regularly carry primary synchronization signals (PSS) and secondary synchronization signals (SSS). PSS and SSS are spanning 12 physical resource blocks (PRBs), each spanning 12 subcarriers, within a single multi-carrier symbol carrying selected sequences[4] being used to identify the respective transmission node. One of the key targets for the design is to enhance one-shot detection by improving the respective sequence characteristics (e.g., by applying a longer sequence and by avoiding time/frequency ambiguities). With achieving this target, the periodicity of the synchronization signals can be reduced without significantly increasing latency. By correlating the received signals with the known sequences, the device is able to perform DL synchronization (both on frame and symbol level) and to identify the transmission point. Once this has been done, the device is able to locate and decode the Physical Broadcast CHannel (PBCH). This channel carries the relevant system configuration (e.g., location and frequency of the Physical Random Access CHannel, PRACH) for the device to be able to continue the access procedure, and it spans 24 PRBs in frequency direction over two successive symbols in time. At this point, the device is aware of the available network and how to access it. The next step is concerned with setting up the physical (involving, e.g., power

---

4 The actual sequence design is defined in [41].

control) and logical connection (involving, e.g., authentication and encryption). This includes various measurements and message exchanges. In Chapter 13, more details are given.

The frequency of transmitting the set of signals given above (i.e., PSS, SSS, PBCH in DL, PRACH in UL) has an impact on the control plane latency (i.e., the time it takes to register to the network) and the signaling overhead (i.e., the more often those signals are transmitted, the more spectral resources are required).

It is foreseen to add high-capacity transmission points operating at mmWave bands, both in a standalone and a non-standalone manner. For the latter, a close-by transmission point from the coverage layer (<6GHz) acts as supporting node for the steps given above, e.g., for all actions related to the messaging. The cell discovery and the alignment of its transmission parameters though, must still be done directly with the mmWave transmission points for both deployment cases. As discussed in Section 11.5, mmWave requires directional and beam-based transmission to overcome the large free-space loss. For cell discovery, the broadcast of synchronization and control signals via beam-based transmissions is a challenge. In [4], different beam sweeping strategies, including time division, frequency division, code division, and spatial division, are compared systematically with respect to cell discovery latency and signaling overhead. It was found out that time division achieves the lowest latency at the price of high signaling overhead, while spatial division allows much lower signaling overhead and provides flexibility to achieve a trade-off between latency and signaling overhead. An auxiliary transceiver based scheme has been proposed to further reduce the signaling overhead and avoid interruption of data transmission (due to hybrid transceiver constraints) during the broadcast of synchronization and control signals.

11.6.3 Control Channel Design

For controlling the system in general and the transmission of data in particular, various physical control message exchanges between the network and the connected devices are required. Some messages carry system-wide settings and are of relevance for all devices, while some messages characterize the connection setup between a specific device (or a sub-group of devices) and the BS and thus are of less relevance for other devices not being part of this sub-group. The former needs to follow similar means as, e.g., applied in 4G for the PBCH design. As any device needs to be aware of the system configuration, the respective transmission needs to be of broadcast type and thus needs to be configured having the weakest possible link in mind to achieve a given minimal reliability. Details related to the PBCH design have already been given in the prior section.

5G will be a packet-switched network, as 3G and 4G have been. Hence, one needs to employ structures for controlling the flow of packets from and to different sources and recipients[5]. In DL, for example, the Physical Downlink Control CHannel (PDCCH) is carrying the instructions (e.g., scheduling grants, resource configurations, and HARQ feedback) from the BS to the connected devices. The first releases of 4G have physically structured the PDCCH in a broadcast manner: The control messages, also known as DL control information elements (DCIs), are multiplexed into a single structure and mapped to the beginning of each TTI, meaning that data and control are

---

5 A more comprehensive treatment of dynamic scheduling can be found in Chapter 12.

multiplexed in a time division multiple access (TDMA) manner. In Release 11, 3GPP has introduced an enhanced PDCCH (EPDCCH) allowing to dedicate single PRBs for the transmission of control messages in an frequency domain multiple access (FDMA) manner.

Discussions in 3GPP indicate that NR will follow a different structure [3]. Instead of separating control and the respective data transmission, so-called *in-resource messaging* is implemented, also referred to as "staying in the box". The main reasons for doing so are:

- "No race to the bottom" (i.e., no need to configure the resources for the weakest possible link);
- Data and control can share reference symbols;
- Control can make use of rank 1 precoding if respective CSI is available;
- Blanking of frequency resources is improved (e.g., for inter-cell interference coordination);
- More degrees of freedom are available for designing the DCIs.

For the actual realisation of the control channels, 3GPP has defined so-called resource element groups (REGs) as the basic control channel building block. These comprise 12 consecutive resource elements (REs) within a single OFDM symbol. Moreover, like in 4G, so-called control channel entities (CCEs), consisting of a number of REGs, are defined as the smallest unit of a scheduled PDCCH transmission. Both localized and distributed variants are foreseen, the latter being able to exploit frequency diversity. As such, each NR-PDCCH is transmitted by using one or several CCEs depending on the respective channel quality. The number of CCEs employed is called aggregation level (AL). Currently, similar to LTE, several ALs, namely 1, 2, 4, 8 or even 16 and 32 are considered. The higher the required coding gain is, the higher the aggregation level has to be chosen.

The transmission of physical control messages is typically required to be very robust to avoid packet loss. The control channel coverage in 5G NR is intended to be at least as good as in 4G. The main building block used to match the transmission to the respective link quality is the already mentioned aggregation level. The higher the aggregation level, the more copies of the control message are transmitted and thus the overall transmission becomes more robust. In addition, if the BS is aware of the channel state, rank 1 precoding may be employed to improve the link quality for the transmission of device specific DCIs. Otherwise, one needs to make use of diversity mechanisms. Two variants, namely space-frequency block codes (SFBC) and per-RE precoder cycling, have been extensively discussed as potential options. Finally, frequency hopping may be applied for the sake of frequency diversity. As coding scheme, PCs are selected, as covered in Section 11.4.2.3.

To enable channel estimation for coherent demodulation, some REs in the REG are dedicated for carrying DMRS. It is envisioned that REGs with configurable DMRS patterns can provide additional trade-offs between channel estimation performance and achieved coding rate for the control channel transmission.

11.6.4 Data Channel Design

While the previous section has covered control plane related aspects, the focus shall now be on the relevant PHY/MAC design choices for the user plane or data channels. The support of a multitude of different use cases is much more emphasized in 5G NR than in earlier generations. To allow the system to be efficient, various means have to be provided for devices to transmit and receive the data.

Most likely, the bulk of the connections accessing the system will still exhibit the following characteristics:

- The overall amount of data to be transmitted is far bigger than a single transmission opportunity is able to accommodate (i.e., a single transmission request requires several transmission opportunities);
- The energy consumption related to the transmission and reception of the data is small compared to other parts of the device (e.g., display);
- The system is able to regularly collect rather concrete context information, such as channel quality.

The listed aspects are mostly, but not exclusively, connected to (e)MBB services. Beside this, 5G NR will include new types of services introducing a new set of relevant aspects either related to the needs of the respective service or the respective device(s):

- In the area of mMTC, some use cases imply the need for transmitting tiny amounts of data (requiring only few transmission opportunities) in a rather sporadic and unpredictable manner. The related devices are typically constrained with respect to cost and energy;
- In the area of URLLC, the allowed transmission latency is required to be very short and the reception of the packet has to be extremely reliable. Both sporadic and regular/periodic transmissions are possible.

Obviously, one needs to make use of various tailored access mechanisms for being able to meet those partly contradicting targets. While these are presented later in this section, initially an overview is given on the most recent progress in 3GPP on the fundamental scheduling concepts and how they are being optimized for multi-numerology operation, followed by a summary of the open issues.

PHY layer processing is not aware of abstract concepts like service-categorization, e.g., URLLC or eMBB. Instead, the differentiation is done via selecting different DCI depending on the current transmission. While initially the research and standards community was heading towards reserving specific types of PHY resources exclusively for some specific services (e.g., so-called mini-slots for URLLC traffic), the prevailing trend in 3GPP now goes towards enhancements of PHY control signaling to indicate the length of a data assignment making the overall system more flexible.

In LTE, if a UE is not in discontinuous reception (DRX) mode, it needs to continuously monitor PDCCH meaning every 1 ms. A potential game-changer on this topic introduced by 3GPP for 5G NR is to allow for more differentiated options. 3GPP has, for instance, agreed that the minimum PDCCH monitoring period may go down to a single symbol, and the consecutive data duration is to be indicated. All that matters is the PDCCH monitoring period itself, especially since data can be scheduled over any number of successive symbols indicated by a respective DCI.

The scheduling delay represents a significant portion of the overall delay occurring in the radio network. In fact, reducing UL scheduling delay has been singled out in 3GPP as a means for achieving the latency required for URLLC applications and for future-proofing the system design. Ideally, URLLC transmissions should get an UL grant right away, i.e., for the very first PUSCH transmission, allocating the service resources with appropriate size and physical layer numerology according to underlying QoS requirements of the data buffered in the UE.

To alleviate the above issues, there are in principle three possible approaches worth exploring:

- Contention-based/grant-free UL data transmission, i.e. without prior scheduling request;
- Enhancements to scheduling requests and buffer status reporting (BSR) mechanisms;
- Semi-persistent scheduling (SPS).

As will be shown, the above are not mutually exclusive. In fact, support for all three is possible and will enable the reduction of the UL data scheduling latency, although optimizing for all three approaches simultaneously within a single system design may not be possible, meaning that compromises will be needed.

11.6.4.1 Contention-based/Grant-free Access

3GPP has agreed to support grant-free, SPS-like, PUSCH transmissions [3]. More specifically, an UL transmission scheme without a grant shall be supported.

Contention-based access in previous generations of cellular systems has been exclusively used as the basis for the initial connection of a device to the network, i.e., where a device would switch from RRC Idle state to RRC Connected, as detailed in Section 13.3. With the introduction of mMTC and URRLC services in 5G, the potential use of this type of access has been extended; namely, by the possibility of a device connecting to the network with minimal signaling overhead and latency. This is especially useful for sporadic small packet transmissions.

Contention-based access protocols are of three types, as depicted in Figure 11-14: (a) multi-stage, (b) two-stage, and (c) one-stage. These can be interpreted very differently, and each can contain several variants.

The one-stage access protocol (c) means that both the access notification and data delivery are done in a single transaction, i.e., using one or several consecutive packets in a single transmission. A two-stage access protocol (b) allows the UE to separate the access notification stage from its data delivery stage, e.g., by allowing for an intermediate feedback message. A multi-stage access protocol (a), for which the current LTE connection establishment protocol is a prime example, is composed of at least three phases, namely, the access, connection establishment phase (including authentication and security), and finally the data phase.

In [42], several proposals have been put forward that realize two-stage and one-stage accesses. In particular, the signaling associated with the connection establishment (i.e., mostly the establishment of mutual authentication and security) is assumed to be reused from a previous session, where the

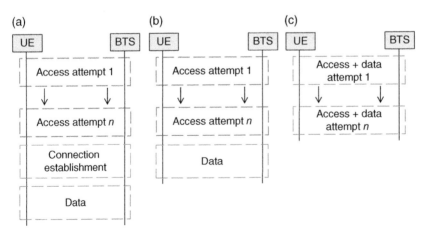

Figure 11-14. High-level description of the three access protocols types considered: (a) Multi-stage access protocol with an access, connection establishment and data phase; (b) Two-stages access protocol with access and data phases; and (c) One-stage access with combined access and data phase.

multi-stage access protocol has been carried out. This reuse of the connection context is achieved through the RRC Extant state introduced in [42]; a new additional state complementing RRC Idle and RRC Connected, as detailed further in Section 13.3. 3GPP has recently applied a similar state named RRC Inactive [43].

In both the one-stage and the two-stage access schemes, contention may occur only in the initial step of the protocol. Designing this type of access is a joint effort between the PHY design (to improve detection and to cope with collisions via advanced signaling processing) and the higher layers (to keep the number of arrivals to the network within the capabilities of the PHY).

A tailored design of the access scheme for mMTC and URLLC services is needed due to the short packet length and the need for supporting a massive number of simultaneously active devices attempting to access the network.

The main PHY functionality requirements for this can be summarized as follows:

- DL synchronization at the device;
- Signature/preamble generation at the device;
- Signature/preamble detection, e.g., via correlation (or compressive sensing) at the BS;
- Repetition and retransmission scheme, eventually based on frequency hopping;
- (Non-orthogonal) multiple access, e.g., through spatial separation;
- Autonomous link adaptation and power control.

The PHY preamble design uses false-alarm probability and missed detection as key metrics. The LTE PRACH is based on Zadoff-Chu sequences [44]. 5G has the following options to adjust the original design of LTE to its needs:

- A higher number of cyclic shifts in at least a part of the PRACH allocations, though reducing the distance between the preambles;
- Simultaneous usage of different root sequences. However, cyclic shifts generated from different root sequences are not orthogonal; hence, residual cross-correlations may increase false alarm rates.
- Usage of m-sequences instead of Zadoff-Chu sequences for higher PRACH capacity, though at the price of a higher PAPR;
- Orthogonal CDMA codes providing much higher capacity, though being more sensitive to weak synchronization;
- Reshaping the arrival distribution of access attempts. Ideally, the access attempts are equally distributed over time, which may be attained by dividing the PRACH opportunities into N slots, and each UE accesses the slot number derived from a modulo N operation of its ID. This solution is limited to delay tolerant use cases and devices with relaxed energy consumption requirement.

Obviously, adopting one or more of the above listed options alleviates the support of higher loads in massive access scenarios compared to the LTE preamble design, as long as the respective downsides can be tolerated.

11.6.4.2 Enhancements to the Design of Scheduling Requests and Buffer State Reporting

In 4G, a SR is used to inform the network that the respective UE has data to transmit, but it does not have enough resources available to transmit the BSR itself, which carries information on buffer status. The buffer status is crucial for the BS scheduler and is typically unknown until the BSR is received. Most SR enhancements proposed recently focus on reducing the delay between initial SR and the first UL data transmission.

Assume a device requiring eMBB data to be transmitted right after having terminated an URLLC session. Without having differentiated SRs, the BS would not be aware of the altered connection requirements, for instance related to latency and reliability, and would thus not be able set up the subsequent data connection as required. This leads either to a connection not meeting the requirements (e.g., if these are more stringent than those of the prior transmission), or the other way round, i.e., a connection overshooting the actual requirements, resulting in inefficient resource usage. Hence, to avoid this, the use of a different SR settings or configurations are required. With allowing for differentiated SRs, the BS is able to allocate the appropriate amount of UL resources, perform link adaptation matched to the use case, and initiate the respective access method. In essence, SR and BSR functionalities are somewhat merged to reduce the latency. The basic idea is to include more details about the BSR into the SR, where some proposals target only to indicate whether the BSR is long or short, while others go further and include even more details, such as, the type of the service the data is originating from.

11.6.4.3   Semi-persistent Scheduling and Grant-free Scheduling

SPS is used in LTE as a scheduling technique with minimal overhead being suitable for traffic with periodic characteristics. SPS is configured (but not activated) via RRC messages with signaling its periodicity. The SPS is then activated via PDCCH-messaging, enabling to re-tune parameters on a faster basis and with less control signaling overhead. More specifically, a UE being configured for SPS-like transmissions waits for a DCI scrambled in relation to a special kind of cell radio network temporary identifier (SPS-C-RNTI), and once received, the UE starts transmitting data with a pre-configured periodicity, as set via RRC signaling.

As with any dedicated resource scheduling scheme, SPS has some inherent inefficiencies. Furthermore, the empty transmissions as needed for implicit release are raising further concerns. Some enhancements to SPS to support 5G are centred around the following items:

- **Configuration/activation split**: Which parameters are configured via RRC signalling, and which are signalled via PDCCH?
- **Sharing of SPS resources**: Should it be allowed, and what sort of collision resolution would need to be introduced?
- **SPS periodicity**: Is support for extreme (i.e., sub-ms range) values needed and justified, in the light of potential usage for URLLC?

As discussed in Section 11.6.4.1, 3GPP is additionally working on standardizing grant-free/contention-based transmissions. It has recently been agreed that UL data transmission without UL grant can be configured by the network to be carried out after semi-static resource configuration in RRC without PHY signaling [3]. If the network configures the system accordingly though, PHY signaling for the activation and deactivation and/or modification of parameters for UL data transmission without UL grant can be applied. Work is ongoing to find a harmonized MAC design for UL SPS and grant-free access based on configured PHY signaling.

11.6.4.4   Pre-emptive Scheduling

As discussed throughout this chapter, 5G will need to support shorter TTI lengths in order to enable lower latencies. On the other side, as long as the latency requirement is not very strict, as for instance for eMBB services, and the traffic volumes to be transmitted are not small, longer TTI durations typically result in higher performance due to lower control overhead. So it is neither reasonable to fix transmission times to be very short (below 1 ms), nor to be long in any case and allow for specific selection instead. Naturally, when allowing transmissions with varying time bases to access the system, special treatments have to be accounted for as outlined next. Let us consider mixed traffic scenarios with both eMBB and URLLC traffic being present. For the DL, in case the BS has already scheduled and indicated eMBB traffic with, e.g., 1 ms TTI duration or longer, and an urgent URLLC packet arrives without respective resources being available, parts of the eMBB DL traffic may be punctured, and the URLLC data symbols inserted instead. Equivalently, in UL the URLLC packet may be allowed to be superimposed to the running eMBB transmission. This solution efficiently embeds URLLC into the frame with avoiding excessive resource reservation [42].

The disadvantage of applying puncturing/superposition as described above, is the negative impact on the eMBB transmission. If the eMBB victim is not aware of the puncturing or superposition, the inserted URLLC symbols will degrade the symbol detection performance significantly. Hence, a first option to lower this effect, is to inform the victim device via control signaling about the particular part of its transmission being punctured. Further solutions for improvement are to use: (i) code-block based HARQ re-transmissions (where only the punctured part is retransmitted) or even to (ii) retransmit the punctured code block immediately in the next scheduling opportunity (i.e., without NACK feedback). For further details, the interested reader is referred to Section 12.3.3.

11.6.4.5 Device-to-Device and Broadcast/Multicast

So far, we have handled unicast transmissions between device and network. For specific use cases, though, it is more efficient to introduce novel access paradigms, related to device-to-device (D2D) and broadcast/multicast (BMS) transmission. The former allows devices to directly communicate without the network being part of the data exchange, though, for controlling the connections, the network may still act in a supporting manner. When payloads are of relevance for a group of devices, BMS is typically more efficient than relying on multiple unicast connections. In the following, we provide some high-level viewpoints. For a deep-dive, the interested reader is referred to the wide range of available publications, e.g., given in [42].

D2D connections can be used for various means:

- Network offloading with the help of content caching within dedicated devices, e.g., [45];
- Coverage extension with the help of data relaying;
- Direct interactions between the respective devices, e.g., between different road users.

The actual transmission can be 'underlaid' or 'overlaid', i.e., either the D2D connections have dedicated resources available, or they use superposition. Before being able to perform the data transfer, though, the related devices need to be aware of the available links and their respective quality. For this, proximity discovery is performed. Various approaches have been discussed in the literature to perform this step. Recently, both strategies with network support [46] and without network support [47] have been analyzed. For the latter, the achievable mean discovery time (i.e., the time until all nodes have been made aware of all potential partners) has been investigated depending on the number of active nodes and with or without the application of FD, as covered in detail in Section 16.2.4. Dedicated discovery messages have been designed, and the transmission probability for each node has been optimized to minimize the mean discovery time. The use of FD during the discovery phase helps to further reduce the mean discovery time. The given references provide further details on specifics of the concepts and performance results. A more thorough treatment of D2D can be found in Chapter 14.

Although being part of early releases of 4G already, services relying on broadcast/multicast are not yet widely applied in 4G networks. Different applications can benefit from this kind of access such as software updates of sensor networks, or multimedia streams to a group of people, e.g., video feeds during concerts or sports events. Recent studies have investigated means to enhance broadcast/multicast transmission:

- Non-orthogonal transmission schemes for stream multiplexing, including both beam-based and multi-level coding based variants;
- The introduction of complementary unicast based feedback schemes.

The former approach allows to increase the spatial reuse by means of multi-antenna precoding in order to superimpose several multi-cast streams on top of a broadcast transmission. A possible usage scenario for such a technique is the transmission of area specific video feeds in a stadium (e.g., the camera feed capturing the area of the game field being far away from the respective multicast group) in addition to a broadcast transmission being of interest for all spectators. An alternative to this beamforming-based approach is to rely on multi-level coding, multiplexing different streams in code domain [42].

A common bottleneck of broadcast systems is the need to configure the transmission according to the weakest possible link. 5G naturally will allow for unicast UL transmissions, which may be used to improve the DL broad-/multicast connections in various ways, such as:

- transmitting channel quality indicators providing context that the system can use to allow for a more efficient initial broadcast transmission, and
- enabling the system to introduce a HARQ mechanism to follow-up the initial broadcast stream by subsequent unicast retransmissions.

For details and performance results, the interested reader is referred to [42].

11.7 Summary and Outlook

5G will more prominently than any of the earlier generations invite new and existing players to improve their products, systems and services by allowing those to obtain access to a wireless communication system. This wide range of new use cases and device classes accessing the network most often requires special treatment of those to ensure proper and efficient functioning. The air interface (AI) is one of the fundamental building blocks to address this. In particular, the AI should be designed in a way to provide more flexibility while avoiding being excessively complicated to build and test. The two possible design extremes are either having a single AI as a "one-fits-all solution", where fixed configurations support each and every single requirement at any point in time, or allowing for an excessive number of solutions tailored to each single family of use cases, where the service provider requires to operate a respective number of RATs concurrently. This chapter has explored various design options within these stated extremes, though ultimately centering on a single multi-service AI supporting some flexible adaptations to allow various kinds of services and device classes to be served according to their specific needs, and also the support of different transmission frequencies with very different transmission characteristics.

The chapter has in particular treated the wide topic of waveform design for 5G NR - a key technology defining many features of the overall AI - and compared a rich set of enhancements improving conventional CP-OFDM in various aspects. Similarly, various coding options to be used in various settings have been identified, e.g., related to the size of the packets being transmitted, and efficient schemes to implement HARQ both for UL and DL have been discussed. While for bands below 6 GHz fully digital beamforming may be applied in the context of massive MIMO, systems operating at higher frequencies might have to rely on hybrid variants adding an analog component, as detailed in the chapter, along with the application of massive MIMO to an example scenario. Finally, various design principles for multiplexing the different service types within a single transmission band have been presented, both taking data and control signaling into account.

References

1 3GPP RP-170379, "Study on New Radio (NR) Access Technology", March 2017

2 3GPP RP-170855, "Work Item on New Radio (NR) Access Technology", March 2017

3 3GPP TR 38.802, "Study on new radio access technology physical layer aspects", V14.1.0, June 2017

4 5G PPP mmMAGIC project, Deliverable D4.2, "Final radio interface concepts for mm-wave mobile communications", June 2017

5 H. G. Feichtinger and T. Strohmer, "Gabor Analysis and Algorithms: Theory and Applications", Springer, 1998

6 5G PPP FANTASTIC-5G project, Deliverable D3.2, "Final results for the flexible 5G air interface link solution", May 2017

7 G. Berardinelli, F. M. L. Tavares, T. B. Sørensen, P. Mogensen, and K. Pajukoski, "On the potential of zero-tail DFT-spread-OFDM in 5G networks", IEEE Vehicular Technology Conference (VTC Fall 2014), Sept. 2014

8 D. Petrovic, W. Rave, and G. Fettweis, "Effects of phase noise on OFDM systems with and without PLL: characterization and compensation", IEEE Transactions on Communications, vol. 55, no. 8, pp. 1607–1616, Oct. 2007

9 5G PPP mmMAGIC project, Deliverable D5.1, "Initial multi-node and antenna transmitter and receiver architectures and schemes", Mar. 2016

10 S. C. Cripps, "Advanced Techniques in RF Power Amplifer Design", Artech House, 2002

11 5G PPP METIS II project, Deliverable D4.2, "Final air interface harmonization and user plane design", Apr. 2017

12 P. Siohan, C. Siclet, N. Lacaille, "Analysis and design of OFDM/OQAM systems based on filterbank theory", IEEE Transactions on Signal Processing, vol. 50, no. 5, pp 1170–1183, Aug. 2002

13 3GPP TR 38.913, "Study on Scenarios and Requirements for Next Generation Access Technologies", V14.2.0, Dec. 2016

14 5G PPP FANTASTIC 5G project, Deliverable D3.1, "Preliminary Results for Multi-Service Support in Link Solution Adaptation", May 2016

15 R. Garzon-Bohorquez, C. Abdel Nour and C. Douillard, "Improving Turbo Codes for 5G with parity puncture-constrained interleavers", Int. Symp. on Turbo Codes and Iterative Information Processing (ISTC 2016), Sept. 2016

16 R. Garzon-Bohorquez, C. Abdel Nour and C. Douillard, "Protograph-Based Interleavers for Punctured Turbo Codes", IEEE Transactions on Communications, Dec. 2017

17 R. G. Gallager, "Low Density Parity-Check Codes", MIT Press, Cambridge, 1963

18 E Arıkan, "Channel Polarization: A Method for Constructing Capacity-Achieving Codes for Symmetric Binary-Input Memoryless Channels", IEEE Transactions on Information Theory, vol. 55, no. 7, pp. 3051–3073, July 2009

19 C. Berrou and A. Glavieux, "Near Optimum Error Correcting Coding and Decoding: Turbo-Codes", IEEE Transactions on Communications, vol. 44, no. 10, pp. 1261–1271, Oct. 1996

20 R. Garzon-Bohorquez, C. Abdel Nour and C. Douillard, "On the Equivalence of Interleavers for Turbo Codes", IEEE Wireless Communications Letters, vol. 4, no. 1, pp. 58–61, Feb. 2015

21 J. Vogt and A. Finger, "Improving the max-log-MAP turbo decoder", Electronics Letters, vol. 36, no. 23, pp. 1937–1939, Nov. 2000

22 D. J. C. MacKay and R. M. Neal, "Near Shannon Limit Performance of Low Density Parity Check Codes", Electronics Letters, vol. 32, no. 18, pp. 1645–1646, Aug. 1996, Reprinted Electronics Letters, vol. 33, no. 6, pp. 457–458, Mar. 1997

23 T. Richardson and R. Urbanke, "Multi-Edge Type LDPC Codes", 2004, see http://citeseerx.ist.psu.edu/index

24 T. J. Richardson, M. A. Shokrollahi and R. L. Urbanke "Design of Capacity-Approaching Irregular Low-Density Parity-Check Codes", IEEE Transactions on Information Theory, vol. 47, no. 2, pp. 619–637, Feb. 2001

25 C. Jones, S. Dolinar, K. Andrews, D. Divsalar, Y. Zhang and W. Ryan, "Functions and Architectures for LDPC Decoding", IEEE Information Theory Workshop, Sept. 2007

26 I. Tal and A. Vardy, "List Decoding of Polar Codes", IEEE Transactions on Information Theory, vol. 61, no. 5, pp. 2213–2226, May 2015

27 B. Li, H. Shen and D. Tse, "An Adaptive Successive Cancellation List Decoder for Polar Codes with Cyclic Redundancy Check", IEEE Communications Letters, vol. 16, no. 12, pp. 2044–2047, Dec. 2012

28 3GPP RAN1 meeting #86bis, Nokia, "Chairman's notes of AI 8.1.3 on channel coding and modulation for NR", Oct. 2016

29 3GPP RAN1 meeting #86bis, AccelerComm, "Complementary turbo and LDPC codes for NR, motivated by a survey of over 100 ASICs", Oct. 2016

30 3GPP RAN1 meeting #87, AccelerComm, Ericsson, Orange, IMT, LG Electronics, NEC, "WF on channel codes for NR short block length eMBB data", Nov. 2016

31 3GPP RAN1 meeting #88, R1-1702856, AccelerComm, "Enhanced turbo codes for URLLC", Feb. 2017

32 L. Zheng and D. N. C. Tse, "Diversity and multiplexing: a fundamental tradeoff in multiple-antenna channels", IEEE Transactions on Information Theory, vol. 49, no. 5, pp. 1073–1096, May 2003

33 T. L. Marzetta, "Noncooperative Cellular Wireless with Unlimited Numbers of Base Station Antennas", IEEE Transactions on Wireless Communications, vol. 9, no. 11, pp. 3590–3600, Nov. 2010

34 M. Hochwald, T. L. Marzetta, and V. Tarokh, "Multiple-antenna channel hardening and its implications for rate feedback and scheduling", IEEE Transactions on Information Theory, vol. 50, no. 9, pp. 1893–1909, 2004

35 O. Elijah, C. Y. Leow, T. A. Rahman, S. Nunoo and S. Z. Iliya, "A Comprehensive Survey of Pilot Contamination in Massive MIMO—5G System", IEEE Communications Surveys & Tutorials, vol. 18, no. 2, pp. 905–923, Q2 2016

36 A. Alkhateeb, O. Ayach, G. Leus and R. W. Heath, "Channel Estimation and Hybrid Precoding for Millimeter Wave Cellular Systems", IEEE Journal on Selected Topics on Signal Processing, vol. 8, no. 5, pp.831–46, Oct. 2014

37 S. Han, C.-L. I, Z. Xu, and C. Rowell, "Large-scale antenna systems with hybrid analog and digital beamforming for millimeter wave 5G", IEEE Communication Magazine, vol. 53, no. 1, pp. 186–194, Jan. 2015

38 5G PPP mmMAGIC project, Deliverable D5.2, "Final multi-node and multi-antenna transmitter and receiver architectures and schemes", June 2017

39 J. Mo and R. W. Heath, "Capacity Analysis of One-Bit Quantized MIMO Systems with Transmitter Channel State Information", IEEE Transactions on Signal Processing, vol. 63, no. 20, pp. 5498–5512, Oct. 2015

40 D.-T. Phan-Huy, P. Ratajczak, R. D'Errico, A. Clemente, J. Järveläinen, D. Kong, K. Haneda, B. Bulut, A. Karttunen, M. Beach, E. Mellios, M. Castaneda, M. Hunukumbure and T. Svensson, "Massive Multiple Input Massive Multiple Output for 5G Wireless Backhauling", IEEE Globecom'2017 ET5GB workshop, Singapore, Dec 2017

41 3GPP TR 38.211, "Physical channels and modulation", V1.0.0, (2017–09)

42 5G PPP FANTASTIC-5G project, Deliverable D4.2, "Final results for the flexible 5G air interface multi-node/multi-antenna solution", May 2017

43 3GPP TR 38.804, "Study on new radio access technology radio interface protocol aspects", V14.0.0, Mar. 2017

44 S. Sesia, I. Toufik and M. Baker, "LTE The UMTS Long Term Evolution: From Theory to Practice", John Wiley & Sons Ltd., 2009

45 A. Masucci, S. E. Elayoubi and B. Sayrac, "Flow level analysis of the offloading capacity of D2D communications", IEEE Wireless Communications and Networking Conference (WCNC 2016), Apr. 2016

46 N. K. Pratas and P. Popovski, "Network-Assisted Device-to-Device (D2D) Direct Proximity Discovery with Underlay Communication", IEEE Global Conference on Communications (GLOBECOM 2015), Dec. 2015

47 M. G. Sarret, G. Berardinelli, N. H. Mahmood, B. Soret and P. Mogensen, "Can full duplex reduce the discovery time in D2D communication?", International Symposium on Wireless Communication Systems (ISWCS 2016), Sept. 2016

12

Traffic Steering and Resource Management

Ömer Bulakçı[1], Klaus Pedersen[2], David Gutierrez Estevez[3], Athul Prasad[4], Fernando Sanchez Moya[5], Jan Christoffersson[6], Yang Yang[7], Emmanouil Pateromichelakis[1], Paul Arnold[8], Tommy Svensson[9], Tao Chen[10], Honglei Miao[7], Martin Kurras[11], Samer Bazzi[1], Stavroula Vassaki[12], Evangelos Kosmatos[12], Kwang Taik Kim[13], Giorgio Calochira[14], Jakob Belschner[8], Sergio Barberis[14] and Taylan Şahin[1,15]

[1] Huawei German Research Center, Germany
[2] Nokia Bell Labs, Denmark
[3] Samsung Electronics R&D Institute, UK
[4] Nokia Bell Labs, Finland
[5] Nokia, Poland
[6] Ericsson, Sweden
[7] Intel, Germany
[8] Deutsche Telekom, Germany
[9] Chalmers University of Technology, Sweden
[10] VTT, Finland
[11] Fraunhofer Heinrich Hertz Institute, Germany
[12] WINGS ICT Solutions, Greece
[13] Samsung Electronics, Republic of Korea
[14] Telecom Italia, Italy
[15] Technische Universität Berlin, Germany
With contributions from Chao Fang and Behrooz Makki.

12.1 Motivation and Role of Resource Management in 5G

One of the main differentiating factors of the 5th generation (5G) mobile and wireless network compared to previous generations is the support of a wide range of services associated with a diverse set of requirements, which are typically grouped under the main service types enhanced mobile broadband (eMBB), ultra-reliable low-latency communications (URLLC), and massive machine type communications (mMTC), as highlighted in Section 2.2. 5G also aims at enabling new business opportunities by addressing the requirements of vertical industries. Accordingly, the resulting 5G system shall be flexible to cope with such diversity in an efficient way. On this basis, resource management (RM) plays a critical role to fulfil the service and slice requirements, such as quality of service (QoS) requirements, while efficiently mapping service flows to appropriate resources in order to optimally adapt to the current traffic and dynamic radio-environment conditions. To this

5G System Design: Architectural and Functional Considerations and Long Term Research, First Edition.
Edited by Patrick Marsch, Ömer Bulakçı, Olav Queseth and Mauro Boldi.
© 2018 John Wiley & Sons Ltd. Published 2018 by John Wiley & Sons Ltd.

end, the resource landscape is extended beyond conventional radio resource management (RRM). In addition to licensed radio bands, the extended realm of resources includes the native use of unlicensed bands, which shall be adaptive and coupled with the changing radio topology, energy, hardware and software resources, such as computational and storage resources, as well as backhaul (BH) and fronthaul (FH) resources.

A fast resource allocation can typically be performed on a radio frame unit basis (e.g., a radio subframe of one millisecond), which is controlled by a scheduler. The scheduler can allocate time and frequency resources considering, e.g., the link qualities and transmission power constraints. A much slower resource allocation can be achieved through balancing the load among different cells and radio access technologies (RATs), which is handled by traffic steering mechanisms and hard handovers with execution times spanning several hundred milliseconds. Nevertheless, the 5G system will include various paradigm changes that will re-shape the RM methods towards more agile solutions.

As also highlighted in Chapter 11, it is envisioned that a flexible frame structure will be employed in 5G, such that physical layer (PHY) numerologies are optimized for specific frequency bands, such as sub-6 GHz, millimeter-wave (mmWave), and for one or more target use cases [1] [2] [3]. Thus, the developed RM schemes need to be built upon this additional degree of flexibility, where it is required to introduce new functionalities, such as dynamic multi-service scheduling (see Section 12.3), considering QoS requirements and classifications (see Section 12.2).

In the 5G era, a tight interworking between novel 5G air interfaces (AIs) and enhanced Long-Term Evolution (eLTE) is targeted, where the integration is done on the radio access network (RAN) level, as detailed in Section 6.5. Along with forms of multi-connectivity where a user equipment (UE) connects to multiple access nodes at the same time, and a new QoS architecture, as introduced in Section 5.3.3 and detailed in Section 12.2, functions that are traditionally slow are envisioned to be operated on a faster time scale. This enables a fast routing of service flows to appropriate access nodes and AIs, as presented in Section 12.4 under dynamic traffic steering. Furthermore, a proactive analysis can be performed to determine new mmWave links to be utilized by traffic steering.

The application of full frequency reuse is an effective way of increasing the network capacity considering the scarcity of available spectrum. This, however, results in inter-cell interference (ICI) which needs to be mitigated to ensure the targeted high capacity and wide coverage. In addition, novel communication modes and new deployment options foreseen in 5G lead to new interference challenges to be overcome. Therefore, interference mitigation schemes, as presented in Section 12.5, are essential and should take into account different deployment options, such as fixed and dynamic radio topologies, flexible duplexing schemes, such as dynamic time division duplex (TDD), and high frequency band operation, while exploiting new interference-resistive designs.

Network slicing is seen as a key enabler for new 5G businesses (e.g., related to vertical industries) that require one or more services with associated service-level agreements (SLAs), as detailed in Chapter 8. In particular, the scope of QoS fulfilment is enhanced toward SLA fulfillment, where a real-time SLA monitoring is crucial to avoid SLA violations. Multi-slice RM in Section 12.6 introduces schemes that are needed to respond to multiple SLAs, where different slices share the same RAN infrastructure.

As detailed in Section 15.3, energy efficiency is one of the design goals of 5G to attain a sustainable system. Hence, active-mode operation needs to be optimized in the 5G network with the help of QoS and channel quality awareness, as presented in Section 12.7. The energy saving gains of the moderated network are obtained due to the unique design of 5G which enables access nodes to be in sleep mode longer in case of no traffic.

Eventually, functional extensions and changes in the device measurement context are needed to enable the aforementioned new functionalities tailored to different 5G use cases, while considering device performance, as covered in Section 12.8. Finally, the chapter is summarized in Section 12.9.

12.2 Service Classification: A First Step Towards Efficient RM

The envisioned 5G services have different requirements in terms of QoS, both regarding delay and throughput as well as reliability and availability aspects. The simultaneous provisioning of these services using a common infrastructure is an issue that should be addressed by the 5G network to have a functional and efficient operation. Toward this direction, in this section, we present briefly the QoS framework for 5G networks, focusing on service classification mechanisms.

12.2.1 QoS Mechanisms in 5G Networks

A first step to be able to allocate efficiently the network resources to the heterogeneous services is the accurate identification of the service types and the corresponding requirements. This knowledge could be provided by the higher layers as in High Speed Packet Access (HSPA) and LTE, where sets of QoS parameters are available for RRM functionalities, such as admission control and packet scheduling decisions. A similar approach to the one followed in LTE, where different bearers are set up within the Evolved Packet System (EPS) to support multiple QoS requirements, also holds for 5G networks. As introduced in Section 5.3.3 and described in detail in [4] [5], the 5G QoS model supports a QoS flow based framework where a QoS flow ID (QFI) is used to identify a QoS flow in the 5G system. The QFI is carried in an encapsulation header without any changes to the end-to-end (E2E) packet header. It can be applied to Protocol Data Units (PDUs) with different types of payload, i.e., Internet Protocol (IP) packets, non-IP PDUs, and Ethernet frames. The QFI will be unique within a PDU session. In the Next-Generation Radio Access Network (NG-RAN), the data radio bearer (DRB) defines the packet treatment on the radio interface (Uu). Particularly, in the downlink (DL), the NG-RAN maps QoS Flows to DRBs based on NG-U (or N3 interface) marking (QFI) and the associated QoS profiles. In the uplink (UL), the UE marks UL packets with the QFI for the purposes of marking forwarded packets to the core network (CN). Specifically, in the UL, the NG-RAN may control the mapping of QoS Flows to DRB in two different ways: Either using reflective mapping, where the UE monitors the QoS flow ID(s) of the DL packets and applies the same mapping in the UL, or using explicit configuration where the NG-RAN may configure by Radio Resource Control (RRC) an UL "QoS Flow to DRB mapping". However, sometimes an incoming UL packet matches neither an RRC configured nor a reflective QoS Flow ID to DRB mapping. In this case, the UE will map that packet to the default DRB of the PDU session, even though this may not correspond to the adequate QoS requirements of the flow. To address this issue, the use of novel service classification techniques should be considered, in which the traffic flows are monitored to extract more detailed service classification information and to identify the service type providing this information as input to RRM functionalities.

Toward this direction, this section focuses on service classification techniques based on machine learning (ML). The proposed classification methods reside in the area of statistical-based classification techniques and are implemented exploiting several flow-level measurements (e.g., traffic volume, packet length, inter-packet arrival time, and so forth) to characterize the traffic of different services.

Other methods of traffic classification, like payload-based classification, need to analyze the packet payload or use deep packet inspection technologies. On the contrary, statistical-based classification techniques are usually very lightweight, as they do not need access to packet payload and can also leverage information from flow-level monitors. It should be noted that these techniques may be used both as independent mechanisms as well as additional mechanisms that will support the already established techniques in order to effectively assign flows to the appropriate bearers and, therefore, fulfill the QoS requirements of the flows. The goal is to further increase the effectiveness of packet scheduling algorithms and other RRM functionalities by assigning the appropriate QoS to each service type or even to different flows of the same service type.

12.2.2 A Survey of Traffic Classification Mechanisms

During the last years, several studies that focus on application and service discrimination based on traffic classification learning techniques have been proposed. Both supervised as well as unsupervised ML mechanisms have been considered [6], as also depicted in Figure 12-1. Regarding the unsupervised techniques, principal component analysis (PCA) based mechanisms [7] and clustering algorithms like K-Means, density-based spatial clustering of applications with noise (DBSCAN) and Autoclass [6] have been investigated. These mechanisms group flows that have similar patterns into a set of disjoint clusters. Their major advantage is that they automatically discover the classes via the identification of specific patterns in the dataset without requiring a training phase like the supervised mechanisms. However, the resulting clusters do not necessarily map 1:1 to services and, even in this case, the clusters still need to be labeled in order to be mapped to the corresponding services.

As far as the supervised ML techniques are concerned, there are various classification schemes that have been proposed for the traffic classification problem like Naïve Bayes, decision trees, random forests and others [8]. A main approach focuses on Bayesian classification techniques [9] in which flow parameters are used to train the Naïve Bayes classifier and create a group of services.

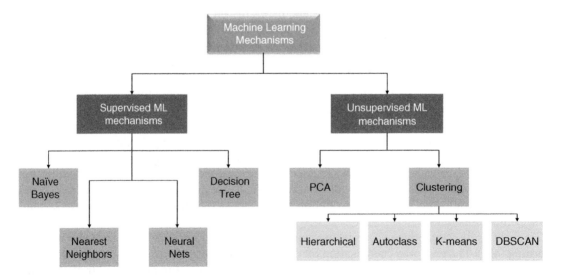

Figure 12-1. Overview of service classification techniques.

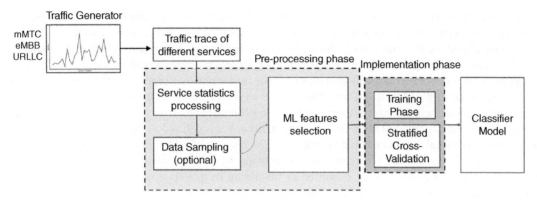

Figure 12-2. Example mechanism for the service classification process.

Then, when new flows arrive, they are assigned to the class for which the class membership probability is maximized. Also, decision tree algorithms [10] represent a completely orthogonal approach to the classification problem, using a tree structure to map the observation input to a classification outcome.

Finally, another concept, which has also been employed for the traffic classification process, is the concept of artificial neural networks (ANNs) that consists of a collection of processing elements that are highly interconnected and transform a set of inputs to a set of desired outputs, inspired by the way biological nervous systems works. Based on this approach, a multilayer perception classification network can be used for assigning probabilities to flows [11]. A set of flow features can be used as input to the first layer of the network, while the output classifies flows into a set of traffic classes by calculating the probability density function of class membership.

12.2.3 ML-based Service Classification Approach

In this section, the problem of service classification is investigated through the example of supervised ML mechanisms. The different steps included in the classification mechanism are presented in Figure 12-2. The first step of the process refers to the collection of traces from different services, whereas the next step focuses on the statistical processing of these traces to separate them in flows. It should be noted that a flow is considered a series of packet transmissions that have the same source and destination, and for which the inter-arrival time is below a specific threshold.

After this processing, a set of features is generated for each flow, including inter-arrival time statistics, packet size statistics and other flow characteristics like the total number of packets, source, destination as well as flow direction. Subsequently, some feature engineering tasks (e.g., data imputation tasks and data cleaning) are performed, completing the pre-processing steps. These tasks include the selection of the most representative features, the transformation of categorical features into numerical values, the normalization of features' values, and other tasks, such as the replacement of missing values, in order to guarantee high data quality. Finally, the implementation of the ML mechanism follows, including two main phases, namely a training phase and a cross-validation phase. In this case, stratification is applied in order to randomly sample the flows' dataset in such a way that each service type is properly represented in both training and testing datasets.

To evaluate the performance of the classification mechanisms and select the most adequate mechanism, several metrics can be used in the train/test sets, like the so-called accuracy metric, the precision, the recall and the F1-score, all of which will now be explained. A very useful tool that illustrates the relationship between the different evaluation metrics and provides a holistic view of each algorithm's performance is the confusion matrix, where the horizontal axis represents the predicted class (i.e., the outcome of the algorithm), and the vertical axis represents the true class. The confusion matrix formulation includes information for the false positives (FP), true positives (TP), false negatives (FN) and true negatives ($TNeg$). In this context, *accuracy* is defined as the percentage of correct predictions to the total number of predictions, i.e. $(TP + TNeg)/(TP + FP + TNeg + FN)$, whereas *precision* represents the percentage of the instances that were correctly predicted as belonging in a class among all the instances that where classified as belonging in this class, as given by $TP/(TP + FP)$. *Recall* is defined as the percentage of the instances of a specific class that were correctly classified as belonging to this class, given by $TP/(TP + FN)$, and *F1-score* is defined as the harmonic mean of the precision and recall.

It should be highlighted that the investigation of a single metric, like accuracy, is not always enough to choose the best mechanism for a classification problem, as the misclassification of a specific class instance may have more significance than the correct classification of others. For this reason, other evaluation metrics have also to be applied to make the most appropriate choice depending on the problem's characteristics.

12.2.4 Numerical Evaluation of Service Classification Schemes

In this subsection, a numerical evaluation of the service classification schemes based on simulation results is provided. In particular, for the simulation scenario, the three main service types URLLC, eMBB and mMTC are considered. The different traces for each service have been generated using specific traffic models, where URLLC and mMTC traces are generated according to the models presented in 802.16p [12], and eMBB traffic is generated based on [13], assuming video streaming traffic (i.e., YouTube). For the classification mechanism, a splitting of 70%-30% for training and testing sets has been considered. For the training set, the label 'service type' of each flow is considered as known; whereas, for the testing set, this label is considered as unknown, and each flow is labeled using the classifier model. The outcome of the proposed mechanism is a classifier model that can be applied to unknown flows to recognize them in an accurate way.

In the simulation scenario, the performance of various ML mechanisms has been investigated including base classifiers as well as ensemble-based classifiers. The goal of ensemble methods is to combine the predictions of several base estimators built with a given learning algorithm to improve generalizability and robustness over a single classifier. To compare the different ML mechanisms, the accuracy metric of each algorithm is presented in Table 12-1, where a Dump classifier that classifies all the flows as type 0 (mMTC service) is also considered. From this table, it can be seen that Decision Tree and Random Forest algorithms lead to the highest accuracy values, outperforming the other ML algorithms.

To provide a more complete view of each classifier's performance, the corresponding confusion matrices are also illustrated in Figure 12-3. Considering that it is desired to eliminate the possibility that a URLLC service is misclassified as another service type, the optimal model should have high values of recall and high accuracy values for the case of mMTC and eMBB services. The confusion matrix shows that the Decision Tree and the Random Forest algorithms result in extremely good

Table 12-1. Accuracy score for each classification mechanism.

Classifier	Accuracy
Dump Classifier	0.51
Naïve Bayes	0.82
Support Vector Machine (SVM)	0.93
Decision Tree	0.99
K Nearest Neighbour Classifier	0.97
Logistic Regression	0.89
Random Forest	0.99
Adaboost Classifier	0.98

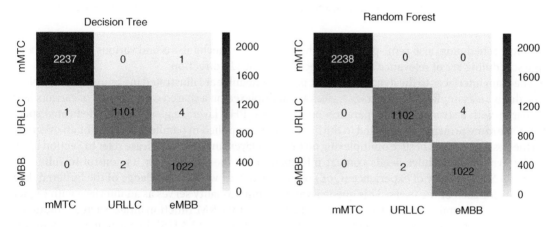

Figure 12-3. Performance evaluation results considering the indicative scenario.

results, as they miss-classify only a few flows, resulting also in high values of recall and precision. Therefore, these two classification mechanisms can be selected for further consideration for the problem of service classification.

12.3 Dynamic Multi-Service Scheduling

Allocation of radio transmission resources is one of the most essential RRM tasks. A major challenge is to balance these resources among different transmission types, e.g., unicast, multicast, and broadcast, as well as scheduled and non-scheduled UL access. Here, we first focus on the network-controlled resource allocation for scheduled unicast transmissions between the network infrastructure and UEs, as handled by the Medium Access Control (MAC) layer dynamic scheduler. In addition to

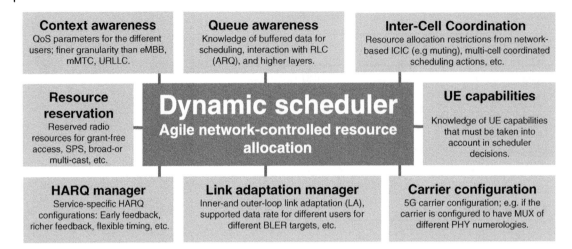

Figure 12-4. High-level illustration of the main interfaces to the dynamic scheduler for unicast transmissions.

dynamic scheduling, also semi-persistent scheduling (SPS) mechanisms and various forms of grant-free scheduling are of relevance for 5G, especially for the UL direction.

The main interfaces to the dynamic unicast packet scheduler are illustrated in Figure 12-4 [14]. Note that this is an example where diverse services are multiplexed on a shared carrier, or set of carriers. For scenarios with a stronger slice separation on MAC and PHY layers, e.g., slices with dedicated spectrum, the same principle as depicted in the figure could be applied in parallel per slice. For an overview on the possible extents of slice multiplexing on different layers in the RAN, please refer to Section 8.2.3.

The dynamic scheduler needs context information for the users under its control to fulfill the users' QoS and quality of experience (QoE) requirements, as well as knowledge of the buffered data amounts for each of the users. This queue awareness functionality serves as an interface with higher layers, e.g., Transmission Control Protocol (TCP) and other RM functionalities, such as multi-AI/multi-slice traffic steering as presented in the following Section 12.4 [15]. In particular, TCP-awareness can help to avoid re-buffering events for streaming and queuing delays for latency-critical traffic. Furthermore, the dynamic scheduler needs to know which radio resources are available for unicast transmissions, and which resources are reserved for other purposes, such as SPS, broadcast, and non-scheduled access. Knowledge of UE categories and related constraints is obviously also needed by the scheduler. UE categories basically express terminal capabilities, such as the maximum supported data rate and multiple-input multiple-output (MIMO) capability which influence on how much data can be scheduled to the UEs, and which formats are supported. As 5G will support cell carriers with a mixture of different PHY numerologies, information about carrier configuration is also needed by the scheduler. The link adaptation manager essentially provides information on the supported data rate for the different users. It also facilitates to operate different users with a service-specific first transmission block error rate (BLER); for example, using an initial BLER target of less than 1% for latency-critical traffic with tight reliability constraints, while using 10% - 20% BLER for best effort eMBB traffic.

As the scheduler also needs to dynamically allocate transmission resources for hybrid automatic repeat request (HARQ) transmissions, it interfaces the HARQ manager as well. Notice here that in

5G, asynchronous HARQ is assumed for both link directions (i.e., UL and DL), giving the scheduler freedom to decide at which point in time it wants to schedule HARQ retransmissions [16]. Similarly, other HARQ enhancements are considered, such as a flexible timing of acknowledged (ACK)/ negative acknowledged (NACK) messages and number of HARQ processes [17], options for early HARQ feedback to reduce latency [18], as well as richer HARQ feedback, allowing the scheduler to optimize the usage of radio resources for HARQ retransmissions [19].

Finally, the scheduler is also subject to constraints from multi-cell cooperation, e.g., the network-based ICI coordination (ICIC) schemes that are further described in Section 12.5.

12.3.1 Scheduling Formats and Flexible Timing

In order to support the diverse service requirements handled by the previously described dynamic scheduler in the 5G New Radio (NR) system, especially variable transmission time interval (TTI) sizes, a flexible resource frame structure and timing are required.

The flexible frame structure is realized by a scalable numerology [3], due to parametrized subcarrier spacing of $2^m \cdot 15$kHz, where m is an element of the set of integer values [0, 1, 2, 3, 4, 5]. This scalable subcarrier spacing allows flexibility of resource units in the frequency domain. It also corresponds to flexible orthogonal frequency division multiplexing (OFDM) symbol lengths in time domain. Similar to LTE, multiple symbols are combined into a slot. The number of OFDM symbols is given by the reference numerology such that the slot time duration is $1/2^m$ ms.

Consequently, with the same control plane (CP) overhead, the number of OFDM symbols is independent of the subcarrier spacing scaled by m. It is further agreed that 5G NR should also support the LTE normal CP duration for each subcarrier spacing. As an example, $m = 0$ and $m = 2$ correspond to 15 kHz and 60 kHz subcarrier spacing, respectively, resulting in a slot duration of 1 ms and 0.25 ms. Similar to the time domain, a minimum scheduling unit for scheduling comprises also multiple subcarriers in the frequency domain, called a physical resource block (PRB) in LTE. Additional to the flexibility by this scalable numerology, further degrees of freedom are provided by the option to schedule slots or mini-slots. Specifically, a slot is defined as 14 OFDM symbols, and the smallest mini-slot is 1 OFDM symbol [4] [16]. Note that, on the other hand, also larger scheduling units can be assigned, spanning multiple slots or mini-slots.

However, to support the diverse traffic and latency requirements mentioned in the previous section, also flexible timing of the scheduled resources is essential. An illustrative example for the envisioned 5G flexible timing is provided in Figure 12-5, where a TDD scenario with three slots for DL followed by two for UL transmissions is shown. Further signaling, such as reference signals, are omitted for visibility.

The flexible timing is applied by the following main design principles. First, DL control and data channel transmissions are multiplexed on a per-user basis and control channels are front loaded, which means that they appear in the start of the transmission before the corresponding DL data transmission, as detailed in [14], Chapter 2. Second, the NR PHY design should allow devices with different bandwidth capabilities to access the same carrier regardless of the NR carrier bandwidth [3]. Third, for an UE scheduled with a resource spanning multiple slots, the corresponding DL control channel signaling is transmitted only once in order to utilize benefits of reduced control channel overhead of longer TTIs. For example, user #3 in Figure 12-5 is scheduled to a resource unit spanning three slots with a single DL grant in slot #1.

While the benefits of flexible timing and scheduling are clear, such a framework requires additional signaling, e.g., transmitted in the DL control information (DCI), which contains among other things

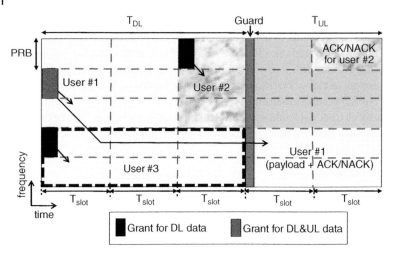

Figure 12-5. Basic illustration of possible scheduling of users on a slot resolution for a TDD scenario, where T_{slot} is the time duration of a slot and UL/DL indicates UL/DL.

resource assignment for UL or DL in an LTE system. Among other fields in the DCI, this includes the timing between a DL assignment and data transmission, the timing between the UL assignment and data transmission, and the timing between DL data reception and acknowledgement. Optionally, a DL grant can also include a future UL transmission allocation, as given in Figure 12-5 by the grant for user #1, and the timing has to be part of the DCI.

These degrees of freedom in scheduling and timing, supported by the flexible numerology and scheduling framework, allow a centralized implementation in one physical location, as seen in Figure 12-4, while RF frontends can be distributed over multiple base station (BS) sites.

12.3.2   Benefits of Scheduling with Variable TTI Size

It is well known from the existing literature that there are fundamental trade-offs between scheduling users to maximize their spectral efficiency, coverage, latency or reliability [20]. This calls for flexible scheduler functionality that allows scheduling each user in accordance with its desired optimization target. One option for this is to design the 5G system to support scheduling with different TTI sizes [21], which is the foundation of the scheme detailed in the following, based on a flexible frame structure that dynamically allows variable TTI size configuration per user and per scheduling instance.

In this way, the following scheduling decisions are possible:

- Use short TTI for low-latency communications users to optimize their latency, at the expense of increased control overhead and lower channel coding gains;
- eMBB users can be scheduled with longer TTIs and wider frequency allocations to cope with the high data rate demands;
- mMTC users can benefit from narrow bandwidth allocation and long TTIs, which are attractive characteristics from cost and coverage perspectives.

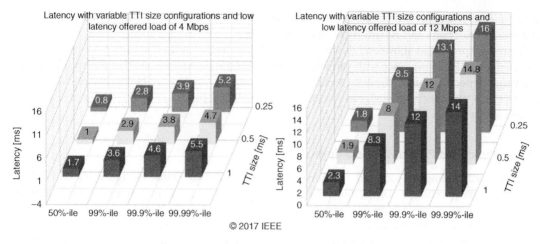

Figure 12-6. Latency values from packet latency cumulative distribution function (CDF) with variable TTI configurations and offered loads for a mix of eMBB and low-latency traffic © 2017 IEEE [25].

In addition, the possibility to set the TTI size per scheduling instance enables optimizing eMBB services that make use of TCP. A short TTI duration can be used in the first transmissions to reduce the round-trip time of the flow control mechanism in the slow start phase of TCP, and later longer TTIs can be configured to maximize the spectral efficiency, when steady operation is reached.

System-level performance evaluations have been carried out to compare the performance of several TTI size configurations, in order to estimate the most suitable TTI size that should be dynamically chosen per UE depending on service requirements, traffic type, radio channel quality and system load. The evaluation is performed in a 3GPP Urban Macro scenario with 7 BSs, each having 3 sectors, 500 m inter-site distance (ISD) and using 10 MHz bandwidth [22]. In-resource control channel (CCH) scheduling grants with link adaptation are assumed, which allows to model different degrees of CCH overhead (i.e., aggregation levels or number of resource elements dedicated to CCH) depending on the UE radio conditions [21] [23].

The packet latency, i.e., the MAC layer one-way UP latency (including scheduling delay) achieved with different TTI sizes and system loads with a mix of eMBB and low-latency services is shown in Figure 12-6. The eMBB traffic is modelled with a single user full buffer download, whereas higher priority low-latency traffic follows a Poisson arrival process with 1 kB payload and varying total cell offered load. More details can be found in [24] [25]. As depicted in Figure 12-6, using a short TTI at low system loads is in general a more attractive solution to achieve low-latency communications. However, looking at the tail values (99.9%-ile and above), a 0.5 ms TTI size offers better latency than the 0.25 ms TTI, even for low loads.

As the load increases, longer TTI configurations with lower relative CCH overhead, and consequently higher spectral efficiency, provide better performance, as these better cope with the non-negligible queuing delay. The 1-ms TTI configuration is beneficial from a latency point of view for high loads and above the 99.9%-ile, due to queuing delay. As the offered load increases, or as UEs with the worst channel conditions are considered, the queuing delay becomes the most dominant component of the total latency; therefore, it is beneficial to increase the spectral efficiency of the transmissions, by using a longer TTI, in order to reduce the experienced delay in the queue. The observed trends are relevant for URLLC use cases requiring latency guarantees of a few milliseconds and reliability levels up to 99.999%.

The results presented above and detailed in [24] [25], as well as related studies performed in [23] (focused on TCP performance with variable TTI size configuration), indicate that the optimum TTI size varies depending on multiple factors. Therefore, it is beneficial to be able to dynamically adjust the TTI size per user's service requirements and scheduling instance, rather than operating the system with a fixed TTI.

12.3.3 Punctured/Preemptive Scheduling

Punctured scheduling, as pictured in Figure 12-7, is proposed to enable multi-service scheduling with diverse requirements [14]. A single block in Figure 12-7 is a scheduling unit in time domain from UE #2 perspective, which can be a mini-slot of one or several OFDM symbols. The symbol duration depends on the subcarrier spacing of the system. Here, UE #1, receiving eMBB traffic, is scheduled by the BS for transmission on the DL shared radio channel. The former is facilitated by the BS sending a scheduling grant, transmitted on a PHY control channel, followed by the actual transmission of the transport block. During the scheduled transmission time of the transport block for UE #1, the DL shared channel for this transmission is in principle monopolized by the UE. As illustrated in Figure 12-7, it may happen that URLLC data for UE #2 arrives at the BS shortly after the transmission towards UE #1 has started. In order to avoid having to wait for the completion of the transport block transmission to UE #1, it is instead propose to immediately transmit the URLLC data to UE #2 by puncturing (i.e., over-writing) part of the ongoing transmission to UE #1. The advantage of this solution is that the latency of the data to UE #2 is minimized, and there is no need to a priori reserve radio resources for transmission of URLLC that may potentially come.

The drawback is in the performance of the transmission towards UE #1. Depending on how large a fraction of the resources for the transmission towards UE #1 is punctured, UE #1 may still be able to correctly receive the data thanks to efficient forward error correction (FEC).

In Figure 12-8, the simulated performance of three transceiver approaches with different extents of knowledge about the puncturing at both transmit and receive sides is shown [26]. In the first approach, assuming prior knowledge about puncturing at both transmitter and receiver sides, rate

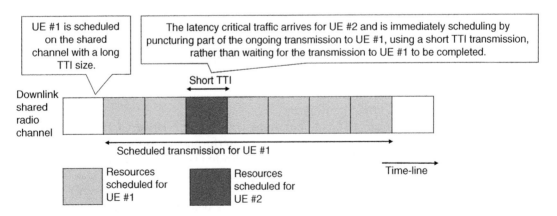

Figure 12-7. Sketch of the basic principles of punctured scheduling on the DL shared data channel.

(a)

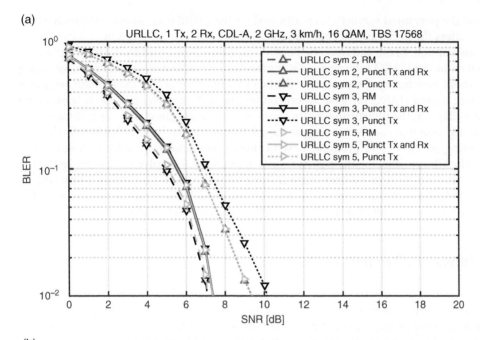

(b)

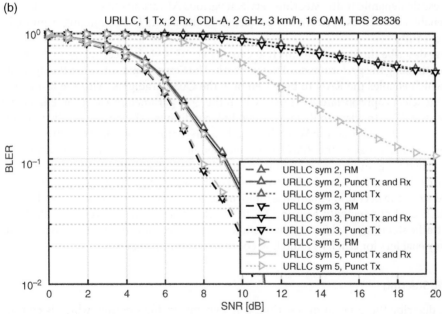

Figure 12-8. Three transceivers: Ideal rate matching, puncturing at transmitter (Tx) only, and puncturing at both Tx and receiver (Rx) with (a) transport block size (TBS): 17568 bits, (b) TBS: 28366 bits [26]. QAM: Quadrature Amplitude Modulation.

matching around the punctured symbol(s) is performed for the eMBB data. Apparently, such transceiver approach offers the performance upper bound, which cannot be achieved in practice due to the unrealistic assumption. In the 2nd case, the puncturing is only performed by the transmitter, but not by the receiver during HARQ retransmission combining. In contrast to the 2nd case, assuming the knowledge about puncturing at the receiver, the 3rd variant assumes that the puncturing is also taken into account during HARQ retransmission combining at the receiver as well. It is observed from Figure 12-8 [26] that the puncturing-aware receiver, i.e., the 3rd compared case, outperforms significantly the 2nd case that uses puncturing at Tx only, especially for large TBS whereby the detrimental puncturing effect is more prominent. With these observations, in order to improve the performance of the UE experiencing the puncturing, it appears to be an advantage to make the UE aware of the puncturing, such that the knowledge that part of the transmission is containing no useful information can be taken into account in the decoding process. The indication of the puncturing is conveyed to the UE using the PHY control channels that also carry scheduling grants.

12.4 Fast-Timescale Dynamic Traffic Steering

With native support for cloud-based deployments and increased availability of computation resources for improving network performance, some of the asynchronous functions in legacy radio access technologies (RATs) could be implemented in a faster timescale in 5G. In this section, we consider the use of fast-timescale dynamic traffic steering between various AI variants (AIVs)[1] within the 5G network as a key enabler of various network key performance indicators (KPIs) ranging from high-capacity to ultra-reliability.

In this section, we describe three mechanisms that can enable 5G RAN in steering traffic between different AIVs for optimizing the performance of the network. The first scheme combines the dynamic QoS framework, which enables the RAN to better enforce the QoS paradigms, with the fast traffic steering scheme that selects appropriate AIVs over which the QoS parameters are enforced. Such a mechanism enables the RAN to adapt to dynamic radio link conditions, which is especially relevant for higher frequency bands, such as mmWave bands. The pre-emptive geometrical interference analysis (PGIA), for instance, enables the RAN to take interference conditions of the mmWave links into account while scheduling traffic over each individual link. The mechanism further considers the formation of resource sharing clusters, which is then used along with traffic steering, to better coordinate the interference in the network. Finally, the multi-node connectivity mechanism considers the enhancement of LTE dual connectivity feature [27] to enable Packet Data Convergence Protocol (PDCP)-level fast traffic steering to achieve higher data rates. The mechanisms described here could be considered as key enablers for efficient radio link coordination and RM in 5G.

12.4.1 Fast Traffic Steering

In this section, we describe the concept of fast-timescale dynamic traffic steering, which is considered to be a key enabler for 5G KPIs. The slower-timescale traffic steering mechanisms were also used for mobility and load balancing in legacy RATs [28]. Such concepts were evaluated from a

---

1 For the definition and examples of the term AIV, the interested reader is referred to Section 6.4.1.

heterogeneous network perspective in [29], with the added constraint of mobility. In 5G, the traffic steering mechanism is considered under the operational constraints of being fast, i.e. in the range of a few milliseconds, or being on a synchronous timescale in relation to the BS or Gigabit NodeB (gNB) operation. The dynamic traffic steering and associated dynamic QoS concept was first proposed in [30] [31] from a throughput enhancement perspective. Means for improving the reliability of the network with the application of dynamic traffic steering and related QoS enhancements were considered in [15]. Another key enabler considered here is the efficient centralization of some key functions, which are not AIV-specific, in the upper layers of the gNB (also called access network - outer (AN-O) layer in [15]). On the other hand, the AIV-specific functions are distributed in the lower-layers of the gNB (called access network - inner (AN-I) layer in [15]), similar to the function splits and related deployment scenarios proposed in 3GPP for 5G NR [32] and detailed in Section 6.6.2.

An overview of the envisioned service flow delivery mechanism [15] [30] in the context of dynamic traffic steering, and the related QoS architecture [4], as detailed in Section 5.3.3 is shown in Figure 12-9. Here, the key idea is that, as compared to the legacy RATs, where the QoS definition is done by the Policy and Charging Rules / Enforcement Function (PCRF/PCEF) and enforcement is done in the packet data network gateway (P-GW), we use the 5G QoS architecture, where the QoS definition for each service flow, called the QoS flow, to be done at the SMF and appropriate mapping of the data packets to the QoS flows by the user plane function (UPF). However, the RAN has advanced capabilities in terms of enforcing and readjusting the QoS paradigms in order to achieve the target KPIs of the network. Such a dynamic QoS framework tightly interworks with the traffic steering framework in the RAN, in order to adjust the QoS parameters of each radio flow over the AI.

The concept can be efficiently mapped to the 5G QoS architecture, in terms of the E2E application flow (i.e., PDU) into multiple QoS flows between the end user device and the UPF, and radio bearers,

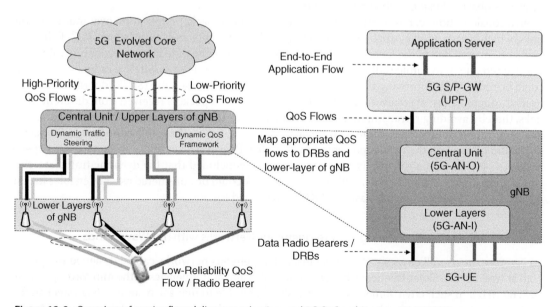

Figure 12-9. Overview of service flow delivery mechanism and 5G QoS architecture [4] [15] [30].

which map multiple QoS flows over the AI. The 5G-RAN has the capability of mapping various QoS flows into one or more DRB sessions, based on how the QoS criteria can be best fulfilled over the radio interface. Such mapping between QoS flows and radio bearers within the 5G-RAN is considered here as dynamic traffic steering function, implemented in the AN-O. The association between E2E application flow and QoS flows is done based on non-access stratum level packet filters in the 5G core (for DL) and in the UE (for UL flows) [4]. Thus, with the dynamic QoS framework and traffic steering function, the 5G-RAN and core have the joint responsibility of ensuring that the 5G KPIs in terms of reliability, latency, throughput, etc., are met by the system.

There are two KPI optimizations that could be achieved using dynamic traffic steering in combination with the dynamic QoS framework - throughput and reliability. The key limitation with LTE networks is the lack of flexibility in the RAN to optimize the mapping of QoS parameters (defined by the P-GW located at the edge of the mobile network) into the DRBs over the AI, depending on the dynamic radio conditions experienced by the user. Thus, for 5G networks with centralized deployments, enhancements are considered mainly focusing on optimizing the throughput KPI in an unreliable network environment comprising of mmWave cells, tightly interworking with LTE small cells. The UE is assumed to have multiple active radio links with different nodes in the AN-I layer, using the multi-connectivity paradigm. Here, the upper layers of the gNB are assumed to monitor the link quality for the end users, based on radio link quality feedback from the lower layers. When a link breakage is detected, the logical AN-O layer or central unit (CU) can deactivate the corresponding radio bearer and re-route the traffic using another active radio link. This ensures that the KPIs related to each QoS flow are constantly monitored and that they are achieved. If the QoS constraints lead to the establishment of guaranteed bit rate bearers with associated resource reservations, such dynamic QoS and traffic steering concept also enables a more efficient resource utilization, since a radio link failure over a link also leads to the bearer and corresponding resource reservations being deactivated [30] [31]. Based on the evaluations presented in [30], up to 20% gain in terms of mean packet delivery delay was observed using such enhancements.

Such considerations can be further enhanced by an efficient packet duplication over multiple radio links, which is seen as a key enabler for achieving higher reliability and lower latency [15]. Reliability could be ensured due to the fact that the UE would still receive the data in case one or more links are interrupted due to dynamic radio conditions, as long as it has at least one active radio link. Latency could be reduced by improving the signal-to-interference-plus-noise ratio (SINR) distribution of the users through the use of receiver-side combining of duplicated packets received from different radio links, thereby, avoiding retransmissions that would increase the latency over the AI. Significant SINR and resulting reliability improvements were observed using two or three link cooperation for packet duplication, based on the results presented in [15]. PDCP layer packet duplication have been specified in 5G [4]. Based on the discussions here, it can be concluded that the dynamic traffic steering concept used in cognition with dynamic QoS framework could be considered as a key enabler for the diverse set of requirements, use cases and KPIs envisioned for 5G.

12.4.2 Proactive Traffic Steering in Heterogeneous Networks with mmWave Bands

To increase the overall capacity of 5G systems, heterogeneous networks can be designed to operate both at traditional cellular frequency bands below 6 GHz and at frequencies in the mmWave spectrum (e.g., 60 GHz). In such environment, a network should be able to manage resources (e.g., steer traffic to the proper transmission link), taking into account the specific characteristics of the various potential transmission links and their actual condition (e.g., interference) at the time of transmission. A proactive analysis, by means of a geometrical-based interference approach, allows to limit transmission collisions (here, meant as transmissions creating such a high mutual interference with neighbor transmission links that makes the communication impossible) and subsequently to avoid signaling overhead required to take the appropriate counteractions to recover from link failures. Such investigation will enable the system to take educated actions before a new transmission link in the mmWave band is initiated: For instance, by performing a pre-emptive traffic steering (e.g., establishing a transmission link on a lower frequency) in the AN-O layer or by implementing a suitable resource partitioning mechanism (e.g., Frequency- or Time-Division Multiplexing, FDD or TDD) in the AN-I layer (scheduler).

As an example of the latter, the considered PGIA allows the network to define and update, prior to the establishment of a new mmWave transmission link, all the existing sets of prospective interfering mmWave transmission links. Each one of these sets, referred to as resources sharing clusters (RSCs), is a set of mmWave transmission links that are coordinated to avoid potential collisions and are revised periodically or on an event-triggered basis. A prerequisite to the algorithm is that the geometrical positions of all the mmWave transmitting and receiving nodes are known by the network at any time regardless of their mobility, as well as some basic information about the mmWave antennas. In the geometrical interference analysis of the PGIA method, described in [15], once an interfering link is found, three variants are considered on how its interference contribution is taken into account and how the merging of RSCs is done. Since it can be assumed that a link coming from a RSC of n links will interfere the interfered link for $1/n$ of the band or $1/n$ of the time, the following variants starting from the originally conceived PGIA apply:

1) Original PGIA: The size of the interfering link cluster is not taken into account in the calculation leading to conservative overestimated interferences and formations of big RSCs;
2) PGIA w/ clusters: The size of the interfering link cluster is considered, nevertheless interference contributions by different links of the same RSC are summed up;
3) PGIA w/ clusters & sum: As in PGIA w/ clusters, plus, to account for all interference contributions, all different interferences coming from different RSCs are summed up and checked, if there is a collision, the most interfering RSC is merged with the interfered.

The set of RSCs emerging from the geometrical step is then used to partition the available resources avoiding transmission collisions.

The simulations are performed under the assumption of a $1\,km^2$ flat area with no obstacles and a random positioning of transmitting and receiving nodes. For receiving nodes, omnidirectional antennas are used. Transmitters are equipped with directional antennas characterized by a main lobe of 4 degrees and a front-to-back ratio (FBR) of 30 dB. An ideal power control mechanism is active: At each receiving node, the received signal power is always of the optimal magnitude. Moreover, a maximum range of a mmWave radio link is set to 200 m with a carrier-to-interference (C/I) threshold of 12 dB. The overhead of the resource sharing mechanisms is negligible, and throughput is divided equally between the links of the same RSC. Simulations are Monte Carlo with 10000 runs for every studied scenario, where the number of mmWave links per km^2 varies from 10 to 200 with 10-link steps.

In Figure 12-10, the fairness of the network with no algorithm or different PGIA algorithms is compared. The different variants of PGIA can significantly reduce the number of interfered links as the number of concurrent mmWave links increases: With no PGIA mechanism, the average percentage

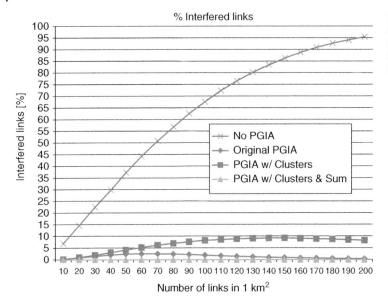

Figure 12-10. Average percentage (%) of interfered links as functions of the number of concurrent links in 1 km², without and with different PGIA algorithms.

of interfered mmWave links rises over 95%, as the number of concurrent links grows to 200, while with the original PGIA, the average percentage of interfered mmWave links is capped around a very low 2.5%. The PGIA w/ clusters variant, since it looks for interference only on a cluster by cluster basis, keeps the interference capped a little higher but below 9.3%. The PGIA w/ clusters & sum variant identifies every possible collision considering the sum of the interfering signals coming from all the different RSCs, with the result being that there are no interfered links regardless of the number of concurrent mmWave links in the area.

In Figure 12-11, the performance of the network with no algorithm or different PGIA algorithms is compared. The different variants of PGIA mechanism not only keep the number of interference links very low, but also achieve a better average throughput per link with a consistent gain in areas with many concurrent mmWave links per area: Both subsequent variants perform better than the original PGIA algorithm, which nonetheless doubles the average throughput in the most loaded scenario. The PGIA w/ clusters variant, by considering potential collision only on a cluster by cluster basis, keeps the size of the RSC quite limited achieving the best performance of all the algorithms.

12.4.3 Multi-Node Connectivity

When deploying new novel 5G AIVs, it can have advantages to leverage existing legacy (i.e., LTE) deployments to improve coverage, as discussed in Section 6.7.2. When a new deployment is rolled out, it is likely that the initial deployments will have coverage limitations, since it will likely not be as dense as an existing mature network. Also, the higher frequencies of novel 5G AIV imply, for example, lower diffraction and higher outdoor/indoor penetration losses, as detailed in Chapter 4, which means that signals will have less ability to propagate around corners and penetrate walls. Thus, to achieve a reliable connection, it may be necessary to rely on LTE for coverage.

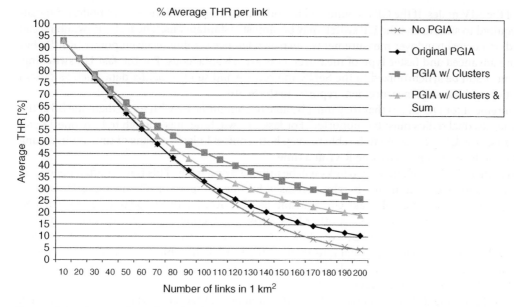

Figure 12-11. Average throughput (marked as THR in the figure) per link as a percentage (%) of the total throughput achievable by one single link as functions of the number of concurrent links in 1 km², without and with different PGIA algorithms.

The dual connectivity (DC) concept from LTE Release 12 [27] is being used as a basis for a tighter integration between LTE and 5G in 3GPP, see also Section 6.5. DC will enable the UE to be connected to LTE and 5G, i.e., both to UP and CP, simultaneously. In DC, the node which is responsible for the CP is the Master eNB (MeNB). The MeNB also routes data to the secondary eNB (SeNB) over the X2 interface. The data rate sent to the SeNB can be based on feedback from the SeNB using flow control. The PDCP layer of the MeNB is common to both AIVs. DC can increase the UE throughput due to UP aggregation (receiving data from both BSs at the same time) and make the connection more reliable via CP diversity. The increased reliability also comes from the case when the UE hands over to another SeNB; in this case, it can still be connected to the MeNB and reliably receive RRC signaling from the MeNB. A disadvantage of DC is that it may be less resource efficient than being connected only to the best cell, since DC will transmit packets also on links with relatively bad quality and, thereby, taking resources from users who could utilize the link better.

With DC, the split is done at the PDCP layer. This means that the protocol layers below PDCP can be more or less independent between the two AIVs. An alternative to DC is to do the split at a lower protocol layer, for example on MAC layer, such as in carrier aggregation. This would have advantages, since it would allow cross-AIV scheduling, which would allow highly dynamic traffic distribution even for short flows. Carrier aggregation also does not require a symmetric UL/DL configuration, i.e., it is possible to have, e.g., fewer UL carriers than what is used on DL. On the downside, MAC-layer aggregation poses very high requirements on BH, or requires the AIVs to be co-located, and may be difficult if the AIVs involve different numerologies.

An alternative to DC is fast UP switching, where the CP is connected to both AIVs at the same time; but, the UP is only transmitted via one of the AIVs, i.e., the PDCP level routes the UP packets to one of the AIV nodes. If the CP is connected to both the LTE node and the 5G NR node, no signaling is required to switch, and the UP switch may be almost instantaneous. The fast UP switching can be based on normal handover measurements, such as reference symbol received power (RSRP), but also more advanced and faster type of measurements are advantageous. For the fast UP switching, the UE must still be synchronized to the SeNB and receive and send signaling data, such as system information and transmit measurement reports, to the SeNB.

A requirement for DC UEs is that they must be able to communicate with more than one BS at the same time, i.e., dual radios must be supported. For fast UP switching, it is enough to have one Tx and one Rx, since the UE is expected to be able to keep both control links by means of time multiplexing operations to listen and measure one RAT at a time.

In cases when DC is configured, two links will be utilized by each UE. This is most often beneficial, since it allows for aggregation and, hereby, increased bit rate compared to if only one AIV is used. However, in many cases, this will make UEs transmit packets over relatively bad links. In case of high load, these links could be better used by other UEs whose connections to the BS are better. In these situations, it is better to employ fast UP switching, where the UEs only utilize their best link. In addition to use best link and lower the total interference, the fast UP switching can also perform rather fast load balancing between the AIVs. Ideally, the switch could in some scenarios be so fast that it can follow the fast fading of a low frequency AIV.

In Figure 12-12, the difference between fast UP switching and DC at varying loads is illustrated. When the AIV is changed, PDCP packets must be moved to the new node. This will need to be done efficiently not to cause delays or interruptions. Also scheduling will be affected, since measurements, e.g., channel quality indicator (CQI) reports, must be available at the new AIV. If the packets cannot be scheduled immediately due to missing measurements, fast UP switching will be subject to interruptions and sub-optimal performance.

How fast UP switching should be done in a more general case can be derived from the wave length of the fast fading for different UE speeds and carrier frequencies. The time for the UE to travel one

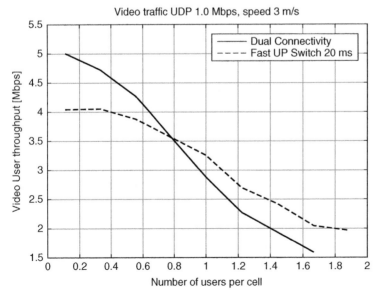

Figure 12-12. The user throughput for each packet for the video service.

fourth of the wave length can indicate the rate that the feedback and switch should be to allow the fast UP switching to follow the fast fading relatively well. From this, it can be deduced that a FS or scheduling decision made every 10 ms would be sufficient to capture the fast fading for UE speeds up to 3 m/s for, e.g., LTE on 2 GHz or up to 0.833 m/s (3 km/h) for an AIV on 6 GHz. Following the fast fading for faster UEs or on higher frequencies would require faster measurement feedback for the fast UP switching. If the fast fading can be followed, even with the same average SINR level for LTE and NR, the fast UP switching would be able to ride on the fading tops and increase the instantaneous SINR and, thereby, the performance.

12.5 Network-based Interference Management

Co-channel interference is an inherent limitation of wireless systems employing universal frequency reuse, i.e., a frequency reuse factor of one. While such feature maintains high spectrum utilization, the interference limits spectral efficiency and, thereby, network and user performance, unless some type of interference management is employed. Overcoming interference is, therefore, essential in ensuring high capacity and wide coverage of high end-user data rates, as well as robust and efficient communication.

A common objective of most interference management schemes relates to improving cell edge user throughputs, coupled with varying levels of network coordination [33] [34]. For instance, centralized scheduling as well as ICIC introduced in LTE Release 8 aim to improve cell-edge SINR through frequency and power allocation. In later LTE releases, the backward compatible enhanced ICIC (eICIC) framework for heterogeneous networks provided the ability to mitigate interference on data and control channels in frequency or time domain. In time domain, interference avoidance within eICIC is facilitated using almost blank subframes (ABS) by scheduling the intended and interfering signals on different subframes [35] [36]. These schemes were complemented by further enhanced ICIC (FeICIC) from Release 11 [37], mainly focusing on interference handling on user side (via interference cancellation schemes). In frequency domain, carrier aggregation (CA) introduced additional degrees of freedom that can be exploited for interference management purposes, as interference is partly avoided by scheduling the control channels of the macro cell and small cells on different carriers. Fast cross-carrier scheduling of the data can also help to reduce interference when there is a strong aggressor cell present while with coordinated multi-point (CoMP), it is possible to take a network-wide approach at interference management by considering a larger both dynamic and fixed set of cooperative radio nodes [38]. The 5G RAN will enable multi-layer multi-node connectivity over one or more AIVs resulting in much higher data rates and additional degrees of freedom for the fast synchronous control. The performance gain of CoMP depends to a large extent on the tightness of the coordination, i.e., the level of synchronization that is needed. It is known that CoMP is sensitive to BH latency for the signaling, knowledge and accuracy of channel state information (CSI), and requires a large communication overhead, which is the reason why much of the theoretical gains are hard to materialize in practice. In the end, designing a proper interference management scheme depends on the use case, deployment scenario, and size of the cooperative set, while maintaining a reasonable degree of flexibility for the RM.

In this section, we present several approaches for interference management requiring some level of network cooperation different to those used in legacy systems. Those include dynamic radio topology schemes, advanced transceiver designs, dynamic TDD and inter-cell pilot interference management, as well as mmWave deployments.

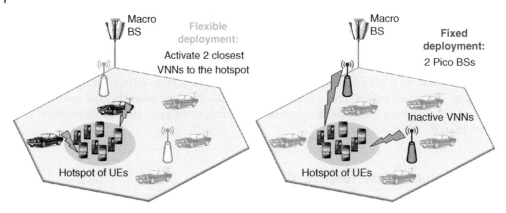

Figure 12-13. NN operation in 5G dynamic radio topology.

12.5.1 Interference Mitigation in Dynamic Radio Topology

In this section, a dynamic radio topology comprising non-static access nodes is examined, which emerges as a promising notion, enabling flexible network deployment and new services [39]. Within the framework of dynamic network topology, which shall be supported as a key design principle in 5G RAN deployments [39], nomadic nodes (NNs) aka vehicular NNs (VNNs) can enable demand-driven service provisioning to increase the network capacity and/or to extend the cell coverage area [40] [41] [42]. NNs can be mounted on cars within a car-sharing fleet, taxi fleet or on privately owned cars, as illustrated in Figure 12-13. Further, NNs can be considered as a complementary enhancement to today's heterogeneous networks, where varying traffic over time and space can be addressed by on demand activation and deactivation of available NNs. However, the large number of unplanned deployments of access nodes in conjunction with the changing environment gives rise to challenges related to significant scheduling needs and backhauling overload among others. Additionally, these dense deployments will face critically increased levels of ICI, since the scarcity and high cost of the spectrum resource inevitably requires intense spectrum reuse. To tackle the detrimental effects of ICI without compromising spectral efficiency, the employment of novel interference management techniques is crucial to be able to perceive increased throughput over the network areas.

In multi-cell cooperative resource allocation, the cells' cooperation takes place at the CP. The fundamental principle in that case is that allocated resources for the users impacted by ICI can be coordinated; thus, they are essentially interference-aware RRM techniques. On the other hand, in multi-cell cooperative transmission and reception, the co-channel inter-node interference management is extended into data transmission and reception enhancement. In such a case, the cells' cooperation takes place at the data plane, envisaging the joint or distributed signal processing.

In the following, UE-centric interference management is analyzed in ultra-dense networks (UDNs) complemented through the notion of the dynamic radio topology. Here, two case studies for a hotspot area and a 5G RAN consisting of NNs under macro BS coverage are evaluated. In the first case study, the flexibility of NNs is exploited by selecting the closest NNs to a hotspot area, which may be formed in response to a street event, for example. The performance of such a dynamic radio topology is compared to a fixed small cell deployment based on pico cells [43]. In this case, a minimum distance of 50 m is set between any active NNs, to reduce the impact of interference among selected NNs and to increase the spatial diversity. In the second case study, a dense NN deployment is considered, where NNs are mounted on cars that are parked along a road side [44]. To overcome increased interference among NNs due to close proximity, coordinated resource allocation and joint transmission (JT) are applied adaptively based on BH conditions (i.e., between access node and serving BS), load constraints, and service type.

In the first case study, there are on average 20 randomly located, inactive and parked NNs present at each macro cell. UEs are distributed in the network as random hotspots (1 per macro cell, on average). Each hotspot is constituted of 50 UEs that are randomly distributed within an annulus region bounded by the radii of 10 and 50 m, centered at a random point in the macro cell. A specific number (1, 2, or 4 NNs) of the closest NNs to the center of each hotspot are activated by the network. UEs are not forced to connect to the activated NNs, but they still attach to the node with the largest RSRP value, as in the conventional cell selection scheme. For performance comparison, a fixed deployment of pico cells is considered, with the number of pico BSs equal to that of the activated NNs. Yet, due to dynamicity of the hotspot, from hotspot perspective pico cells are randomly located and there is no correlation between the hotspot and pico cell locations. The DL throughput gains at 10 %-ile and 50 %-ile user throughput with respect to the pico cell deployment are shown in Figure 12-14 [43]. It can be seen, for example, that by activating one NN closest to the hotspot, the 10 %-ile throughput gain is around 150 % compared to the case of one pico cell. In addition, NNs are assumed to be stationary during their operation, while their availability changes with respect to time and space according to their battery state or driver needs (hence, the term "nomadic"). Furthermore, as NNs are integrated into vehicles, due to low height of 1.5 m like the one of UEs, severe fading characteristics can be expected on the wireless BH link as opposed to well-elevated small cells (e.g., at 5-10 m height for fixed relay nodes). Accordingly, to ensure the expected benefits of NN operation, active NNs shall be properly selected such that the BH link quality is optimized.

In the second case study, the key interference management mechanisms applied are JT between the access links of NNs (i.e., between NNs and UEs) when ideal BH is provided. The selection of candidate UEs for JT is based on the difference of RSRP measurements from serving and neighboring NNs. Given the number of UEs experiencing low channel quality, a number of resource blocks (RBs) are reserved for JT, and resource allocation between different NNs is done. For remaining UEs, interference management is applied, where dynamic frequency partitioning or muting of resources for some NNs is performed [31].

A particular deployment of 5 NNs is investigated. The NNs operate mainly in licensed carriers at 3.5 GHz, though they can also operate in unlicensed bands at 5 GHz subject to access conditions and the availability of Wi-Fi access points. For the channel models in the simulation scenarios, 3GPP-compliant parameters were used, corresponding to the urban micro, UMi, scenario 4 for licensed and outdoor Licensed-Assisted Access (LAA) [45]. More details on the simulation parameters for this study can be found in [31].

The results for the licensed carrier are demonstrated in Figure 12-15. The bars show the mean user throughput in case NNs are activated, or if interference management is performed on top. The blue bar (solid fill) indicates the baseline, where all the users are attached to the macro. On top of that, the dotted white bar is the gain we see when we activate a number of NNs and offload some traffic from the macro. Finally, the top bar (pattern fill with diamonds) shows the gains when we also employ coordination or cooperation between the activated NNs. Interference management is crucial since as the

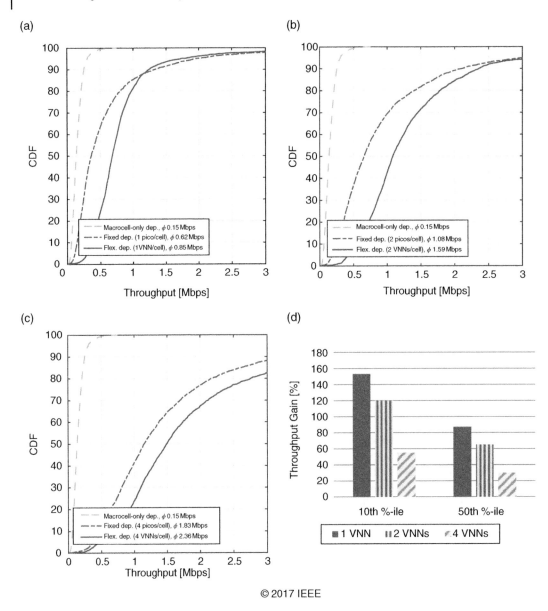

© 2017 IEEE

Figure 12-14. User throughput CDFs on DL with different deployment scenarios (a), (b), and (c), together with the throughput gain of the NN deployments compared to pico cell deployments at the 10th and the 50th percentiles of the CDFs (d). Mean user throughput (denoted as ϕ) levels are provided in the legends © 2017 IEEE [43].

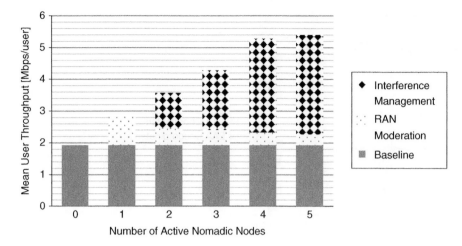

Figure 12-15. Mean user throughput for different NN activations © 2016 IEEE [44].

number of NNs increases, the performance is degraded due to interference from surrounding NNs. Hence, adaptive interference coordination and cooperation (e.g., coordinated Scheduling, JT, and dynamic NN selection) mechanisms improve spectral efficiency and user throughput in dynamic radio topologies.

As an extension of this study, the coordinated scheduling of the RAN resources utilizing unlicensed spectrum, aka LAA, is examined as a candidate solution for boosting the achievable gains in interference limited scenarios, such as UDNs [46]. To do this, one needs to identify which RAN nodes are utilizing shared unlicensed bands and coordinate them in time and frequency domain so as to minimize interference and maximize the gain. In this study, the activation of NNs with capability to operate in unlicensed bands is assumed in a hotspot scenario, as a mean to boost performance due to the interference limitations in licensed carrier.

A key challenge in LAA is the effect of *listen-before-talk* (LBT) on the actual performance, since the dense activation of LAA nodes might lead to severe collisions if there is no centralized LAA coordination. The users provide measurements for the unlicensed channels (per carrier) using the licensed bands, also providing the RSRP of dominant interferers. Based on the unlicensed channel availability, a subset of users and NNs is selected on demand, for which either uncoordinated resource allocation or interference management is performed, which can be adapted based on the neighborhood of activated NNs. As illustrated in Figure 12-16, the more NNs appear in unlicensed bands without coordination, the more the gains get limited. On the other hand, if coordination is performed (i.e., interference coordination by muting at cell edges or multi-connectivity) one can observe huge gains in terms of DL throughput. The figure mainly shows the trend of capacity gains when no interference from Wi-Fi is assumed within the given scheduling period. Note that since one of the limitations in LAA is the foreseeable delay due to HARQ, it is assumed that the re-transmissions are handled by the primary cell (e.g., macro or NN) on a licensed carrier.

The next potential step for dynamic network topologies would be to allow the vehicular nodes to assist the network also while on the move [47]. Large gains have been shown for the network throughput and the fairness for the vehicular users on-board [48], but recently also for serving out-of-vehicle users [49].

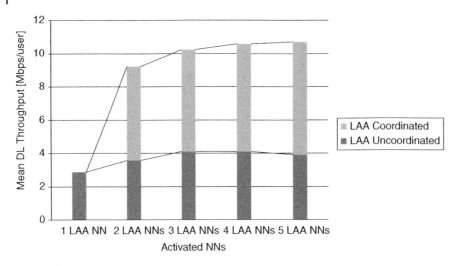

Figure 12-16. Mean user throughput for different LAA NN Activations © 2017 IEEE [46].

12.5.2 Interference Management Based on Advanced Transceiver Designs

As a first step mitigating the adverse effects of co-channel interference on the receiver side, conventional linear receivers were improved by utilizing information on the interference covariance matrix and, thus, evolved into minimum mean square error - interference rejection combining (MMSE-IRC) receivers. Network-assisted interference cancellation and suppression (NAICS) in 3GPP Release 12, meanwhile, further introduced standardized support to facilitate ICI cancellation and suppression on the UE receiver side. One scheme that received much attention during the NAICS study and work items was symbol-level interference-aware detection (IAD). While IAD provides superior performance over MMSE-IRC receivers, it has been verified in information theory that sequence-level interference-aware techniques can always outperform IAD. Advanced transceiver designs along this line have been proposed for 5G interference management with the help of network coordination via Xn interface between BSs determining on which resources they will be applied.

One representative technique is sliding-window coded modulation (SWCM), the first low-complexity systematic approach at the physical layer with theoretically optimal performance guarantee for interference channels [50] [51] [52] [53] [54]. In the SWCM scheme, a single communication block is used to communicate messages in multiple, say b, subblocks, each consisting of n symbols. Each transmitted symbol is decomposable to multiple layers: For example, a 16-QAM symbol can be viewed as the combination of two 4-QAM layers, two 4-pulse-amplitude modulation (PAM) layers, or four binary phase shift keying (BPSK) layers. Each message is sent over multiple subblocks and multiple layers. Figure 12-17 illustrates the basic SWCM encoder structure in which each message is sent over two subblocks and two layers. To be specific, it is assumed that a 16-QAM signal is transmitted, defined as the weighted superposition of two 4-QAM layers. The message $m(j)$ is encoded via a binary channel code of length $4n$ and then scrambled and/or interleaved into a codeword $c(j)$ for $j = 1, 2, \ldots, b-1$. For transmission in subblock j, the first half of the codeword $c(j)$ from the current message $m(j)$ is mapped to one 4-QAM layer, while the second half of the codeword $c(j-1)$ from the previous message $m(j-1)$ is mapped to the other 4-QAM layer. Superimposing the two layers forms a sequence of n 16-QAM symbols to be transmitted in subblock j, conveying partial information

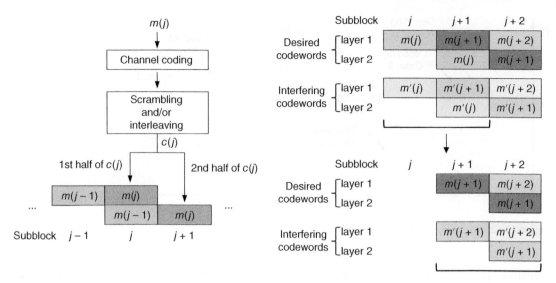

Figure 12-17. Basic SWCM encoder and decoder structures © 2016 IEEE [54].

of two messages $m(j-1)$ and $m(j)$. In total, $b-1$ messages are sent over b blocks, resulting in a slight rate loss from the actual code rate.

Each message is sent over multiple subblocks and multiple layers in a staggered manner. As its name suggests, desired messages are recovered by sliding a decoding window spanning over multiple subblocks and applying successive cancellation decoding within and across decoding windows. Figure 12-17 illustrates an example of the basic SWCM decoding operation. In the decoding window spanning over subblocks j and $j+1$, a receiver in the SWCM scheme recovers the desired codeword $c(j)$ and the interfering codeword $c'(j)$ successively, each using a conventional point-to-point decoder. It then slides the decoding window to subblocks $j+1$ and $j+2$, and again applies successive cancellation decoding to recover the next desired message $m(j+1)$. If both the desired and interfering transmitters use the two layer SWCM encoder structure illustrated in Figure 12-17, the resulting achievable rate region of (R, R') is characterized by a nonconvex polytope region with different trade-offs between the rate R of the desired codeword and the rate R' of the interfering codeword according to a particular decoding order:

1) The desired codeword $c(j)$ only;
2) $c(j)$, then the interfering codeword $c'(j)$;
3) $c'(j)$, then $c(j)$;
4) The next interfering codeword $c'(j+1)$, then $c(j)$.

as illustrated in Figure 12-18. Figure 12-18 also illustrates the achievable rate region of the entire system with two sender-receiver pairs communicating over an interference channel, which is determined by the intersection of the achievable rate regions at both receivers, and compares it with the optimal maximum likelihood decoding (MLD) achievable rate region; see [50] and [51] for the detailed information-theoretic analysis.

As promised from network information theory, the SWCM scheme can improve spectral efficiency over conventional schemes by using layered signaling and a staggered transmission structure. Layered

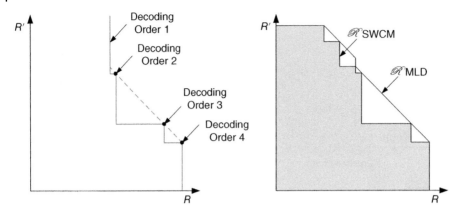

Figure 12-18. SWCM and MLD achievable rate regions © 2016 IEEE [54].

signaling enables more flexible decoding for some layers of desired and interfering codewords, achieving higher rates than non-layered schemes. Staggered transmission allows for a longer code-word to be sent over multiple layers, providing diversity and robustness for decoding either desired or interfering codewords at different receivers. Both link-level and system-level simulations are presented to demonstrate these benefits quantitatively. The basic SWCM encoder and decoder structures are implemented by off-the-shelf codes, such as 3GPP LTE Turbo codes for OFDM MIMO systems. Note that any off-the-shelf codes, such as low density parity check (LDPC) codes adopted in 3GPP NR can be used for implementation without structural modification. The link-level performances of SWCM and conventional schemes are evaluated in the 2×2 MIMO Ped-B interference channel with an average interference-to-noise ratio (INR) of 15 dB. As shown in Figure 12-19, simulation results for the signal-to-noise ratio (SNR) gain at a target BLER of 10% demonstrate that SWCM outperforms MMSE-IRC (now widely used in industry) by 10.1 dB and IAD (i.e., the 3GPP Release 12 NAICS receiver) by 6.1 dB. Also the system-level performance was evaluated under 3GPP Release 12 NAICS evaluation assumptions, indicating that SWCM outperforms MMSE-IRC by 71.5% and IAD by 38.6% at the cell edge under a resource utilization of 60%. See [53] and [54] for the detailed practical implementation, results, and analysis.

As for any other advanced coding and modulation technique, the statistics of the ICI can be designed via the modulation of interfering signals. An active interference design scheme using a novel modulation, namely frequency and quadrature amplitude modulation (FQAM), is proposed as a physical layer technique for interference management. Employing FQAM, the statistical distribution of the ICI becomes non-Gaussian, achieving higher rates than Gaussian noise, since Gaussian noise is the worst-case distribution of the ICI as additive noise, when a receiver treats interference as noise for decoding. See [55] for a detailed analysis and results.

12.5.3 Interference Mitigation in Massive MIMO Dynamic TDD Systems

The use of very large arrays at BSs has been proposed as a very promising way of drastically improving the network capacity [56], as also discussed in Section 11.5. A number of very appealing advantages can be potentially achieved out of this technology: Dramatic capacity improvements can be

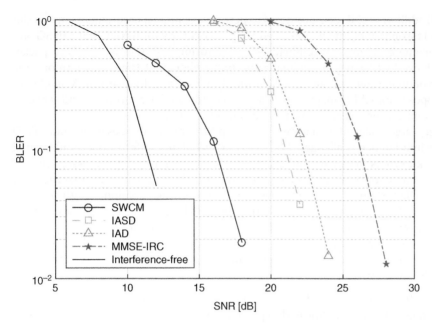

Figure 12-19. Link-level performance for the 2/2-layer SWCM MIMO scheme, interference-aware successive decoding (IASD), IAD, and MMSE-IRC in the 2×2 MIMO Ped-B interference channel with average INR of 15 dB © 2016 IEEE [54].

obtained, especially when it is used to serve multiple users; the effects of uncorrelated noise and multipath fading vanish when the number of antennas grows to infinity as proven by random matrix theory; the directivity of the beams can be greatly improved, hence reducing the side-lobe interference; and finally, the radiated energy can be significantly reduced due to the large array gains.

However, the applicability of FDD in massive MIMO systems has been shown to be challenging because of the amount of pilot overhead and feedback that would be needed for channel estimation [57]. Although there exists an interesting debate in the community on the feasibility of FDD for massive MIMO systems, the feature of channel reciprocity makes TDD very appealing for massive MIMO systems [58]. Furthermore, flexible TDD designs are appropriate as a means to modify the capacity split between UL and DL and increase spectrum flexibility. However, such designs introduce the problem of strong cross-link interference (CLI) when a DL transmission happens at the same time of an adjacent UL transmission. CLI may for instance take the form of BS-to-BS interference from the DL transmission in one cell to the UL reception in an adjacent different cell, where the likely line-of-sight (LoS) between the BSs and the strong transmit power may cause substantial performance degradation. In addition, two UEs in different cells but in proximity of each other may be subject to CLI as interference from the UL transmission of one user may affect the DL reception of the other. It can be more difficult to deal with the UE-to-UE interference compared to the BS-to-BS interference, as the interference situation can change continuously due to the UE mobility.

In LTE Release 12, a technique known as enhanced interference mitigation and traffic adaptation (eIMTA) was introduced to support UL/DL configuration changes according to UL/DL traffic situations in a cell to improve the user average packet throughput. Solutions to combat CLI in that framework included: i) additional CSI measurements to perform separate measurements in two slot types, namely fixed and flexible slots, and ii) enhanced UL power control for flexible slots. In 5G NR, dynamic TDD has become even more prominent as a way to improve the user average packet throughput and to achieve low latency by reducing the frame alignment time. Furthermore, the transmission direction of time resources can be semi-statically or dynamically changed. A number of solutions are being proposed for CLI management based on coordinated beamforming and scheduling, advanced transceivers, power control, sensing mechanisms, etc [16].

The utilization of very large scale antenna arrays with beamforming capabilities in dynamic TDD systems brings further interference challenges: Massive MIMO systems suffer from a fundamental interference limitation known as pilot contamination that has its origin in the necessary reuse of non-orthogonal pilot signals across cells. Hence, the above described problem of badly managed CLI generated in dynamic TDD systems may become even more severe when combined with the pilot contamination effect typical of a a massive MIMO setting, potentially creating a problem of "beamformed CLI" that must be avoided via selecting the right TDD configuration for overlapping cells.

It was shown in [59] and summarized in Figure 12-20 that the flexibility offered by a dynamic TDD architecture can be leveraged to combat interference in massive MIMO dynamic TDD systems by finding answers to the following two questions:

- Which transmission path (UL/DL) should be used for training at the small cell tier?
- In which order should UL/DL slots be allocated to prevent both beamformed CLI while matching the load distribution?

To answer the previous questions, the following observations have been made: i) The pilot overhead introduced by employing DL pilots (S_D in Figure 12-20) in a TDD massive MIMO system is very

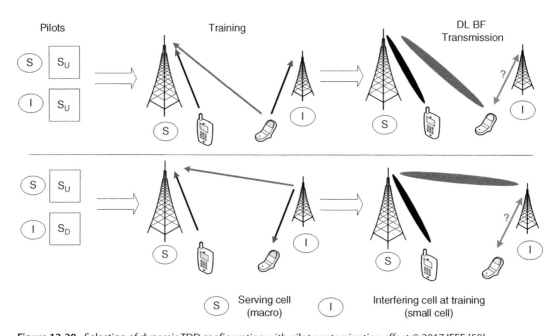

Figure 12-20. Selection of dynamic TDD configuration with pilot contamination effect © 2017 IEEE [59].

Table 12-2. Interference-based TDD configuration classification.

Cell		Interference avoidance		Beamformed interference	
PCR-D (DL training)	S	D	U	D	U
	I	D	U	U	D
PCR-U (UL training)	S	D	U	D	U
	I	U	D	D	U

high. Hence, when using a large number of antennas, channel training should be performed in the UL (S_U in Figure 12-20) to avoid the costly overhead; ii) the rest of the single-antenna BSs or those equipped with a smaller number of antennas may choose either UL or DL to perform channel training in TDD systems; iii) beamformed interference to users at other cells caused by pilot contamination effect may be avoided by the right selection of transmission paths at each time slot; and iv) strong CLI (in particular, BS-to-BS) generated by flexible TDD deployments must be prevented as LoS channels between BSs would cause interference to be particularly problematic.

Hence, following the above observations, the following classification can be established when a pilot contamination regime (PCR) is in place, i.e., when pilot contamination occurs due to the TDD configuration of the network:

- According to the reception of beamformed interference:
  - Interference avoidance mode: Node not listening when beamformed interference is present
  - Beamformed interference mode: Node listening when beamformed interference is present

- According to the contaminating pilot:
  - PCR-Downlink (PCR-D): Massive MIMO pilots contaminated by interfering DL training sequence
  - PCR-Uplink (PCR-U): Massive MIMO pilots contaminated by interfering UL training sequence

The problem of possible beamformed interference adds to the problem of CLI, hence making a joint solution essential. Table 12-2 shows how the above observations can be implemented in simple TDD design recommendations. The two columns within each of the configuration (namely interference avoidance and beamformed interference) indicate how the interfering cell should be configured when the serving cell is operating in downlink (D) or in uplink (U).

12.5.4 Multi-Cell Pilot Coordination for UL Pilot Interference Mitigation

Accurate channel knowledge is crucial for the efficient operation of MIMO systems. In practical systems, channel estimation is performed by sending predefined symbols called pilots, based upon which the receiver can process the received signal to estimate the channel. Current systems use orthogonal pilots in each sector/cell and non-orthogonal pilots across cells. This leads to the UL inter-cell pilot interference problem in systems with frequency reuse factor one, i.e., the pilots that the BS receives from a target user suffer from pilot interference coming from users in neighboring cells scheduled on the same time-frequency resources as the target user. This problem is detrimental in UL multi-user MIMO scenarios which rely on accurate channel knowledge to perform space division multiple access (SDMA). Additionally, in TDD systems, the BS can acquire the channel knowledge necessary for DL multi-user MIMO precoding via UL pilots sent by users. In this case, pilot interference leads to imperfect channel knowledge, which degrades the precoding quality and ability to perform SDMA in the DL.

LTE-A systems assign non-orthogonal yet distinguishable UL sequences to users across cells. These sequences are cyclic extended Zadoff-Chu (ZC) sequences spread over the subcarriers of interest. A cyclic extension is used to maximize the number of distinguishable sequences. The available sequences in each cell are constructed by phase rotating a root sequence identified by a root index, and are mutually orthogonal. The root sequences (and corresponding root indices) across cells are different. Different root sequences or phase rotations thereof are not orthogonal, though they are distinguishable via their root indices.

A small number of solutions exist on mitigating inter-cell pilot interference within the multi-cell OFDM context. One work is [60], where the authors propose a multi-cell assignment of ZC sequences such that the channel estimation worst-case mean square error (MSE) is minimized. However, the authors assume that user pilots occupy all available subcarriers, making the used framework not suitable for LTE-A or 5G systems. Because of the need to pad guard subcarriers and unused subcarriers with zeros, the resulting sequences after the inverse fast Fourier transform (IFFT) neither have the ideal auto-correlation nor the cross-correlation properties of ZC sequences, as assumed in [60]. Further, their approach assumes that BSs treat pilot interference as noise, which is suboptimal at high UL SNRs occurring in, e.g., small-cell scenarios. In such scenarios, a better approach would be the suppression of pilot interference at the BS to recover the desired pilots with as little interference as possible.

Mainly for CoMP applications, LTE-A allows for pilot orthogonality among multiple cells, where a BS assigns, from its pool of available orthogonal sequences, pilot sequences for users in neighboring cells. Such a solution is not feasible for 5G systems, as the number of users a BS can serve within its cell on the same time-frequency resources decreases. One possible solution that suppresses pilot interference and leaves the number of served users within a cell unchanged can be realized by exchanging ZC root indices among BSs in a distributed scenario. Alternatively, a central controller can signal to each BS the root indices of sequences used by neighboring BSs/cells. Both of these implementations allow a given BS to construct the sequences used in neighboring cells and perform channel estimation which includes not only the channel of the desired user, but also that of users in neighboring cells [61]. The channel of the former is then effectively estimated with little but existing residual interference (due to the non-orthogonality of sequences across the cells), while the estimated channels of the latter can be dropped or used according to the desired application (e.g., CoMP coordinated beamforming or joint transmission rely on the channel knowledge of users in neighboring cells). Channel estimation is performed in time domain and exploits the fact that in practical OFDM systems the number of taps is (much) smaller than the number of subcarriers, which results in reduced number of variables in time domain (i.e., taps) that need to be efficiently estimated. Even better results can be obtained by first carefully choosing the sequences that are assigned over multiple cells before exchanging the sequence information. Indeed, [61] proposes a greedy selection and assignment of sequences to further reduce the channel estimation MSE. A possible implementation of this scheme is a central one, where a central controller first optimizes the sequence choice over the network and then sends the indices of all concerned BSs to each BS, as shown in Figure 12-21 (left) for a two-cell system. In this example, BS 1 informs all users within its cell of the sequence root index

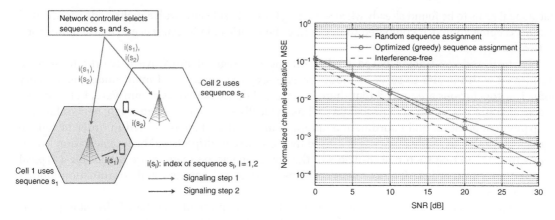

Figure 12-21. Left: Signaling steps in a centralized implementation, and Right: Greedy sequence assignment gains over random assignments.

(here, sequence 1) to be used as chosen by the central controller. Additionally, BS 1 uses the information about sequence 2's root index that is sent by the central controller to mitigate pilot interference by users in cell 2. Figure 12-21 (right) shows the gains of greedy sequence assignment for a four-cell system consisting of one cell-edge user per cell, where pilot sequences occupy 180 OFDM or OFDM-like subcarriers, and with 40 independent and identically distributed channel taps per user. Clear gains of greedy sequence assignment can be seen in the medium to high SNR regime where sequence non-orthogonality becomes the system bottleneck. The framework in [61] naturally extends to other types of sequences, e.g., pseudo-noise sequences, which are currently discussed in 3GPP.

By generalizing the idea of UL CoMP data reception to pilot sequence reception in [61], the inter-cell pilot interference effect can be mitigated to a large extent. The channel estimation quality for non-CoMP applications improves, while efficient CoMP operation can take place without reducing the number of users that can be served in a cell. This observation is especially important for 5G eMBB and URLLC services.

12.5.5 Interference Mitigation in mmWave Deployments

In Section 11.5, the need for highly directional transmission in mmWave systems was discussed. For this reason, it is often argued that the interference in mmWave cellular systems might not be so detrimental as compared to current deployments at lower frequencies. However, whether interference plays a significant role or not, actually depends on the deployment, i.e., on the BS density as well as on the capabilities of the antenna array at the users. Due to the limited range of mmWave communication, a high BS density might be required to achieve an acceptable coverage.

For ultra-dense deployments, users might have a line-of-sight to several BSs, eventually leading to higher interference, in particular if we need to accept a less precise hardware-constrained beamforming approach, for instance due to phase noise or limited digital-to-analog converter (DAC) or analog-to-digital converter (ADC) resolution. In fact, in Chapter 3 of [62] it was shown that there are substantial gains to be found already when searching for the best beams with a multi-node coordinated beam sweeping approach compared to neglecting the multi-node interference in dense multi-node networks.

Analytical and ray tracing based studies on the coverage of mmWave networks have been performed with the aim to serve as input for assessing the potential of multi-node connectivity [63]. The conclusions are indeed that there is a need of dense networks at mmWave, but that also means that there is a high LOS probability. Thus, typically several BSs are in reach, hence allowing to exploit cooperation and coordination among multiple BSs. This is needed in order to reduce signal outage due to sudden blockages caused by the sparse multipath and high penetration loss at mmWaves. Exploiting such macro-diversity the coverage can be made more robust for a typical UE, and intermittent interference can also be avoided by cooperation and coordination among multiple BSs.

Further improvements of the coverage can be achieved, in particular in the UL, by taking advantage of the sparse multipath property of mmWave channels with BSs employing partial-zero-forcing, in which an optimal fraction of the antennas is devoted to cancel the few strongest interferers. However, the coverage for narrow streets and for users blocked by buildings or foliage is still a challenge, to which the use of multi-node relaying schemes might be an important solution, as also discussed in the context of self-backhauling in Section 7.4.

In addition to more robust coverage and interference mitigation in dense networks, there are several other potential benefits of multi-node cooperation at mmWave frequencies, also in less dense networks, in contrast to traditional multi-node cooperation at lower frequencies:

- Increased aggregated power gains in cases when the system is peak power limited per node;
- Artificially increase multipath and thus (distributed) MIMO rank to better support (distributed) spatial multiplexing and massive MIMO gains at sparse mmWave channels;
- Support integrated multi-node/multi-beam soft handover;
- Efficient load balancing for energy efficient operations.

To illustrate the potential gain by tight multi-node cooperation at mmWave frequencies, in the following, a case study on joint transmission hybrid precoding for energy-efficient mmWave networks is presented.

As an example test case, in what follows, a joint transmission hybrid precoding for energy-efficient mmWave networks is presented. A mmWave hybrid beamforming design, maximizing spectral efficiency, has been shown to provide performance close to a fully digital precoding scheme [64], [65]. However, only a few works have studied the energy efficiency of the precoder design taking the total energy consumption into account, i.e., also hardware and not only radio frequency (RF) power. When joint transmission is allowed, a major research problem is to jointly design the analog and the digital precoder such that the total power consumption is minimized and the QoS of each user is satisfied. In this work, a joint transmission hybrid precoding structure scheme is designed and compared to the fully digital joint transmission precoding scheme, from multiple BSs to each user. The total power required by the BSs is minimized under the constraint to satisfy a spectral efficiency requirement for each user.

In the following, a mmWave network is considered consisting of M BSs and K users. BS v is assumed to be equipped with N_v antennas and S_v RF chains, where $S_v < N_v$. Each user is assumed to be equipped with a single antenna and is not pre-associated with any BS. Due to the fact that strict synchronization among BSs is difficult to achieve, each user is assumed to use successive interference cancellation to sequentially detect multiple streams from different BSs.

Due to the unit modulus constraint, the optimization problem for hybrid precoding is usually too complex to find a global optimal solution; and hence most research tries to find a suboptimal solution that gives performance close to the fully digital precoding scheme. In a similar manner as [66], the above problem can be reformulated as a semidefinite program, conditioned on fixed analog precoders. The algorithm starts by solving for the optimal fully digital precoder without the analog constraint. Then, the analog precoder is initialized as the element-wise normalization of the digital precoder and conditioned on the analog precoder, and the digital precoder is obtained by solving the optimization problem. For further details, see [62], (3-19)-(3-22).

The two left-most curves of the simulation results in Figure 12-22 are obtained for 3 BSs (1 macro BSs and 2 small BSs) and 4 users. The macro BS is assumed to have 32 antennas and 4 RF chains. The small BSs are assumed to have 8 antennas and 4 RF chains. The maximum transmit power is set to 46 dBm for the macro BS and 43 dBm for small BSs. Note that, even though the maximum transmit RF power of macro BSs and small BSs is assumed to be 46 dBm and 43 dBm, respectively, the average sum transmit RF power needed for 3 BSs to achieve a target per-user spectral efficiency of 2.5 bit/s/ Hz is approximately 1 W (30 dBm), and the average sum transmit RF power increases with the target per-user spectral efficiency.

Figure 12-22 shows that the total power consumption of the BSs when using the joint transmission hybrid beamforming scheme is substantially lower than that of the joint transmission digital beamforming scheme due to the reduced number of RF chains. The two right-most curves of the simulation results in Figure 12-22 show that when the number of coordinated BSs decreases, the total power consumption increases. This is because more coordinated BSs provide a better chance for a user to be jointly served by BSs with good channel conditions, while decreasing the number of BSs would result in increased transmit RF power from the remaining BSs in order to serve users experiencing

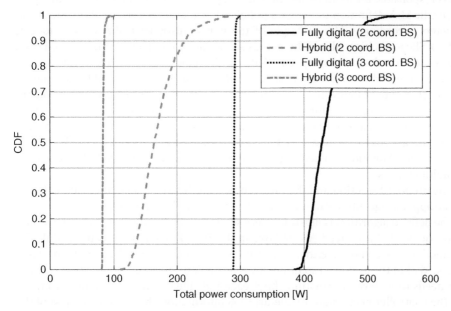

Figure 12-22. CDF of the total power consumption for 3 and 2 coordinated BSs employing joint transmission hybrid beamforming and joint transmission fully digital beamforming, respectively. The target spectral efficiency is 4 bit/s/Hz.

blockages or deep fading. That is, the increase in the transmit RF power outweighs the decrease in the hardware power consumption under our parameter settings. In addition, the distribution of the power consumption over the cooperative nodes becomes more even with several cooperating nodes, which might be useful for meeting potential effective isotropic radiated power (EIRP) restrictions of mmWave nodes.

12.6   Multi-Slice RM

Network slicing is an emerging concept based on the idea of running multiple logical networks as virtually independent business operations on a common physical infrastructure [39]. Network slicing is in detail introduced in Chapter 8. With respect to RRM, especially the management of the scarce radio resources is a critical issue. Thus, pooling and sharing these resources among the logical networks (the so-called network slices) in an efficient manner is the main target. For 5G NR, a new access stratum (AS) sublayer above PDCP level is introduced by 3GPP [67] which handles the mapping of QoS flows to certain DRBs, see also Section 6.4.2.1. To this end, potential architectures to implement a slice-aware RM have been introduced in Section 8.2.3 and 8.3.4 as well as in [68]. Based on that, an algorithm which dynamically adjusts QoS class identifiers (QCIs) of the individual data streams to meet target SLAs has been designed. To satisfy defined SLAs between potential business partners, it is essential to introduce the idea of network slicing to a mobile radio network. In the following, the principle idea and the algorithm are described in detail concluded by simulation results, a possible RAN implication and derived necessary signaling information.

To realize slice-aware RM in an efficient manner, additional functionality needs to be introduced to the RAN, which is responsible for monitoring and enforcing SLA for individual slices by mapping the abstract slice-specific SLA definition to the QoS policies. It monitors the status of the SLAs and adapts QoS parameters accordingly. The principle of the newly introduced SLA control loop is shown in Figure 12-23. Slice-specific multiple thresholds are defined to protect the SLA of each slice. The thresholds are related to relevant slice-specific KPIs, which have to be satisfied. Typical QoS-related KPIs, such as guaranteed throughput or latency, but also business-driven additional agreements have to be protected. If the monitoring entity detects a threshold violation it will give a higher prioritization to the corresponding data flows. Using data-stream-specific QCI adaptation aims to meet SLA requirements, while realizing data-flow-specific QoS is still a task of the AIV-specific MAC scheduler. With this approach, slice-specific QCI adaptation is transparent to the MAC scheduler. Therefore, no further adaptation is needed to the scheduler, and a common MAC scheduler could be used for multiple slices.

More details on the proposed solution and the interaction with the MAC scheduler can be found in [31] [15], while the applied algorithm is described and analyzed in the following.

In Figure 12-24, the defined algorithm is illustrated as a sequence chart. The SLA controller would be a functionality of the newly introduced AS sublayer in the 5G NR 3GPP architecture, while the SLA monitoring functionality could be either handled in the central unit (CU) or in the Next Generation (NG) CN. Within the proposed solution, it would be done in the CU. In the first step (1), the SLA controller needs to receive information about each individual data flow coming from the CN. On one hand, the controller needs state-of-the-art QoS information about the QCI class of the data flow, as, e.g., defined in [69], while on the other hand it needs information about which data flow corresponds to which slice. Therefore, a slice ID [32] has to be transmitted for each data flow from

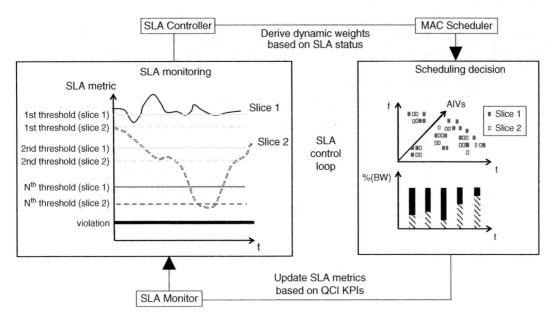

Figure 12-23. SLA control loop.

the CN to get the information how to handle data flows of different slices. The SLA controller is aware of the existing network slices and their defined SLAs. Based on a defined SLA metric, the SLA controller assesses the current status of the network slice (2). The SLA metric consists of radio specific KPIs, such as data rate, latency and BLER. However, in addition, some business-driven parameters could influence the metric, such as an additional weight for prioritized services, e.g., for public safety. Multiple thresholds are defined as depicted in Figure 12-23. The more thresholds are violated, the more aggressive the prioritization will be adapted. A time window needs to be defined for how long the monitoring entity should observe the slice-specific data flows before the SLA controller takes a decision. After the decision was taken in (2), the SLA controller adapts the data flows with the derived QCI classes and forwards it to the destination BSs (3). The adapted data packets are now received and sorted into the new corresponding QCI-related buffers as an input source for the MAC scheduler. The millisecond based scheduling decisions are taken based on the new buffer statuses in step (4). The AIV-specific RRM functionality needs to monitor the statistics of the flow based QoS KPIs and send it back to the SLA monitor (5). For a certain amount of time, the SLA monitor needs to observe the situation again and then send an indicator to the controller showing whether a certain threshold of a network slice is exceeded or not (6). After a certain amount of time is exceeded, the SLA metric is compared with multiple thresholds and a decision is taken again on which data flows are adapted (7). This could result in a down-prioritization, if the SLA metric is recovering. Even a more aggressive prioritization is possible, if the SLA metric still exceeds a threshold (8). If the most critical situation appears which is indicated by violating the most critical threshold (9), e.g., defined as the Nth threshold in Figure 12-23, feedback information has to be sent to the core network to react on the forthcoming SLA violation (10).

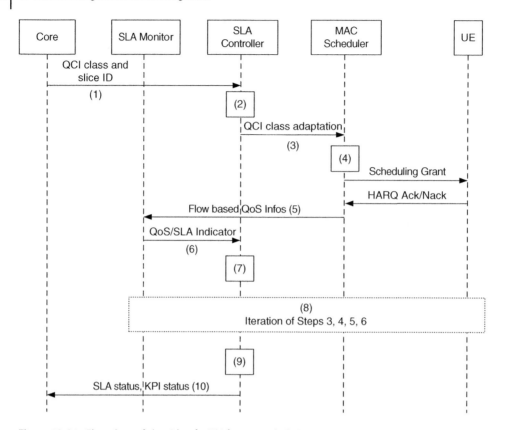

Figure 12-24. Flow chart of algorithm for RM for network slicing.

In a today's scenario, 2 dedicated RANs (subnetworks) may be operated in parallel for independent businesses, each covering a channel bandwidth of 10 MHz, Subnetwork 1, representing a special purpose network which is overprovisioned to guarantee a high quality of service, serves 100 users with low data demand resulting in a low network load. In contrast, subnetwork 2 represents a low-cost best effort network that serves 710 users causing a fully loaded system with lower performance per user. Figure 12-25 shows the probabilities for the achievable user throughput in both subnetworks and the total one.

With the slicing concept, both subnetwork types (now called slices 1 and 2) may run as logical networks on a common RAN infrastructure which also allows sharing their frequency resources (resulting total bandwidth of 20 MHz). Detailed simulation assumptions can be found in [15]. For slice 1, an SLA is assumed to guarantee the same overall network capacity as it was the case with the dedicated subnetwork 1, whereas users of slice 2 are still served via best effort. For this setup, it is not expected that the slices will achieve the same user-specific throughput performance as the dedicated subnetworks for two reasons: Different RRM approaches (subnetwork-specific scheduling vs. joint scheduling with prioritization of slice 1) and different interference conditions (especially the low interference in subnetwork 1 vs. the fully loaded shared network).

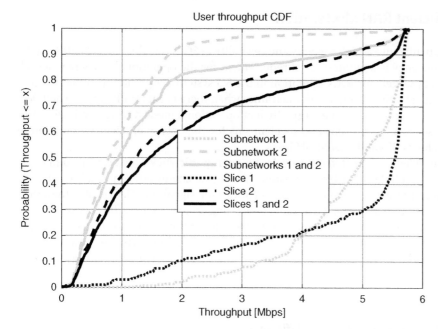

Figure 12-25. Simulation results of RM for network slicing.

The results are depicted in Figure 12-25. For slice 1, the user throughput distribution has changed because of the higher interference occurring when both slices are served on the shared band. The scheduler compensated the users of slice 1 with lower channel quality by allocating more resources to keep the target SLA of slice 1. Users in slice 2, even when served via best effort only, profit from increased resource space. In this way, also the overall performance is strongly improved by the proposed concept, compared to the scenario with 2 dedicated networks (see solid lines).

As demonstrated, multi-slice RM can achieve performance gains due to resource pooling while protecting the performance of individual slices. For simplicity, the example was only related to network capacity as KPI, but the concept also allows guaranteeing a mix of different KPIs like throughput, latency, and/or reliability.

The final design recommendation to support network slicing in the RAN when radio resources are shared can be found in [1]. The following data exchange in the RAN is required:

- The SLA control loop functionality acts as a centralized logical entity;
- A slice ID of each data flow and the corresponding SLA KPIs need to be known by the SLA controller;
- The SLA monitor needs information of the MAC scheduler about the status of the flow-specific QCI KPIs;
- The SLA controller needs to get an SLA status indicator of the monitoring entity and basic parametrization of thresholds based on the defined SLAs from the CN or management and orchestration (MANO);
- The SLA controller needs to feedback information to the CN, if a certain SLA cannot be protected.

12.7   Energy-efficient RAN Moderation

One of the key drivers behind ultra-high capacity and reliability in 5G is the ultra-densification of the RAN. Thus, energy efficiency is a quintessential requirement for the operational efficiency of such networks. One of the key targets for RAN operation in the 5G system design has been that the network should be operational only during those timeframes when there is data traffic to be scheduled, thereby enabling the network to apply an energy-efficient operational mode during low-load conditions. In this chapter section, we present two mechanisms for the energy-efficient moderation of RAN nodes, which includes BSs and related BH links.

The first mechanism called coordinated sleep cycles for energy efficiency considers the joint operation of radio access and BH links, in order to maximize the overall energy efficiency of the network. This mechanism intelligently adapts the optimization of always-on signals in 5G, to enable the access and BH links to jointly enter an active/inactive state, depending on real-time traffic conditions. The second mechanism called cell on/off coordination enables the optimal operation of RAN nodes through load-based activation / deactivation of cells. The mechanism proposes the use of a coordination layer above layer-2 of the RAN to control the operational parameters of the lower layer nodes, with the help of traffic steering, described in Section 12.4.

12.7.1   Coordinated Sleep Cycles for Energy Efficiency

One of the key design disruptors in 5G is the lean system design, which enables the gNBs to enter sleep mode when there is no active data to be scheduled [70]. Such a fundamental design assumption also enables the application of novel technology enablers for achieving higher power savings in the 5G RAN. BH power consumption has always been a significant contributor to the total power consumption of the network. A detailed study of the BH power consumption for different link types was done in [71], and it was shown that for legacy networks the BH network for the access link consumed approximately 70% of the total power. With the ultra-densification of 5G-RAN, such considerations become relevant and important due to the probability of a significant increase in the total network power consumption, and solutions addressing these need to be developed. A key solution targeting such a problem is the joint operation of RAN and BH links, so that when the RAN node is inactive, the corresponding BH link could enter sleep mode as well, in order to maximize the achievable power savings in the network. In [72], the technology potential in terms of power savings of such joint RAN-BH operation have been considered using various BH link types. Based on the evaluations, it was shown that significant reductions in the total network power consumption could be achieved using such mechanisms.

The self-backhauling (sBH) concept, as introduced in Section 7.4, where the 5G radio is reused over the access and BH links, thereby enabling an integrated access and BH scenario is gaining increasing relevance in 5G [73]. The use of the mmWave band for 5G deployments enables the full reuse of frequency bands used for access and BH, with the help of directional antennas used for the access and BH links. Such deployment scenarios shown in Figure 12-26 (a), linked to the possible upper and lower layers of the gNB concept, provide an attractive solution for network operators to provision the last leg of connectivity for the end users. Network operators could also use the multi-connectivity paradigm to improve the reliability of such deployments, especially

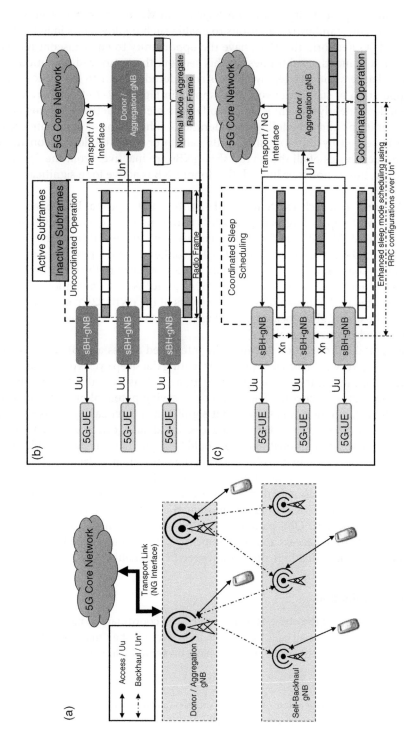

Figure 12-26. (a) Integrated access and BH deployment scenario using sBH gNBs [74], (b) Normal RAN-sBH operation, (c) Coordinated RAN-sBH active-mode operation [15] [70], see also Section 7.4.

due to the relatively higher probability of link breakage in mmWave bands. 3GPP has also agreed to study such scenarios and standardize solutions to enable such deployments in 5G [74].

In a partially loaded 5G network, the gNBs have the opportunity to enter discontinuous transmission modes during the inactive subframes. In this section, the optimized operation of such a network for integrated access and BH deployments is considered, similar to the work done in [15] [70]. The normal operation of such a network is as shown in Figure 12-26 (b), where each of the sBH gNBs operate independently and enter sleep mode during each inactive subframe depending on real-time load conditions and scheduling considerations. The donor or aggregate gNB which aggregates the traffic served by multiple sBH-gNBs can enter sleep mode only on those time instances where the sleep cycles of all the served sBH-gNBs are inactive. From the figure, we can observe that if the operation is not coordinated, the donor gNB is active for a significant period of time, consuming energy.

As described in Section 7.4, in such a system the sBH-gNBs would appear as UEs to the donor gNB over the Un* interface[2], while appearing as regular gNBs to the end users over the NG-Uu interface. Since sBH-gNBs have the additional role of providing connectivity to the end users, the conventional discontinuous reception (DRX) paradigm (where the BS configures the UE to enter inactive state based on the scheduling constraints, until there is data to be scheduled to the UE [BI09]) cannot be applied here. In order to address this issue and for enabling joint RAN-BH operation, it is proposed that the sBH-gNBs, located in the logical AN-I / lower layers of the gNB [32], coordinate their sleep mode operation with the donor gNB, located in the logical AN-O / CU or upper layer of gNB [32], using enhanced DRX mode request signaling.

Such a coordinated mode of operation between the integrated access and BH links is shown in Figure 12-26. From the figure, we can observe that coordinating the sleep cycles of the sBH nodes can minimize the operational time of the donor or aggregation gNB, as well, thereby, minimizing the total power consumption of the network. The total power saving potential of such enhancements was presented in [70] [15], where it was shown that significant power saving gains of up to 50% (relative to LTE) and 17.5% (relative to 5G with fiber access backhauls), can be achieved using such a coordinated operation. An additional constraint that could be applied here is the possible extra delay that is induced by such an operation by possibly delaying the scheduling of data by a few subframes. It is assumed that such coordinated mode of operation takes the QoS paradigms of the flow into account while optimizing the operation of donor and sBH nodes. With the use of the dynamic QoS framework (see Section 12.4.1), 5G-RAN has the additional capability of enforcing such enhancements without impacting the E2E QoS parameters. While the coordinated operation is mainly described here from an integrated access and BH perspective, it could be applied to any type of BH link with the coordination done using implementation specific mechanisms.

12.7.2 Cell On/Off Coordination

This sub-section discusses one of the other aspects of inter-cell coordination, namely cell switch on/off for network energy saving. As small cells have much less coverage than macro- and micro-cells, it is highly likely that certain small cells in the network have no served users or traffic in long periods

---

2 The Un* interface is envisioned to be the interface between a serving node and the sBH-gNBs (as an evolved Un interface, where the Un interface has been defined for relaying in LTE-A).

and can be switched off or go to low power mode for energy saving. These kinds of techniques are called network energy saving technologies in mobile networks.

The cell switch-off approach is a system level approach, which works in an area covered by multiple cells with same or different RATs [75]. It has been introduced in LTE networks. In this approach, there is no need to modify the lower-layer components in BSs. When the traffic load in a certain area is low, some cells can be shut down and the served users can be handed over to neighboring cells. The inactive cells can be turned on during the busy time, signaled by neighboring working cells or the operations, administration and maintenance (OAM) functionality of the system. There are two ways to switch on/off cells. One is the signaling directly between BSs. The other one is the dedicated control from the OAM system.

The popular cell switch-off structure is called the hierarchical cell structure [76], in which always-on macrocells are deployed for basic coverage and micro/pico-cells are planned for capacity boost. Cells for capacity boost only operate when the traffic load is high in macrocells. Otherwise, they are switched off to save energy if the macrocell is able to provide enough capacity.

Clearly, the cell switch-off approach needs the awareness of lower-layer states of the network and the coordination of BSs. In the literature, both centralized and distributed approaches have been proposed, though centralized solutions are preferred for performance [77]. The above-mentioned cell switch-off approaches are applied in LTE networks, in which the locations of macro and micro/pico-cells are well planned for optimized throughput and coverage. If only considering the moderate number of macro and micro/pico-cells in a region, the switch-on/off coordination can be implemented on top of the current LTE system. In 5G networks, the coordination would be a challenge due to the presence of an extremely large number of cells and different types of RAT. The inter-cell coordination will be a key design for cell switch-on/off and other network coordination functions in 5G networks.

One idea is to introduce the coordination layer on top of the layer-2 of the RAN to coordinate the lower-layer behaviors of the RAN network entities [78]. The coordination layer provides logically centralized control and coordination for heterogeneous RAN including different RATs and flexible implementation of RAN. Logically-centralized coordination does not mean global coordination of an operator's entire mobile network, but a scalable solution adapted to the coordination needed by a geographic region. The key element in the coordination layer is the RAN coordinator implemented flexibly, i.e., it can be implemented in a single device, inside a BS, or in a cloud environment.

A key concept in the coordination layer is that of radio network abstraction. The term "abstraction" in this context means the application of a set of principles to extract key but simplified information from the layers 1 and 2 of the RAN, so that abstracted network status reflecting reality and multi-dimensional relationships of low-layer entities can be generated and exploited for efficient coordination at RAN. Radio network abstraction can shield the lower layer implementations and enable a scalable coordination solution in 5G networks. Said abstraction makes a logically-centralized coordination approach feasible as it contributes to the reduction of signaling overhead.

The cell switch-on/off as an application in the coordination layer is shown in Figure 12-27. The coordination layer collects the network status from lower layers through defined interfaces. By further processing the network status information at the coordination layer, the necessary information is aggregated and extracted to generate the so-called network graph used for the cell switch-on/off. Different centralized algorithms can be applied to decide the optimized cell setting based on current network conditions and system policies. Simulation results for the performance of coordination

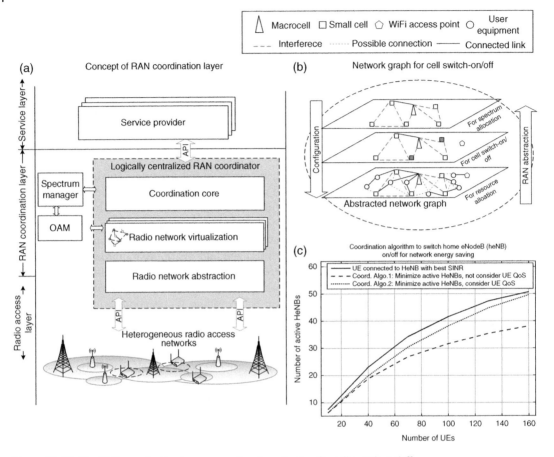

Figure 12-27. The RAN coordination layer concept and application for cell switch-on/off.

algorithms are shown in Figure 12-27c. More details of the algorithms can be found in [79]. The advantages of this approach are several-fold: a) thanks to the signaling overhead reduction by network abstraction, it can apply to large-scale networks; b) network abstraction can abstract away sufficient differences of RATs and facilitate the implementation of cell switch-on/off across different RATs; c) the same network status information can be abstracted and used for different RAN coordination applications and thus provide flexibility.

While the cell switch-off approach tries to make a good balance between performance and energy saving, it has several limitations to overcome. Firstly, frequently switching on/off cells may lead to service interruption and worsen the user experience. Its usage should be limited to a semi-static manner. Secondly, switching off cells may reduce the battery life of served UEs, as they have to connect to other cells far away. A joint optimization may hence need to take into account users' conditions when the switching off decision is made. Thirdly, switching off a cell may create a coverage hole and require remaining active cells to increase their coverage, which may compromise the energy saving gain.

12.8 UE Context Management

New fast RM for new 5G AIVs will take into account data from services, applications down to physical layer and will have an impact on BS and UE measurements gathering context data for efficiently controlling spectrum resources. The reference [80] defines context awareness as delivering context information in real-time on the network, devices, applications and the user and his environment to application and network layers in the context of International Mobile Telecommunications-2020 (IMT-2020). This context could be classified on device state level (e.g., battery state level and processing load), user status (e.g., QoE preferences, activities, location, and mobility status), current environment (e.g., devices in the neighborhood, topology, background activities, and weather) and network status (e.g., load, throughputs, reliability, supported radio technologies, interference and spectrum availability). While 3GPP Release 13 already provides context data as information elements, for instance, UE power preference indicator (PPI) enabling the BSs to configure properly DRX values, RSRP, and reference signal received quality (RSRQ) measurements of serving and neighboring cells, there are still many challenges in particular in the exploitation of user or device information for radio resource allocation in heterogeneous 5G networks deploying dense and widespread small cells. Therefore, the identification of the context information required from different sources (e.g., from the UE) is of key importance to support efficient RRM required by 5G.

While designing the UE context in 5G networks, the amount of data to be gathered and the complexity of RM algorithms need to be designed carefully between the network performance enhancements they enable and the load they impose on both the BS and the UE in terms of data gathering, signaling, processing and storage [81]. For example, since multiple use cases, possibly with contradicting KPIs, are identified in 5G, different AIVs may be used in different use cases. Consequently, more frequent inter-AIV switches (i.e., the switch from a certain AIV to another AIV) are foreseen. For example, an inter-AIV switch can occur due to a change in running applications and the related AIVs, or due to the switching between bands and related AIVs. So, to enable the efficient switching from one AIV to another, the UE may need to perform separate measurements for each AIV.

To address the above challenges, a novel UE context management framework envisioned for 5G is proposed in [15] [81], and it can be decomposed to the following essential logical entities:

- Measurement Functions, in which the UE and the BS perform measurements;
- Communication Function, which sends the UE (BS) measurements to the BS (UE); and
- Configuration Function, which selects the most suitable UE measurement configuration profile.

In what follows, we describe how the different logical entities interact with each other. Firstly, a set of the so-called measurement configuration profiles (MCPs) are defined and stored at both the BS and the UE. Each MCP contains a predefined set of UE measurement configurations (e.g., UE measurement intervals, measurement sampling rate, and maximum number of measured cells). The framework allows the UE and the BS to select the best suitable MCP according to a variety of parameters. Those parameters can be categorized into different groups as following:

- UE-calculated parameters: All the parameters calculated by the UE (e.g., UE mobility state, UE power state, and UE capability) and then reported to the BS;
- Infrequently-changing BS-calculated parameters: All the infrequently changing parameters calculated by the BS (number of neighbor cells, BS served cell size, BS capabilities, etc.), and then sent to the UE (either through dedicated or broadcasted signaling);
- Frequently-changing BS-calculated parameters: All the frequently changing parameters calculated by the BS (current active radio bearers, load of neighbor BSs, etc.).

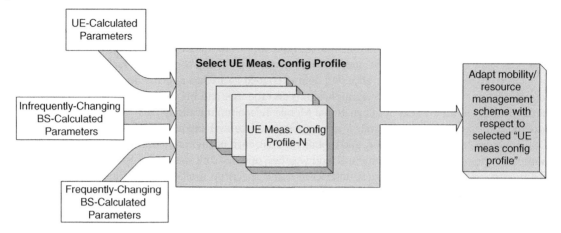

Figure 12-28. Framework for adaptive UE measurement configuration process © 2017 IEEE.

The algorithm defining the interaction among the logical entities consists of 3 main steps, as explained below and also summarized in Figure 12-28. The introduced framework provides BSs with the flexibility to extend the defined MCPs by adding new MCPs. The BS shall send the new MCPs to the UE (either through dedicated or broadcasted signalling).

STEP-1: The UE selects the best suitable MCP according to the "UE-calculated parameters" and the "Infrequently-changing BS-calculated parameters". Subsequently, the UE shall adopt the RRM scheme indicated by the selected UE-MCP, e.g., adjusting measurement intervals according to the selected profile.

STEP-2: The BS reselects or fine-tunes the "active UE-MCP". When a UE establishes a connection with the BS, it shall transmit to the BS the "UE-calculated parameters". Therefore, the BS may reselect the UE-MCP taking into consideration the "UE-calculated parameters", "Infrequently-changing BS-calculated parameters", as well as the "Frequently-changing BS-calculated parameters". As a result of this reselection or fine-tuning of the suitable UE-MCP, the BS may command the UE to adjust the current active UE-MCP.

STEP-3: UE and BS both update each other with latest calculated parameters. Whenever the UE detects that the "UE-calculated parameters" are different from the values transmitted to the BS, it shall inform the BS with the updated parameter set. Similarly, the BS shall inform the UE when the BS detects that the values in the last calculated parameter set differ from the ones which have been provided by the BS to the UE. Consequently, the best suitable UE-MCP shall be reselected accordingly.

12.9 Summary and Outlook

This chapter has provided an overview of the fast traffic steering and RRM mechanisms envisioned for 5G. These include powerful mechanisms for multi-service co-existence, where users or traffic flows with highly diverse QoS requirements are efficiently multiplexed and managed by the RAN. An important pre-requisite for QoS-aware RRM is to have knowledge of the QoS attributes for different data flows. Toward the former, novel service classification techniques have been developed. Such service classification techniques help identify characteristics of different data flows, and thereby complement the QoS parameters that potentially are made available to the different RAN protocol layers, i.e., those that come with the new 3GPP-defined QoS architecture for NR. At the MAC and PHY layers, 5G NR comes with a large number of options for scheduling and multiplexing of users, enabled through, e.g., a flexible PHY design with dynamic frame structure, and configurable PHY numerologies. Among others, dynamic scheduling should be made compatible with variable TTIs. The notion of a dynamic radio topology is highlighted as a promising complementary enhancement to fixed heterogeneous networks. For users associated with multiple cells, it was described how either intra-site carrier aggregation or multi-node connectivity can enable fast traffic steering actions. Fast traffic steering is a powerful instrument for efficient operation and optimization of multi-cell 5G networks. As for any cellular systems, also 5G will be subject to co-channel interference, and hence a set of ICIC schemes are suggested. The 5G ICIC methods explores the new opportunities that come with the enhanced radio design, as well as architecture options, to further boost the performance. The latter include also ICIC mechanisms for CLI mitigation for TDD scenarios. Finally, it was described how the application of slicing for different service categories in the RAN can help orchestrate the multiplexing of multiple services.

The majority of the discussed RRM options in this chapter are anticipated to be supported for the 5G NR specifications from 3GPP. However, it should be emphasized that the 3GPP NR specifications only set the boundaries for the various RRM options, while there remains a high degree of freedom for vendor-specific RRM algorithms, such as, scheduling policies, traffic steering algorithms, ICIC, and slice orchestration. The options for novel RRM innovations are not yet preempted, so further enhancements are possible to be introduced in forthcoming 5G NR 3GPP releases, as well as novel product-specific algorithms.

References

1 5G PPP METIS-II project, Deliverable D2.4, "Final Overall 5G RAN Design", June 2017

2 5G PPP METIS-II project, White paper, "Preliminary views and initial considerations on 5G RAN architecture and functional design", Mar. 2016

3 3GPP TR 38.913, "Study on Scenarios and Requirements for Next Generation Access Technologies; (Release 14)", v14.2.0, March 2017

4 3GPP TS 38.300, "Technical Specification Group Radio Access Network; NR and NG-RAN Overall Description; (Release 15)", v.15.0.0, Jan. 2018

5 3GPP TS 23.501, "Technical Specification Group Services and Systems Aspects: System Architecture for the 5G system; (Release 15)", v.15.0.0, Dec. 2017

6 T. T. T. Nguyen and G. Armitage, "A survey of techniques for internet traffic classification using machine learning", *IEEE Communications Surveys & Tutorials*, vol. 10, no. 4, pp. 56–76, Fourth Quarter 2008

7 R. Yan and R. Liu, "Principal Component Analysis Based Network Traffic Classification", *Journal of Computers*, vol. 9, no. 5, pp. 1234–1240, May 2014

8 J. S. Aafa and S. Salim, "A Survey on Network Traffic Classification Techniques", *International Journal of Engineering Research & Technology*, vol. 3, no. 3, Mar. 2014

9 A. Moore and D. Zuev, "Internet traffic classification using Bayesian analysis techniques", in *ACM International Conference on Measurement and Modeling of Computer Systems (SIGMETRICS)*, June 2005

10 N. Williams, S. Zander and G. Armitage, "A preliminary performance comparison of five machine learning algorithms for practical IP traffic flow classification", *ACM SIGCOMM Computer Communication Review*, vol. 36, no. 5, pp. 5–16, Oct. 2006

11 T. Auld, A. W. Moore and S. F. Gull, "Bayesian neural networks for Internet traffic classification", *IEEE Trans. Neural Networks*, no. 1, p. 223–239, Jan. 2007

12 IEEE, "IEEE 802.16p Machine to Machine (M2M) Evaluation Methodology Document (EMD)", May 2011

13 P. Ameigeiras, J. J. Ramos-Munoz, J. Navarro-Ortiz and J. Lopez-Soler, "Analysis and modelling of YouTube traffic", *Transactions on Emerging Telecommunications Technologies*, vol. 23, no. 4, pp. 360–377, June 2012

14 5G PPP FANTASTIC-5G project, Deliverable D4.2, "Final results for the flexible 5G air interface multi-node/multi-antenna solution", Apr. 2017.

15 5G PPP METIS-II project, Deliverable D5.2, "Final Considerations on Synchronous Control Functions and Agile Resource Management Framework for 5G", March 2017

16 3GPP TR 38.802, "Study on New Radio (NR) Access Technology Physical Layer Aspects; (Release 14)", v14.0.0, Mar. 2017

17 S. Khosravirad, G. Berardinelli, K. I. Pedersen and F. Frederiksen, "Enhanced HARQ design for 5G wide area technology", *IEEE Vehicular Technology Conference (VTC Spring 2016), 5G Air Interface Workshop*, May 2016

18 G. Berardinelli, S. Khosravirad, K. I. Pedersen, F. Frederiksen and P. Mogensen, "Enabling early HARQ feedback in 5G networks", *IEEE Vehicular Technology Conference (VTC Spring 2016)*, May 2016

19 S. Khosravirad, K. I. Pedersen, L. Mudolo and K. Bakowski, "HARQ Enriched Feedback Design for 5G Technology", *IEEE Vehicular Technology Conference (VTC Fall 2016)*, Sept. 2016

20 B. Soret, P. Mogensen, K. I. Pedersen and M. C. Aguayo-Torres, "Fundamental Tradeoffs among Reliability, Latency and Throughput in Cellular Networks", *IEEE Global Communications Conference (Globecom 2014)*, Dec. 2014

21 K. Pedersen, G. Berardinelli, F. Frederiksen and A. Szufarska, "A Flexible 5G Frame Structure Design for Frequency-Division Duplex Cases", *IEEE Communications Magazine, vol. 54, no. 3*, pp. 53–59, Mar. 2016

22 3GPP TR 36.814, "Further advancements for E-UTRA physical layer aspects; (Release 9)", v9.0.0, Mar. 2010

23 K. I. Pedersen, M. Niparko, J. Steiner, J. Oszmianski, L. Mudolo and S. R. Khosravirad, "System Level Analysis of Dynamic User-centric Scheduling for a Flexible 5G Design", *IEEE Global Communications Conference (GLOBECOM 2016)*, Dec. 2016

24 G. Pocovi, K. I. Pedersen, B. Soret, M. Lauridsen and P. Mogensen, "On the impact of multi-user traffic dynamics on low latency communications", *International Symposium on Wireless Communication Systems (ISWCS 2016)*, Sept. 2016

25 P. Karimi, F. Sanchez Moya, O. Bulakci, D. M. Gutierrez-Estevez, M. Ericson, A. Prasad and G. Fodor, "Holistic Resource Management and Air Interface Abstraction Models", *IEEE Conference on Standards for Communications & Networking (CSCN) 2017*, Sept. 2017

26 3GPP R1-1700374, "Downlink multiplexing of eMBB and URLLC transmissions", Intel Corporation, Jan. 2017

27 H. Wang, R. C and K. I. Pedersen, "Dual connectivity for LTE-advanced heterogeneous networks", *Wireless Networks, The Journal of Mobile Communication, Computation and Information.*, vol. 21, no. 6, Aug. 2015

28 N. T. K. Jørgensen, D. Laselva and J. Wigard, "On the potentials of traffic steering techniques between HSDPA and LTE", *IEEE Vehicular Technology Conference (VTC Spring)*, Budapest, May 2011.

29 P. Munoz, R. Barco, D. Laselva and P. Mogensen, "Mobility-based strategies for traffic steering in heterogeneous networks", *IEEE Communications Magazine*, vol. 51, no. 5, pp. 54–62, May 2013

30 A. Prasad, F. S. Moya, M. Ericson, R. Fantini and Ö. Bulakci, "Enabling RAN Moderation and Traffic Steering in 5G Radio Access Networks" *IEEE Vehicular Technology Conference (VTC Fall 2016)*, Sept. 2016

31 5G PPP METIS-II project, Deliverable D5.1, "Draft Synchronous Control Functions and Resource Abstraction Considerations", May 2016

32 3GPP TR 38.801, "Study on new radio access technology: Radio access architecture and interfaces; (Release 14)", v14.0.0, Mar. 2017

33 E. Pateromichelakis, M. Shariat, A. U. Quddus and R. Tafazolli, "On the Evolution of Multi-Cell Scheduling in 3GPP LTE / LTE-A", *IEEE Communications Surveys & Tutorials*, vol. 15, no. 2, pp. 707–717, July 2012

34 E. Hossain, M. Rasti, H. Tabassum and A. Abdelnasser, "Evolution Toward 5G Multi-tier Cellular Wireless Networks: An Interference Management Perspective", *IEEE Wireless Communications*, vol. 21, no. 3, pp. 118–127, June 2014

35 A. Damnjanovic, J. Montojo, Y. Wei, J. Tingfang, L. Tao, M. Vajapeyam, T. Yoo, O. Song and D. Malladi, "A survey on 3GPP heterogeneous networks," *IEEE Wireless Communications*, vol. 18, no. 3, pp. 10–21, June 2011

36 R. Madan, J. Borran, A. Sampath, N. Bhushan, A. Khandekar and T. Ji, "Cell Association and Interference Coordination in Heterogeneous LTE-A Cellular Networks", *IEEE Journal on Selected Areas in Communications*, vol. 28, no. 9, pp. 1479–1489, Nov. 2010

37 3GPP TS 36.300, "Technical Specification Group Radio Access Network Evolved Universal Terrestrial Radio Access (E-UTRA) and Evolved Universal Terrestrial Radio Access Network (E-UTRAN), Overall description; (Release 13)", v13.2.0, Dec. 2015

38 3GPP TR 36.819, "Coordinated multi-point operation for LTE physical layer aspects; (Release 11)", v11.2.0, Sept. 2013

39 Next Generation Mobile Networks (NGMN) Alliance, "NGMN 5G White paper", Feb. 2015

40 Z. Ren, S. Stanczak and P. Fertl, "Activation of nomadic relay nodes in dynamic interference environment for energy saving", *IEEE Global Communications Conference (Globecom 2014)*, Dec. 2014

41 Ö. Bulakci, Z. Ren, C. Zhou, J. Eichinger, P. Fertl, D. Gozalvez-Serrano and S. Stanczak, "Towards flexible network deployment in 5G: Nomadic node enhancement to heterogeneous networks", *IEEE International Conference on Communications (ICC 2015)*, June 2015

42 G. Tsoulos, Ö. Bulakci, D. Zarbouti, G. Athanasiadou and A. Kaloxylos, "Dynamic Wireless Network Shaping via Moving Cells: The Nomadic Nodes Case", *Wiley Transactions on Emerging Telecommunications Technologies (ETT)*, vol. 28, no. 6, Jan. 2017

43 T. Şahin, Ö. Bulakci, P. Spapis and A. Kaloxylos, "Flexible Network Deployment in 5G: Performance of Vehicular Nomadic Nodes", *IEEE Vehicular Technology Conference (VTC Spring 2017)*, June 2017

44 E. Pateromichelakis, H. Celik, R. Fantini, Ö. Bulakci, L. M. Campoy, A. M. Ibrahim, J. Lorca, M. Shariat, M. Tesanovic and Y. Yang, "Interference Management Enablers for 5G Radio Access Networks", *IEEE Conference on Standards for Communications & Networking (CSCN) 2016*, Nov. 2016

45 3GPP TR 36.889, "Feasibility Study on Licensed-Assisted Access to Unlicensed Spectrum; (Release 13)", v13.0.0, July 2015

46 E. Pateromichelakis and Ö. Bulakci, "LAA-as-a-Service as Key Enabler in 5G Dynamic Radio Topologies", *IEEE Conference on Standards for Communications & Networking (CSCN) 2017*, Sept. 2017

47 Y. Sui, A. Papadogiannis, J. Vihriälä, M. Sternad, W. Yang and T. Svensson, "Moving Cells: A promising solution to boost performance for vehicular users", *IEEE Communications Magazine*, vol. 51, no. 6, June 2013

48 Y. Sui, I. Guvenc and T. Svensson, "Interference Management for Moving Networks in Ultra-Dense Urban Scenarios", *EURASIP Journal on Wireless Communications and Networking*, no. 111, April 2015

49 X. Tang, X. Xu, T. Svensson and X. Tao, "Coverage Performance of Joint Transmission for Moving Relay Enabled Cellular Networks", *IEEE Access: The New Era of Smart Cities*, July 2017

50 L. Wang, E. Sasoglu and Y.-H. Kim, "Sliding-window superposition coding for interference networks", *IEEE Inernational Symposium on Information Theory*, June 2014

51 L. Wang, "Channel coding techniques for network communication, Ph.D. thesis," UCSD, 2015

52 H. Park, Y.-H. Kim and L. Wang, "Interference management via sliding-window superposition coding", *IEEE Global Communications Conference (Globecom 2014), International Workshop on Emerging Technologies for 5G Wireless Cellular Networks*, Dec. 2014

53 K. T. Kim, S.-K. Ahn, Y.-H. Kim, H. Park, L. Wang, C.-Y. Chen and J. Park, "Adaptive sliding-window coded modulation in cellular networks", *IEEE Global Communications Conference (GLOBECOM 2015)*, Dec. 2015

54 K. T. Kim, S.-K. Ahn, Y.-S. Kim, J. Park, C.-Y. Chen and Y.-H. Kim, "Interference management via sliding-window coded modulation for 5G cellular networks", *New Waveforms and Multiple Access Methods for 5G Networks, IEEE Communications Magazine*, vol. 54, no. 11, pp. 82–89, Nov. 2016

55 S. Hong, M. Sagong, C. Lim, S. Cho, K. Cheun and K. Yang, "Frequency and quadrature-amplitude modulation for downlink cellular OFDMA networks", *IEEE Journal on Selected Areas in Communications*, vol. 32, no. 6, pp. 1256–67, June 2014

56 F. Rusek, D. Persson, B. K. Lau, E. Larsson, T. Marzetta, O. Edfors and F. Tufvesson, "Scaling Up MIMO: Opportunities and Challenges with Very Large Arrays", *IEEE Signal Processing Magazine*, vol. 30, no. 1, pp. 40–60, Dec. 2012

57 J. Jose, A. Ashikhmin, T. L. Marzetta and S. Vishwanath, "Pilot Contamination and Precoding in Multi-Cell TDD Systems", *IEEE Transactions on Wireless Communications*, vol. 10, no. 8, pp. 2640–2651, June 2011

58 J. Hoydis, K. Hosseini, S. t. Brink and M. Debbah, "Making Smart Use of Excess Antennas: Massive MIMO, Small Cells, and TDD", *Bell Labs Technical Journal*, vol. 18, no. 2, pp. 5–21, Sept. 2013

59 D. Gutierrez-Estevez, "Interference-Aware Flexible TDD Design for Massive MIMO 5G Systems", *IEEE WCNC* 2017, Mar. 2017

60 J. W. Kang, Y. Whang, H. Y. Lee and K. S. Kim, "Optimal pilot sequence design for multi-cell MIMO-OFDM systems", *IEEE Transactions on Wireless Communications*, vol. 10, no. 10, pp. 3354–3367, Aug. 2011

61 S. Bazzi and W. Xu, "Mitigating inter-cell pilot interference via network-based greedy sequence selection and exchange", *IEEE Wireless Communications Letters*, vol. 5, no. 3, pp. 236–239, June 2016

62 5G PPP mmMAGIC project, Deliverable D5.2, "Final multi-node and multi-antenna transmitter and receiver architectures and schemes", June 2017

63 5G PPP mmMAGIC project, Deliverable D5.1, "Initial multi-node and antenna transmitter and receiver architectures and schemes", Mar. 2016

64 C. Rusu, R. Mendez-Rial, N. Gonzalez-Prelcic and R. W. Heath, "Low complexity hybrid precoding strategies for millimeter wave communication systems", *IEEE Transactions on Wireless Communications*, vol. 15, no. 12, pp. 8380–8393, Dec. 2016

65 O. E. Ayach, S. Rajagopal, S. Abu-Surra, Z. Pi and R. Heath, "Spatially sparse precoding in millimeter wave MIMO systems", *IEEE Transactions on Wireless Communications*, vol. 13, no. 3, pp. 1499–1513, Mar. 2014

66 J. Li, E. Bjornson, T. Svensson, T. Eriksson and M. Debbah, "Joint precoding and load balancing optimization for energy-efficient heterogeneous networks", *IEEE Transactions on Wireless Communications*, vol. 14, no. 10, pp. 5810–5822, June 2015

67 3GPP TR 38.912, "Study on New Radio (NR) access technology; (Release 14)", v14.0.0, Mar. 2017

68 P. Arnold, N. Bayer, J. Belschner and G. Zimmermann, "5G Radio Access Network Architecture Based on Flexible Functional Control / User Plane Splits", *European Conference on Networks and Communications (EuCNC) 2017*, June 2017

69 W. Fu, Q. Kong, W. Tian, C. Wang and L. Ma, "A QoS-Aware Scheduling Algorithm Based on Service Type for LTE Downlink", *International Conference on Computer Science and Electronics Engineering (ICCSEE 2013)*, Mar. 2013

70 A. Prasad, M. Uusitalo and A. Maeder, "Energy Efficient Coordinated Self-Backhauling for Ultra-Dense 5G Networks", *IEEE Vehicular Technology Conference (VTC Spring 2017)*, June 2017

71 J. Salmelin and E. Metsala, "Mobile Backhaul", John Wiley & Sons, 2012

72 A. Prasad and A. Maeder, "Backhaul-Aware Energy Efficient Heterogeneous Networks with Dual Connectivity", *Telecommunication Systems*, vol. 59, no. 1, pp. 25–41, May 2015

73 X. Ge, H. Cheng, M. Guizani and T. Han, "5G Wireless Backhaul Networks: Challenges and Research Advances", *IEEE Network*, vol. 28, no. 6, pp. 6–11, Nov. 2014

74 3GPP RAN meeting #75, AT&T, Qualcomm, Samsung, RP-170821, "New SID Proposal: Study on Integrated Access and Backhaul for NR", Mar. 2017

75 3GPP RAN3 meeting #66bis, Huawei, R3-100162, "Overview to LTE Energy Saving Solutions to Cell Switch Off/on", Jan. 2010

76 3GPP RAN3 meeting #65bis, Ericsson, R3-092478, "Considerations on Energy Saving Solutions in Heterogeneous Networks", Oct. 2009

77 G. D. González, J. Hämäläinen, H. Yanikomeroglu, M. García-Lozano and G. Senarath, "A Novel Multiobjective Cell Switch-Off Framework for Cellular Networks", *IEEE Access*, vol. 4, pp. 7883–98, Nov. 2016

78 5G-PPP COHERENT project, Deliverable D2.2, "System Architecture and Abstractions for Mobile Networks", July 2016

79 T. Chen, X. Chen and R. Riggio, "A Network Graph Approach for Network Energy Saving in Small Cell Networks", *IEEE Vehicular Technology Conference (VTC Spring 2016)*, May 2016

80 International Telecommunication Union – Radio (ITU-R) Working Party WP 5D, "Draft New Recommendation "IMT Vision - Framework and overall objectives of the future development of IMT for 2020 and beyond", Doc. R12-SG05-C-0199 (approved by Study Group SG5), June 2015

81 Ö. Bulakci, D. M. Gutierrez-Estevez, M. Ericson, A. Prasad, E. Pateromichelakis, G. Calochira, J. Belschner, P. Arnold, F. Sanchez Moya, A. M. Ibrahim, F. Bronzino, H. Celik and G. Fodor, "An Agile Resource Management Framework for 5G", *IEEE Conference on Standards for Communications & Networking (CSCN) 2017*, Oct. 2017

13

Initial Access, RRC and Mobility

Mårten Ericson[1], Panagiotis Spapis[2], Mikko Säily[3], Klaus Pedersen[4], Yinan Qi[5], Nicolas Barati[6], Tommy Svensson[7], Mehrdad Shariat[8], Marco Giordani[9], Marco Mezzavilla[10], Mark Doll[11], Honglei Miao[12] and Chan Zhou[2]

[1] *Ericsson Research, Sweden*
[2] *Huawei German Research Center, Germany*
[3] *Nokia Bell Labs, Finland*
[4] *Nokia Bell Labs, Denmark*
[5] *Samsung Electronics R&D Institute, UK*
[6] *New York University, USA*
[7] *Chalmers University of Technology, Sweden*
[8] *Samsung Electronics R&D Institute, UK*
[9] *University of Padova, Italy*
[10] *New York University, USA*
[11] *Nokia Bell Labs, Germany*
[12] *Intel, Germany*
With contributions from Hao Guo, Nima Jamaly, Behrooz Makki, Mikael Sternad and Anna Wigren.

13.1   Introduction

This chapter covers the control plane (CP) procedures for the access of a user equipment (UE) to the network. In general, the system access is a rather complex procedure, and depending on the used mechanisms and the requirements it may be an inefficient process resulting in waste of resources. The 5th generation (5G) CP functions for the system access have the same purpose as in older generations of mobile systems, such as Wideband Code Division Multiple Access (WCDMA) and Global System for Mobile Communications (GSM). However, new requirements in 5G, for instance related to latency and energy efficiency, raise the need for different and more efficient approaches.

To easier understand the role of the CP functions, consider the events in Figure 13-1, which depicts a UE's life as a 5G terminal. When the UE is not active, it is in a battery saving mode. To be able to wake up, the UE must periodically listen to system information (SI) as well as the paging channel (to detect a connection which is network-initiated) from the cell it is camped on. This enables the UE to perform a so called initial access to the cell and enter the Connected state. When the UE enters the Connected state, the UE can transmit and receive data. As can be seen in Figure 13-1, the nodes are connected to the 5G core network (CN) via the Xn interface and the N2 and N3 interfaces for CP and

5G System Design: Architectural and Functional Considerations and Long Term Research, First Edition.
Edited by Patrick Marsch, Ömer Bulakçı, Olav Queseth and Mauro Boldi.
© 2018 John Wiley & Sons Ltd. Published 2018 by John Wiley & Sons Ltd.

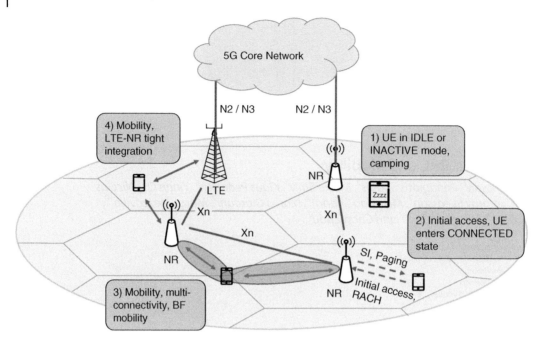

Figure 13-1. A day in the life of a 5G UE, showing basic control plane functionality.

user plane (UP), respectively, as detailed in Section 6.2. In the Connected state, the CP must support security, mobility and radio bearer establishment. Some key functionalities of New Radio (NR) are the multi-connectivity (MC) ability for higher throughput and/or higher reliability, and to handle advanced beamforming (BF) techniques, see also Section 11.5, including BF mobility. In addition to this, 5G will support a tight integration with enhanced Long Term Evolution (eLTE).

In the 3rd Generation Partnership Project (3GPP), the CP functions related to radio-specific functionality are standardized in the form of the Radio Resource Control (RRC) protocol. The RRC messages utilize the same protocol stack as the UP. That is, the same protocol layers, headers and coding are used. The RRC protocol is used by the base station (BS), in 3GPP referred to as enhanced Node-B (eNB) for 4G and Gigabit Node-B (gNB) for 5G, to control the UE behavior both in Connected and in Idle state. Essentially, the responsibilities of the RRC protocol are the handling of system information and RRC connection and measurement configuration.

However, this chapter is not only about the RRC protocol, but also handles other aspects of initial access functionality such as synchronization and system access using the Random Access CHannel (RACH). The way that these processes have been handled in previous generations was optimal for certain use cases, but the introduction of new and stringent requirements in 5G imposes the need for certain enhancements. To understand this need in more detail, Table 13-1 shows the main 5G service types as detailed in Section 2.2, their corresponding requirements (compared to LTE), and the high-level implication on the CP functions.

It becomes evident that an enhanced initial access in 5G plays a very important role for supporting a massive amount of connections, as needed for massive machine-type communications (mMTC), but also for supporting ultra-reliable low-latency communications (URLLC) services. For

Table 13-1. Implications of the main 5G services and their requirements on the 5G control plane functions.

Main 5G services and their requirements	Implications on CP functions
Enhanced mobile broadband (eMBB): The user experience shall be improved in terms of vastly higher data rates and almost seamless connectivity	Both the higher data rates and almost seamless connectivity pose new requirements on the mobility functionalities of 5G, and imply tighter integration between (e)LTE and NR.
Ultra-reliable low-latency communications (URLLC): Very high requirements for capabilities such as latency, reliability, and availability.	This service poses very high requirements on the initial access CP functions to be able to perform very fast setup and prioritize between different services.
Massive machine-type communications (mMTC): Very large number of connected devices typically transmitting a relatively low volume of non-delay-sensitive data.	This service poses extreme requirements on efficient initial access and state handling.
All service types: Very high energy efficiency for both network and users	Puts requirement on more efficient system information transmission compared to LTE, as well as an efficient use of UE state transitions.

the latter, for instance, methods to prioritize users and services already during the random access phase can strongly improve latency and reliability. Since massive beamforming will play a prominent role in a 5G system, as mentioned before, the methods to synchronize to the network will also differ compared to previous mobile communication systems, where the exact design is again very important especially in mMTC scenarios. Finally, also improved mobility, UE states and state handling play an important role in addressing the requirements listed in Table 13-1.

The remainder of this chapter is structured as follows. Section 13.2 covers the initial access, including the transmission of system information, optimized RACH access, and various enablers for initial access in the context of beam-based connectivity. Section 13.3 then describes the fundamentals of UE state handling and why 5G requires a new state to support better battery savings and faster setup for data transmissions. Section 13.4 then covers mobility, discussing both normal handover (HO) with enhancements, and mobility in the context of multi-connectivity. Another important topic for 5G mobility is how to handle beamforming and mobility, which is also treated in Section 13.4. Finally, the chapter is summarized in Section 13.5.

13.2   Initial Access

13.2.1   Initial Access in General

Initial access comprises a set of functions across multiple layers of the air interface in the radio access network (RAN) protocol stack and, to some extent, the CN/RAN interface as in the case of paging and state transition. In LTE, these functions are:

1) **Acquisition and cell search**, allowing a UE to synchronize to a cell. This includes the following steps: (a) acquisition of frequency and synchronization to a cell, (b) acquisition of frame timing of the cell, (c) determination of the physical layer cell identity of the cell. This is performed using the Primary and Secondary Synchronization Signals (PSS and SSS, respectively). The PSS allows the

UE to identify the cell ID and the generic timing information, enabling it to identify the position of the SSS, and the SSS provides the frame timing and the cell identity group to which the cell ID belongs.

2) **System information distribution**, enabling the UE to acquire the cell system information to be able to access the network. The system information includes the downlink and uplink cell bandwidths and configurations, etc. This information is transferred using the Master Information Block (MIB) and the Secondary Information Blocks (SIBs). The MIB is needed by the UE to be able to read the SIB information. The SIB includes permissions and configurations of the UE, such as cell selection and reselection, random access parameterization, access barring information provision, warning messages, etc.

3) **Random access**, allowing the UE to connect to the network for initial access, handover purposes, scheduling requests, UE positioning, etc. This is done using a shared channel, namely the aforementioned Random Access CHannel (RACH). The UEs may access the RACH either on a contention basis, or contention free, depending on the purpose; the former approach being mainly used for initial access and RRC connection establishment, and the latter mainly for handover, uplink resynchronization, and positioning.

4) **Paging** is used for network-initiated connection setup when the UE is in RRC Idle state. The UE is notified for downlink data using the paging channel in LTE.

While the initial access in 5G will in principle consist of exactly the same steps and functions, some of the listed functions should be significantly enhanced for 5G. In the following subsections, improvements are hence discussed that increase the ability of the system to handle more demanding service requests or an increased number of access requests.

13.2.2 System Information and 5G RAN Lean Design

The LTE system information (SI) comprises information such as access information for the UE, node specific information, system wide information, public warning system (PWS) information, etc. One of the drawbacks with LTE was the rather low possibility for the cell to enter a so-called micro sleep, also referred to as cell discontinued transmission (DTX). There are several reasons for this related to the system information distribution such as:

- Reference symbols are transmitted even if there is no data;
- System information is transmitted periodically regardless of the load in the system;
- The Physical Downlink Control CHannel (PDCCH) is transmitted across the full system bandwidth, i.e., the same number of PDCCH symbols are used for all resource blocks (RBs);
- For synchronization signals, LTE uses a fixed periodicity of 5 ms, which decreases BS sleep efficiency at low load.

Since system information is always transmitted, it is difficult to make the LTE system more energy efficient at low load. If there is no, or very limited, possibility for BSs for micro sleep, the total power consumption of the network is likely to increase when the BS density is increased [1]. In addition to the economic and environmental reasons to minimize energy consumption, there are also many reasons related to engineering. Today, the main obstacle for miniaturization of radio BSs is heat. Energy consumption directly drives product weight and volume through heat dissipation. Reducing energy consumption will have nonlinear impacts not only on weight and volume, but also on what a BS is, how it is built, what it can do, where it can be deployed, what energy sources it utilizes, etc.

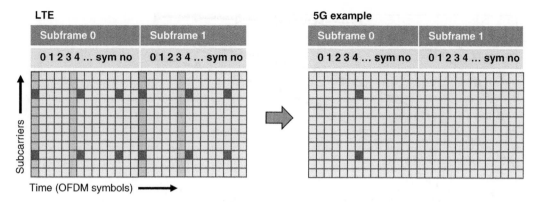

Figure 13-2. Example of 5G lean design compared to LTE for one PRB. Since there are no active users in the cell, the 5G cell can turn off the PDCCH symbols (yellow/light grey dots), decrease the periodicity of the synchronization signals (green/grey dots) and decrease the RS transmissions (red/black dots).

How can 5G be made more energy efficient? Below we go through how the system information can be designed to be more lean in 5G, diving in more detail into the bullets listed before:

Reference Signals

The reference signals (RS) necessary for channel estimation should in 5G typically only be transmitted in the same sub-frame, over the same bandwidth, and in the same beam, i.e., based on the same beam-forming as the corresponding data. This is different compared to LTE, where also the cell-specific reference signals (CRS) in previous sub-frames can be used to aid channel estimation, as illustrated in Figure 13-2, where the red or dark slots are RS. The exact procedure for NR is currently under discussion in 3GPP [2][3].

System Information Transmission Using user On-demand Approach

In the 3GPP NR discussions, the system information is divided into minimum SI and "other SI". Minimum SI is periodically broadcasted (as in LTE today). The minimum SI comprises basic information required for initial access to a cell and information for acquiring any other SI broadcast periodically (as in LTE) or provisioned on a demand basis, which is a novel concept compared to LTE [4].

PDCCH

In LTE, PDCCH is transmitted across the full system bandwidth, i.e., the same number of PDCCH symbols are used for all RBs. This is not very resource and energy efficient. For 5G, a more efficient PDCCH transmission is foreseen. It will probably be more limited to the resources used by the user data. Again, see an example in Figure 13-2, where the yellow or grey slots are PDCCH symbols.

Synchronization Signals

LTE uses a periodicity of 5 ms for synchronization signals. However, if the periods between the synchronization signals can be increased, the BS sleep efficiency can be increased [5]. The reason is that it takes some time to deactivate and reactivate certain components. For longer sleep durations, more components can be put to sleep, and the sleep power usage becomes lower. The current agreement

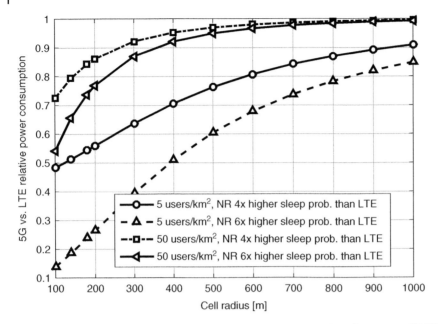

Figure 13-3. Example of the relative 5G power consumption vs. LTE for different NR cell DTX probabilities.

in 3GPP [6] enables different periodicities of the synchronization channels (5, 10, 20, 40, 80 or 160 ms), depending on, e.g., traffic load. See Sections 11.6 and 13.2.3 for more information on this topic.

Figure 13-3 shows the relative power consumption per cell for 5G compared to LTE for a range of cell radii for two fixed user densities of 5 or 50 users/km². When the network is densified, the cell radius shrinks. For small cell radii, there is hence a higher probability to have no active users in the cell, which may enable the cell to enter the cell DTX. The major difference between 5G and LTE is the ability to actually utilize the cell DTX when there is no active traffic. In Figure 13-3, it is assumed that NR has 4 and 6 times higher probability to enter cell DTX if the cell is empty, as detailed in [7]. The figure shows that if 5G is designed so that it allows better cell DTX sleep probabilities than LTE, the power consumption can be decreased substantially.

13.2.3 Configurable Downlink Synchronization for Unified Beam Operation

As mentioned in Section 13.2.1, downlink synchronization is the first task of the initial access for UEs to acquire initial timing and frequency synchronization with the network, and to detect the physical cell ID (PCI), which is used for the reception of most other channels and signals. One of the major targets of the 5G radio system is to support significantly higher carrier frequencies than LTE, e.g., up to 100 GHz. Moreover, the 5G system is envisioned to support different deployment scenarios including heterogeneous networks with ultra-densely deployed small cells operating in high frequency bands, e.g., millimeter-wave (mmWave) bands. Additionally, 5G can operate in standalone and non-standalone mode, as introduced in Section 1.2. While the synchronization signaling requirements for all these scenarios are very different, it is highly desired to design a unified synchronization signal transmission framework to support all these envisioned scenarios. It should also be noted that a 5G radio system endeavours to be highly configurable so that the system can be flexibly configured according to its operation scenario.

As described in Section 13.2.2, a lean system design to achieve much better forward-compatibility has been considered as one of the main design targets of the 5G radio system. One part of achieving this is to minimize the "always-on" signals, such as cell specific-reference signals in LTE. As a result, the synchronization signals (SS) together with MIB channels are the only remaining "always-on" signals transmitted periodically in the 5G radio system [8]. It is therefore very important to keep the SS transmission frequency on a proper level so that on one hand, UEs can acquire initial synchronization very quickly, and on the other hand, the overhead of SS transmission is kept to a minimum level for the sake of network energy efficiency. To accomplish this, a configurable SS transmission framework has been developed and standardized in 3GPP [9].

As illustrated in Figure 13-4, in the configurable SS transmission scheme, a series of synchronization signal burst sets are transmitted, where each set contains one or several synchronization signal blocks (SSB). Each SSB consists of 4 OFDM symbols, one PSS, one SSS and two PBCH OFDM symbols carrying MIB. All OFDM symbols in the SSB are transmitted from the same antenna port, i.e., by using the same beam direction. This means that the beam index is implicitly signaled by the time location of the SSBs. To ensure timely acquisition of initial timing and frequency synchronization, regardless of periodicity of synchronization burst set, a certain number of SSBs shall be transmitted within a time window of 5 ms. Since each SSB is transmitted in a particular beam direction, the number of SSBs in 5 ms essentially determines the number of beams used for SS transmission in a cell. Consequently, configurable SS transmission incorporates single-beam and multi-beam transmission into a unified solution. Specifically, when single beam operation is applied, there is only one SSB transmitted in a 5 ms time window like in LTE. To realize multi-beam operation, there are more than one SSB to be transmitted in 5 ms. Due to different beamforming gain requirements in different frequency bands, the maximum number of SSBs in a 5 ms time window is defined as follows: 4 for frequency bands below 3 GHz, 8 for frequency bands above 3 GHz and below 6 GHz, and 64 for frequency bands from 6 GHz to 52.6 GHz [10]. These values set the maximum number of beams that can be used for SS transmission. When more beams (i.e., spatial repetitions) are used, beam width

Figure 13-4. Example of configurable SS burst set transmission.

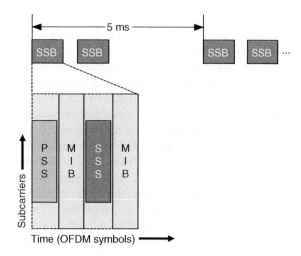

becomes narrower and the beamforming gain increases, there will be less opportunities to put the BS into (micro-) sleep. As a result, such configurable SS transmission provides a great optimization space for the trade-off among the different aspects mentioned above.

13.2.4 Digital Beamforming in the Initial Access Phase

The previous sections discussed system information distribution and the synchronization signals transmitted from the BSs for enabling the UE to synchronize to the respective cell and beam. This task is very complicated in cases of highly directional links, since in addition to the time and frequency domains, the 5G network elements (i.e., BS, UE, or relays) have to discover each other in an additional dimension, namely that of space. For instance, each receiver must determine the angles of arrival for the incoming signals before a data link is established. Depending on the beamforming architecture, scanning all possible angles of arrival may entail high delays in the CP. These delays, as we show below, may be well above what is acceptable in the current 4G LTE systems, and will definitely not meet the stringent delay requirements of the 5G system. Note that at the early stages of deployment, the mmWave coverage is projected to be "spotty", and handovers between small mmWave cells and LTE macro cells will be quite frequent, leading to even higher delays. Therefore, a fast discovery procedure to find the point of attachment in the angular space is needed.

Assuming that mmWave initial access will follow the same five-step process of LTE, we can identify one major difference: at the end of the synchronization phase, the UE learns where the BS is, and at the end of the random access phase, the BS learns where the UE is.

The main beamforming architectures are analog, digital and hybrid beamforming, as detailed in Section 11.5. The analog architecture relies on inexpensive phase-shifters in radio frequency (RF) to create and steer a beam towards a desired direction. The digital architecture relies on digital samples obtained by the analog-to-digital converters (ADCs) attached to each antenna element of a multi-antenna array. With analog beamforming, each receiver can steer the beam in only one direction at a time. Therefore, the receiver will have to direct the beam in all the available angles one by one. With digital beamforming on the other hand, thanks to all the available digital samples, the receiver can simultaneously steer beams into as many directions as the number of its antenna elements. This however occurs at the expense of increased power consumption, due to the usage of a higher number of ADCs than in the analog beamforming case. Hence, angular scanning with digital beamforming becomes M times faster, where M is the size of the angular domain the receiver needs to scan. Another factor affecting network discovery is the mode of signal transmission, which is here assumed to be based on single-stream analog beamforming or omnidirectional analog transmission. This analysis captures the most extreme cases, since multi-stream transmission is expected to perform somewhat in between omnidirectional and single-stream analog BF [11]. The total size of the angular space is a function of the signal transmission mode and the beamforming architecture at the receiver: it is the product of the size of the angular space at the receiver and at the transmitter. For example, in the synchronization phase, if both the BS and the UE employ analog directional transmission and beamforming, and have array sizes of 64 and 16 respectively, 1024 angle combinations have to be scanned one by one. Note that if a UE is at the edge of the cell, it may further need to scan the angular space more than once to accumulate enough energy to detect the synchronization signal in a low signal-to-noise ratio (SNR) regime.

In Figure 13-5, we focus on the synchronization phase and assume a transmission scheme where at each synchronization slot the BS randomly picks a direction and transmits the synchronization signal in analog directional way or omnidirectionally (where the transmission angle is fixed). In this example,

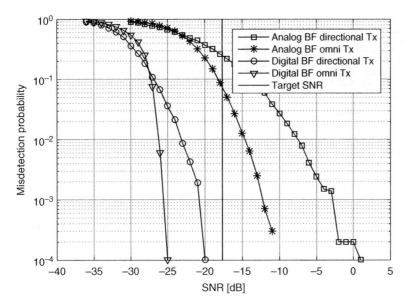

Figure 13-5. Misdetection probability vs data SNR.

Table 13-2. Initial access options nomenclature. O: analog omni-directional, D: analog directional, Dig: digital directional.

Option	Sync BS Tx	Sync UE Rx	RA BS Rx
DDO	Analog Directional	Analog Directional	Analog Omni
DDD	Analog Directional	Analog Directional	Analog Directional
ODD	Analog Omni	Analog Directional	Analog Directional
ODDig	Analog Omni	Analog Directional	Digital Directional
ODigDig	Analog Omni	Digital Directional	Digital Directional

digital receiver beamforming exhibits undisputable superiority. It can be detected even for SNRs below -17 dB. This SNR value was chosen as a threshold, since at this SNR a mmWave system operating at 1 GHz bandwidth can offer data rates of 10 Mbps. Next, we move to comparing five discovery design options where signal transmission is performed sequentially rather than randomly [12]. These options are differentiated and named based on the synchronization signal transmission mode (analog directional or analog omni-directional), synchronization signal reception architecture (analog directional, analog omni-directional or digital directional), and random access reception mode (analog directional, analog omni-directional or digital directional). Note that since at the end of the synchronization phase the BS's location is known to the UE, except in one case, the random access preamble is always assumed to be transmitted in analog directional, so that all the directivity gain of the antenna array is exploited. The nomenclature of these different schemes and options is explained in Table 13-2.

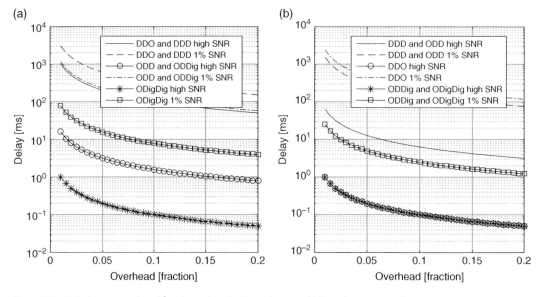

Figure 13-6. Delay vs. overhead for a) synchronization phase and b) random access phase.

The comparison is made in terms of delay and overhead for both edge users, defined as the 1st percentile of SNR, and also high-SNR users. The overhead is defined as the ratio of the synchronization signals duration over the transmission period. For example, if the signal duration is 100 μs, and it is transmitted every 5 ms, then the overhead is 2%. Naturally, as the overhead is increased, i.e., as more resources are used for these signals, the delay falls accordingly. Figure 13-6 shows the delay for the synchronization and random access vs. the overhead. As can be seen, digital beamforming dramatically reduces the synchronization and random access times.

Nevertheless, the issue of energy consumption is an important one. To offset the high power consumption of digital beamforming, one could employ ADCs of low quantization, say 2-3 bits, as discussed also in Section 11.5.3. Since the power consumption of ADCs scales exponentially with the bit resolution, employing low quantization ADCs will bring the power consumption of digital beamforming to the same level as its analog counterpart, as detailed in Section 16.2.3.2. Note that the effect of reducing the quantization resolution is negligible in low SNR regimes of the edge user.

13.2.5 Beam Finding for Low-Latency Initial Access

The previous subsection has analyzed the use of digital beamforming (BF) in the initial access phase so as to reduce the synchronization delay. This is particularly important in scenarios with mobility, where the high channel dynamics would necessitate fast mechanisms to find alternative communication links. In this section, we study the performance of large-but-finite mmWave networks using codebook-based analog beamforming. We consider an efficient genetic algorithm (GA) based approach for initial access beamforming. With the proposed algorithm, the appropriate beamforming matrix is selected from a set of predefined matrices such that the network end-to-end (E2E)

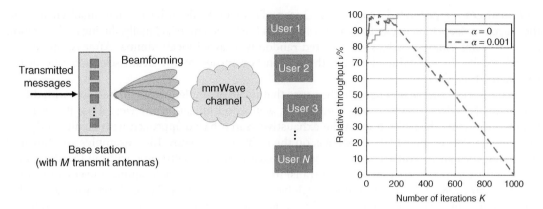

Figure 13-7. On the left: the considered mmWave multi-user system. On the right: an example of the convergence process of the GA-based beamforming for systems with ($\alpha = 0.001$) and without ($\alpha = 0$) delay cost of the algorithm.

throughput is optimized. The considered system model is depicted on the left side of Figure 13-7, where a base station with M antennas serves N single-antenna users. Specifically, the E2E throughput at the end of the K-th iteration of the genetic algorithm is defined as

$$R(K) = (1 - \alpha K) \sum_{i=1}^{N} \log_2 \left(1 + \text{SINR}_i^K \right), \tag{13-1}$$

$$\text{SINR}_i^K = \frac{\dfrac{P}{M} g_{i,i}^K}{BN_0 + \dfrac{P}{M} \sum_{i \neq j}^{N} g_{i,j}^K}, \tag{13-2}$$

where $g_{i,j}^K$ is the *(i, j)*-th element of the matrix $G^K = \left| \mathbf{H} \mathbf{V}^K \right|^2$ with $\mathbf{H}$ denoting the channel matrix and $\mathbf{V}^K$ being the selected precoding scheme in the K-th iteration of the algorithm. Also, P is the transmit power, B is the system bandwidth and N_0 is the power spectral density of the noise. Finally, α represents the relative delay cost for running each iteration of the algorithm. As seen in the above equations, there is a trade-off between the running delay of the algorithm and improving the beamforming efficiency. As the number of iterations K increases, more accurate beams are selected by the algorithm, and the users' signal-to-interference-and-noise ratio (SINR) increases. On the other hand, with $\alpha \neq 0$, the delay cost of the algorithm reduces the E2E throughput. Thus, as seen in the following, there is a finite optimal number of iterations maximizing the E2E throughput in delay-constrained applications.

The details of the proposed scheme can be found in [13]. Shortly, the algorithm is based on the following procedure. The algorithm starts by getting L possible beam selection sets randomly, and each of them means a certain beam formed by transmit antennas, i.e., a submatrix of the codebook. During each iteration, the best selection result is determined, named as the *Queen*, based on the objective metrics. For instance, the beamforming matrix with the highest E2E throughput is chosen if the above expression for $R(K)$ is considered as the objective function. Next, the Queen is kept and

$S < L$ matrices are generated around the Queen. This can be done by making small changes to the Queen such as changing a number of columns in the Queen matrix. Finally, during each iteration $L - S - 1$ beamforming matrices are selected randomly to avoid local minima. After N_{it} iterations, considered by the algorithm designer, the Queen is returned as the beam selection rule in the considered time slot.

The proposed scheme is generic in the sense that it can be implemented in the cases with different channel models, beamforming methods as well as optimization metrics. Also, it can reach (almost) the same throughput as in the exhaustive search based approach with significantly less implementation complexity. For instance, on the right side of Figure 13-7, examples are shown of the GA performance in different iterations in the cases with ($\alpha = 0.001$) and without ($\alpha = 0$) costs of running the algorithm, assuming $M = 32$ transmit antennas, $N = 8$ single-antenna users, precoding codebook size 128, SNR = 10 dB, and Rayleigh fading channels. In the figure, the relative achievable throughput compared to the maximum throughput achieved by exhaustive search is plotted. We observe that very few iterations are required to reach the maximum throughput, if the running delay of the algorithm is taken into account. That is, considering the cost of running the algorithm, the maximum throughput is obtained by finding a suboptimal beamforming matrix and leaving the rest of the time slot for data transmission. On the other hand, as the number of iterations increases, the cost of running the algorithm reduces the end-to-end throughput converging to zero at $K = 1/\alpha$. With no cost for running the algorithm, on the other hand, the system performance improves monotonically with the number of iterations. However, the developed algorithm leads to (almost) the same performance as the exhaustive search based scheme with very limited number of iterations (note that with the parameter settings of Figure 13-7, an exhaustive search implies testing in the order of 10^{30} possible beamforming matrices). For example, with the parameter settings of Figure 13-7 and $\alpha = 0$, the proposed algorithm reaches more than 95% of the maximum achievable throughput with less than 200 iterations, thus making it particularly suitable for delay-constrained systems.

13.2.6 Optimized RACH Access Schemes

After receiving system information and synchronizing to the cell, the UE should indicate its desire to receive uplink resources. In LTE, the UE randomly selects one of the random access preambles (64 preambles) configured in the broadcasted system information. The current design may create problems, due to collisions, when large numbers of devices simultaneously attempt to access the system. These potential collisions will lead to additional access delays which may impact services. Up to now, several schemes have been proposed in the literature for handling the random access procedure. These schemes can be classified into two large groups, namely *pull-based* and *push-based* [14]. In the first set of solutions, the RACH preambles are being split in prioritization groups, and the more delay sensitive devices compete against fewer devices for accessing the system. Additionally, the time for a second attempt to access the system could be fine-tuned according to the collision rate. In the second set of solutions, a push-based procedure is used to achieve a small data transmission (SDT) from the MTC devices. However, these schemes are designed mainly for prioritizing access based on the transmission requirements and are not aiming at solving the collision rate problem. Also, these focus mainly on traditional MBB use cases, and not on URLLC or mMTC use cases.

Based on the description above, we conclude that there are two key challenges for machine type communication: the massive access and the latency requirement.

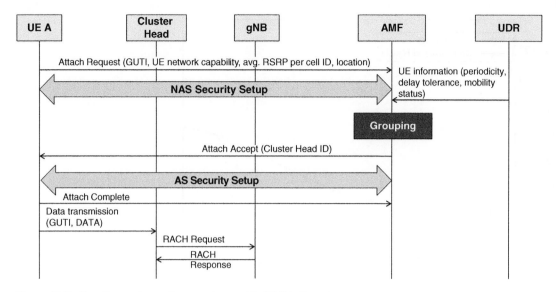

Figure 13-8. Signaling exchange for grouping and for RACH attempt.

For the random access of a vast amount of devices, a solution based on the grouping of the devices seems to be appropriate. Instead of having all the group members performing random access using one of the 64 preambles when they have to transmit, one could aggregate the transmission requests and only one device (i.e., the group head) could perform the RACH request. This will result in a significant reduction in the collision rate in the RACH. A slotted access scheme, where each device transmits according to its needs will further benefit the system.

The proposed solution is shown in Figure 13-8:

1) The devices are grouped by the network based on their mobility and their communication characteristics (e.g., data to be transmitted, packet delay requirements).
2) The network schedules the cluster heads' transmission opportunities based on their transmission requirements. The scheduling information includes how many timeslots each device should attempt to access the network and which preambles should be used.
3) The intra-cluster communication may take place either via a different interface or via scheduled device-to-device (D2D) communication.

In Figure 13-9, it is shown that using the group-based system access reduces the collision rate significantly, since fewer devices (only the group heads) compete for RACH preambles. This also has a direct impact on the average initial access delay (related to the whole process including random access, random access response, terminal identification, and contention resolution), since the devices are accessing the system with fewer collisions and thus experience fewer retransmissions. For a low number of devices, the collision rate is small since the preambles are enough for the random access. As the number of devices increases, the collision rate and the CP latency increases for both approaches, but in the case of the group-based access it is considerably lower.

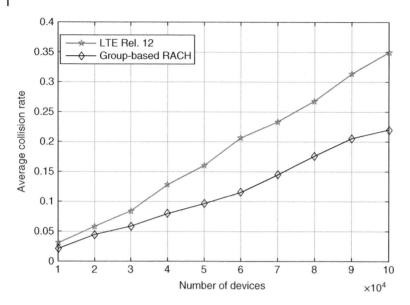

Figure 13-9. Average collision rate for the group-based system access compared with LTE.

The previous solution can handle the first of the two the key challenges introduced in this section, but it fails to successfully handle the strict latency requirements. One possible solution could be to reserve a set of dedicated preambles for the use of devices with high priority. This solution, however, is not efficient, since the number of RACH preambles is very small (i.e., 64 preambles) and has to be used both for random access and for handover purposes. What could be done is to:

1) maintain the preamble use from the devices without stringent delay requirements as is, and
2) the delay-sensitive requests apply a combination of preamble signatures at a given random access time slot.

The approach would enable requests with more restrict delay requirements to have higher priority, since combinations of preambles can always be identified by the receiver. The combination of the preambles may take place either in time or frequency domain, thus providing a large set of potential combinations. The network will inform the UE about its reserved preamble combinations and the respective time and frequency shifts for each service priority level after the initial UE attach.

As depicted in Figure 13-10 for high priority initial access assessment, the preamble coding outperforms the other two approaches, since the combination of preambles reduces the probability of a collision when accessing the system. Collisions of high-priority transmissions will occur only if two low-priority devices select the same preambles as the combination dedicated for a high-priority device that attempts to access the system at the same time. As can be seen in the figure, the proposed scheme provides ~1000 times lower retransmission probability for high-priority users compared to LTE without resource split, whereas it provides ~100 times lower retransmission probability compared to LTE with resource split.

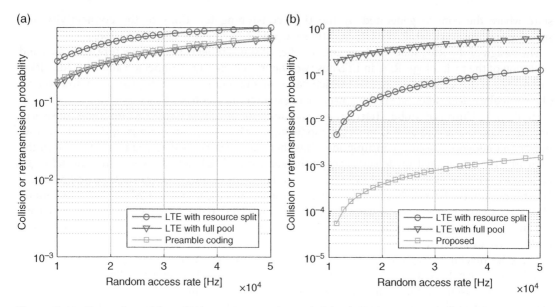

Figure 13-10. Comparison of the collision or retransmission probability for high- and low-priority requests.

13.3 States and State Handling

13.3.1 Fundamentals of the RRC State Machine for 5G

The RRC state machine for the 5G RAN will consist of three RRC states [3][15]: RRC Idle, RRC Connected and a new inactive state, which in this section is called RRC Connected Inactive.

The RRC state machine is designed by mapping the operational functions of the UE to the radio resource control and management, so that the different functions of RRC and RRM procedures can be supported efficiently. RRC states are managed the entire time from UE power-on until the UE is powered off. For example, after power-on, the UE starts to search for a public land mobile network (PLMN). When the network has been found and the UE has registered to the network, the network takes over the state management, and the state management process continues depending on the UE data activity, network coverage and user mobility.

Let's assume that a UE should establish a data connection for example to download a newsletter with attached figures. The UE transitions from RRC Idle to RRC Connected state to download the newsletter. When the newsletter is being read, the traffic activity becomes low between UE and network. If the UE remains in RRC Connected state, there would be unnecessary power consumption at the transceiver and allocation of radio resources, even though power and resources could be saved. The simplest option is to disconnect the UE from the network and go to RRC Idle state to save power. This is a valid approach if the data transmission activity remains low for a relatively long period. However, if the user soon clicks another link on the newsletter, then the UE must again request the whole RRC setup, and the network must re-create the radio bearers from scratch, which would take a lot of signaling and energy. This type of traffic is best served with an intermediate state between active mode and idle mode, or more precisely, a low-activity state where the context generated by the time-consuming functions of the connectivity and security are preserved while the UE can still save power during low-activity periods.

The design of the RRC state machine and number of states reflects the architecture, use cases and evolution of the technology. RRC in LTE has two states, namely RRC Connected and RRC Idle. RRC Connected is the state for active UEs, which can transmit and receive user plane data and control plane signaling. RRC Idle state is the power saving state for low-activity UEs.

Similarly, in Universal Mobile Telecommunications System (UMTS), the state machine consists of one Idle mode state and four Connected mode states. The Idle mode state is optimized for low-power and network resource consumption, thus there is no UE context stored in the UE nor in the network. The Connected mode states are optimized for high UE activity where the normal high-data-rate traffic for active UEs and most of voice and data traffic are being transmitted and received. For this reason, the UE's RAN context is stored in the UE and the network. UMTS Connected mode also contains low-power states where the UE is still connected to the network and listens to the paging and broadcast channels, while the uplink data transfer is not supported without state transition to the active Connected state.

The state machine design for 5G RAN is challenging due to the high number of 5G use cases, which have a high diversity and even contradictory requirements. The state handling mechanism for 5G needs to consider all the 5G use cases and therefore is a key component for 5G design. The user plane and control plane latency should be low to reflect the 5G use cases, and is required to be significantly reduced compared to existing cellular systems. This enables a good user experience and improves the battery life of UEs, since it enables a fast transition from a power-efficient state to an active state, which means that the devices can spend more time in the low-power state.

Figure 13-11 presents the 5G state machine with the three states and their related state transition procedures.

The RRC Connected state is optimized for high UE activity and has similar characteristics as RRC Connected in LTE, e.g., the UE context is stored in the network and in the UE, UE mobility is network

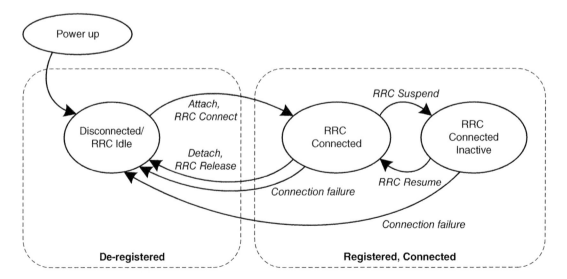

Figure 13-11. 5G RRC state machine [16].

controled, and the UE location is known at cell level. In RRC Idle, the UE context is not stored, and the use of this state may be limited to power-up and to fault recovery procedures. RRC Connected Inactive is proposed as the low-activity UE state. The new RRC Connected Inactive state will keep the UE context in the UE and the network to avoid a setup of radio bearers and security, and the connection between RAN and CN is kept active. When transitioning from RRC Connected to RRC Connected Inactive, the configured Access Stratum (AS) state information is retained by the UE and network. When transitioning back to RRC Connected, the AS state can be restored without the entire configuration signaling (e.g., state transition from RRC Idle to RRC Connected). Only the radio resources exclusively assigned to the UE are released and can be used by other UEs when the UE enters RRC Connected Inactive. In RRC Connected Inactive, the UE identifies itself to the RAN by a unique UE ID provided by the BS. The UE can move within the pre-defined area in the RAN by performing cell reselections and without notifying the network. Considering the large number of use cases, the UE behavior can be configured based on the service requirements of the UE. The configurability of the RRC Connected Inactive state is a key feature to differentiate the UE behaviour for different service requirements, while using only a single but flexible low-activity state.

The identified characteristics of the RRC state model are shown in Table 13-3. The mobility and system access procedures of the new state model are configured based on different aspects of use cases, device capability, access latency, power saving, security requirements and privacy.

13.3.2 Mobility Procedures for Connected Inactive

The mobility procedures in RRC Connected Inactive state are related to cell selection and reselection, location tracking, location update, and reaching the UE by paging. Mobility during Connected Inactive state is based on cell selections and reselections performed by the UE, similar to cell (re) selections in LTE RRC Idle state. The 5G CN-RAN connection, for both control and user plane, remains established for the UE in RRC Connected Inactive state. To avoid frequent path switching on the CN–RAN interface, the cell reselections within the allowed RAN tracking area (RTA) are not visible to the CN. Therefore, the CN–RAN interface path and UE context remains at the last serving BS,

Table 13-3. RRC states in 5G.

5G State	Mobility procedure	Monitoring dedicated physical channels	Allowed mode for DL channel monitoring	UE location known on	Uplink activity allowed	Storage of RAN context information
RRC Idle	Cell selection & reselection	No	Discontinuous with DRX	Tracking area list level	No	No
RRC Connected Inactive	Cell selection & reselection	Configurable, yes/no	Discontinuous with DRX	RAN tracking area level	Configurable, Contention-based UL data	Yes
RRC Connected	Network-controlled handover	Yes	Both continuous and discontinuous with DRX	Cell level	Yes	Yes

i.e. the gNB which suspended the UE's active connection to RRC Connected Inactive for the coming low-activity period. The last serving gNB can now take the role of a mobility anchor, which allows the CP and UP of the CN-RAN interface to be kept unmodified towards the CN. The overall mobility procedure for RRC Connected Inactive is presented in Figure 13-12, including the state transitions between RRC Connected and RRC Connected Inactive [16].

If the UE in Connected Inactive state moves out of the RTA, there is a need for location update, either in the form of a periodical update or a RTA update. The gNB, where the UE is currently located, sends the UE's current cell or RTA (i.e., location) to the anchor gNB, which can initiate the relocation on the CN-RAN interface and hand over the anchor gNB role to the current gNB.

The anchor gNB initiates the paging in the RAN when receiving downlink packets. Upon receiving the paging response from the UE, the anchor gNB delivers the UE context data and the buffered DL packets to the current gNB where the UE is located. This then becomes the new serving gNB, re-configures the UE to RRC Connected state and performs the required path switching procedure.

In the UE-initiated connection, the currently visited gNB will retrieve the UE context from the anchor gNB based on the UE's reported last serving Cell ID, and the visited gNB will buffer UL packet(s) until it has performed the required path switching. In case the UE has not moved from the anchor gNB, the UE context is instantly available and there is no need to perform any path switching.

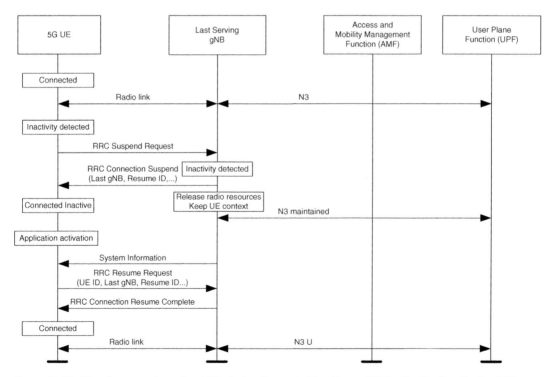

Figure 13-12. Signaling procedure of mobility during Connected Inactive and RRC activation/inactivation [16].

In the distributed RTA management, the anchor gNB may detect UE movement out of the current RTA based on its received UE registrations. The anchor gNB should deliver the UE context to the new gNB to let it take the role of anchor gNB, to trigger the RTA update procedure, and to perform CN-RAN interface path switching.

13.3.3 Configurability of the Connected Inactive State

In LTE, there are two RRC states: RRC Idle and RRC Connected. This reduced the standardization effort and simplified the state machine operation for LTE. RRC Idle is optimized for low UE power consumption. The UE Access Stratum (AS) context is neither stored in the UE nor in the network. The RRC Connected state is optimized for high UE activity where the UE AS context is stored in the UE and the network, and the CN-RAN interface is configured and active. For power saving in the RRC Connected state, discontinuous reception (DRX) is adopted to enable different levels of UE power saving when there is no continuous data transmission. LTE's two-state model works well for example in MBB use cases.

However, the model is shown to be inefficient for handling use cases that involve a frequent transmission of small data packets, such as sensor measurements or smart phone keep-alive messages. The inefficiency comes from signaling procedures, which are used to transition the UE from RRC Idle to RRC Connected, create the UE AS context and configure the CN-RAN interface. With a large number of UEs which frequently transmit small data packets, this becomes inefficient for the network. It is also costly to keep the UEs in RRC Connected without active data transmission due to dedicated resources, handovers and measurement reports. The configurable RRC state machine was developed to handle the mobility state of UEs with diverse requirements of the UE services, such as battery life, latency, bandwidth, mobility, etc., during RRC Connected Inactive.

The UE state model should fulfil all the requirements coming from different 5G use cases and deployments. All use cases share some common characteristics and default procedures, such as suspending and resuming the connection, tailoring the UE mobility procedures to reflect the UE speed, security and privacy, data rate, latency and reliability, expected UE battery life, etc. A power-optimized configuration might be used for a UE with a requirement for a long battery life. On the other hand, a configuration optimized for latency can be used for UEs with applications requiring low latency. The configurability options can be classified into commonly used functions between UE and network [17]:

- Configurable discontinuous reception (DRX);
- LTE-5G tight integration;
- State transitions;
- Measurement configuration;
- Paging and location tracking;
- Synchronization.

A widely configurable DRX at the UE is needed for different traffic patterns and battery requirements. For example, very small packets may be transmitted infrequently by sensors, and those types of devices also need a long battery life. In addition, there may be use cases requiring quite low control plane latency without stringent power consumption requirements, such as vehicles sending and receiving safety and traffic related information. Therefore, the RRC Connected Inactive supports configurable DRX where some devices will monitor the system control channels frequently, while some may stay in low-activity state for several hours, but also benefit from quick connection resumption.

In case of 5G stand-alone deployment or a tight integration of LTE and 5G, the dynamic usage of available resources in available radio access technologies (RATs) will enable a wide range of services with the best coverage. The tight integration of LTE and 5G demands common state handling for multi-radio UEs, especially when connected to the same CN. Non-integrated state transitions between LTE and 5G would lead to significant signaling load especially on LTE side, since the state transition between 5G Connected Inactive to LTE Idle would result in a UE context release for the UE and network. In RRC Connected Inactive state, the multi-radio UE can be configured to camp either on LTE or 5G, preserve the UE context between inter-RAT cell reselections, and monitor the paging from both RATs. Yet another alternative would be to configure multi-radio UEs to camp simultaneously on both RATs and possibly try to simultaneously access both RATs. This would be beneficial for use cases with a fast establishment of multi-connectivity after low-activity periods, but requires more battery power, since two systems are monitored for paging and location tracking.

State transitions can be configured and optimized for different service characteristics. The applications or services transmitting small packets frequently can be configured with different state transition characteristics compared to applications transmitting and receiving large volumes of data. The UE with small data packets can send it as a multiplexed payload of the RRC connection resume request message, since the security context is available in the network. A UE with large data volumes should do a fast state transition to RRC Connected to benefit from maximum data throughput and low latency.

The measurement configuration reflects the characteristics of the requested service. For example, a static device or sensor can be configured to monitor neighbor cells less frequently compared to moving a UE. A UE supporting ultra-reliable communication should monitor the neighbor cells and control channels more frequently and possibly from multiple RATs to establish fast the suspended multi-connectivity at the time of connection resumption. Another example is measurement configuration and beamforming at higher frequencies, where UEs can be configured to monitor dedicated reference signals at specific beams or to measure common signals transmitted in wider beams.

In 5G, the RAN controls the UE paging and location tracking, where the gNB that terminates the CN-RAN interface operates as the paging initiator and mobility anchor. The RAN-based paging area may be configured to be small to ensure that the inter-gNB Xn interfaces are available to all the gNBs within the tracking area. In this case, the RAN-controlled location tracking and paging is mostly suitable for low-mobility UEs to avoid frequent RTA updates. For UEs that require high mobility, a larger tracking area is beneficial, and the paging can be initiated from CN. In LTE RRC Idle state, the synchronization between UE and network is not maintained. In RRC Connected Inactive state, the UL synchronization is not needed for all UEs. However, UEs requesting low-latency system access, such as in the context of industry automation, may maintain the UL synchronization. Those UEs can monitor a dedicated channel and therefore enable seamless resumption of the RRC connection for low-latency system access.

Figure 13-13 illustrates a procedure where the UE in RRC Connected state has an active connection towards the network. After detecting the inactivity, the UE requests suspension of the connection. Alternatively, the network may notice the end of incoming data and thus detect the potential inactivity period for the UE. The connection suspension using the RRC Suspend command will include the UE service-specific characteristics, which will fulfil the low-activity period requirements in terms of power consumption vs. system access latency and granularity of location tracking.

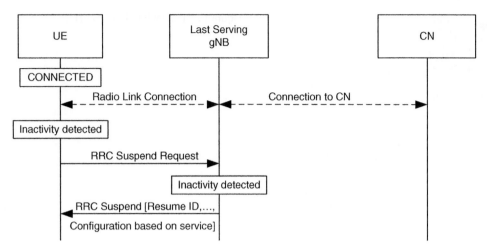

Figure 13-13. Inactivation with service specific configuration [16].

13.3.4 Paging in Connected Inactive

Locating and tracking the low-activity UEs in the RAN means that the RAN is having a distributed management function, where each cell belongs to an RAN tracking area (RTA). The cells in an RTA have information available about their adjacent RTAs in order to create and deliver up-to-date lists on allowed RTAs to the UEs that are in RRC Connected Inactive state.

In the example in Figure 13-14, the 5G gNBs serve three cells. For example, gNB1 is serving 5G cells 13, 14 and 22. Each gNB connects to a CP entity which is hosting the mobility management function in the core network by using the RAN-CN CP interface N2, and to the UP function by using the RAN-CN UP interface N3, see also Section 6.2. The neighboring gNBs are inter connected in the RAN by using the Xn interface.

When a UE is configured to enter Connected Inactive state, it receives from its serving gNB a list of cells belonging to the allowed RTAs. In Figure 13-14, the UE gets configured with RTA1 and 2, and the UE can move within cells in these RTAs during Connected Inactive state without informing the network about the performed cell reselections. The UE location is known by the RTA, thus the serving gNB does not need to perform the serving gNB change even if the UE performs cell reselection and informs the network about it. A serving gNB change can be done after the UE goes back to RRC Connected for active data transmission.

Each 5G cell advertises in broadcasted information its Cell ID, and the list of cells belonging to a RTA can be signaled to the UE using dedicated signaling. The allowed RTAs are handled transparently to the UE with assistance of the current gNB, which takes care of reporting the UE's current location to the UE's last serving gNB.

In the RRC Connected Inactive state, the CN/RAN connection, i.e. both control and user planes remain established for the UE. If the UE reselects a cell that is not in its RTA list, it initiates a RTA update procedure, as shown in Figure 13-15. After a RACH procedure, the UE sends a RTA update request message that is integrity-protected using the AS security context that is stored in the UE and the anchor gNB. The message includes shortMAC-I for UE authentication, the current UE's

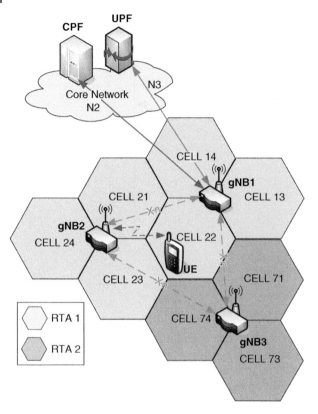

Figure 13-14. RAN Tracking Areas [16].

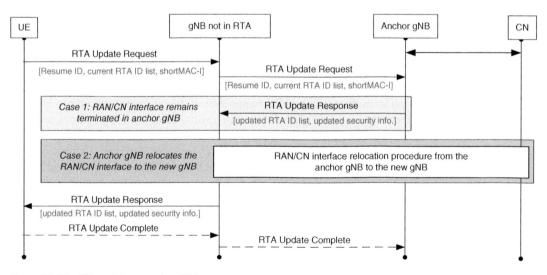

Figure 13-15. RTA update procedure [18].

RTA list and the resume ID that contains the address of the anchor gNB. If a gNB other than the anchor gNB received the RTA update request message, it forwards the message to the anchor gNB. The anchor gNB may keep the RAN/CN interface and respond with an RTA update response message (case 1) or initiate the RAN/CN interface relocation to the gNB that received the RTA update request message (case 2). The RTA update response message includes a new RTA list. After receiving the RTA update response message, the UE may send a RTA update complete message as an acknowledgement of the successful procedure.

When mobile terminated (MT) data arrives at the next generation core user plane (NGC-UP), it forwards the data to the anchor gNB of the UE. The anchor gNB buffers the received MT data and initiates the paging procedure to reach the UE, see Figure 13-16. The anchor gNB sends the paging message to all gNBs in the RTA list of the UE. The gNBs then can page the UE through the cells that are in its RTA list. Here, the anchor gNB needs to keep a gNB-gNB Xn interface relationship with all the gNBs in the RTA list of the UE.

Upon the reception of the paging message, the UE proceeds to resume its RRC connection from RRC Connected Inactive to RRC Connected state. The RRC resumption procedure may include a context fetching procedure if the UE was camping in a gNB other than the anchor gNB.

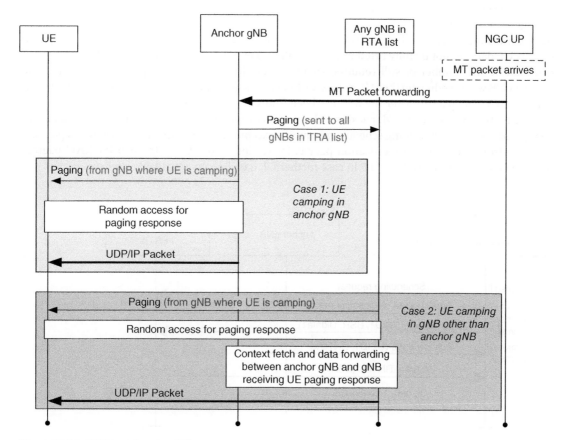

Figure 13-16. RAN-based paging [18].

13.3.5 Small Data Transmission in RRC Connected State

The new 5G radio needs to support efficient and low-latency small packet transmission. This is crucial since many applications, sensors and mMTC devices are expected to generate small packets. If the state transition is needed for every small packet, there will be high protocol overhead due to state transitions. Even if the transition from RRC Connected Inactive to RRC Connected is a light-weight signaling procedure, there is still some overhead and latency which cannot be avoided. Thus, the transmission of packets directly from RRC Connected Inactive would be beneficial for optimized small packet transmission, since the transceiver in a device is active only for a short period. When a UE wants to transmit UL data, it could use a small packet transmit procedure (SPTP). The SPTP, depending on the procedure, may involve just a single contention-based transmission (one step) or may be preceded by a contention-based scheduling request and grant (two step) [19]. The SPTP resembles a standard RACH operation, but allows user data transfer already during scheduling request and scheduling grant. This procedure can support many concurrent and HARQ-enabled transmissions of variable-sized packets and with minimal control overhead. If there is a downlink packet for the UE, RAN-based paging [16][20] can be done, and this in turn triggers the UE to initiate the SPTP.

When the UE has been configured for small data transmission during low-activity periods, a UE in RRC Connected Inactive state can move without any mobility-related signaling as long as its SPTP configuration is usable, possibly restricted to a set of cell IDs or tracking area IDs. In Figure 13-17, the UE transmits its small UL data to any gNB in the RAN-based tracking area (RTA) in SPmsg3. In case its UL data cannot be fully fitted into the small packet block reserved to it with SPmsg2, the UE can indicate the number of still required small packet blocks or the transport block length. The current gNB will schedule further UL grants for RNTI1x, which allow the UE to transmit the remainder of the UL data packet. After the BSR has been fully consumed, gNB1 stops providing UL grants. The UE is still able to receive DL packets as long as its assigned RNTI1x is valid, but the UE has no configured resources for sending scheduling requests, except those for SPmsg1. Accordingly, a possible response from the UE's communication peer is forwarded to the UE without prior RAN paging, while the UE again utilizes the SPTP in case further UL data becomes available later on.

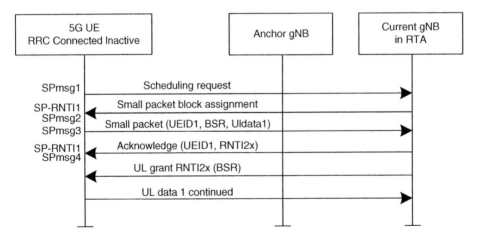

Figure 13-17. UE in RRC Connected Inactive state transmits small packet in UL data.

In Figure 13-17, it is assumed that the LTE security architecture is retained, and only the anchor gNB is able to decipher and possibly check the received data packet for integrity. Accordingly, the current gNB forwards the ciphered ULdata1 packet to the anchor gNB and receives the deciphered (and decompressed) ULdata1 packet back in return. The current gNB can take the role of the new anchor gNB and update the downlink path towards the UE and trigger the mobility management procedures.

13.4 Mobility

13.4.1 Introduction

As discussed earlier in Section 2.3, the reliability and interruption delay requirements for some 5G services are significantly more stringent compared to 4G. In addition to this, 5G is expected to operate in a wider range of frequencies (1-100 GHz) than 4G. This also means that beamforming techniques may be needed to compensate for the higher propagation loss at high frequencies. This section presents several different methods for addressing the 5G active mobility requirements. One of the agreements made so far is that 3GPP "will aim to define HO for NR with an interruption as close to zero as possible while only having single Tx/Rx in the UE, and 0 ms interruption at least for the case that the UE supports simultaneous Tx/Rx with source cell and target cell during HO" [21]. This is a major improvement compared to LTE, which always involves a data interruption during handover. Basically, there are two different alternatives to solve this:

- Multi connectivity with role switch;
- Normal handover with enhancement.

In this section, both alternatives are investigated. Another important new topic will be mobility in the context of beamforming. Due to the requirements of a lean design of the 5G DL reference symbols, the UL measurements can complement the DL measurements for beamforming mobility. One method to perform UL measurements for beamforming mobility is described in Section 13.4.2. The beamforming mobility design should support a fast switching and tracking of the communication beam to combat rapid changes in link quality. Also, the design should be able to exploit the availability of multiple overlapping beams that can be used for the communication with a single UE. One solution to fulfil these requirements is a multi-connectivity solution called cluster-based mobility, see Section 13.4.3, which is a set of nodes that the UE can detect and which are prepared in advance for a fast re-routing of the signaling and user data. Section 13.4.4 discusses ways to minimize signaling for multi-connectivity.

Enhanced normal handover is treated in Sections 13.4.5 and 13.4.6. More precisely, Section 13.4.5 presents ways to avoid data interruption at a handover, and Section 13.4.6 analyzes ways to improve the performance during high velocity by improving the channel state information.

13.4.2 Mobility Management via UL-based Measurements

Next generation cellular systems must provide a mechanism by which user equipments (UEs) and mmWave gNBs establish highly-directional transmission links, typically formed with high-dimensional phased arrays, to benefit from the resulting beamforming (BF) gain and to compensate for the increased pathloss experienced at high frequencies, as detailed in Section 11.5. In this context, directionality requires a fine alignment of the transmitter and the receiver directional paths, an operation which might dramatically increase the time it takes to access the network. Moreover, the dynamics of the mmWave channel imply that the directional path to any cell can deteriorate rapidly, necessitating the need for intensive tracking of the mobile terminal [22].

Therefore, periodical monitoring of the channel quality between each UE and mmWave gNB pair, in order to perform a variety of control tasks (including handover, path selection, radio link failure detection and recovery, beam adaptation, etc.), is fundamental to provide efficient mobility management schemes. However, while channel tracking and reporting is relatively straightforward in cellular systems at conventional frequencies, the mmWave bands present several significant limitations: (i) the high variability of the channel in each link due to blockage; (ii) the need to track multiple directions for each link; and (iii) reports from the UE back to the cells must be made directional.

To address these challenges, a novel multi-connectivity UL measurement framework has been proposed, as depicted in Figure 13-18, that, with the joint effort of the legacy LTE frequencies, enables fast, fair and robust cell selection [23].

Unlike in traditional LTE schemes, the proposed framework is based on the channel quality of uplink rather than DL signals. This eliminates the need for the UE to send measurement reports back to the network and thereby removes a possible point of failure in the control signaling path. Moreover, if digital beamforming or beamforming with multiple analog streams is available at the mmWave cell, the directional scan time can be dramatically reduced when using UL-based measurements. Finally, mobile terminals are the most energy-constrained network entities, due to their limited battery capacity, contrary to the BS nodes which are always power-connected and do not suffer from strict energy requirements. Therefore, a UL measurement framework, in which digital beamforming, being the most power-consuming beamforming architecture, is used at the gNB side, should be preferred to enable a more efficient mobility management scheme.

In detail, each UE directionally broadcasts a sounding reference signal (SRS) in a time-varying direction that continuously sweeps the angular space. Each potential serving cell scans all its angular directions and monitors the strength of the received SRS, building a report table (RT) based on the channel quality of each receiving direction, to capture the dynamics of the channel. Once the RT of each mmWave gNB has been filled for each UE, each mmWave cell sends this information to a centralized coordinator, for instance residing in the LTE eNB, which, due to the knowledge gathered on the signal quality in each angular direction for each gNB-UE pair, obtains complete directional knowledge over the cell it controls. Hence, it is able to match the beams of the transmitters and the receivers to provide maximum performance.

Therefore, the coordinator reports to the UE, on a legacy LTE connection, which mmWave gNB yields the best performance, together with the optimal direction in which the UE should steer its beam to reach the candidate serving mmWave gNB in the optimal way. The choice of using the LTE control link is motivated by the fact that the UE may not be able to receive from the optimal mmWave link if not properly configured and aligned. Moreover, since path switches and cell additions in the mmWave regime are common due to link failures, the control link to the serving mmWave cell may not be available either. Finally, the coordinator notifies the designated mmWave gNB, through a high-capacity backhaul link, about the optimal direction in which to steer the beam, for serving each UE.

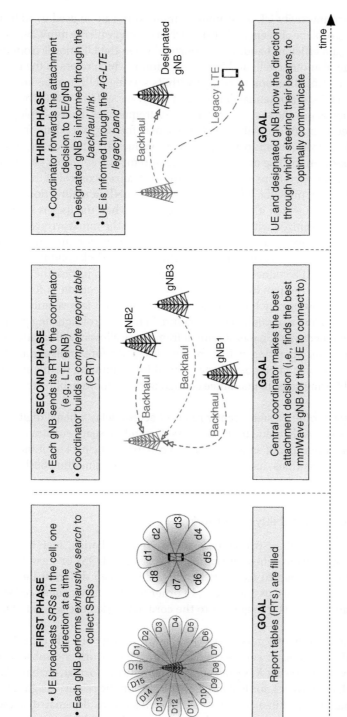

Figure 13-18. Slot scheme for the presented multi-connectivity uplink measurement framework. Green and red dashed lines refer to the control messages exchanged via the legacy communication link and the high-capacity backhaul connections, respectively.

The proposed multi-connectivity UL measurement framework can be used to address some of the most important 5G control plane challenges that arise when dealing with the mmWave frequency bands.

In particular, robust and efficient handover (performed when the UE moves from the coverage of one cell to the coverage of another cell) and beam adaptation (which refers to the need for a user to periodically adapt its steering direction to realign with its serving gNB) can be improved. Frequent handover, even for fixed UEs, is a potential drawback of mmWave systems due to their vulnerability to random obstacles, which is not the case for lower carrier frequencies. Dense deployments of short-range gNBs, as foreseen in mmWave cellular networks, may further exacerbate frequent handovers between adjacent gNBs. A loss of beamforming information due to channel change is another reason for handover and re-association. The presented UL measurement framework can ensure more efficient mobility management operations by exploiting the centralized control of the LTE eNB over the network, to periodically determine the UE's optimal mmWave gNB (and direction) to associate with, or the new direction through which it should steer the beam, when the user is in connected mode, i.e., when it is already synchronized with both the LTE eNB and the mmWave gNBs. The use of both the sub-6 GHz and the mmWave control planes is a key functionality for efficient handover management. In fact, especially when considering highly unstable mmWave link conditions or initially scarce mmWave deployments, the LTE connectivity ensures a ready backup in case the mmWave links suffer an outage. Furthermore, the handover and beam adaptation decision is forwarded to the UE through the controller, whose legacy link is much more robust and less volatile than its mmWave counterpart, thereby removing a possible point of failure in the control signaling path [24].

In addition, a multi-connectivity UL measurement framework allows the final attachment decision to be made by the controller operating at LTE frequencies. Therefore, unlike in traditional attachment policies, the association can be possibly performed by accounting for the instantaneous load conditions of the neighboring cells, to guarantee enough fairness and reliability to the whole cellular network.

Moreover, an UL design offers a significantly reduced access delay when a digital architecture is preferred. The main reason is that, due to the BS's less demanding space constraints with respect to a mobile terminal, a larger number of antenna elements can usually be packed at the gNB side, resulting in a larger number of directions that can potentially be scanned simultaneously through a digital beamforming scheme.

13.4.3 Cluster-based Beam Mobility Framework

5G communications can be outage-prone for higher frequency bands, such as mmWave bands. In these bands, communication relies on strong line-of-sight (LOS) or near-LOS components via beamforming. This may lead to frequent handovers between different beams or cells that support a specific location, as discussed before. In order to support frequent cell switching in an agile way, and in a manner that is transparent to the core network, access point clustering can be instrumental.

A cluster is defined as a group of access points (APs) in the vicinity of a UE, capable of serving that UE. The APs included in the cluster can be configured by the network and subsequently reconfigured when the UE moves. Clustering APs can happen in a multi-tier format, enabling better exploitation of environment characteristics, for instance being based on the relative height of the APs.

To coordinate the mobility within a cluster, one of the APs can be designated as the cluster head (CH), which is connected to the core network through the N2 and N3 interfaces [25], see also

Section 6.2. The CH is also connected to all other APs in the cluster. To enable the CH to coordinate the inter-AP switching in a fast and efficient manner, it is assumed that a limited number of hops exist between the CH and the APs in the cluster. Depending on the topology of the network, the capacity and latency of the transport links, and the network position and role of the CH may vary.

The functional split for a cluster with sufficiently well-dimensioned transport links can be made quite similar to the functional split of coordinated multi-point systems. Here, all the intelligence may be located in a central node (i.e., the CH), which is responsible for all control and user plane protocol handling, including how the mmWave beams are tracked at different APs. The CH decides which APs serve the UE or which APs stay in stand-by mode.

For non-ideal transport or wireless (self-)backhauling cases, see Section 7.4, the CH can handle Packet Data Convergence Protocol (PDCP)-level functionalities, whereas other cluster APs handle the Radio Link Control (RLC), Media Access Control (MAC) and PHY layers, corresponding to the 3GPP split option 2 [26], as detailed in Section 6.6. This way, it is possible to maintain a single PDCP entity in one gNB, while switching the RLC, MAC and PHY from one transmission and reception point (TRP) to another. This protocol split also allows for arranging UE-specific BS clusters [27], as opposed to static clusters in the case of a higher centralization of functionality.

Intra-AP beam switching is triggered by the UE measurement feedback to the AP. In case of inter-AP beam switching, the measurement report will be forwarded to the CH from the current serving AP. The CH will request the target AP for beam switching. If positive feedback is received from the target AP, the UE will eventually be informed (via the CH and the serving AP) to switch its beam, and will be served via the target AP after the switch.

The high path loss, and high susceptibility to blockage are two main factors affecting the mmWave systems' coverage, especially outdoors. Densely deployed mmWave small cells with multi-node coordination seem to be a feasible solution to both these issues. The use of coordinated BSs to enhance the data rate and coverage of the network has been widely studied in the context of 4G LTE/LTE-A networks. Various techniques, e.g., joint transmission, coordinated beamforming, and cooperative communication, have been considered.

The height of the APs has a significant impact on the probability of LOS links, as shown in Figure 13-19 a) below, where the LOS link between a low-rise AP and the UE is blocked, but a LOS link between a high-rise AP and the UE is still available.

Here, we target UEs requiring immersive 5G experiences and thus demanding extremely high data rates. In addition to the mmWave APs installed to street furniture with relatively low height, a small group of significantly simplified APs with beamforming capability are installed on high buildings to provide a beam resource pool (BRP) as shown in Figure 13-19 b). The high-rise APs can be implemented as remote radio heads (RRHs) in order to reduce the complexity and cost. When the LOS links between low-rise APs and those UEs with extremely high data rate requirements are blocked, the beams in the BRP can be used to establish a LOS connection. Therefore, consistent user experiences without interruption can be supported by this two-tier deployment. In Table 13-4, it is shown how many high-rise APs (30 m high) need to be installed together with each 100 low-rise APs (3 m high) for a target overall LOS probability of 95%, based on the results in [28]. It can be seen that the LOS probability can be significantly enhanced with only a few high-rise APs.

The high- and low-rise APs can be connected to switches via high-rate and low-latency fibre connections. A CH is connected to all high- and low-rise APs via a switch. The functionalities of the CH can be integrated into low-rise APs so that each low-rise AP is connected to the high-rise APs and acquires beam resources from the BRP when needed.

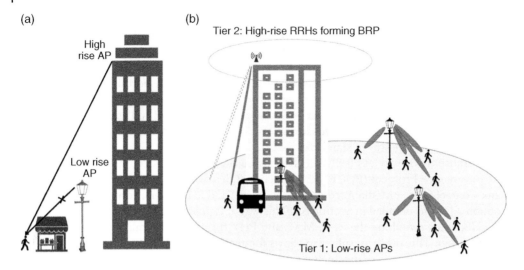

Figure 13-19. a) High- and low-rise APs, b) two-tier deployment.

In both architectures, macro cells operating below 6 GHz might be potentially available and offer additional help, but are not essential to the proposed architecture. Two possible system architectures are illustrated in Figure 13-20.

The UE context will be saved and updated in both the UE-associated low-rise APs and the CH. Based on the UE context, the CH maintains a list of candidate beams from the pool in a dynamic manner. Once a low-rise AP detects that its associated UE is experiencing a low data rate because of LOS link blockage, it sends the request to the cluster head to check if the available beam list is empty or not. If there are no available high-rise beams, the low-rise AP continues the transmission to the UE, but with degraded performance. Otherwise, the CH sends the list to the low-rise AP.

It should be noted that even if the list is not empty, the CH should only choose from those high-rise beams where the associated high-rise RRHs are close enough to the target UE, so that the signal strength can be kept at a high level. In the meantime, the CH also requests the available high-rise APs in the list to measure the pathloss to the target UE to select the best high-rise AP. If the best high-rise AP significantly enhances the SINR of the target UE, the CH controls the switch to connect the low-rise AP to the selected high-rise AP.

Meanwhile, the low-rise APs could still maintain the uplink channel between themselves and the UE, since, for 5G immersive experiences, the very high data rate required is normally only on the DL, and the UL connection could still be sufficient even if it does not experience LOS. Another reason that the original UL connection should be maintained is that the low-rise APs must monitor the UL channel and measure the signal strength of the received signal. Once it is above a certain threshold, e.g. if LOS is available to the low-rise APs again, the transmission will go back to the low-rise APs and the high-rise APs will be released back to the pool. Note that the DL and UL connections could be asymmetric as described above.

There is no need to do handover between low- and high-rise APs. Local traffic steering can easily be employed in the proposed framework to steer traffic from the low-rise AP to high-rise AP once the LOS link of a low-rise AP is lost. Consequently, the signaling procedure can be significantly reduced.

Table 13-4. Number of high rise APs.

Blocking building height [m]	Probability of LOS with low-rise APs only	Number of high-rise APs	Probability of LOS with joint low-/high-rise AP deployment
3	0.93	1	0.9990
5	0.44	3	0.9578
10	0.02	25	0.9636
15	0.0005	100	0.9513

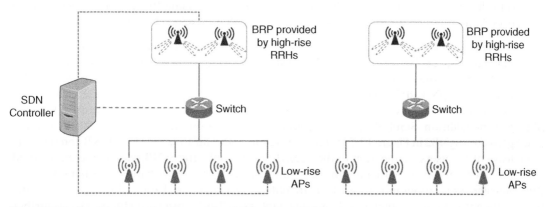

Figure 13-20. Possible system architectures involving cluster heads.

Another option is to integrate the CH with the low-rise APs as illustrated in Figure 13-20, so that each of the low-rise APs can control the switch to select candidate RRHs or beams from the BRP. The operation procedure is similar to the previous case, but the functionalities of the CH are now performed by the low-rise APs instead. The benefit of such architecture is that the signaling procedure can be further simplified. However, the signaling simplification comes with a cost: If each low-rise AP must be able to perform CH functionalities, the complexity of the low-rise APs is increased, whereas in the architecture with a dedicated CH, only the CH node needs to have the capability to perform the relevant functions.

13.4.4 Partly UE-autonomous Cell Management for Multi-Connectivity Cases

One of the challenges for UEs in multi-node connectivity (or multi-cell connectivity, carrier aggregation, etc.) is that multiple cell associations need to be managed. A fully network-controlled connectivity management approach will lead to increases in RRC and backhaul signaling if adopted. For multi-connectivity scenarios with the primary cell on the macro layer and the secondary cell on the small cell layer, numerous studies have shown that the mentioned signaling overhead is dominated by cell management events for the small cell layer, see e.g. [29][30][31].

It is therefore proposed to adopt a UE-autonomous cell management approach for the small cell layer for cases with multi-connectivity [32], while the primary cell management (handover) on the macro layer is still fully network-controlled. Since primary cell management actions happen less frequently, and typically require interaction with the core network, such actions are assumed to continue to be fully network-controlled and UE-assisted [30]. The fundamental principles of the UE-autonomous secondary cell management proposal can be summarized as follows:

1) The network prepares a set of small cells among which the UE is allowed to perform autonomous secondary cell addition and removal, and cell change;
2) The network signals a list of these prepared small cells to the UE, including the measurement that it shall use for performing autonomous secondary cell addition, removal, and change;
3) Once the UE fulfils the criteria for secondary cell addition, removal, or change for the prepared small cells, it takes the corresponding action without first sending RRM measurements to the network, and waiting for the corresponding RRC signaling messaged in the downlink.

The basic principles of UE-autonomous secondary cell management actions are further exemplified in Figure 13-21. Here, the notation of master gNB (MgNB) for the macro layer and secondary gNB (SgNB) for the small cell layer is adopted, similar as in LTE-Advanced. The procedure for UE-autonomous SgNB addition simply means that the UE sends a RA message to that cell when it wants to add this link (i.e., without involving RRC signaling). The involved SgNB informs the UE's MgNB of the addition. Similarly, for the SgNB change operation, the UE only sends a RA to the new SgNB (i.e., SgNB2 in Figure 13-21). The SgNB2 thereafter informs the UEs MgNB and requests the release of the users' connection to the SgNB1. In principle, the SgNB release operation could also have been conducted with a RA message, but it is most efficiently handled with a scheduled RRC message, as pictured in Figure 13-21. In addition, signaling is also required for configuring the UEs for autonomous secondary cell management, as well as for preparing the SgNBs, as further detailed in [29].

The performance of UE-autonomous secondary cell management has been evaluated by means of extensive system level simulations. For the highway scenario studied in [29], it is found that the average number of required RRC messages per UE per second is reduced from 4.9 to 0.35 by using UE-autonomous secondary cell management. Similarly, the associated Xn signaling is reduced by approximately 50%. See also [33] for results on the data interruption time for multi-node connectivity.

13.4.5 Enhanced Synchronous Handover without Random Access

The basic mobility functionality for LTE is based on UE-autonomous cell re-selections for RRC Idle mode and network-controlled handovers—with UE-assistance—for RRC Connected state. Numerous measurements from live LTE networks have confirmed that the mobility performance is generally good, observing close to 100% handover success rates and low percentages of ping-pongs, i.e. undesirable handovers between pair of cells within a short time [34][35]. However, due to the asynchronous nature of the LTE handover functionality with random access (RA) at every cell change, there is an undesirable temporary data interruption gap at every handover. Field measurements reveal that the data interruption time at each handover ranges from, at best, 20-30 ms and up to 100 ms (or even more) for some networks [34][35]. To overcome this problem for 5G, it is proposed to adopt a time-synchronized and RA-less

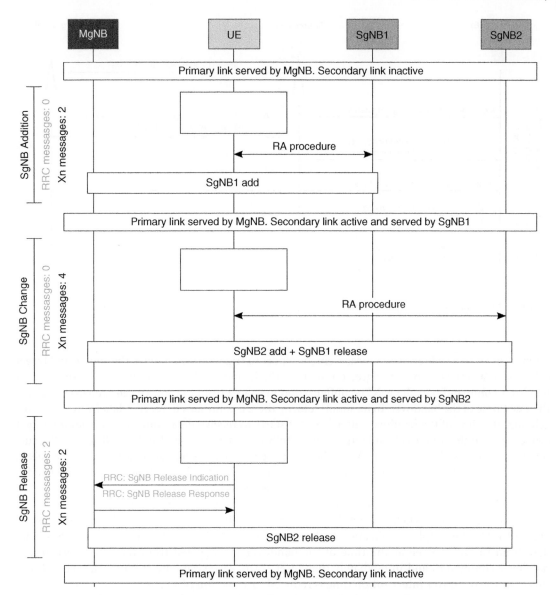

Figure 13-21. Summary of signaling procedure for UE-autonomous SgNB addition, change, and release.

handover functionality [36]. Similar solutions have also been discussed in 3GPP, see for example [37] and [38], but no decision has been taken at this moment. This is a natural evolution as many other features will require time-synchronized BSs for the 5G-era.

Following [39], the basic principle of RRC Connected mode synchronous RA-less handover between a source and a target cell is illustrated in Figure 13-22.

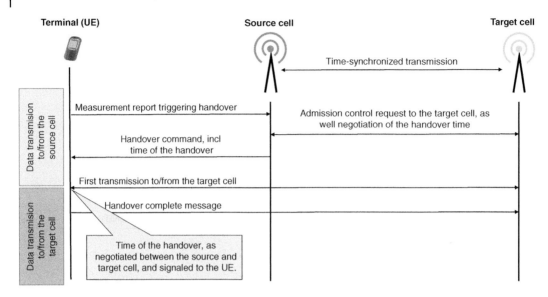

Figure 13-22. Signaling flow diagram illustrating the basic principles of synchronized RA-less handover.

Once the source cell receives a measurement report to trigger the handover, the source and target cells exchange information and agree on the time the handover should take place. The subsequent handover command (i.e., the RRC reconfiguration message) from the source cell informs the UE of the exact time of the handover. In its simplest form, the UE continues to receive data from the source cell until the time of the handover, after which it starts to receive data immediately from the target cell. Given that the source cell and target cell are fully time-synchronized and the UE knows the current value of the timing advance (TA) for the source cell, the UE is capable of measuring the received time-offset from the two cells and compute the TA for the target cell, as outlined in [36]. The handover command from the source cell may also include an uplink pre-scheduling command for the UE to immediately transmit in the uplink towards the target at the time of the handover.

The timing of the synchronized RA-less handover functionality is further illustrated in Figure 13-23. The simplest realization is shown in Figure 13-23 a), where the UE stops receiving data from the source cell at the time of the handover (as signaled from the source cell in the handover command), and immediately thereafter starts receiving data from the target cell. For this case, the handover interruption time is reduced to a fraction of a subframe (or TTI), accounting for the received time differences of the signals from the two cells and potential UE processing times for performing the switch from source to target. In a more advanced version of the synchronous RA-less handover functionality, the UE continues to listen to the source cell for a short time period, while in parallel also receiving data from the target cell. For the latter case, there is no data interruption time on the physical layer between the source and target cell. During the time where the UE receives data from both cells, those data packets could be the same (i.e., duplicated) or different data packets. It should also be noticed that for achieving a virtually zero data interruption time during handovers, there needs to be corresponding network support for fast and efficient data forwarding and flow switching between the two involved cells. In [40], more

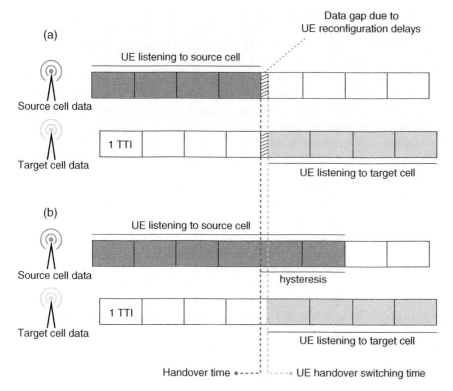

Figure 13-23. Simple illustration of the timing diagram for synchronous RA-less handover: (a) case where the UE receives data only from a single cell at time, (b) option with hysteresis time where the UE receives data from both cells.

details related to this are reported, where the UE processing times, BS processing times, and backhaul signaling latencies between the source and target cell are taken into account.

The benefit of the proposed synchronous RA-less handover functionality is a significant reduction of the data interruption time during handovers, approaching virtually zero. Moreover, the handover execution process is reduced compared to that of LTE, as there is no RA at the target cell at every handover. Measurements from LTE networks shows that the RA procedure typically takes on the order of ~10 ms [37]. The faster handover execution process translates to increased mobility robustness since the system is able to react faster, resulting in even lower handover failure probabilities [36]. Finally, the fact that the handover process no longer requires RA translates to savings in the required RA access resources for the system. For example, the handover rate per UE is found to typically vary from few handovers per minute to handovers every second depending on the network topology and velocity of the device [33][35][41][42].

13.4.6 RAN Design to Support CSI Acquisition for High-Mobility Users

Channel state information (CSI) at transmitters is fundamental in many advanced transmission schemes. However, feedback delays in FDD, framing delays in TDD and transmission control delays of multiple milliseconds result in severe outdating of this information for terminals at vehicular velocities. Backhaul delays increase the problem when using coordinated multi-point transmission (CoMP). Channel prediction based on extrapolation of the short-term fading has proven to be inadequate at vehicular velocities and high carrier frequencies.

For this reason, a new scheme was proposed in [43], which may radically extend the prediction horizon when used on vehicles, and which is based on the usage of an additional antenna, a "predictor antenna", placed in front of the transmission antennas in the direction of travel. This approach can provide an order-of-magnitude improvement in channel prediction performance compared to Kalman or Wiener-extrapolation of previous measurements based on the channel statistics. The principle can be used to improve downlink transmissions that require CSI at transmitters (CSIT) in FDD as well as TDD systems.

There have been investigations in [43] on how different types of antenna designs on vehicles affect the attainable cross-correlation between the channel measured by the forward predictor antenna and the channel later experienced by the main antenna, when it has moved to the position previously occupied by the predictor antenna. The attainable precision in the prediction of complex channel gains is directly related to this cross-correlation. The use of monopole antennas placed on flat and uncluttered vehicle roofs was shown to provide the highest correlations, resulting in average cross-correlations of 0.97-0.98 for antenna separations of 0.5-3 wavelengths, as shown in Figure 13-24.

Those experimental results were based on measured data obtained in cooperation with TU Dresden, in their testbed using a 20 MHz OFDM signal working at 2.68 GHz, at 50 km/h vehicle

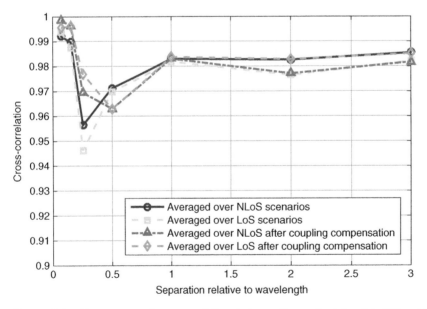

Figure 13-24. Mean measured cross-correlations between the received signals in LOS and NLOS scenarios, as a function of the spacing between the forward predictor antenna and a rearward main antenna on the roof of a vehicle moving at 50 km/h. Results are shown without and with the use of a pre-compensator of the mutual electromagnetic coupling between the antennas.

velocity. The useful antenna correlation is however reduced due to mutual electromagnetic couplings with very closely spaced antennas, around 0.25 wavelengths apart. In [44], a simple and efficient scheme has been found that counteracts this effect. It is based on using an open-circuit antenna decoupling method. The effect on the average antenna correlation of using such a pre-compensator is also shown in Figure 13-24.

Figure 13-25 shows the attained normalized mean square prediction error (NMSE) for this case (two monopole antennas on a flat vehicle roof), when predicting the channels for 10 kHz wide OFDM subcarriers at 2.68 GHz, at 45–50 km/h vehicle velocities. The statistics summarize the variability of the prediction accuracy over 1958 subcarriers, for different settings of the spacing between the prediction antenna and the main antenna. The statistics are collected over all predicted subcarriers and separate measurements.

The benefits of the predictor antenna concept to alleviate beamforming mispointing in massive MIMO backhauling of high speed vehicles, even when local scattering and multipath propagation around the vehicle generate very fast fading, was discussed and demonstrated in [45] and [46]. Additional application areas that would benefit from CSIT are coordinated multi-point (CoMP) with soft handover and robust backhauling for delay and/or mission critical services to fast moving vehicles.

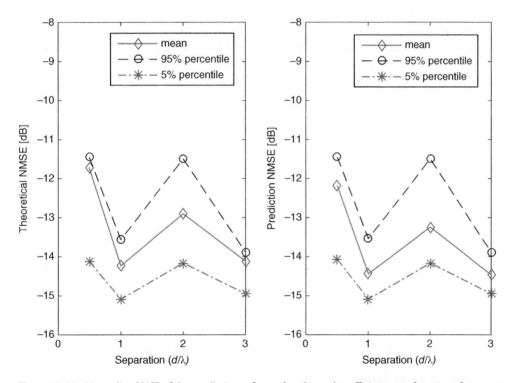

Figure 13-25. Normalized MSE of the predictions of complex channel coefficients as a function of antenna separation, using the predictor antenna scheme – Left: theoretical limits for the NMSE, calculated from the measured correlations between the two antenna signals. Right: corresponding measured prediction performance.

To summarize, the predictor antenna concept has been shown to be feasible, and to be able to provide accurate channel state information for large prediction horizons in time, corresponding to multiple wavelength distances in space.

13.5 Summary and Outlook

This chapter provided an introduction to the latest research on 5G (NR) initial access, RRC state handling and mobility. The 3GPP specification for NR at the time of writing of this book is under development, with a good overview available in [47] and the ultimate specification expected to be available under [48]. Nevertheless, various decisions on NR have already been taken, and other novel concepts as described in this chapter have been proposed and are likely to eventually be standardized, possibly in some modified form.

For instance, the NR initial access will likely be extended with support for beamforming, which will play an important role for NR especially at higher frequencies. In addition, ways to prioritize between different services already at initial system access may be introduced. An energy-efficient and lean design of NR system information transmission is also important for various 5G use cases, and was hence also covered in detail in this chapter.

A novel aspect in 5G will be the introduction of a new RRC state that will allow for a longer device battery life-time and also a faster switch to RRC Connected mode than the transition from RRC Idle to RRC Connected. Further, mobility in NR will be improved with the tighter integration with LTE. Work is also ongoing to improve mobility in general in 3GPP by using, e.g., UL measurements and RACH-free handover. Finally, NR multi-connectivity together with beamforming mobility will give almost seamless connections and improved user experience.

References

1 M. Ericson, "Total Network Base Station Energy Cost vs. Deployment", IEEE Vehicular Technology Conference (VTC Spring 2011), Sept. 2011
2 3GPP TR 38.912, "Study on New Radio (NR) access technology", June 2017
3 3GPP TS 38.211, "NR; Physical channels and modulation", V15.0.0, Dec. 2017
4 3GPP R2-168858, "Text Proposal to TR 38.804 on on-demand SI provisioning for NR", NTT DOCOMO, Nov. 2016
5 B. Debaillie, C. Desset and F. Louagie, "A Flexible and Future-Proof Power Model for Cellular Base Stations", IEEE Vehicular Technology Conference (VTC Spring 2015), May 2015
6 3GPP R1-1708890, Final Report of 3GPP TSG RAN WG1 #88bis, Apr. 2017
7 5G PPP METIS-II project, Deliverable D6.2, "5G Asynchronous Control Functions and Overall Control Plane Design", Apr. 2017
8 3GPP TR 38.802, "Study on New Radio Access Technology, Physical Layer Aspects (Release 14)", V14.1.0, June 2017
9 3GPP RAN1 #88, "Chairman meeting notes", Feb. 2017
10 3GPP RAN1 #89, "Chairman meeting notes", May 2017

11 C. Nicolas Barati, S. Amir Hosseini, Sundeep Rangan, Pei Liu, Thanasis Korakis, Shivendra S. Panwar and Theodore S. Rappaport, "Directional Cell Discovery in Millimeter Wave Cellular Networks", IEEE Transactions on Wireless Communications, vol. 14, no. 12, pp. 6664–6678, Dec. 2015

12 C. N. Barati, S. A. Hosseini, M. Mezzavilla, T. Korakis, S. S. Panwar, S. Rangan, and M. Zorzi, "Initial Access in Millimeter Wave Cellular Systems", IEEE Transactions on Wireless Communications, vol. 15, no. 12, pp. 7926–7940, Dec. 2016

13 H. Guo, B. Makki, and T. Svensson, "A genetic algorithm-based beamforming approach for delay-constrained networks", International Symposium on Modeling and Optimization in Mobile, Ad Hoc, and Wireless Networks (WiOpt 2017), May 2017

14 K. Chatzikokolakis, A. Kaloxylos, P. Spapis et al., "On the Way to Massive Access in 5G: Challenges and Solutions for Massive Machine Communications", EAI International Conference on Cognitive Radio Oriented Wireless Networks (CrownCom 2015), Apr. 2015

15 I. Da Silva, G. Mildh, M. Säily and S. Hailu, "A Novel State Model for 5G Radio Access Networks", IEEE International Conference on Communications (ICC 2016), Workshop on 5G RAN Design, May 2016

16 5G PPP METIS-II project, Deliverable D6.1, "Draft Asynchronous Control Functions and Overall Control Plane Design", July 2016

17 S. Hailu, M. Säily and O. Tirkkonen, "Towards a configurable state model for 5G radio access networks", Global Wireless Summit 2016, Nov. 2016

18 S. Hailu and M. Säily, "Hybrid paging and location tracking scheme for inactive 5G UEs", European Conference on Networks and Communications (EuCNC 2017), May 2017

19 5G PPP FANTASTIC-5G project, Deliverable D4.1, "Technical Results for Service Specific Multi-Node/Multi-Antenna Solutions", Mar. 2016

20 D. Aziz, H. Bakker, A. Ambrosy and Q. Liao, "Signalling Minimization Framework for Short Data Packet Transmission in 5G", IEEE Vehicular Technology Conference (VTC Fall 2016), Sept. 2016

21 3GPP R2-1702451, "Report of 3GPP TSG RAN WG2 meeting #97", Apr. 2017

22 M. Giordani and M. Zorzi, "Analysis of the User Tracking Performance in 5G Millimeter Wave Mobile Networks", Annual Mediterranean Ad Hoc Networking Workshop (Med-Hoc-Net'17), June 2017

23 M. Giordani, M. Mezzavilla, S. Rangan and M. Zorzi, "Multi-Connectivity in 5G mmWave cellular networks", Annual Mediterranean Ad Hoc Networking Workshop (Med-Hoc-Net'16), June 2016

24 M. Polese, M. Giordani, M. Mezzavilla, S. Rangan and M. Zorzi, "Improved Handover Through Dual Connectivity in 5G mmWave Mobile Networks", IEEE Journal on Selected Areas in Communications (JSAC), vol. 35, no. 9, pp. 2069 – 2084, June 2017

25 M. Shariat et al., "5G Radio Access above 6 GHz", Transactions on Emerging Telecommunications Technologies, Wiley, vol. 27, no. 9, pp. 1160–1167, 2016

26 3GPP 38.801, "Study on new radio access technology: Radio access architecture and interfaces", Apr. 2017

27 5G PPP mmMAGIC project, Deliverable D3.2, "Evaluations of the concepts for the 5G architecture and integration", June 2017

28 Y. Qi, M. Hunukumbure and Y. Wang, "Millimeter Wave LOS Coverage Enhancements with Coordinated High-Rise Access Points", IEEE Wireless Communications and Networking Conference (WCNC 2017), Mar. 2017

29 L.C. Giménez, P. H. Michaelsen and K. I. Pedersen, "UE Autonomous Cell Management in a High-Speed Scenario with Dual Connectivity", IEEE International Symposium on Personal, Indoor and Mobile Radio Communications (PIMRC 2016), Sept. 2016

30 K.I. Pedersen, S. Barbera, P-H. Michaelsen and C. Rosa, "Mobility Enhancements for LTE-Advanced Multilayer Networks with Inter-Site Carrier Aggregation", IEEE Communications Magazine, vol. 51, no. 5, pp. 64 – 71, May 2013

31 3GPP R2-132339, "Autonomous SCell Management for Dual Connectivity Cases", NSN, Nokia Corporation, Aug. 2013

32 5G PPP FANTASTIC-5G project, Deliverable D4.2, "Final results for the flexible 5G air interface multi-node/multi-antenna solution", May 2017

33 L. C. Gimenez, P. H. Michaelsen and K. I. Pedersen, "Analysis of data interruption in an LTE highway scenario with dual connectivity", IEEE Vehicular Technology Conference (VTC Spring 2016), May 2016

34 A, Elnasher and M.A. El-Saidny, "Looking at LTE in Practice: A performance analysis of the LTE system based on field results", IEEE Vehicular Technology Magazine, vol. 8, no. 3, pp. 81–92, Sept. 2013

35 L.C. Gimenez, M. Carmela; M. Stefan, K.I. Pedersen and A. Cattoni, "Mobility Performance in Slow- and High-Speed LTE Real Scenarios", IEEE Vehicular Technology Conference (VTC Spring 2016), May 2016

36 S. Barbera, K.I. Pedersen, C. Rosa, P.H. Michaelsen, F. Frederiksen, E. Shah and A. Baumgartner, "Synchronized RACH-less Handover Solution for LTE Heterogeneous Networks", IEEE International Symposium on Wireless Communication Systems (ISWCS 2015), Aug. 2015

37 3GPP TR 36.881, "Study on latency reduction techniques for LTE", V.14.0.0, June 2016

38 3GPP R2-1706626, "RACH-less HO in NR when UE is in CA or DC", June 2017

39 5G PPP FANTASTIC-5G project, Deliverable D4.2, "Final results for the flexible 5G air interface multi-node/multi-antenna solution", May 2017

40 L.C. Gimenez, P-H. Michaelsen, K.I. Pedersen and T.E. Kolding, "Towards Zero Data Interruption Time with Enhanced Synchronous Handover", IEEE Vehicular Technology Conference (VTC Spring 2017), June 2017

41 S. Barbera, P. Michaelsen, M. Saily and K.I. Pedersen, "Improved Mobility Performance in LTE Co-Channel HetNets Through Speed Differentiated Enhancements", IEEE Global Communications Conference (GLOBECOM 2012), Dec. 2012

42 S. Barbera, P.H. Michaelsen, M. Saily and K.I. Pedersen, "Mobility Performance of LTE Co-Channel Deployment of Macro and Pico Cells", IEEE Wireless Communications and Networking Conference (WCNC 2012), Apr. 2012

43 M. Sternad, M. Grieger, R. Apelfröjd, T. Svensson, D. Aronsson and A. B. Martinez, "Using Predictor Antennas for Long-Range Prediction of Fast Fading for Moving Relays", IEEE Wireless Communications and Networking Conference (WCNC 2012), 4G Mobile Radio Access Networks Workshop, Apr. 2012

44 N. Jamaly, R. Apelfröjd, A. Belen Martinez, M. Grieger, T. Svensson, M. Sternad and G. Fettweis, "Analysis and Measurement of Multiple Antenna Systems for Fading Channel Prediction in Moving Relays", European Conference on Antennas and Propagation (EuCAP 2014), Apr. 2014

45 D.-T. Phan-Huy, M. Sternad and T. Svensson, "Adaptive Large MISO Downlink with Predictor Antenna Array for Very Fast Moving Vehicles", International Conference on Connected Vehicles (ICCVE2013), Dec. 2013

46 D.-T. Phan-Huy, M. Sternad and T. Svensson, "Making 5G Adaptive Antennas Work for Very Fast Moving Vehicles", IEEE Intelligent Transportation Systems Magazine, vol. 7, no. 2, pp. 71–84, 2015

47 3GPP TS 38.300, "Technical Specification Group Radio Access Network; NR; NR and NG-RAN Overall Description; Stage 2, (Release 15)", Sept. 2017

48 3GPP TS 38.331, "Technical Specification Group Radio Access Network; NR; Radio Resource Control (RRC); Protocol specification (Release 15)", Sept. 2017

14

D2D and V2X Communications

Shubhranshu Singh[1], Ji Lianghai[2], Daniel Calabuig[3], David Garcia-Roger[3], Nurul H. Mahmood[4], Nuno Pratas[4], Tomasz Mach[5] and Maria Carmela De Gennaro[6]

[1] Industrial Technology Research Institute, Taiwan
[2] University of Kaiserslautern, Germany
[3] Universitat Politècnica de València, Spain
[4] Aalborg University, Denmark
[5] Samsung Electronics R&D Institute, UK
[6] Magneti Marelli, Italy

14.1 Introduction

Device-to-device (D2D) communication refers to the "direct mode" of communication between two or more user equipments (UEs) that are within the transmission range of each other, where "direct mode" refers to the fact that communication takes place without any traffic going through network entities, such as base stations (BSs). The D2D concept has been continuously evolving from the academia as well as the industry perspective. There are various existing D2D standards such as Wi-Fi Direct, Bluetooth, etc. The 3rd Generation Partnership Program (3GPP) in Release 12 and onwards provided enhancements to Long Term Evolution (LTE) systems supporting proximity or D2D-based services. D2D work within 3GPP has been ongoing and will continue to evolve as part of the 5th generation (5G) releases. This chapter provides a detailed summary of the work, the technical details of the envisioned D2D evolution, and 5G-specific D2D challenges and solution approaches.

Vehicle-to-vehicle (V2V) communication also exploits the direct mode of communication between two or more vehicles, in contrast with vehicle-to-anything (V2X) communication that includes V2V, vehicle-to-pedestrian (V2P), vehicle-to-infrastructure (V2I), and vehicle-to-network (V2N) information exchanges. Future V2X connectivity aims to enable various value-added services. The automotive industry is working with 3GPP with the aim of designing technical enablers in LTE and New Radio (NR) to support services including vehicle platooning, advanced driving, remote driving, and sensor information sharing.

Contrary to the unicast communication typically involved in mobile broadband (MBB) connectivity, V2X heavily relies on multi-cast communication, i.e., the transmission of information to multiple recipients, for example to disseminate warning and neighboring vehicles related

5G System Design: Architectural and Functional Considerations and Long Term Research, First Edition.
Edited by Patrick Marsch, Ömer Bulakçı, Olav Queseth and Mauro Boldi.

information. Besides, multicasting can enable short reception delays as required in V2X. This chapter discusses further details related to broadcasting in V2X, including problems due to simultaneous message broadcasting by multiple distributed entities, distributed and centralized radio resource management, etc.

This chapter is structured as follows. The remainder of Section 14.1 provides further details on D2D and V2X application scenarios and challenges. Section 14.2 provides an overview on the current status in different standardization bodies w.r.t. D2D and V2X. Section 14.3 then ventures into a particular question w.r.t. 5G design, namely the choice of appropriate waveforms for D2D and V2X support in 5G. Sections 14.4 and 14.5 cover two main enablers related to D2D and V2X, namely the device discovery over the so-called sidelink between communicating entities, and sidelink mobility. Sections 14.6 and 14.7 then delve more into the specific application and implementation of D2D and V2X and practical scenarios, while Section 14.8 provides an outlook to possible D2D extensions in 5G. Finally, the chapter is summarized in Section 14.9.

14.1.1 Application Scenarios

In this section, key D2D and V2X applications and use cases are listed. Some of them are already available in 4G systems, such as public safety related use cases, while others are envisioned to be supported by 5G systems, e.g., related to wearable devices and the Internet of Things (IoT):

- **Public safety communication**: In situations such as fire outbreak, earthquake, tsunami, and other natural and unexpected disasters, often the existing network infrastructure can be destroyed and is therefore not available. In such situation, a D2D-based network that is quick to setup for emergency services is being designed and deployed;
- **Proximity-based applications**: This set includes applications where devices are in range of direct communication with each other and thus can discover each other and locally communicate, e.g. for local advertisements, social apps enabling users to locate nearby like-minded users, content sharing and/or playing games etc.;
- **Traffic offloading**: In order to save network resources and also to minimize end-to-end (E2E) delay, D2D-based local data sharing can be enabled;
- **Wearable devices and IoT**: One of the important goals of 5G is to provide connectivity to various types of devices and not just focus on broadband services. For example, D2D devices can act as relays to in-building and out of coverage sensor devices, and home appliances can communicate in D2D mode, avoiding the need for all devices to directly access a BS. Moreover, wearable devices such as smart watches and health monitoring devices can save power by sparsely connecting to handheld devices which can act as relays;
- **V2X**: V2X applications encompass an overall system architecture and impose several design challenges; some of them (e.g., ultra-low latency and high reliability requirements) are expected to be inherently supported by the 5G system. Envisioned V2X applications in 5G are broadly classified into two areas:
  - **Safety-related V2X applications**, such as fully-autonomous or tele-operated driving, high-density car platooning, map services and precise positioning, sensor/camera/radar information sharing among vehicles, augmented reality (e.g., see-through behind truck), or using vehicles as relays for traffic forwarding, etc.;

– **Non-safety-related or commercial V2X applications**, such as connected vehicles, mobile high data rate entertainment (e.g., virtual reality, audio/video streaming, online gaming, 3D/UHD video telepresence), mobile hot-spots, software updates, etc.

14.1.2 Technical Challenges from 5G Design Perspective

Going forward into the 5G era, D2D and V2X functionalities are expected to be natively supported into the control and user plane protocol stacks of 5G radio and next generation network systems, rather than as add-on functions, both from device as well as network perspective. This requires addressing several challenges from a control and user plane design perspective. Some of the important D2D and V2X aspects considered for the 5G framework are as follows:

- **Channel sounding among pairs of devices**. One key requirement in the context of D2D communications is the need to estimate the link quality between devices. In 5G, this should be done based on the reuse of the same sounding reference signals (SRS) that are also used for the cellular uplink. This is different to the LTE Release 12 approach, where dedicated signals for D2D link estimation are used. A key challenge is then that a device can either send its own SRS or receive SRS transmitted from another device, but not both at the same time. Hence, it is required to design SRS muting patterns such that devices can ultimately estimate the links to other devices in proximity over time;
- **Control signaling among devices**. Another design challenge is how to enable control signaling between directly communicating devices (e.g., ACK/NACK, channel quality indicator, CQI feedback). In particular, this is expected to build upon the same control channels as designed for cellular communications. If, for instance, there are certain signals foreseen for uplink control signaling, and others for downlink control signaling, then if the uplink control signals are reused for the control signaling between D2D pairs, a device can only transmit uplink control signals or receive these from another device, but not both at the same time;
- **Challenges in cooperative D2D communication**. Cooperative D2D communication is when D2D pairs are utilized as relays to facilitate the transmission between a cellular user and its BS to improve spectral efficiency. In this case, the interface supporting communication between two devices is enhanced to support unicast D2D communication and/or one-to-many or one-to-all D2D communication among pairs of devices, where one of these devices can be the source or D2D transmitter (DT), while other device(s) can be the destination or D2D receivers (DRs). Besides, such D2D devices facilitate cellular user transmission by acting as relay devices. Some of the challenges are related to how the 5G radio access network (RAN) dynamically allows cooperative D2D mode selection and communication, while at the same time ensuring interference mitigation, for instance in case of simultaneous D2D and cellular UE-to-BS communication over shared radio resources;
- **Interference management for the simultaneous dissemination of safety messages in vehicular networks**. One of the main motivations for the use of V2X communications is the road safety improvement that these communications can bring. In this context, it is expected that vehicles transmit periodic messages with their current position, speed and acceleration, among other parameters. These messages need to be received by the vehicles near the transmitter to be aware of their presence and avoid potential vehicle collisions. In a real-world scenario with many vehicles

spread in rural and urban areas, the resources used to disseminate these messages need to be spatially reused, leading to interference and message collisions. The high vehicle mobility and message transmission frequency make the interference management of these scenarios extremely difficult.

14.2 Technical Status and Standardization Overview

In this section, a view on the current technical status for D2D and V2X is given, together with an overview of the standardization of these solutions.

14.2.1 D2D: 3GPP Standardization Overview

3GPP started working on LTE-based D2D in Release 12. The initial study phase was mostly driven by operators and vendors trying to come up with use cases, scenarios and system requirements related to the support of D2D services over LTE networks. Agreed outcomes were then included in the System Architecture group SA1 specification TS 22.278 [1]. The SA1 outcome specified service requirements for proximity based discovery, also referred to within the 3GPP community as Proximity Service (ProSe) discovery, as well as ProSe communication over evolved Universal Mobile Telecommunications System (UMTS) Terrestrial Radio Access Network (E-UTRAN). These are specified for users within network coverage as well as outside network coverage. In Releases 12 and 13, outside network coverage specifications were mostly focused on public safety services. Another aspect of D2D communication specified as part of Releases 12 and 13 is the Enhanced Packet Core (EPC) support of ProSe communication over wireless local area networks (WLANs). Service requirements to enable the EPC to provide network support for connection establishment, maintenance and service continuity for WLAN direct communication are specified. However, direct control of the WLAN link is not within the scope of 3GPP specifications.

Corresponding to SA1 requirements, stage 2 of the ProSe features in the Evolved Packet System (EPS) were specified by SA2 and are available in TS 23.303 [2]. This stage includes specifications for ProSe direct or EPC-level discovery and ProSe direct communication using E-UTRAN or WLAN direct. ProSe discovery identifies ProSe-enabled UEs that are in proximity. Discovery may be EPC-assisted or direct discovery. ProSe direct communication enables establishment of communication paths between two or more ProSe-enabled UEs that are in direct communication range, again based on either E-UTRAN or WLAN. Specific to public-safety usage, ProSe-enabled public safety UEs can establish the communication path directly among each other, regardless of whether these are served by E-UTRAN. For out-of-coverage scenarios, UE-to-UE as well as UE-to-network relays are supported.

The UE-to-UE interface for discovery and communication is defined as PC5 interface in TS 23.303 [2] and referred to as *sidelink* in RAN specifications. 3GPP RAN1 aspects including synchronization signal design and synchronization procedures, device discovery, resource allocation, etc., are specified in 3GPP TS 36.211 [3].

TS 36.331 [4] specifies the Radio Resource Control (RRC) protocol aspects of sidelink discovery and communication. Among other aspects, it specifies enhancements to the RRC protocol to cover certain control plane design aspects for the sidelink, such as sidelink discovery and communication monitoring, and the sidelink synchronization process. Overall security aspects of proximity-based services are in the scope of SA3, and Release 12 specifications and beyond are captured in 3GPP TS 33.303 [5]. ProSe security is broadly specified according to two aspects. First, security for common procedures, i.e. security of those ProSe features which share common procedures, such as service authorization procedures for ProSe discovery and ProSe communication. Second, security of the ProSe features, including security for ProSe direct discovery, security for one-to-many ProSe direct communication, and security for EPC-level discovery of ProSe enabled UEs.

Future and ongoing work in 3GPP specific to Release 14, Release 15 and onwards is focused on commercial use cases such as wearable devices and IoT. As part of this work, out-of-coverage scenarios are discussed for commercial applications and aim to specify layer-2 based relay support. Also, both 3GPP and non-3GPP interfaces are required to be supported on the PC5 interface. From physical layer (PHY) design perspective, as part of future releases, new radio specification is expected to natively support sidelink transmission, as will be detailed in Section 14.3.

14.2.2 V2X: 3GPP Standardization Overview

3GPP V2X related specification so far has been focused on enhancements to Long Term Evolution – Advanced (LTE-A). However, 3GPP has an agreed roadmap to support V2X as part of its ongoing 5G effort, i.e., as part of New Radio (NR) and Next Generation (NG) work items. Use cases and system requirements specific to V2X services over LTE have come from the automotive industry as well as operators and vendors. These are specified in TS 22.185 [6]. Going forward for Release 15 and onwards, the 3GPP roadmap aims to support V2X services corresponding to use cases and requirements being studied and included in SA1 TR 22.886 [7]. These services could use 3GPP systems for the transport of their communication based on the new radio and/or LTE. SA1 TR 22.886 [7] and subsequently TS 22.186 grouped all the use cases into five categories: vehicle platooning, advanced driving, remote driving, extended sensors, and general aspects.

The system level architecture aimed to address use cases and requirements in TS 22.185 is specified in TS 23.285 [8]. This specification includes architecture enhancements to facilitate vehicular communications for V2V, V2P, V2I, and V2N communications. V2X communication over the PC5 interface as well as V2X communication over the LTE-Uu interface are specified. A new study item corresponding to 5G use cases and system requirements in TS 22.186 was agreed by the SA plenary. Corresponding RAN aspects of V2X services are specified in TS 36.331 (Radio Resource Control) [4], TS 36.201 (physical layer) [9]; TS 36.212 (multiplexing and channel coding) and TS 36.213 (physical layer procedures) [10].

14.2.3 ETSI ITS Communications Architecture and Protocol Stack

The European Telecommunication Standards Institute (ETSI) Intelligent Transport Systems (ITS) Technical Committee (TC), established in 2007, is responsible for Cooperative ITS (C-ITS) standards development, interoperability and conformance testing. The standardization activity is European-focused but recognized worldwide. The ETSI ITS standard [11] defines four ITS station (ITS-S) sub-systems (personal, vehicle, roadside and central) which communicate in an ITS environment.

ETSI ITS V2X communication is based on ITS-G5 access layer technology [12] and allows 6-12 Mbps default data rates in a 10 MHz radio channel. It uses Institute for Electrical and Electronics

Engineers (IEEE) 802.11 physical and data link layers (IEEE 802.11 MAC plus ANSI/IEEE 802.2 Logical Link Control). Three dedicated ITS bands are defined in the 5.9 GHz frequency range: G5A (road safety), G5B (non-safety), and G5D (future). Due to short-lived V2X communication links, data exchange is enabled without the need for association or authentication. Apart from the detailed radio communication requirements in the 5.9 GHz ITS band, protocols and message specifications, the standard also defines the following C-ITS applications and corresponding use cases: road safety, traffic control and efficiency, fleet and freight management, and location-based services.

The ITS communications architecture is based on the Open System Interconnection (OSI) model extended for ITS, as shown in Figure 3 in [13]. The higher layers are intended to be access technology agnostic to enable forward-compatibility. The protocol stack is defined for basic ITS-S functionality and includes the following layers:

- **Applications layer** – support for road safety, traffic efficiency, infotainment and other use cases;
- **Facilities layer** – applications, information services and session support, e.g., Global Positioning System (GPS) positioning, state monitoring (car engine, lights), messages and time management;
- **Networking and transport layer** – protocols for data delivery and routing between ITS stations (e.g., IPv6 support, handover between access technologies, etc.);
- **Access layer** – internal and external ITS-S communication using various media (ITS-G5, Wi-Fi, Ethernet, Cellular, Bluetooth, GPS, etc.);
- **Management entity** – configuration of ITS-S station and cross-layer information exchange;
- **Security entity** – privacy and security services.

In addition, the following ITS-related protocol stack extensions are defined:

- **Decentralized congestion control** [14] – used to avoid unstable ITS-G5 radio behavior. It is a cross-layer function to enable time-critical road safety applications with high reliability and low latency. It controls ITS-G5 radio channels' load, maintains network stability, throughput efficiency and fair resource allocation to ITS stations. In addition, it uses high-priority messages resource reservation and fast adoption to changing radio channels. In the access layer, it includes transmission power control, rate control, adaptive clear channel assessment and probing;
- **Cooperative awareness basic service** [15] – facilities layer service to allow road users and infrastructure to be informed about object states (e.g., vehicle time, position, motion state, activated systems) and attributes (e.g., dimension, type, role). Information is exchanged using periodic Cooperative Awareness Messages (CAM) using V2V or V2I with a predefined syntax, semantics and handling;
- **Decentralized environmental notification** [16] – a facilities layer service triggered by road hazard warning applications. Decentralized environmental notification (DEN) messages contain road hazard (e.g., icy road) or abnormal traffic conditions (type, position, duration). Notification dissemination (via V2V or V2I) uses pre-defined geographic areas of the receiving ITS stations close to a detected event. Information is presented to the driver to react, or forwarded to a central ITS for traffic management purposes. The protocol (including messages syntax and semantics) is designed to manage situations when a detected event persists after the originating ITS station is far from its location;
- **GeoNetworking** [17] – ITS networking and transport layer protocol. This provides ad-hoc networking based on geographical addressing and geographical routing between ITS stations using short-range wireless technology (i.e., ITS-G5). ITS stations addressing is based on network

addresses and also geographical areas addressing (e.g., GeoBroadcast, involving the distribution to all nodes in a geographical area). Packets are forwarded based on the geographical location of network nodes and packet destinations;

- **Local dynamic map** [18] – a database of time- and location-referenced objects influencing or influenced by road traffic. A digital map data store maintains useful moving or stationary objects info, e.g.:
  - Lane specific info (e.g., curbs);
  - Pedestrian walking;
  - Bicycle paths;
  - Road furniture (e.g., traffic signs or lights);
  - Dynamic objects sensed by other users.

A facilities layer function describes object dependencies, relationships and timestamps. CAM and DEN messages are data sources for the local dynamic map (LDM). The standard describes LDM functional behavior, functions, interfaces and data objects for safety applications.

14.2.3.1 The Role of 5G in ETSI ITS

Work on ETSI ITS Release 1 focusing on the standard foundation including required messages was finalized in 2013 (currently in maintenance mode) to enable "Day 1" deployments and use cases. The basic set of applications was defined in Table 6.1 in [19]. Release 1 focuses on V2X ITS-G5 single radio channel operation, though not excluding cellular or broadcast as communication architectures that can accommodate the applications.

Further work on Release 2 started in Oct 2014 with planned completion in 2018. Among other topics, it introduces multi-channel ITS-G5 radio operation, service advertisement and new messages to support new use cases. Some of the Release 2 work items are:

- **Cooperative adaptive cruise control** – here, vehicles follow closely and safely, with braking and accelerating done cooperatively using V2V;
- **Vehicle platooning** – linking vehicles (e.g., trucks) into a train-like group to save fuel, fit more cars on the road, and improve safety;
- **Collective perception** – vehicle sensors data sharing;
- **Vulnerable road users** – involving the usage of V2P for the detection of vulnerable road users such as pedestrians, cyclists and motor cycles, and the prevention of accidents;
- **New applications in the ITS band 63 - 64 GHz**.

The current ETSI ITS standard architecture allows an easy integration of new communication technologies including 5G. From the standardization point of view, 5G radio technologies will play an important role in developing the ETSI C-ITS ecosystem, given its native support for automotive verticals, high throughput and capacity, low-latency and high-reliability communications. From the deployment point of view, a better cost efficiency may be an important long-term benefit of using only 5G or even earlier cellular V2X (C-V2X) infrastructure, addressing communications requirements from different verticals along V2X. Nevertheless, there are different views in the industry, for instance among telecommunications and automotive players, how to position 5G in C-ITS and what the evolution path should be. The long-term impact of 5G on V2X depends on the answers to some key questions, which are currently discussed between the industry and regulatory partners related to the C-V2X technology positioning vs. ITS-G5, their coexistence in the 5.9 GHz ITS band, and potential regulatory mandates.

14.2.4 IEEE Wireless Access in Vehicular Environments – WAVE

The IEEE Wireless Access in Vehicular Environments (WAVE) system is a radio communication system intended to enable communications between vehicles and infrastructure, and communications among vehicles, for the U.S. The reference standard is 1609 with its sub sections [20]. The term "dedicated short range communications" (DSRC) is typically used in the U.S. to refer to the radio spectrum or technologies associated with WAVE, while the Society of Automotive Engineers (SAE) has specified messages in SAE J2735 [21] to be used by applications which use DSRC for WAVE.

The protocol stack used for WAVE is shown in Figure 2 of Std 1609.0 [20], with all layers related to the standards provided. Going from the lower to the upper layers of the stack, we find:

- The **Access layer**, which is divided in Medium Access Control (MAC) and PHY layer. IEEE Std 1609.4 [20] (Multi-Channel Operations) specifies extensions to the IEEE 802.11 MAC layer protocol. In particular, an IEEE 802.11 amendment was used for the WAVE protocol stack, called 802.11p, which specifies maximum delays of ten milliseconds for high-priority messages;
- The **Network and Transport layer** is standardized in the 1609.3 document [20], showing how existing protocols, like Logical Link Control (LLC) and Internet Protocol version 6 (IPv6), are used together with new services, like the WAVE service advertisements and the WAVE Short Message Protocol (WSMP);
- The **Higher layers** include several standards, related to the specific applications to be implemented and the corresponding facilities. Focusing on safety applications, the facility layer in WAVE is captured in the SAE J2735 standard [21], which describes the basic set of messages to be used among vehicles and with road side units (RSUs).

The set of messages used in WAVE is composed by:

- **Basic safety messages (BSMs)**. These are used in multiple safety applications in each vehicle. These applications are largely independent of each other, but all make use of the incoming stream of BSMs from surrounding (nearby) vehicles to detect potential events and dangers. In the basic system defined by the SAE J2735 standard, each moving vehicle updates and sends its own BSM every 100 ms. Other nearby devices (typically vehicles, but potentially also RSUs) detect this broadcast and process it as they see fit, i.e., running whatever safety applications they wish;
- **Roadside alert (RSA) messages**. These are the messages used in various traveller information applications. For instance, an Emergency Vehicle Alert message is used to inform mobile users of nearby emergency operations. The types of data in the roadside alert messages can include information such as travel delays, incident and diversion data, construction messages, and other data that the traffic management centre (TMC) may wish to deliver to drivers;
- **Probe vehicle messages (PVMs)**. These are used by multiple applications. Vehicles gather data on road and traffic conditions at intervals. The probe data provides information on traffic conditions, weather, and road surface conditions;
- **Signal phase and time (SPAT) messages**. SPAT messages are used to convey the current status of one or more signalized intersections. Along with map messages (see below), the receiver of such messages can determine the state of the signal phasing and when the next expected phase will occur.

It will be used from applications that notify to the driver which would be the best speed to be kept to overcome the traffic light (e.g., in the context of GLOSA – Green Light Optimal Speed Advice);

- **Map messages.** Map data messages are used to convey many types of geographic road information. Currently, its primary use is to convey one or more intersection lane geometry maps within a single message.

The *security module* is standardized in 1609.2 [20], while the *management module* is proposed in [22]. The set of applications designed for the US standard WAVE includes both V2V applications defined in the ETSI-ITS standard "Day 1", see Section 14.2.3, and a set of V2I applications, like in-vehicle signage, speed limit and traffic signal violation, which focus on safety-related applications with information coming from the infrastructure.

14.2.5 Other Industry Organizations

5G Automotive Association (5GAA) – A global cross-industry organization formed in September 2016 with the intention to address connected mobility and road safety needs, promote and develop cellular V2X communications and also support standardization and commercial adoption. Its members include car manufacturers, suppliers, telecom operators and mobile equipment vendors. Among others, its work focuses on influencing use cases harmonization, business models, technology roadmap evolution, spectrum allocation, and regulatory, certification and approval processes. In its first White Paper [23], the organization argues the case for cellular V2X technology applicability for road safety and cooperative driving, and compares it with IEEE 802.11p based radio access. According to 5GAA, the major advantages of LTE V2X 3GPP Release 14 include [23]:

- Improved system performance, including higher spectral efficiency;
- Ability to concurrently address short-range V2V, V2I, and V2P communication requirements with long-range network-based V2N;
- Radio spectrum flexibility - based on D2D communication modes using the 3GPP PC5 interface, C-V2X technology can operate without network infrastructure and operator subscription in the 5.9 GHz ITS bands, while device-to-tower and device-to-network modes may use commercial mobile communication bands;
- Leveraging device-to-network communications and the V2N mode to enable cloud-based services and edge computing;
- Technology scalability and backward and forward compatibility based on 3GPP releases;
- Ability to use existing higher layer automotive industry standards;
- Quick introduction and commercial availability of C-V2X planned in 2018, i.e., based on 3GPP Release 14 technology.

More recently, the organization highlighted the benefits of a technology-neutral approach and forward-compatibility for V2X, and emphasized the benefits of cellular V2X technology with respect to DSRC in comments submitted to the U.S. Department of Transportation [24], regarding the planned mandate of introducing DSRC in new vehicles.

Car2Car Communication Consortium – A non-profit organization started in 2002 by European vehicle manufacturers and supported by automotive suppliers, research organizations and others. It supports the creation of standards for communicating vehicles and an increase of road traffic safety and efficiency by deployment of C-ITS and V2X. Currently, the organization comprises of

89 members, with 18 vehicle manufacturers, 40 equipment suppliers and 31 research organizations. The consortium is one of the key players supporting the deployment of C-ITS across Europe. In 2014, the organization started working on various C-ITS deployment pilots and corridor projects. Its members focus on developing the ETSI ITS-G5 standards family to ensure interoperability across borders and brands. More recently, the organization publicly presented its high-level V2X technology and applications roadmap including increasing cooperation and automation levels. In the recent press release in February 2017 [25], the consortium declares support for the European strategy on C-ITS deployment published in November 2016 [26], which favors ETSI ITS-G5 for direct vehicle communication. The press release indicates ITS-G5 market readiness, citing the recent ITS-G5 integration into the AUTomotive Open System Architecture (AUTOSAR) Release 4.3 industrial standard for vehicle electronics, and that hardware compatibility to IEEE 802.11p is becoming mandatory in the USA, and also mentions planned commercial V2X rollouts in 2019.

C-ITS Platform – This was established in 2014 to address barriers and enablers for C-ITS deployment in the EU. Its work focuses on knowledge exchange and policy cooperation, shared vision and roadmap development between the European Commission, public and private stakeholders from member states and local authorities. In its first phase report published in 2016 [27], the organization defined a list of Day 1 C-ITS services prioritized for deployment. Furthermore, it advocates a hybrid communications approach for C-ITS, spanning various radio access communication technologies, and recommends that C-ITS message transmission, depending on the application, should be access-layer agnostic, including the IEEE 802.11p/ETSI ITS-G5 usage for short-range or cellular 4G/5G for long-range communications, amongst others. The platform work has continued in phase two [28], but a working group focusing on hybrid communications and spectrum allocation was put on hold.

GEAR 2030 – A group launched in January 2016 by the European Commission to increase competitiveness and sustainable growth in the EU automotive industry, and also serve as a forum for discussion and strategic advice, building consensus, political support and developing policies. Apart from member states, it includes stakeholders from different sectors, including vehicle manufacturers, component suppliers, the aftermarket sector, insurance providers, road safety, telematics, and environment and consumers' organizations. The main objective is to analyze key trends and develop recommendations to address challenges and opportunities for the automotive industry. Along with global competitiveness and value chain adaptation, autonomous and connected vehicles and their smooth roll-out are one of its three focus areas.

14.3 5G Air Interface Candidate Waveforms for Sidelink Support

The focus of this section is on a "baseline" D2D scenario where devices are synchronized via timing advance (TA) to their BS, but where there is no explicit coordinated synchronization of D2D links. As we will see, this may lead to performance degradation on the sidelink, which may be alleviated by the choice of waveforms that are less sensitive to synchronization errors.

14.3.1 Synchronization Problems and Possible Solutions

We consider a D2D scenario as depicted in Figure 14-1, where devices synchronize with one common BS via a continuous closed-loop control mechanism enforced for each device in the uplink. The TA is expressed in time units and represents the degree of anticipation in the transmission, to achieve perfect synchronization of the signals arriving at the BS. However, if opportunistic sidelink communications are allowed, devices are free to transmit data among each other at will. Under such conditions, time-aligning each device before the transmission can begin could be inacceptable because of the increment in latency incurred by the synchronization mechanism that should be executed before each transmission. Thus, being able to support multiple concurrent D2D transmissions precludes perfect synchronization between such devices. In fact, an asynchronous operation may even be necessary to reach an acceptable level of spectral efficiency.

A straightforward approach could be to ignore such issues, i.e., using the same TA for the sidelink communications. In the scenario depicted in Figure 14-2, for example, if device #1 and device #2 do not modify their TA, then the timing offset (TO) experienced at device #2 with respect to that of device #1 is the difference between their TA plus the additional offset term introduced by the distance d_3. $TO = TA1 - TA2 + (d_3/c)$, where c is the speed of light. As it turns out, the orthogonality property of Orthogonal Frequency Division Multiple Access (OFDMA)-based technologies like IEEE 802.11p and LTE [29], which divide the spectrum in multiple and orthogonal parallel sub-bands, is only verified under ideal conditions, that is to say, under perfect frequency synchronization and precise time alignment within the cyclic prefix. These ideal conditions mean that the receiver

Figure 14-1. D2D scenario with synchronization toward one BS.

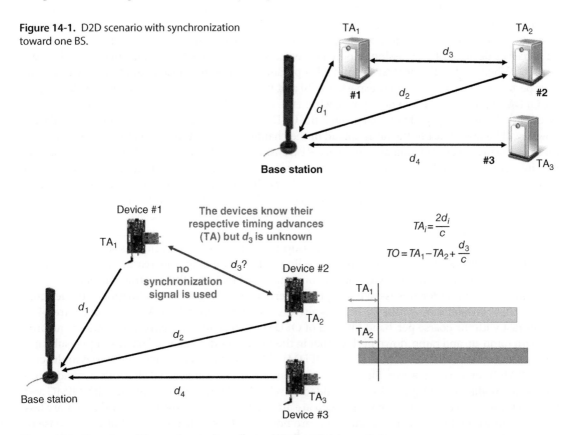

Figure 14-2. Overview of the synchronization effects of D2D sidelink transmission.

(RX) clock should be phase-locked to the transmitter (TX) clock in order to avoid phase error, and additionally, the data beginning mark after the training sequences should be perfectly identified for sampling purposes, which is not true if there is no time alignment.

For example, Figure 14-3 a) shows an orthogonal frequency division multiplexing (OFDM) waveform in time domain corresponding to a communication where the device only uses the first half of the carriers of the available frequency band. It is also shown in Figure 14-3 a) that the TO of the transmission of an OFDM waveform with an initial cyclic prefix may be interpreted as a τ -sample delay in time domain. Figure 14-3 b) shows the same waveform in frequency domain without TO (continuous line) and overlaps it with plots of the OFDM waveform with increasing sample delays (dashed lines) of 50 ns each. Theoretically, the first half of the band should contain data, and the second half should contain zeroes, which is true for the perfectly aligned OFDM waveform. However, when a TO is present, the supposedly empty sub-bands show increasing levels of interference, until reaching a point where definitely non-zero samples will be obtained. Ultimately, a large TO leads to inter-carrier interference (ICI) and inter-symbol interference (ISI), which can only be avoided if the two communicating entities perform an additional synchronization optimized specifically for the sidelink communication, or if the BS centrally controls the TAs of the devices such as to keep TOs on sidelinks within constraints.

Returning to Figure 14-3, let us assume a sidelink communication where device #1 and device #3 are the transmitters and device #2 is the receiver. Device #1 and device #3 transmit in the same carrier frequency but on two distinct non-overlapping sub-bands (upper and lower). It is assumed that device #2 perceives device #1 transmission as perfectly synchronized and device #3 transmission with a certain TO. When the fast Fourier transform (FFT) operation of the joint transmission from device #1 and #3 is performed, the device #2 may experience ISI and ICI from device #3. This is due to the fact that if the OFDM symbol of device #3 does not lie within the FFT window of device #2, the FFT duration may be extended over the symbol boundary, and if the OFDM symbol of device #3 lies within the FFT window, it will be less robust to the delay spread of the channel. Note that the ISI and ICI from device #3 cause performance degradation also to device #1 transmission, which is perfectly synchronized. This multiple TO issue in OFDM may be solved by increasing the cyclic prefix duration, so that it covers both the longest transmission delay and the channel's maximum delay spread; however, this may require the use of TO estimation methods for OFDM. Assisted by the cyclic prefix, the receiver may absorb small positive values of TO (i.e., the signal arrives later than expected by the receiver), as long as the TO is roughly within the duration of the cyclic prefix, because the cyclic prefix is discarded prior to frequency sampling. However, note that OFDM performance degrades significantly as soon as the TO is such that the channel transient exceeds the cyclic prefix duration, and increasing the cyclic prefix decreases the maximum throughput, a direct consequence of reducing the spectral efficiency.

A potential solution for this issue is adopting alternative multi-carrier modulation schemes currently under consideration for the 5G PHY, as also discussed in Section 11.3. As shown in Figure 14-4, in contrast with the coarse per-band filtering of OFDM, such candidates have a smoother time profile, with ramp-up and ramp-down periods that in the frequency domain yield narrower per-subband filtering as in universal filtered OFDM (UF-OFDM) [30], also known as universal filtered multi-carrier (UFMC), or per-subcarrier filtering as in filterbank multi-carrier (FBMC)/offset quadrature amplitude modulation (OQAM) [31], [32] aided by selective filters [33]. As a result, with FBMC and UF-OFDM the sidelobes in frequency domain are attenuated, and thus the ICI and ISI issues are less critical than with OFDM when transmissions are not perfectly synchronized, but at the expense of increased implementation complexity.

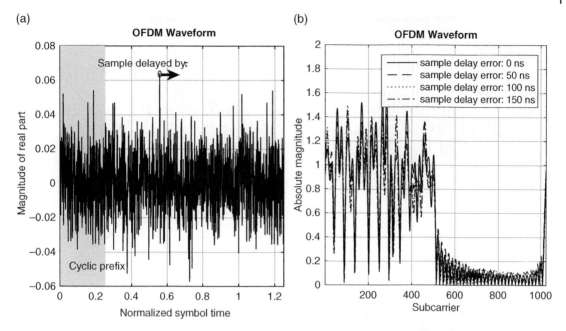

Figure 14-3. a) OFDM waveform in the time domain affected by timing offset. b) Received OFDM spectrum for different values of sampling delay error.

14.3.2 Enhancements for V2X

IEEE 802.11p includes short-range communications for data exchange between vehicles and between vehicles and roadside infrastructure (V2I). However, in the current design of LTE-V2X, the synchronization requirements are becoming a crucial issue. For example, while vehicles may be synchronized with elements of the roadside infrastructure (i.e., BSs), it is not expected that the synchronism between vehicles in a V2V sidelink communication scenario would be supported by such elements. Thus, specific synchronization signals are now under discussion for the support of this direct V2V communication. Moreover, as explained in previous sections, the concept of TA may not be reused, because it only assists in the determination of the distance to a common element (i.e., the BS) and is useless when determining the distance between two vehicles with varying relative distance. Consequently, it is of the utmost benefit to assess improvements of OFDM or alternatives which should be able to tolerate a lower degree of synchronization without introducing or increasing ICI or ISI.

In this sense, an evaluation of the impact of timing synchronization for OFDM, FBMC/OQAM and UF-OFDM is found in [34], tested and validated through a real channel under real transmission conditions, in a wireless communication system using readily available radio frequency (RF) communication hardware boards [35] and implementing the IEEE 802.11p protocol. The findings in [34] are in agreement with previous analytical and numerical contributions, complementing the state of the art by assessing the impact of the TO from the hardware platform viewpoint. The authors adjust

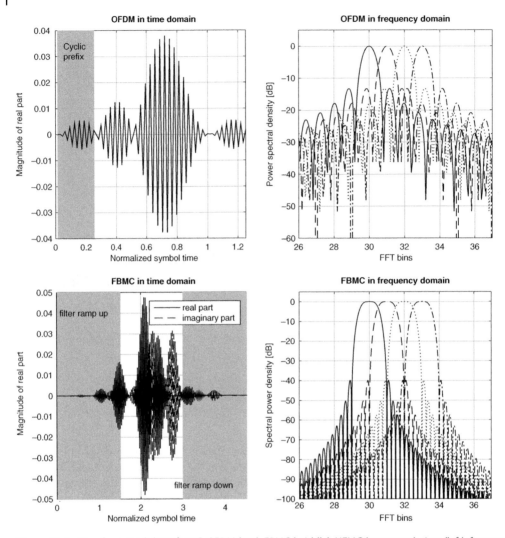

Figure 14-4. Waveform candidates for 5G, OFDM (top), FBMC (middle), UFMC (next page), time (left), frequency (right).

the TO in successive 50 ns steps as shown in Figure 14-5, which corresponds to a degree of dis-synchronization of 1 sample, and evaluate the performance of the waveforms.

Although implementing an OFDM waveform is less complex, the results confirm that OFDM is less robust against the lack of synchronization, i.e., its performance degrades significantly already for small timing offsets. However, note that there exist mechanisms that improve OFDM robustness, such as using a longer cyclic prefix or filtering, as in the case of UF-OFDM. On the other hand, FBMC/OQAM and UF-OFDM are more robust against the lack of synchronization, protecting the target bit error rate (BER) for a wider range of timing offsets. However, FBMC is not as easily applicable to MIMO, and implementations of FBMC are more complex than for OFDM, while UF-OFDM has a complexity similar to OFDM. In any case, [34] highlights the need for a software-defined radio

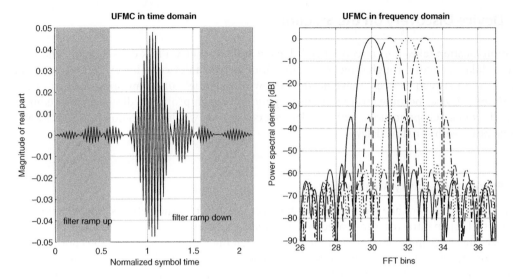

Figure 14-4. (Continued)

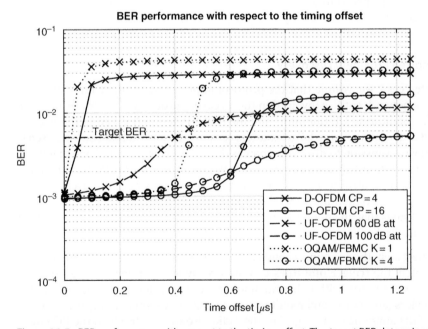

Figure 14-5. BER performance with respect to the timing offset. The target BER determines the highest tolerable TO.

approach to provide more flexibility to accommodate sidelink communication into a frame-structured cellular system within a V2X environment. This can be done via a dynamic selection of the most proper waveform according to the particularities of the communication mode (i.e., uplink or sidelink) by means of a joint and harmonized simultaneous implementation of all such waveforms, as detailed in Section 11.3.4.

14.4   Device Discovery on the Sidelink

D2D communication takes the concept of cell-densification further via link densification, and enables new cellular services through direct communication between devices. However, to allow direct communication between devices, these need to know of potential communicating devices in the neighborhood. Devices become aware of their neighbors through proximity discovery procedures. Additionally, some 5G applications with low-latency constraints would also require devices to discover each other with very low delay.

14.4.1   Proximity Discovery Architecture

D2D communications can take place either with direct assistance of the network infrastructure or autonomously [36]. In infrastructure-based systems, such as LTE-Advanced (LTE-A) and LTE-A Pro, the discovery mechanism involves communication through the BS, thus increasing the control overhead and consequently the discovery time. The involvement of the BS is required as the D2D resources are shared with the cellular resources. In the following, we consider a network-supported scenario (as currently done in LTE), and a scenario with limited network support. Furthermore, we consider the case of orthogonal spectrum dedicated to D2D, as well as the shared spectrum model (limited to the in-coverage case).

14.4.2   Network-supported Proximity Discovery

In the network-supported case, we assume that the protocol in place follows the LTE-A discovery protocol. There are three network entities involved in the discovery protocol, namely the announcing UE (announcer), the monitoring UE (monitor) and the ProSe function (located in the network infrastructure at the EPC). The announcing UE announces certain information that could be used by UEs in proximity that have permission to discover, whereas the monitoring UE is the one that monitors certain information of interest in the proximity of announcing UEs. The ProSe function is responsible for the admission control to ProSe, coordination of the proximity discovery and the monitoring of the ongoing D2D communication links.

There are four essential steps that an announcer takes to complete the proximity discovery protocol, when authorized to use ProSe:

- In the **discovery request** step, the announcer sends a discovery request to the ProSe function;
- If this request is accepted, the **trigger monitors & assign resources** step takes place where the ProSe function triggers the monitors within the announcer proximity region, assigning and informing the intervening devices (announcer and monitors) which system resources have been allocated to the transmission of the announcer message, i.e., the radio resources over which the PC5 interface will take place;
- In the **E-UTRAN discovery** step, the announcer transmits its message to the monitors over the PC5 interface in the E-UTRAN;
- Finally, in the **match report** step, the devices that were able to decode successfully the announcer message, transmit the received information to the ProSe function for confirmation of the announcer identity. After this last step, the direct D2D communication link can be established.

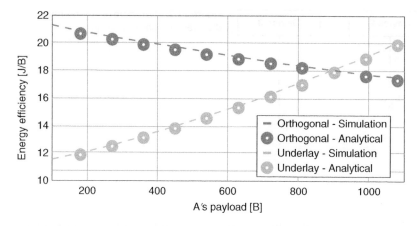

Figure 14-6. Energy efficiency comparison between orthogonal and underlay discovery according to the announcers payload (A's payload) [37].

In contrast with the state-of-the-art cellular network proximity discovery, we assume that the communications associated with the proximity discovery occur underlaid to the cellular communications. Specifically, we assume that the proximity discovery transmission shares the same time and frequency resources as the downlink transmission. The underlay transmission is enabled through receivers capable of multi-user decoding coupled with power control. The results show that this scheme is more energy efficient, as shown in Figure 14-6, than the traditional orthogonal transmission, specifically for low transmission payloads [37].

14.4.3 Out-of-Coverage Proximity Discovery

Specific to out-of-coverage cases, a fast-discovery mechanism for 5G networks aided by full-duplex (FD) technology will now be presented. FD technology, which allows a device to transmit and receive simultaneously in the same frequency band, as discussed in more detail in Section 16.2.4, has been considered for the design of the future 5G system [38]. Autonomous device discovery (i.e., device discovery without network assistance) procedures consist of two key steps, namely: i) transmitting a discovery beacon signal with device identification information for neighbor nodes to be aware of its presence, i.e., *announcing*, and ii) listening to neighbors' beacon signals to discover the neighbors in proximity, i.e., *monitoring*. Once a node becomes aware of the presence of a neighbor, a positive acknowledgement is transmitted. As shown in Figure 14-17, such acknowledgements can be piggy-backed in the discovery beacon, in order to save radio resources. Autonomous device discovery involves a trade-off between the announcing step and the monitoring step, since conventional devices cannot announce and monitor simultaneously. Moreover, transmitting beacons too frequently results in collisions of the beacon signal and leaves little room for neighbor discovery. On the other hand, too sporadic beacon transmission leads to a prolonged discovery time.

FD transmissions provide a favorable solution to this trade-off. By allowing continuous reception (i.e., always listening), a device does not have to trade-off beacon transmission with

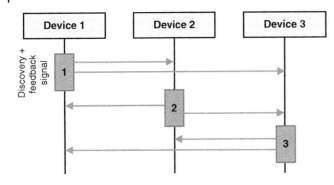

Figure 14-7. Procedure of autonomous discovery (broadcast message).

listening for neighbors' beacons. Thus, the number of collisions is reduced, consequently decreasing the discovery delay. In scenarios with low-latency requirements, FD may considerably reduce the discovery delay, while reducing collisions and control overhead and improving energy efficiency.

The discovery beacon signal includes the piggy-backed feedback of decoded discovery signals. Such feedback may contain more than one acknowledgement, since several discovery signals may be successfully decoded by the time when the beacon is transmitted. Figure 14-7 shows an example of the procedure. Device 1 is the first at transmitting the discovery signal. Such signal is received by devices 2 and 3. Then, device 2 transmits the discovery signal, which includes the feedback indicating that device 2 has discovered device 1. Finally, device 3 transmits the discovery signal, including the feedback that it has discovered both devices 1 and 2.

14.4.4 Performance Evaluation of Device Discovery with Full-Duplex Nodes

14.4.4.1 System Model

The potential latency benefits of FD communication in D2D discovery in a $100 \times 100\text{m}^2$ area network will now be evaluated in detail. The available bandwidth is divided into four orthogonal frequency slots. At each transmit time interval (TTI), the beacon is transmitted across one of these slots chosen randomly. A simple linear receiver that treats the interference as noise, here referred to as interference-as-noise (IAN) receiver, and an ideal advanced interference cancelling (IC) receiver that can cancel three strongest interference streams (owing to the fact that the devices are equipped with four antennas) are considered. The system model and simulation parameters are further detailed in [39].

14.4.4.2 Performance Evaluation Results

Figure 14-8 displays the average discovery time in the network as a function of the transmission probability, for different numbers of nodes in the network considering IAN receivers. Both cases of half-duplex (HD) and FD nodes are compared. A low transmission probability translates to long average delays due to the large idle time, whereas a high transmission probability reduces the correct beacon detection due to overwhelming collisions. The optimal transmission probability

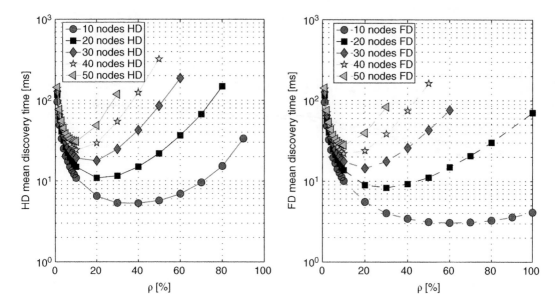

Figure 14-8. Single cluster mean discovery time with HD and FD, assuming the IAN receiver and four frequency resources [39].

corresponds to the minimum average discovery time for each network size. As expected, FD leads to lower discovery time and allows a higher optimal transmission probability. For example, in the case of a network with 10 nodes, the optimal transmission probability is increased from 40% with HD to around 65% with FD communication. Moreover, the optimal discovery time is also reduced by a factor of approximately 40% from around 5 ms with HD to around 3 ms with FD. The minimum discovery time as a function of the number of users per resource slot (θ) is further highlighted in Figure 14-8, for both receiver types. As expected, IC receivers significantly outperform IAN receivers, leading to a faster discovery. The performance improvements of FD are relevant for a small network, especially when IC receivers are used, while the gains tend to vanish for significantly large networks, e.g. with more than 25 nodes, due to the overwhelming effect of the additional interference. Note that such results are obtained by assuming the transmission probabilities corresponding to the minimum average discovery times, as taken from the findings depicted in Figure 14-9.

14.5 Sidelink Mobility Management

14.5.1 General Considerations

Sidelink mobility is an important issue which needs to be addressed and natively supported within the overall 5G framework. In the recent past, this aspect has been very sparsely studied and researched. However, a common and standardized solution needs to be agreed for wider deployment

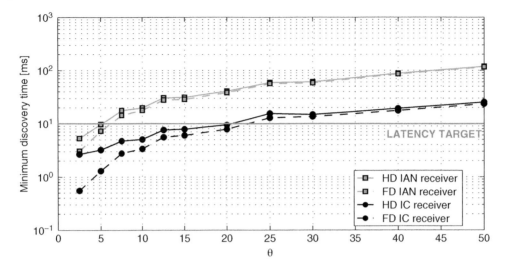

Figure 14-9. Minimum discovery time for HD and FD assuming the IAN and the IC receiver [39].

of various D2D applications. In general, mobility solutions aim to ensure reliable and maximally uninterrupted connection to the network even in the context of devices changing their location as well as their point of attachment to the network. The challenges due to device mobility in the case of D2D communication are very different, since now these involve more than one UE, i.e., also the mobility of the link between two devices with an ongoing D2D communication needs to be supported. This for instance implies, in the case of LTE terminology, a solution to support mobility of the PC5 interface.

Another aspect of D2D mobility involves group scenarios where two or more devices belonging to the same group are involved. For example, a platoon of vehicles may consist of a leader device and several other devices connected to it. Another example is a group of wearable or IoT devices. Mobility issues arise when moving groups of devices approach the cell edge, but not all group members satisfy handover conditions simultaneously. In all such cases, procedures to handover a D2D group with two or more devices are required, so that the established D2D and/ or V2V link within the group or between a pair of D2D devices is not interrupted. Moreover, each UE in the group of D2D and/or V2V devices needs to be handed over to the target cell as a group rather than in an individual fashion, to avoid unnecessary signaling overhead, among other issues.

Yet another aspect that needs to be considered in 5G is the mobility among multiple radio interface variants. In 5G it is widely envisioned that multiple RATs, such as NR and enhanced LTE (eLTE), and also multiple 5G air interface variants (AIVs) will co-exist and operate on a wide range of frequencies, which also has to be taken into account in the context of D2D mobility.

To address the above mentioned sidelink mobility aspects, suitable mobility management schemes targeted to D2D devices should be developed. New functionalities and procedures should be introduced to efficiently handover a group of two or more D2D communicating devices.

14.5.2   D2D Mobility Management Schemes

D2D mobility management procedures can be divided into four parts:

- **Signal quality measurement**: As a first step, UEs involved in D2D communication measure side-link quality, i.e. the link quality of their PC5 link over which D2D communication is ongoing. In 5G, these measurements can be performed when needed and requested, e.g., using on-demand system information blocks (SIBs) or dedicated signaling. Thus, the timing as well as target of measurement can be decided by the 5G RAN using an on-demand approach. Alternatively, UEs may proactively send such reports to the BS;
- **Coordination between the source and target BSs**: The 5G RAN may also communicate with BSs of other AIVs, within the same Public Land Mobile Network (PLMN) or beyond. Enhancements to the X2 interface (or Xn interface in NR terminology) are required to support this in order to achieve inter-RAT D2D compatibility and service continuity;
- **Resource allocation by the target BS**: After the source 5G RAN decides upon the target 5G RAN to which D2D UEs should be handed over, the make-before-break rationale needs to be adopted. This rationale requires that the D2D context information is transferred from the source 5G RAN to the target 5G RAN before the actual handover execution. Context information here refers to various aspects including D2D-specific information such as communication types, radio resources, traffic type, D2D group, etc. This enables the target 5G RAN to prepare customized D2D resources for the D2D unicast or group-cast communication. The handling of D2D resources across different AIVs is another issue. One approach may be that all D2D UEs use the original D2D resources for D2D communications until all D2D UEs hand over to the new target RAN/AIVs. This approach makes D2D communications continuous when D2D UEs perform the handover procedure. Another way is to design temporary D2D resources used for D2D UEs performing the handover procedure. The temporary D2D resources, beside the D2D resources provided by the original RAN or by the target RAN/AIVs, can be only used by D2D UEs performing the handover procedure for data transmission;
- **Packet switch from the source BS to the target BS**: After the handover decision is completed and a handover command is sent to the D2D UEs, they execute the handover and might use original, temporary or newly assigned D2D resources in order to continue the ongoing D2D communications.

Figure 14-10 shows the impact of mobility on D2D communication reliability, where reliability is defined as the percentage of packets properly received within the given maximum E2E latency. During the simulation, the handover delay of each of the UEs was kept at either 2 ms or 200 ms. These handover delay values were chosen to reflect an acceptable range of handover delay requirements in various cases. Simulation results show that the D2D mobility management scheme helps to achieve significantly higher D2D communication reliability. The main reason is that for the LTE handover scheme, the ongoing D2D communication terminates whenever one of the D2D UEs satisfies the handover condition, and the D2D service does not resume until both UEs successfully handover to the same target cell and re-establish the D2D link.

The discussion has so far considered two D2D UEs involved in the handover procedure, though this can be extended to group scenarios with more than two UEs, with possible enhancements and signaling optimizations. A group-based tracking area update (TAU) design as well as a group-based

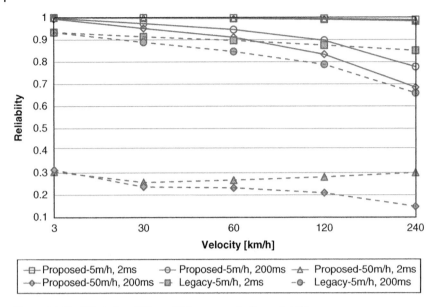

Figure 14-10. Effect of mobility on D2D communication reliability.

Random Access CHannel (RACH) access are suitable enhancements and optimization options for group-based communication.

One has to stress, though, that the aforementioned considerations and simulation results were based on the assumption that D2D communication takes place among UEs served by the same BS. In principle, it would of course be possible to schedule D2D transmissions across cells, though with the need for additional coordination among the involved BSs.

14.6 V2X Communications for Road Safety Applications

This section will now explore very practical considerations and constraints related to the usage of V2X communications in specific road safety scenarios.

14.6.1 General System Design Aspects

For certain safety services, like vehicle collision avoidance, it is important to ensure a certain level of awareness, warning dissemination and coordination among neighboring vehicles or ITS stations. In this context, CAMs as introduced in Section 14.2.3 provide information on the presence, position, and basic status of communicating ITS stations to neighboring ITS stations [15]. By receiving CAMs, the ITS stations are aware of other stations as well as their positions, movement and other basic attributes. The C-ITS standard defines that the CAM transmission frequency is between 1 and 10 Hz. With respect to warning dissemination, DEN messages provide information about a specific environment or traffic warning [16]. DEN transmission can be periodic while the event that caused the warning is active. In this case, the transmission frequency is between 0.1 and 25.5 Hz. In addition to the previous messages already standardized in the C-ITS standard, other types of messages can be required for, e.g., platoon management. In this particular case, coordination among platoon members may be required to leave space for a new platoon member, change lanes, divide the platoon in a diverging highway, etc. These three types of messages have to co-exist in a shared channel with a potential high density of ITS stations.

The destinations of the previous messages depend on the particular service associated with it. In general, they are vehicles in close vicinity. This fact together with the need of short reception delays have motivated the distribution of these messages using broadcasting techniques by the generator of the messages, i.e., without the support of an infrastructure. This is the approach used in the C-ITS standard, in which the ITS-G5 [40] technology is used for the lower protocol layers. ITS-G5 is based on the IEEE 802.11p standard, inheriting the IEEE standard 802.11 based multiple access method, i.e., carrier sense multiple access with collision avoidance (CSMA-CA). In this method, stations transmit after sensing that the channel is free for a random period of time. CSMA-CA in congested networks might force stations to wait for very long times before transmission as a result of high channel usage [41].

Motivated by the strict requirements on the reliability and latency of safety-related ITS applications, the ITS-G5 technology includes decentralized congestion control (DCC) methods to regulate the channel load. The DCC methods on the access layer are the transmit power control (TPC), the transmit rate control (TRC), the transmit datarate control (TDC), the DCC sensitivity control (DSC), and the transmit access control (TAC). The TAC and TRC mechanisms are relevant for non-safety applications only, since they constrain the minimum packet transmission periodicity and, in the case of the safety-related packets, the periodicity is established by the application and should not be restricted by lower layers [42][43]. The effects of the other three methods are somewhat tangled (cf. Table 14-1), being possible to counteract one with each other. In general, tuning the DCC parameters is a very complex problem [44]. This issue is highlighted in [45], where the authors argue that the use of the DCC mechanisms can be counterproductive.

Table 14-1. The effect of DCC methods on certain key performance indicators. The arrows indicate that the magnitudes increase (↑) or decrease (↓) when the methods are active.

	TPC	TRC	TDC	DSC	TAC
Packet transmission frequency	=	↓	=	=	↓
Collision probability	↓	↓	↓	↑	↓
Interference level	↓	↓	↓	↑	↓
Coverage area	↓	=	↓	=	=
Interfered area	↓	=	=	=	=
Reliability	↓	↑	↓	↓	↑
Delay	↓	↑	↓	↓	↑

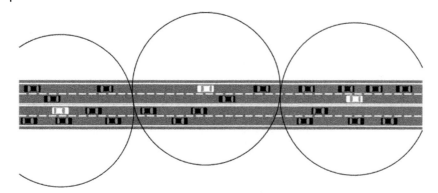

Figure 14-11. Sphere packing in a motorway to select simultaneous transmitters. White vehicles transmit simultaneously, and black vehicles wait for other transmission opportunities.

The problem of tuning different transmission parameters is not only linked to ITS-G5 and DCC. This problem is a consequence of the simultaneous message broadcasting by multiple points distributed in space. In fact, a transmitter cannot increase the range of its transmissions without influencing the interference. At the same time, the packet transmission frequency is inversely related with the distance at which the closest simultaneous transmitter is located. This problem can be viewed as a certain form of sphere packing, as depicted in Figure 14-11. Considering all this, the following system design guidelines aim at ensuring both good reliability and delay:

1) It is important to ensure that messages are received with good enough signal-to-interference-and-noise ratio (SINR) levels at the desired distances to the transmitters. This imposes a lower bound to the distance between closest simultaneous transmitters;

2) ITS stations should transmit alternatively ensuring that the previous lower bound is satisfied. Therefore, stations have to wait for a certain amount of time before being able to use the channel again. For instance, if there is one ITS station every 10 m, and if the distance between closest simultaneous transmitters should be greater than 300 m, a station has to wait for the transmissions of the other 29 ITS stations before using the channel again. In other words, the ITS station is allowed to use the shared channel 1/30 of the time;

3) This fraction and the packet transmission frequency can be used to compute the maximum available time for transmitting one message. For instance, following with the previous example, if ITS stations should transmit one message every 10 ms, these messages should be transmitted in, at most, 1/3 ms, in order to allow 30 transmissions in 10 ms;

4) This time should be enough to ensure the SINR levels of the first step with the available resources, e.g., channel bandwidth. If not, the system could become unreliable.

14.6.2 Impact of the Existence of Several Message Ranges on the System Design

The guidelines presented in the previous section aim to provide a good service performance in a worst-case scenario. The real word is, however, not so extreme. For instance, all messages do not need to have the same requirements in terms of the maximum distance at which an ITS station

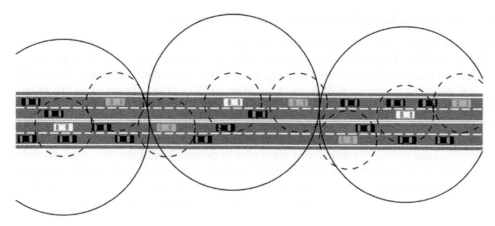

Figure 14-12. Selection of simultaneous transmitters for two different ranges.

Figure 14-13. First ring of neighbours.

should be able to decode it. As opposed to Figure 14-11, Figure 14-12 shows two packings in a motorway for two different message ranges. These ranges can be achieved by using, e.g., different transmission power levels. The white and grey vehicles can simultaneously transmit short range messages, whereas only white vehicles can simultaneously transmit larger range messages. By using this approach, resources are reused more frequently in space with short range messages, which allows higher packet transmission frequencies, since ITS stations have to wait less time to use the channel.

As commented before, shorter ranges can be achieved by using lower transmit power. This may, however, decrease reliability if noise is an important factor of the SINR. This motivates the study of other approaches, like the use of the millimeter wave (mmWave) bands. These bands are characterized by large path and obstruction losses, as detailed in Section 4.2.1, which limit interference. In addition to this, obstruction losses or shadowing created by neighboring vehicles can be beneficial to confine the transmissions to the first ring of neighbors as illustrated in Figure 14-13. Indeed, these neighbors are those in physical collision risk and, therefore, are those stations that may require communications with higher packet transmission frequency. This fact renders the combination of mmWave and centimeter wave (cmWave) bands a very suitable solution for road safety applications. In particular, cmWave bands can be used for some kind of awareness and warning dissemination using low packet transmission frequencies. On the other hand, mmWave bands can be used for collision avoidance and coordination of vehicles in the close vicinity using high packet transmission frequencies.

14.6.3 Distributed versus Centralized Radio Resource Management

The C-ITS standard has promoted the use of the distributed radio resource management inherited from IEEE 802.11p. More specifically, the CSMA-CA mechanism controls the channel usage, i.e., a station uses the channel after sensing it to be free for certain amount of time. This mechanism simplifies the resource management and does not need the support of an infrastructure. It has, however, a major drawback known as the *hidden node problem*. Hidden nodes are ITS stations that cannot hear each other and, therefore, may falsely use the channel simultaneously in a CSMA/CA system. If these nodes are close enough to each other, they may cause destructive interference to other stations that are trying to decode one of the messages. In order to mitigate this effect, the transmit power should be large enough to be heard at long distances, in such a way that hidden nodes are too far away to cause destructive interference to stations that should receive the message. This solution limits both the packet transmission frequency and the use of mmWave bands due to related propagation characteristics. In some cases, the required power increment to avoid the hidden node problem might be prohibitive, like in urban scenarios where buildings significantly obstruct the signal to vehicles in perpendicular streets, cf. Figure 14-14.

In scenarios where hidden nodes reduce the reliability or packet transmission frequency to worrisome values, distributed resource management should be avoided. 5G systems with the support of a cellular infrastructure can make use of a centralized resource management. In this case, the cellular network can use vehicle location information to manage the channel usage in a more efficient way. Moreover, this type of management allows the use of the mmWave band.

14.7 Industrial Implementation of V2X in the Automotive Domain

Car manufacturers, automotive suppliers, semiconductor manufacturers and communication module manufacturers have worked a lot in the last ten years to integrate their competencies and provide to the vehicles a platform which allows the V2X communication to extend the concept of

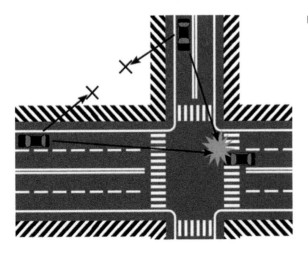

Figure 14-14. The hidden node problem.

"connected car". While 4G communication allows vehicles to connect to the Internet, to download videos, maps, and to connect to many applications useful for the trips and for the entertainment of the passengers, V2X communication is more focused on guaranteeing a safe and comfortable journey by informing the driver about possible dangers on the road, with a large amount of space-time to react. By participating to the EU corridors organized by the Car2Car Communication Consortium, see also Sections 14.2.5 and 14.7.2, many V2X providers have tested several use cases to analyze the benefits of V2X communication, and now industry is moving toward the production of the "V2X platform", trying to understand which would be the best way to introduce V2X communication to vehicles. Some car manufacturers already started to introduce V2X communication to vehicles, e.g. as General Motors has done this for the Cadillac CTS in 2017 [46].

The introduction of V2X communications to vehicles raises several issues. For instance, security is a major aspect to be investigated and solved to guarantee an effective and safe communication among vehicles and with other road users [47]. The second aspect to be analyzed and solved is related to the privacy of vehicle passengers [48].

Both the US and EU are working on the themes of security and privacy through several research projects [49][50], funded respectively by the US department of transport (DOT), the Car2Car Communication Consortium, and the EU community. In the US, the standard IEEE 1609.2 has to be followed to guarantee the requirements of security, while in the EU, ETSI has provided the standard TS 103 097 [51] for security header and certificate formats, TS 102367-3 [52] for DEN messages, and TS 102637-2 [53] for CAM messages.

Other issues related to V2X communication specifics for vehicles are the physical placement of the V2X platform and antennas on the vehicle, and the costs of the equipment for V2X communication and related retro-fit for vehicles that are already in use. These issues have been partially solved for V2X, and are currently some of the aspects to be covered by the future 5G platform for vehicles. They are described in the following paragraphs.

14.7.1 Placement of the V2X Platform within the Vehicle

One of the issues to be solved related to the introduction of V2X in vehicles is where to place related functionality, i.e., whether this should be integrated into a unique platform with other external communication related functionality, or as a separate board. In the following, several configurations are presented:

- **Roof-mounted integrated smart antenna**: In a unique antenna, there are all the possible communication solutions: V2X, 4G, BT, Wi-Fi, GPS, and eventually radio AM/FM. An additional human machine interface could be provided inside the vehicle to show to the driver the advertisement coming from V2X, otherwise this information could be integrated in the cluster or in the dashboard, by using respectively the same area of the body computer messages, or of the navigator. This solution, all in one, has the advantage to cover with one hardware all the possible communication sources, so it minimizes the cabling, which reduces the latencies. However, it has also some disadvantages: The presence of several antennas in the same area could cause interferences, and the diversity of cellular V2X w.r.t. 802.11p is not fully usable because antennas are too close to each other. Moreover, the smart antenna is external to the vehicle, so it is easiest to be removed by a third person. It could reach very high temperatures (more than

100 °C), which are not acceptable for some modules, like the 4G chipset. The presence of the 4G module on the smart antenna requires a big space for integration in the global hardware. Moreover, the computational capabilities cannot be scaled or upgraded, since they are on the same hardware as the antennas;

- **RF front-ends under the roof, antennas on the roof, computational unit in the cluster or dashboard**: This solution has the advantage of being more robust to the temperatures, which are usually under 85° under the roof, and to be scalable/upgradable, since the computational unit could be exchanged with new versions without affecting the roof side. The presence of all antennas and RF front-ends respectively up and under the roof reduces the cabling inside the vehicle, which is limited to the cable going from the RF front-end to the computational unit. Less cables inside the passenger compartment will reduce the weight of the vehicle. Moreover, when the RF front-end is connected to the computational unit just with a digital connection, there are no signal losses, which are instead present when the cabling is analog. The disadvantage of this solution is the presence of multiple hardware components, which could be hacked from outside;
- **Standalone box installed in the dashboard, and antennas on the roof**: The standalone box includes both RF front-end and computational unit for V2X. It is a solution that allows the retrofit to all previous vehicles already sold, while the first two solutions were not ready for retro-fit. Moreover, this solution has the advantage of dedicating a computational unit exclusively to V2X, so it does not have the constraints of the first solution, where the computational unit is used also from the other communication sources. The disadvantage is the connection between the standalone box and the antennas, which requires RF cables all around the vehicle, with higher costs and weight and large signal loss. Moreover, the standalone box does not usually guarantee all connectivity functionalities together (4G, BT, Wi-Fi, V2X), so it would be possible to have in the vehicle multiple sources of connectivity from outside, which could be cyber-attacked;
- **Smart corner**: The RF front-end is placed within the rear and/or front lamps, while the computational unit is placed in the dashboard. With this placement, the maximum radio diversity can be exploited. The computational unit, internal to the vehicle, is connected with a digital link to the corners. Inside the lamp, also the global positioning system (GPS) antenna could be integrated. This distributed solution improves the reliability since there are 4 RF front-ends, and it can be easily upgraded since the computational unit is separated from the front-ends.

14.7.2 Test Deployments and Outcomes

Tests on V2X communication have been conducted in the EU and the US, with several goals:

- Test V2V applications in some use cases (mainly in the US);
- Test V2I applications for safety and road information;
- Test of the first prototypes of security systems developed for V2X.

In the EU, there is the C-ITS corridor project Rotterdam-Frankfurt-Vienna, involving the Netherlands, Germany and Austria. In this project, the road works application is tested, safety applications are tested, and security is experimented [54]. Another pilot in Europe is Scoop@France [55], involving Renault and PSA group as original equipment manufacturers (OEMs) with 2000 vehicles and several hundreds of RSUs. The objective of the pilot is to test the "Day 1" applications. In France, also the Transpolis urban mobility lab has been built [56]. 20 million Euros are invested from both the private and the public sector. Among the several topics that this site has to cover, there are the applications of "Day 1", but also autonomous vehicles and connected cars. These pilots all start from the previous field tests already experimented in the EU, driven by European projects like SimTD [50], Fot-Net [57], TEAM [58], Drive-C2X [49], etc.

In the US, there have been several pieces of news about possible mandates from the government related to the introduction of V2X on all vehicles, starting from new ones, and using some after-market product to equip also old vehicles [59]. For this reason, car makers, suppliers and chipset provides accelerated the tests and many pilots were developed, and many others are currently in development phase. Some of them are:

- **Southeast Michigan testbed** [60]: This was started in 2007 and used to develop and test the proof-of-concept engineering project conducted by the US DOT and the auto industry. It was essential to determine the feasibility and technical limitations of DSRC operating at the 5.9 GHz bands. It was further focused on security, cooperative adaptive cruise control (CACC), V2I safety applications, and environment-related applications. The same test bed has been extended for a large deployment of connected vehicles and infrastructure that will address real-world transportation problems in this region. This will be the first of its kind in the world. 20,000 vehicles will be involved, most of them will be company-owned and employee-driven fleets. Up to 500 infrastructure nodes will be installed, located based on safety and congestion needs, and near OEM facilities. Up to 5,000 safety devices will be distributed and installed: after-market safety devices, retro-fit safety devices, pedestrian safety devices, and nomadic seed devices. It will have a fully functional security system. It is expected to be operational starting from 2017 until 2019;
- **Safety Pilot - Ann Arbor testbed** [61]: This is a research initiative managed by the National Highway Traffic Safety Administration (NHTSA) and the Research and Innovative Technologies Administration (RITA) Intelligent Transportation Systems Joint Program Office. It is operational since 2012. More than 2800 DSRC-equipped vehicles including passenger cars, trucks, buses, motorcycles and a bicycles are used, 25 RSUs are installed, several V2V and V2I safety applications are tested, and many stakeholders have the possibility to access and use the operational environment;
- **Mcity- Ann Arbor testbed** [62]: In Ann Arbor, a unique closed testbed for evaluating the capabilities of connected and automated vehicles and systems has been established. Operational from July 2015, it includes approximately 5 lane miles of roadway with intersections, traffic signs and signals, sidewalks, benches, simulated buildings, robotic pedestrians, street lights and obstacles such as construction barriers. The environment and road geometry varies to simulate the different situations that connected and automated vehicles will face on live roads;
- **Wyoming I80 Corridor** [63]: In Wyoming, 170 miles of roads have been equipped with infrastructure, starting from 2015, to provide fleets with: advisories, roadside alerts, parking notifications and dynamic travel guidance. This test site would be used to give to commercial vehicles and fleets continuous information related to traffic and weather (wind, snow, fires, visibility) to reduce the road congestion caused by trucks moving slowly or stopping the traffic, which is a big problem in that area;
- **New York & Brooklyn** [64]: Three pilots have been started in the New York area, where there is a high accident rate due to the many intersections, to the difference among day/night conditions, and the mix of residential and commercial buildings, which concentrate in a unique area many drivers of very different attitudes.

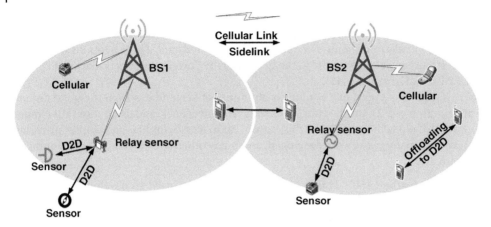

Figure 14-15. Application of D2D communication in non-public-safety scenarios.

The drawback of the deployment and the introduction of DSRC infrastructure is of course the costs: the equipment of all the road infrastructure, for highways, for cities, for all vehicles, requires high cost, which cannot be shared with the costs related to the other technologies. This is one of the reasons why some geographical areas like China and Europe are evaluating to use cellular V2X (C-V2X) to obtain the same safety results as with DSRC/ITS-G5, but allowing for the reuse of cellular infrastructure that is already deployed. 5G communication will become the natural evolution for the V2X applications, because it will be able to meet requirements of latencies more restrictive than the legacy cellular technologies, while as new cellular technology it is expected to ultimately provide a high penetration rate. This can overcome the problem of the costs for the installation of V2X equipment.

14.8 Further Evolution of D2D Communications

As one of the critical technical enablers for 5G networks, D2D communication not only provides an efficient approach to enable V2X communication to improve safety and transport efficiency, but also an efficient alternative to cope with massive machine-type communications (mMTC) and enhanced mobile broadband (eMBB) services. For instance, as can be seen from Figure 14-15, some remote sensor devices with high pathloss values to the BS can transmit their reports in the uplink through a relay sensor. In this way, the uplink coverage can be extended and the power consumption of the remote sensors can be improved for mMTC services. Figure 14-15 also describes the use case where D2D communication is exploited to offload the cellular network traffic and ultimately provide a higher system capacity. In this section, we provide insights into the application of D2D communication to serve non-public-safety services.

14.8.1 Exploitation of D2D to Enhance mMTC Services

In 3GPP, the possibility to exploit D2D communication for mMTC applications is also studied in [65] where one cell phone acts as a relay for a group of mMTC sensors. Since the D2D pairing procedure is performed in a distributed manner without help from a BS, it brings a loss in global awareness. Additionally, very few cell phones will appear in deep indoor or rural areas, especially at night. Thus, the solution proposed in [65] has limited efficiency.

In order to improve the performance of mMTC services, a new solution has been proposed where D2D communication among different mMTC devices is enabled and exploited. To achieve high efficiency for D2D communication, a clustering approach is hence required at the BS to make sure that a relay UE only serves remote UEs in its proximity. Afterwards, the BS needs to configure the transmission mode (TM) of each UE, taking into account context information such as channel state information (CSI), location and battery level information. Once a D2D link is successfully established, it is used to relay the report of remote UEs. Therefore, the context-aware D2D communication can be divided into two steps:

- Device clustering and TM selection;
- Uplink report of remote UEs through D2D.

14.8.1.1 Device Clustering and TM Selection

As aforementioned, a device clustering algorithm should be performed in the BS to allow only intra-group sidelink communication. When a remote UE is initially attached to the network, it needs to synchronize to the BS at first. Due to a higher available transmission power at BS, there is a better coverage in the downlink compared to the uplink, and therefore the remote UE can obtain the synchronization signals from the BS. After synchronization, the remote UE receives the system information blocks (SIBs) where the D2D-related configuration information is provided. Upon receiving the system information, the remote UE reports its context information (e.g., channel state information, location and battery level) to the BS. Since the deployment of the new sensor is performed manually by technicians, such information can be submitted by the equipment of technicians. Based on the received information, a TM update procedure can be executed. This procedure also applies to the case where the BS updates the TM of its served UEs. In this procedure, the BS first performs the context-aware device clustering and transmission mode selection (TMS) algorithm. After that, both the relay and remote sensors will be paged and configured by the BS. Besides, other dedicated control information is also transmitted (e.g., IDs of the sensors involved in sidelink communication, resources for sending sidelink discovery messages, and conditions that the sidelink should fulfill). The relay UE will then send a discovery message to the remote UE, in which the reference signals are embedded. The remote UE determines whether the discovery request should be accepted based on the estimated reference signal received power (RSRP). After that, the remote UE sends back an ACK/NACK message. And in case the discovery is accepted, a security association is established between the relay and remote UEs. At the end, the result of the discovery process is further fed back from the relay UE to the BS. The BS will avoid to set up the same D2D link if a NACK message is received and the remote UE stays in cellular TM until the next TM update procedure.

14.8.1.2 Uplink of Remote UEs

Once a D2D link is established, it is exploited for uplink packet transmission of the remote UE. The corresponding configuration information is carried by the SIBs and downlink control information (DCI), which should be stored at the UE when the D2D discovery procedure is successful. When the remote UE is paged by the network or a new data packet arrives at the device buffer, the remote UE transmits its packet to the relay UE. In this step, both D2D ends should be aware of the time and

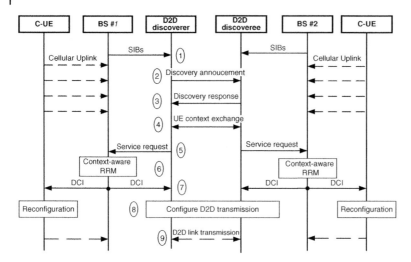

Figure 14-16. Signalling diagram in multi-cell scenario.

frequency resource for the D2D transmission. After successfully receiving packets from remote UE(s), the relay UE replies with acknowledgment message(s) to the remote UE(s). Otherwise, a NACK message is transmitted back and triggers the re-transmission procedure. Afterwards, the relay UE further forwards the received packets to the serving BS. This process can be performed as a normal cellular uplink transmission where a control plane (CP) connection needs to be established. Another alternative is that the received packets will be buffered in the relay UE and then transmitted together with its own packet. In this case, an advantage in power saving for the relay UE can be achieved since the relay UE only needs to perform the CP connection establishment procedure once. Upon successfully receiving packets from the relay UE, the BS sends an acknowledgment message to the relay UE. It should be noted that, along with this communication procedure, real-time context information of both the relay and remote UEs should be transmitted in the data reports, in order to update the context information at the BS.

14.8.2 Radio Link Enabler in Reuse Mode to Improve System Capacity

In order to reuse the uplink spectrum of the cellular network, context information [66][67] can be used to efficiently control the mutual interference. Therefore, signaling schemes supporting context-aware RRM for D2D communication should be designed in 5G. In Figure 14-16, a proposed signaling scheme is illustrated where the two D2D UEs are served by two different cells.

At first, UEs receive SIBs broadcasted by their serving BSs and dedicated to D2D operation. In 3GPP, the currently standardized SIB18 and SIB19 [4] provide information not only on the resource pools used for the transmission of D2D discovery and communication messages in the serving cell, but also on resource pools used in neighboring cells. With this information, a UE is able to receive the discovery message sent by another UE from a different cell. Therefore, the D2D discoveree can successfully receive the discovery request sent by the D2D discoverer. Along with the discovery message, reference signals for D2D channel estimation are carried and the D2D discoveree decides whether the discovery request should be accepted based on the radio quality. If the D2D discovery request is accepted, the D2D discoveree replies an ACK message to the D2D discoverer. Then, the context information of the two D2D UEs are exchanged and they send the service request with the collected information to their serving BSs. After that, both BSs run the RRM algorithm jointly, and a configuration message is sent from each BS to its D2D UE. A coordination between different BSs is required, in order to guarantee that a D2D UE does not receive its D2D and cellular transmission in the same time and frequency resource. Moreover, the configuration message contains information about which resource is used for each D2D UE to transmit and receive D2D data. Based on the acquired information, each of the D2D UEs configures its transmission with the allocated resource. Finally, the bi-directional D2D transmission can start. Please note that the D2D discovery procedure does not require the two UEs to enter RRC Connected state. In addition, one D2D UE only needs to enter RRC Connected state to send the D2D service request message to the BS. In a scenario where one D2D UE is out of the cell coverage, only the D2D user within the cell coverage can send the D2D service request to the serving BS. After receiving the configuration information from the BS, this UE will forward it to the other UE which is out of cell coverage.

14.8.3   Radio Resource Management for D2D

14.8.3.1   Context-aware Clustering and TMS to Optimize D2D-based mMTC

From efficiency perspective, D2D communication should only be applied between two nearby devices. An approach to ensure this property is to apply clustering among devices and allow only intra-cluster D2D communication. Additionally, a clustering algorithm can assist remote sensors to find proper relay sensors. The design and implementation of an efficient clustering algorithm should take into account the context information (e.g., location information, traffic type, battery life requirement, etc.) which can help to improve system performance. For instance, in a dense urban scenario, sensors located deeply indoor or in basements experience high penetration loss. Therefore, relay sensors may be the sensors located on the higher building floors.

After clustering, in order to configure the TM of each device, the BS needs to estimate its battery life by analyzing the context information. For example, the propagation loss can determine the power consumption during one uplink transmission period, and the traffic type is directly related to the uplink report frequency of the device. The following approach is used to both configure the relay and remote UEs and generate the numerical results reported in the next section:

1) Sensors with good cellular links and enough battery capacity are considered as relay sensor candidates;
2) Sensors out of the network coverage and sensors with battery life lower than requirement try to establish a D2D link with the relay sensors;
3) If the propagation loss in between two D2D ends is below a threshold, the D2D link can be successfully established, and it is used to relay the uplink reports of the remote sensors.

In Figure 14-17, the system performance of the proposed context-aware D2D communication to assist mMTC services is given, together with the performance of the cellular network without using D2D communication as a baseline. Regarding the clustering algorithm, the total number of

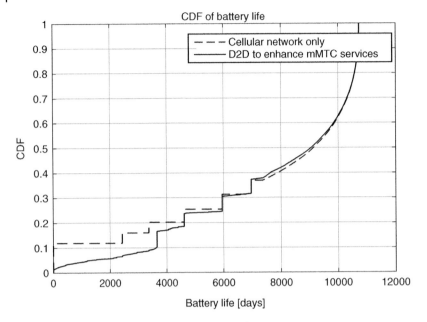

Figure 14-17. System performance w.r.t. battery life.

clusters is set to be $K = 1000$. At first, the BS selects K devices with the cellular pathloss values lower than a predefined threshold. Then, these K devices are considered as centroids of the K clusters, and each of the other devices is attached to the cluster, the centroid of which has the shortest distance from the considered device. The power consumption parameters in [68] are used in this work. Additionally, the Madrid grid model proposed in [69] is used as the environment model, see also Section 15.2.3. Moreover, an inter-site-distance (ISD) of 1732 m [70] is applied. To capture the propagation characteristics, 3D channel models proposed in [70] and [71] are implemented for cellular and D2D channels, respectively. Considering the traffic model, each device tries to send an uplink packet of 1000 bits every five minutes. The other parameters are aligned with [70] and [72].

In Figure 14-17, the cumulative distribution function (CDF) of the battery life of UEs is plotted. As it can be seen from the figure, 15% of MTC UEs cannot transmit uplink reports to the BS by using only the cellular uplink, while only 3% of UEs cannot connect to the BS if the D2D communication is exploited. Thus, the service availability is improved from 85% to 97%. Moreover, 76% of UEs can meet the battery life requirement of ten years by using the cellular uplink, while this value can be improved to 89% by exploiting the D2D communication.

14.8.3.2 Efficient Reuse of Cellular Resources for D2D Communication

In order to improve the overall system capacity, certain cellular traffic can be offloaded to D2D communication. In this scheme, a D2D underlay transmission mode refers to the case where the resource of cellular links is reused by the D2D transmission. In order to foresee whether the uplink resource of a cellular link can be efficiently reused by a D2D link, the BS needs to be aware of the

Table 14-2. System performance w.r.t. system capacity (parameter: $\dfrac{d_m}{d_{(m,n)}} \geq 1$).

	Cellular	D2D	Cellular+ D2D	Capacity w/o D2D
Capacity	30.84 Mbps	27.36 Mbps	58.20 Mbps	36.05 Mbps

mutual interference between the D2D link and cellular links. Considering a scenario with M cellular links and N D2D links, $M \times N$ channel gain information is required to characterize the interference power for the D2D links, where each value stands for the link quality between a cellular UE and a D2D receiver. Considering that the values of M and N can be large, the collection of these interference power levels can introduce heavy signaling overhead. Therefore, in order to achieve an efficient signaling scheme, other context information can be exploited to quantify the interference. For example, a BS may consider that the interference from a cellular link to a D2D receiver is under control if the following equation is fulfilled:

$$\frac{d_m}{d_{(m,n)}} \geq Z,$$

where d_m is the distance between the two D2D ends, and $d_{(m,n)}$ is the distance from the cellular UE to the receiver of the D2D link. Z is a predefined threshold value and it is set to be 1 to generate the numerical result discussed later. Moreover, if the interference from the D2D transmission causes a SINR deterioration lower than 1.5 dB for the cellular link, the BS considers that the cellular resource can be reusable.

To evaluate the performance of the proposed D2D communication scheme, outdoor UEs with a maximal transmission power of 23 dBm are deployed in a Madrid grid scenario defined in [69], with a density of 1000 users/km^2. Macro cells are deployed and operate at 800 MHz with 10 MHz uplink bandwidth. Moreover, the number of D2D links is statistically the same as the number of cellular uplink UEs. The 3D channel models proposed in [69] are applied to model both the cellular and D2D links. More detailed information regarding the simulation models and assumptions are captured in [69].

Table 14-2 shows the system performance of the proposed D2D communication. The target SNR values for the open loop power control are randomly generated in intervals from 10 dB to 15 dB for cellular links and 7 dB to 10 dB for D2D links. In Table 14-2, the notation "Cellular+D2D" represents the overall capacity of both cellular and D2D links. As a comparison, the cellular capacity without D2D communication is denoted as "Capacity w/o D2D". As seen from this table, due to the additional interference from D2D links in reuse mode, the capacity of cellular links in the proposed scheme has approximately decreased by 15% compared with the case where cellular link resources are not reused for D2D. However, thanks to the contribution from the D2D links, the overall capacity is increased by 61% compared with the cellular network where no D2D communication is allowed. The above results demonstrate the performance gain by applying smart D2D communication. On the other side, another key aspect for this technology is how to efficiently collect the required context and maintain this in the central unit.

14.8.4 Cooperative D2D Communication

Cooperative D2D communication where D2D pairs implement relay functionalities to facilitate transmission between a cellular user and its BS is a way to improve spectrum efficiency. In cooperative D2D communication, multiple transmitters or the source nodes are allowed to transmit simultaneously in the same and limited spectrum resources whether they are within the BS coverage or outside coverage. Thus, such approach increases the spatial spectrum utilization of the system. Through cooperation, transmitting terminals can together form a virtual antenna array to increase their transmit reliability or throughput. However, this typically requires data sharing transmission and coordinated joint transmission among cooperating terminals, which can be costly, especially under long-term resource balancing considerations. Besides, some fairness consideration is also important to allow each D2D pair to achieve at least the same performance as that with no cooperation. Cooperative D2D communication may involve unicast D2D communication and/or one-to-many/all D2D communication among pairs of devices over the PC5 interface. A cooperative communication scheme enables the 5G RAN to dynamically allow cooperative D2D mode selection and communication, at the same time ensuring interference mitigation, e.g., in case of simultaneous D2D communication and cellular user to BS communication over the shared radio resources, etc. To enable cooperative D2D communications, among others, approaches for cooperative mode selection, relay selection, cooperative transmission, and resource allocation are discussed in [73].

In D2D communication, interference management is one of the key topics to ensure high spectral efficiency. Various techniques involving MIMO signal processing, power control, and transmission mode selection have been proposed to reduce the interference between the D2D pair and cellular users, especially when multiple D2D pairs are allowed to share the same channel. Some mechanisms need to be designed to further mitigate the interference both among D2D pairs and between the D2D transmitters (DTs) and the cellular system. By allowing cooperation among DTs, more D2D pairs can be allowed to transmit simultaneously in the same and limited spectrum resource, increasing the spatial spectrum utilization of the system. This approach is studied in detail in [74].

Performance comparisons of the average sum throughput between the considered cooperative D2D transmission method and non-cooperative approaches have been done for different D2D pair numbers. We consider a scenario, as shown in Figure 14-18a, where the BS is located at the origin, and a cellular user is randomly distributed in a circular cell of radius r_1 meters. DTs are randomly distributed with uniform distribution in a circular area with radius d_1 meters, and D2D receivers (DRs) are randomly distributed similarly in a ring with an inner and outer radius equal to d_2 and d_3 meters, respectively.

In Figure 14-18b, we show the average sum rate versus the transmit SNR of the individual pairs for the case where d_1=1 meter, d_2=30 meters, d_3=40 meters and r_1=300 meters for both the considered cooperative D2D transmission method and the non-cooperative method. This is the case where the DTs are close to each other. The performance of the considered algorithm is compared with the non-cooperative case. From the simulation results, we can see that the average sum rate increases with the number of D2D pairs regardless of whether cooperation exists because more D2D pairs are considered for data transmission. However, the improvement will saturate when the number of D2D pairs is large enough. Besides, the average sum rate for the considered cooperative D2D transmission method is better than that for the non-cooperative method because resource balancing and fairness (considering a rate-gain constraint) are considered in the investigated method.

(a)

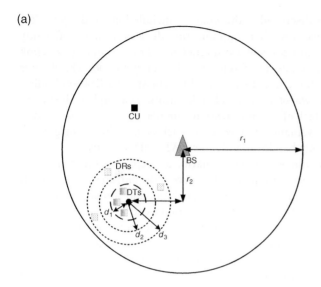

(b)

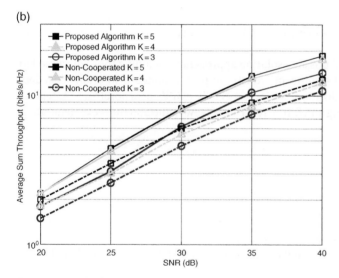

Figure 14-18. (a) Illustrations of D2D pairs, BS and cellular user distributions in the simulation (b) Performance comparisons between proposed cooperative D2D transmission method and without any cooperative method at different D2D pair number.

14.9 Summary and Outlook

In 5G, the direct communication mode between two nearby devices is expected not only to contribute toward performance improvement, but also to enable new applications and services. This chapter has introduced various aspects of D2D and V2X communications. Related applications encompass all three envisioned 5G services, i.e., ultra-reliable low-latency communications (URLLC), eMBB and mMTC. D2D and V2X design challenges, as discussed in this chapter, include how to design reference signals and control signals between two directly communicating devices, how to efficiently utilize cooperative D2D communication to improve spectrum efficiency, how to manage sidelink mobility challenges, and how to control and mitigate interference in vehicular networks. The chapter has provided a comprehensive summary of the current status in different standardization bodies, including 3GPP and ETSI. Further, candidate waveforms for sidelink communication have been discussed, for example in order to mitigate the lack of synchronization. Further, an efficient device discovery approach has been proposed and evaluated, followed by sidelink mobility management functionalities and procedures, including group handover and inter-RAT D2D mobility.

From the perspective of a traffic safety and efficiency scenario, a deep inspection on V2X communications has also been carried out. Shortcomings of the current decentralized congestion control method and system design guidelines aiming at a better reliability and transmission delay have been discussed. It is proposed to use different transmission ranges to meet the diverging requirements of the real world and at the same time improve spectrum efficiency. Further, insights on the industrial implementation of V2X communication have been provided, including placement options of the V2X platform and up-to-date testbed deployments. Section 14.8 has evaluated the highlights and benefits of using D2D communications for mMTC and eMBB services. It can be seen that by assigning some devices as relays for other remote users, battery life and availability in mMTC services can be efficiently improved. Additionally, enabling D2D communication to reuse resources of cellular links and applying cooperative D2D communication can improve the system throughput.

There are several design challenges and possible solution approaches, including those discussed in this chapter. For a wider deployment and significant industry impact, a standard solution is needed. It will be interesting to see how standardization bodies, especially 3GPP, incorporate and integrate various D2D and V2X aspects into ongoing and future releases of New Radio and the Next Generation system architecture.

References

1 3GPP TS 22.278, "Service requirements for the Evolved Packet System (EPS); stage 1", March 2017
2 3GPP TS 23.303, "Proximity-Based Services (ProSe) (Release 14)", Dec. 2016
3 3GPP TS 36.211, "Evolved Universal Terrestrial Radio Access (E-UTRA); Physical channels and modulation", June 2017
4 3GPP TS 36.331, "Evolved Universal Terrestrial Radio Access (E-UTRA); Radio Resource Control (RRC); Protocol specification (Release 14)", April 2017
5 3GPP TS 33.303, "Proximity-based Services (ProSe); Security aspects", June 2017
6 3GPP TS 22.185, "Service requirements for V2X services", March 2017
7 3GPP TR 22.886, "Study on enhancement of 3GPP support for 5G V2X services", March 2017
8 3GPP TS 23.285, "Architecture enhancements for V2X services", June 2017
9 3GPP TS 36.201, "Evolved Universal Terrestrial Radio Access (E-UTRA); LTE physical layer; General description", March 2017
10 3GPP TS 36.213, "Evolved Universal Terrestrial Radio Access (E-UTRA); Physical layer procedures", June 2017

11 ETSI TR 101 607, "Intelligent Transport Systems (ITS); Cooperative ITS (C-ITS); Release 1", V1.1.1, May 2013

12 ETSI EN 302 663, "Intelligent Transport Systems (ITS); Access layer specification for Intelligent Transport Systems operating in the 5 GHz frequency band", V1.2.1, July 2013

13 ETSI EN 302 665, "Intelligent Transport Systems (ITS); Communications Architecture", V1.1.1, Sept. 2010

14 ETSI TS 102 687, "Intelligent Transport Systems (ITS); Decentralized Congestion Control Mechanisms for Intelligent Transport Systems operating in the 5 GHz range; Access layer part", V1.1.1, July 2011

15 ETSI TS 102 637-2, "Intelligent Transport Systems (ITS); Vehicular Communications; Basic Set of Applications; Part 2: Specification of Cooperative Awareness Basic Service", V1.2.1, March 2011

16 ETSI TS 102 637-3, "Intelligent Transport Systems (ITS); Vehicular Communications; Basic Set of Applications; Part 3: Specifications of Decentralized Environmental Notification Basic Service", V1.1.1, Sept. 2010

17 ETSI EN 302 636-3, "Intelligent Transport Systems (ITS); Vehicular Communications; GeoNetworking; Part 3: Network Architecture", V1.2.0, Oct. 2014

18 ETSI EN 302 895, "Intelligent Transport Systems (ITS); Vehicular Communications; Basic Set of Applications; Local Dynamic Map (LDM)", V1.1.1, Sept. 2014

19 ETSI TR 102 638, "Intelligent Transport Systems (ITS); Vehicular Communications; Basic Set of Applications; Definitions", V1.1.1, June 2009

20 IEEE Std 1609, and sub sections: 1609.0, 1609.2, 1609.3, 1609.4.

21 SAE, "DSRC Implementation Guide - A guide to users of SAE J2735 message sets over DSRC", Feb. 2010

22 IEEE P1609.6, "Draft Standard for Wireless Access in Vehicular Environments (WAVE)—Remote Management Services"

23 5G Automotive Association, White Paper, "The Case for Cellular V2X for Safety and Cooperative Driving", Nov. 2016

24 5G Automotive Association, "Comments Of The 5G Automotive Association In the Matter of Proposed Changes in the Commission's Rules Regarding Federal Motor Vehicle Safety Standards; V2V Communications", April 2017

25 CAR 2 CAR Communication Consortium Statement, "Strong support for the European Strategy on C-ITS Deployment", Feb. 2017

26 European Commission, COM (2016) 766, "A European strategy on Cooperative Intelligent Transport Systems, a milestone towards cooperative, connected and automated mobility", Nov. 2016, see https://ec.europa.eu/transport/themes/its/c-its_en

27 European Commission, "C-ITS Platform - Final Report", Jan. 2016

28 European Commission, "C-ITS Platform Phase II – Final Report", Sept. 2017

29 D. Martín-Sacristán, J. F. Monserrat, J. Cabrejas-Penuelas, D. Calabuig, S. Garrigas, and N. Cardona, "On the way towards fourth-generation mobile: 3GPP LTE and LTE-advanced", EURASIP Journal on Wireless Communications and Networking, Aug.2009

30 V. Vakilian, T. Wild, F. Schaich, S. Ten Brink and J.-F. Frigon, "Universal filtered multi-carrier technique for wireless systems beyond LTE", IEEE Global Communications Conference (GLOBECOM 2013) Workshops, Dec. 2013

31 P. Siohan, C. Siclet and N. Lacaille, "Analysis and design of OFDM/OQAM systems based on filterbank theory", IEEE Transactions on Signal Processing, vol. 50, no. 5, pp. 1170–1183, Aug. 2002

32 M. Bellanger, "FBMC physical layer: a primer", PHYDYAS, Jan. 2010

33 K. W. Martin, "Small side-lobe filter design for multitone datacommunication applications", IEEE Transactions on Circuits and Systems II: Analog and Digital Signal Processing, vol. 45, no. 8, pp. 1155–1161, Aug. 1998

34 D. Garcia-Roger, J.F. de Valgas, J.F. Monserrat and N. Cardona, "Hardware Testbed for Sidelink Transmission of 5G Waveforms without Synchronization", IEEE International Symposium on Personal, Indoor and Mobile Radio Communications - (PIMRC 2016), Sept. 2016

35 Mango Communications, "WARP v3 Kit", see https://mangocomm.com/products/kits/warp-v3-kit

36 A. Asadi, Q. Wang and V. Mancuso, "A survey on device-to-device communication in cellular networks", IEEE Communications Surveys Tutorials, vol. 16, no. 4, pp. 1801–1819, Q4 2014

37 N. K. Pratas and P. Popovski, "Network-assisted device-to-device (d2d) direct proximity discovery with underlay communication", IEEE Global Communications Conference (GLOBECOM 2015), Dec. 2015

38 N. H. Mahmood, G. Berardinelli, F. Tavares and P. Mogensen, "On the potential of full duplex communication in 5G small cell networks", IEEE Vehicular Technology Conference (VTC Spring 2015), May 2015

39 M. G. Sarret et. al., "Can Full Duplex reduce the discovery time in D2D Communication?", IEEE International Symposium on Wireless Communication Systems (ISWCS 2016), Sept. 2016

40 ETSI EN 302 665, "Intelligent Transport Systems (ITS); Communications Architecture", V1.1.1, Sept. 2010

41 K. Bilstrup, E. Uhlemann, E. G. Ström and U. Bilstrup, "On the ability of the 802.11p MAC method and STDMA to support real-time vehicle-to-vehicle communication", EURASIP Journal on Wireless Communications and Networking, vol. 2009, no. 5, Jan. 2009

42 K. Bilstrup, E. Uhlemann, E.G. Strom and U. Bilstrup, "Evaluation of the IEEE 802.11p MAC Method for Vehicle-to-Vehicle Communication", IEEE Vehicular Technology Conference (VTC Fall 2008), Sept. 2008

43 J. J. Blum, A. Eskandarian and L. J. Huffman, "Challenges of intervehicle ad hoc networks", IEEE Transactions on Intelligent Transportation Systems, vol. 5, no. 4, pp. 347–351, Dec. 2004

44 A. Alonso and C. Mecklenbraeuker, "Dependability of decentralized congestion control for varying VANET density", IEEE Transactions on Vehicular Technology, vol. 65, no. 11, pp. 9153–9167, Nov. 2016

45 A. Vesco, R. Scopigno, C. Casetti and C.-F. Chiasserini, "Investigating the effectiveness of decentralized congestion control in vehicular networks", IEEE Global Communications Conference (GLOBECOM 2013) Workshops, Dec. 2013

46 Safe Car News Press Release, "Cohda Wireless powers V2X technology in 2017 Cadillac CTS", May 2017, see http://safecarnews.com/cohda-wireless-powers-v2x-technology-in-2017-cadillac-cts/

47 H. Krishnan and A. Weimerskirch, "Verify-on-Demand – A Practical and Scalable Approach for Broadcast Authentication in Vehicle-to-Vehicle Communication", SAE 2011, World Congress, April 2011

48 A. Weimerskirch, J.J. Haas, Y.-C. Hu and K.P. Laberteaux, "Chapter 9 - Data Security in Vehicular Communication Networks", in H. Hartenstein and K.P Laberteaux (editors), "VANET Vehicular Applications and Inter-Networking Technologies", John Wiley & Sons, March 2010

49 Drive C2X European project, see http://www.drive-c2x.eu/project

50 SimTD project, see http://www.simtd.de/index.dhtml/enEN/index.html

51 ETSI ITS 103 097, "Security Header and Certificate Formats", V1.1.1

52 ETSI ITS 102 637-3, "Decentralized Environmental Notification Basis Services", V1.1.1

53 ETSI ITS 102 637-2, "Cooperative Awareness Message", V1.2.1

54 C-ITS Corridor, see http://c-its-korridor.de/?menuId=1&sp=en

55 Scoop@F, see http://www.scoop.developpement-durable.gouv.fr/

56 Transpolis, see http://www.transpolis.fr/en

57 Fot Net Data project, see http://fot-net.eu/

58 Tomorrow's Elastic Adaptive Mobility (TEAM) project, see https://www.collaborative-team.eu/

59 National Highway Traffic Safety Administration (NHTSA), Press Release, "US DOT to Move Forward with Vehicle to Vehicle Communication Technology for Light Vehicles", Dec. 2016

60 Mcity Infrastructure (Michigan), see https://mcity.umich.edu/research/mcity-infrastructure-data-collection-and-management-system-development/

61 Safety Pilot, see http://safetypilot.umtri.umich.edu/index.php

62 Mcity - Ann Arbor Test Bed, see https://mcity.umich.edu

63 US DOT pilot in Wyoming, see http://www.its.dot.gov/pilots/pdf/04_CVPilots_Wyoming.pdf

64 US DOT pilot in Ney York City, see https://www.its.dot.gov/pilots/pdf/02_CVPilots_NYC.pdf

65 3GPP RP-151948, "New WI proposal: D2D Based MTC", Dec. 2015

66 J. Lianghai, A. Klein, N. Kuruvatti, R. Sattiraju and H.D. Schotten, "Dynamic Context-aware Optimization of D2D Communications", IEEE Vehicular Technology Conference (VTC Spring 2014), International Workshop on 5G Mobile and Wireless Communication System, May 2014

67 J. Lianghai, A. Klein, N. Kuruvatti and H.D. Schotten, "System Capacity Optimization Algorithm for D2D Underlay Operation", IEEE International Conference on Communications (ICC 2014), Workshop on 5G Technologies, June 2014

68 J. Lianghai, B. Han, M. Liu and H.D. Schotten, "Applying Device-to-Device Communication to Enhance IoT Services", IEEE Communications Standards Magazine, vol. 1, no. 2, pp. 85–91, 2017

69 5G PPP METIS project, Deliverable D6.1, "Simulation guidelines", Oct. 2013

70 3GPP TR 45.820, "Cellular System Support for Ultra-Low Complexity and Low Throughput Internet of Things (CIoT) (Release 13)", Nov. 2015

71 3GPP R1-132030, "Channel models for D2D performance evaluation", May 2013

72 ITU-R M.2135, "Guidelines for evaluation of radio interface technologies for IMT-Advanced", 2008

73 5G PPP METIS II project, Deliverable D6.1, "Draft Asynchronous control functions and overall control plane design", June 2016

74 5G PPP METIS II project, Deliverable D6.2, "5G Asynchronous control functions and overall control plane design", April 2017

Part 4

Performance Evaluation and Implementation

15

Performance, Energy Efficiency and Techno-Economic Assessment

Michał Maternia[1], Jose F. Monserrat[2], David Martín-Sacristán[2], Yong Wu[3], Changqing Yang[3], Mauro Boldi[4], Yu Bao[5], Frederic Pujol[6], Giuseppe Piro[7], Gennaro Boggia[7], Alessandro Grassi[7], Hans-Otto Scheck[8], Ioannis-Prodromos Belikaidis[9], Andreas Georgakopoulos[9], Katerina Demesticha[9] and Panagiotis Demestichas[10]

[1] *Nokia, Poland*
[2] *Universitat Politècnica de València, Spain*
[3] *Huawei, China*
[4] *Telecom Italia, Italy*
[5] *Orange Labs, France*
[6] *iDATE, France*
[7] *Politecnico di Bari, Italy*
[8] *Nokia, Sweden*
[9] *WINGS ICT Solutions, Greece*
[10] *University of Piraeus, Greece*

15.1   Introduction

In the race towards the 5th generation (5G) of mobile and wireless communications, many stakeholders are involved with their own proposals for technologies, algorithms and procedures. Especially in the current phase of definition of the new system, the right choice between these proposals is essential to secure the success and widespread adoption of 5G. In particular, it is expected that the new generation will enable novel business opportunities, as stressed in Chapter 2, and provide a performance leap that will justify expenses related to its development and deployment.

The introduction of the new standard is a very long and expensive process and, additionally, major design agreements are extremely difficult to revert once the system is mature. Hence, the need to quantify the performance of key design concepts long before any type of hardware implementation is available (e.g., prototypes or trial equipment) is extremely important. Such evaluation is often done by means of 'pen and paper' analysis or through various computer simulations. In order to provide an accurate and unbiased assessment, the right evaluation procedures, metrics and models need to be discussed and agreed among the stakeholders involved in the process. Additionally, an economic evaluation of the introduction of 5G needs to be taken into account, as also stressed in Chapter 2.

5G System Design: Architectural and Functional Considerations and Long Term Research, First Edition.
Edited by Patrick Marsch, Ömer Bulakçı, Olav Queseth and Mauro Boldi.
© 2018 John Wiley & Sons Ltd. Published 2018 by John Wiley & Sons Ltd.

This chapter is organized as follows. Section 15.2 provides information on major performance evaluation frameworks that were developed for legacy and 5G systems. Additionally, this section presents performance assessments for some of the representative 5G use cases, as introduced in Section 2.5. Section 15.3 focuses on network energy efficiency, i.e., the capability of the network to operate with certain (preferably low) electric power consumption. In recent years, optimization of this factor has become one of the most important criteria for mobile network operators (MNOs) and a major target for telecom equipment vendors. Section 15.4 then introduces a techno-economic analysis of potential 5G deployments, giving insight into costs and benefits of running the 5G network. Finally, Section 15.5 summarizes this chapter and highlights the key performance and economic evaluation results that 5G brings.

15.2 Performance Evaluation Framework

In recent years, wireless telecommunications have been at the forefront of the technological development of the modern society. As explained in Chapter 2, comparing to legacy solutions, the 5G system is designed not only to push to the extreme the broadband access for humans, but it is also expected to focus on different kinds of machine-type communications and related services. This diversity of services motivates the evolution and extension of the existing evaluation methodology, mainly focused on human-based mobile broadband communication, to a comprehensive and unbiased **5G evaluation framework** allowing for a fair comparison of proposed concepts, which will give an insight into the achievable performance in most relevant 5G use cases.

15.2.1 IMT-A Evaluation Framework

Even if the 5G system will introduce a set of brand new functionalities, the evaluation framework does not need to be developed from scratch. There is a clear experience of success in the process followed for the evaluation of International Mobile Telecommunications Advanced (IMT-A). IMT-A is recognized as the 4th generation (4G) of mobile and wireless networks, the one generally recognized as the enabler of the widespread success of mobile broadband applications. In the 2007-2008 timeframe, the radio sector of the International Telecommunication Union (ITU-R) Working Party (WP) 5D developed the evaluation guideline for IMT-A for its performance assessment, and defined performance metrics and technical performance requirements. A summary of these can be found in two ITU-R reports, i.e., the report on guidelines for the evaluation [1], and the report on radio interface requirements [2]. The former report contains the detailed simulation assumptions and the evaluation methodologies for IMT-A. This document represents a significant reference ensuring the proper harmonization of the tools used by the independent external evaluation groups for the performance evaluation of IMT-A technology candidates. It includes three components of the evaluation framework:

- *Test environments*, which consist of:
  - A *traffic model* based on the service to be evaluated;
  - A *deployment scenario*, which provides the geographical characteristics where the service is deployed (e.g., indoor hotspot, dense urban area);
  - An *evaluation configuration*, i.e., the assumed evaluation parameters applied to the selected traffic (service) and deployment scenario.

- *Evaluation methodology and procedures* for each key performance indicator (KPI):
  - High-level assessment method, e.g., inspection, analysis or simulation, as defined in more detail in Section 15.2.3;
  - Detailed evaluation method and procedure.
- *Evaluation models*, e.g., channel model, etc.

In 2009 and afterwards, the 3rd Generation Partnership Project (3GPP) has taken the IMT-A evaluation method (with minor enhancements) as the baseline for the assessment of LTE and LTE-A. The main assumptions for system level simulation are captured in [3], while link level simulation considerations (needed, e.g., for performance evaluation of advanced receivers) can be found in [4].

On the other hand, the IEEE body has also elaborated a methodology to evaluate IEEE 802.16 m, also referred to as Worldwide Interoperability for Microwave Access (WiMAX), which was its IMT-A proposal. This methodology is captured in [5], and has many commonalities with those of 3GPP, although it is more detailed in some parts, such as link-to-system mapping or traffic models.

15.2.2 IMT-2020 Evaluation Process and Framework

The formal evaluation framework for 5G, known as IMT-2020 evaluation framework, is the official process of ITU-R, as also described in Section 17.2.2. The most obvious candidate to meet the IMT-2020 requirements is the New Radio (NR) standard developed by 3GPP.

ITU-R has developed reports on minimum technical requirements [6] and evaluation guidelines [7] for IMT-2020 technology proposals, which will be used in the IMT-2020 evaluation and submission process, also detailed in Section 17.2.2. In addition, Task Group 5/1 in ITU-R is tasked with conducting the sharing and compatibility studies for World Radio Conference 2019 (WRC-19), in order to secure 5G spectrum globally.

The main steps of the ITU-R process and alignment with 3GPP work are shown in Figure 15-1. As an initial step, a Circular Letter to invite technology proposals was released in March 2016. Further steps consist of the submission of technology proposals, followed by their official evaluation. It should be noted that ITU-R itself doesn't develop the technical specifications nor the evaluation of candidates. It instead announces a call for external evaluation bodies and requests the contribution from the scientific world to complete this task.

Figure 15-1. ITU-R process and its alignment with 3GPP specifications.

In [8], the detailed IMT-2020 submission and evaluation process is defined. Nine steps are described to approve a candidate radio interface technology (RIT) or a set of RITs (SRIT) as a part of the IMT-2020 specification. Here, step 2 and step 7 define the "entry criteria" and the "exit criteria" for the process, respectively. Any candidate RIT/SRIT needs to fulfil the requirements defined in step 2 according to the following submission process:

- A RIT needs to fulfil the minimum requirements for at least three test environments: two test environments related to enhanced mobile broadband (eMBB), and one test environment for massive machine-type communications (mMTC) or ultra-reliable low-latency communications (URLLC), representing the 3 main 5G service types defined in Section 2.2;
- A SRIT consists of a number of component RITs complementing each other, with each component RIT fulfilling the minimum requirements of at least two test environments and together as a SRIT fulfilling the minimum requirements of at least four test environments comprising the three service types.

If it fails, the proposed RIT/SRIT cannot enter the ITU-R submission process for IMT-2020. In step 7, a candidate RIT/SRIT that successfully passed the step 2 needs to further fulfill the following requirement to be approved as (part of) IMT-2020: the RIT/SRIT must meet the requirements for all five test environments defined in [7], comprising the three main service types eMBB, mMTC and URLLC.

This process allows the proposals to target an initial capability as defined in step 2, and then later achieve the full capability with respect to all 5G usage scenarios, for instance through a further development or consensus building. By these means, a powerful and unified 5G standard that gains wide industry and regional support is expected.

Just like in the case of IMT-A evaluation, 3GPP has committed to submit its proposal to the IMT-2020 process in the beginning of 2020. First considerations on scenarios, requirements and models for (but not limited to) this process can be found in [9].

15.2.3 5G PPP Evaluation Framework

Before the official IMT-2020 evaluation framework was established, several 5G Public Private Partnership (5G PPP) projects have investigated the topic of 5G evaluation to satisfy the needs and challenges set for 5G in different fora, all identifying the need to update the evaluation methodologies considered in previous standards. Firstly, because of the novel requirements posed by new services, scenarios or system configurations, and secondly because of the new technologies that are foreseen to fulfill these requirements, as for instance outlined in [10].

Concerning the 5G services, as explained in Section 2.2, the focus of 5G is not only on mobile broadband, but also on massive connectivity and reliable and low-latency communication. This implies new devices for machine-type communication as well as new traffic models and KPIs (e.g., reliability or connection density). Additionally, the scenarios investigated for 5G are not just based on regular homogeneous base station (BS) placements, but are increasingly closer to realistic heterogeneous deployments. Moreover, the 5G system is assumed to be based on the integration of multiple radio access technologies (RATs), such as evolved LTE, NR, Wi-Fi, etc., and involve multi-connectivity among these or among multiple transmission points, as covered in Section 6.5. Finally, 5G will also involve operations at higher frequencies with large carrier bandwidths, and the usage of massive MIMO, as detailed in Section 11.5. These aspects also require extensions of the channel models to properly capture radio propagation at higher frequencies and the correlation of channel characteristics at different carrier frequencies, as described in Section 4.3.4.

With regard to the new technologies that are being proposed to fulfil the 5G requirements, the use of new waveforms, as detailed in Section 11.3, has gained quite some attention. There is a clear impact of this change on the 5G evaluation in the link-level modeling and link-to-system mapping, requiring the development of new models. 3D beamforming is another technology that influences the simulation models and especially the channel models, as 3D extension becomes a must for an accurate evaluation. In addition, device-to-device (D2D) communication capabilities, or the formation of moving networks, should be integrated in the 5G evaluation methodology and also have implications on the channel modeling needs, as detailed in Section 4.3.7.

In response to these factors, a new evaluation framework has been detailed by 5G PPP in [11]. This document analyzes 5G use cases investigated in several 5G PPP projects, defines the appropriate KPIs, and proposes a set of performance evaluation models.

The first set of KPIs are the so-called **inspection KPIs**, whose evaluation is based on the examination of statements from each specific 5G proposal. The inspection KPIs are basically questions that can be answered with a yes or no. 5G PPP considers six of these:

- **Bandwidth and channel bandwidth scalability**, referring to the ability of the system to operate with different bandwidths (at least supporting 1 GHz) and carrier frequencies;
- **Deployment in IMT bands**, i.e., allowing to deploy 5G in at least one identified IMT band;
- **Operation above 6 GHz**;
- **Spectrum flexibility**, i.e., the capability to accommodate different downlink (DL) and uplink (UL) transmission patterns in both paired and unpaired frequency bands;
- The support of **inter-system handover** between 5G and at least one legacy system;
- Efficient **support of a wide range of services** over a continuous single block of spectrum.

All inspection KPIs were evaluated positively in [12] for the 5G system proposed in METIS-II. For **analytical KPIs**, namely those KPIs that are evaluated through calculations based on available technical information, the 5G PPP framework assumes the following:

- **Mobility interruption time**: a time span during which a user equipment (UE) cannot exchange user plane packets with any BS during transitions between the cells. This KPI relates to the capability of 5G to provide a continuous connectivity for devices on the move. It was shown that 0 ms interruption time is possible, if multi-connectivity solutions are employed, as discussed in Section 6.5. This is in line with ITU-R requirements;
- **Peak data rate**: the highest theoretical single-user data rate, assuming error-free transmission and utilization of all radio resources for the corresponding link direction. This value is linked to the maximum supported number of MIMO streams, modulation order, coding scheme and transmission bandwidth. Although peak data rates are unlikely to be experienced in realistic operations (a KPI of *experienced* user data rate is a far more accurate approximation, as explained later on), they show a potential of the cellular system to cater for the needs of broadband services. For peak data rates, values of 21.7 Gbps and 12.4 Gbps were assessed for DL and UL, respectively, which is above the ITU-R target;
- **mMTC device energy consumption**: reflected through the device battery lifetime without recharging and using a single 5 Wh battery, under the assumption that the device is stationary and the energy consumption is related only to communication aspects. This KPI reflects the ability of

the 5G system to provide an energy efficient procedure for emerging Internet of Things (IoT) services. Assuming a sporadic data transfer of low payloads, a lifetime of 10 years or more can be reached;

- **Control plane latency**: represents the transition time from an inactive and energy efficient mode (e.g., when devices do not exchange any user data with the network) to an active mode. Low values are necessary for energy efficiency reasons and to provide an always-connected experience. For the new generation of cellular devices, when using the newly introduced Radio Resource Control (RRC) Connected Inactive state as detailed in Section 13.3, control plane (CP) latency can be as low as 7.125 ms, i.e. far below the target of 20 ms set by ITU-R;
- **User plane latency**: defined as the one-way transmission time of a packet between the transmitter and the receiver. This KPI not only relates to the efficiency of the radio interface, but also, e.g., to the handling of data buffers at the device side. User plane latency is assumed to comprise the following steps: (1) transmitter processing delay at the BS, (2) frame alignment, (3) synchronization, (4) transmission of a packet over a number of transmit time intervals (TTIs), (5) Hybrid Automatic Repeat reQuest (HARQ) retransmission probability, and (6) receiver processing delay in the UE. Taking into account all these factors and assuming a 0.125 ms TTI for a single packet transmission, a value of 0.763 ms can be obtained, which is below the 1 ms target of ITU-R.

In order to complete the evaluation of 5G system concepts, link-level and system-level simulations are required to assess the KPIs that depend on the actual propagation conditions or system load. The **simulation KPIs** considered in the 5G PPP framework are:

- **Experienced user throughput**: the instantaneous data rate measured separately for DL and UL;
- **Traffic volume density**: the total number of bits correctly received by the infrastructure (in UL) or UE (in DL), measured over a certain geographical area and a period of time divided by the considered area and period;
- **End-to-end (E2E) latency**: one trip time or round-trip time in a packet transmission. In each case, the time is measured at the interfaces between layers 2 and 3;
- **Reliability**: the percentage of packets successfully received in a system within the maximum E2E latency. In this context, **availability** is typically defined both as the percentage of locations where the user gets the desired quality of experience, and the probability of a service not being blocked;
- **Retainability**: the percentage of time when transmissions fulfill the experienced user throughput or reliability requirements;
- **mMTC device density**: the maximum density of users supported in a spatial area with a minimum percentage of messages correctly received;
- **RAN energy efficiency**: defined as the overall energy consumption in the 5G RAN compared to the energy consumption in the RAN of legacy systems;
- **Supported velocity**: an estimate of the maximum velocity for which a certain data rate can be achieved;
- **Complexity**: a KPI that may refer to the size or volume of an analog component, the number of operations of a digital process, or the cost of a certain implementation;
- **Coverage**: While different definitions exist, one common definition in the context of broadband services [11] is that this is calculated as the experienced user throughput over the target value and expressed as a percentage. More precisely, a user is assigned a value of 100% if its throughput is equal or higher than the target, and a proportionally lower percentage otherwise. It is averaged over all realizations.

The novelty in the 5G PPP framework does not only come from the definition of new KPIs, but also from the proposal of new simulation deployment scenarios and models. These scenarios and models are covered in [12], and detailed in several deliverables from METIS-II [13], FANTASTIC-5G [11], mmMAGIC [14] and SPEED-5G [15]. In addition, some characteristics of the 3GPP evaluation framework for 5G covered by [9] have been incorporated in the 5G PPP framework.

On one hand, proposed models and configurations consider synthetic deployment scenarios, namely indoor hotspot, urban macro, outdoor small cells, and rural macro or long-range communications. Several possible configurations are provided for each one. On the other hand, the major novelty of the deployment scenarios in the 5G PPP framework comes from the definition of realistic deployment scenarios, such as an indoor office scenario or the so-called *Madrid grid* [13]. The definition of these aims at providing more realistic conditions for the evaluation of 5G.

The deployment scenarios definition is complemented with the specification of the user, traffic, channel and mobility models for individual use cases. Concerning the channel modeling, the basic references are the IMT-A channel models [1][3], extended to support higher frequencies, bandwidths, numbers of antennas and 3D models [16]. Moreover, additional aspects have been included for 3D modeling [17], for the support of high speeds [18], for propagation in small cells [19], and for vehicular communication links including direct communication between vehicles [20], as also covered in Section 4.3.7. With regard to traffic models, both full buffer and bursty File Transfer Protocol (FTP)-like traffic is considered. The latter has different parameterizations ranging from the simplest case with fixed packet sizes and packet inter-arrival times to complex random values of these parameters generated according to exponential or Poisson distributions.

Compared to analytical KPIs, evaluation results for simulation KPIs vary strongly between use cases, deployment scenarios and traffic models, therefore it is impractical to discuss them without a wider context. The following section gives exemplary evaluations of simulation KPIs and a basic background information derived from selected 5G PPP projects.

15.2.3.1 FANTASTIC-5G

The goal of FANTASTIC-5G was to define a flexible multi-service air interface for the 5G system [21]. It focused on 7 different use cases (i.e., 50 Mbps everywhere, high speed train, sensor networks, tactile Internet, automatic traffic control/driving, broadcast-like services, and dense urban society, cf. Section 2.5) and developed a number of technical solutions allowing the 5G air interface to significantly improve the performance of the baseline 4G technology, as well as to satisfy the respective 5G KPIs.

Wide Area Coverage Scenario

Mobile users require high-speed Internet connection for advanced interactive services. To provide a satisfactory experience, a minimum data rate should be consistently provided to all the users, even at the cell-edge. Most studies set this limit to at least 50 Mbps in DL and 25 Mpbs in UL [22]. To reach this goal, FANTASTIC-5G enhanced the typical macro-cell deployment with the usage of massive MIMO, based on a 2-stage precoding strategy [23]. The first stage uses a grid-of-beams configuration, which uses a beamforming matrix to create a regular grid of highly directional signals in the angular domain. The second stage is a regularized zero-forcing precoder working on top of the first stage, which can be seen as a virtual array of high-gain antennas. Additionally, FANTASTIC-5G also formulated a coordinated beamforming method for inter-cell interference reduction, derived from [24]. It involves a sub-sectorization of the cells and the activation of specific patterns of sub-sectors, which reduces the average interference among adjacent cells. To demonstrate the performance gain provided by developed technical components, various system level simulations were conducted. First of all, three main scenarios were taken into account: (1) rural, with 100 users/km^2 and the inter-site distance (ISD) set to 1000 m, suburban with 400 users/km^2 and an ISD of 600 m, and urban with 2500 users/km^2 and an ISD of 200 m. At the physical layer, an array of 16x8x2 antennas with half-wavelength spacing over a bandwidth of 100 MHz was used [25].

Evaluation results for a rural scenario are depicted in Figure 15-2 and demonstrate the traffic density going from 1.38 Gbps in legacy LTE-A up to 12.4-14.7 Gbps in the new solutions proposed for 5G. For the other scenarios, the absolute values change depending on the ISD, but the gain is similar. Figure 15-3 shows the service coverage, defined here as the ratio of the experienced user throughput over the target throughput (50 Mbps in DL), limited to 100% when the throughput is exceeded. Moreover, in this case it is possible to observe that the new techniques can greatly outperform LTE-A, as the coverage is increased from 25-50% to more than 95%.

Video Broadcasting Scenario

In this scenario, the aim is to investigate the efficiency of one-to-many transmission techniques able to allow a large number of mobile users to receive the same real-time video stream. As a baseline approach, the 4G technology proposes the multicast-broadcast single frequency network (MBSFN) technique [3], where multiple BSs transmit the exactly same signal under tight synchronization, and all the replicas add up in power at the mobile users. To improve efficiency and reliability, FANTASTIC-5G enhanced the simple MBSFN approach with two new technical components, namely an adaptive selection of the modulation and coding scheme (MCS) and a HARQ retransmission of broadcast packets [26]. Specifically, the MCS is chosen based on the

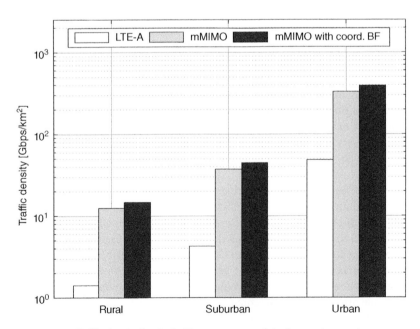

Figure 15-2. Traffic density for the "wide area coverage" deployment scenario.

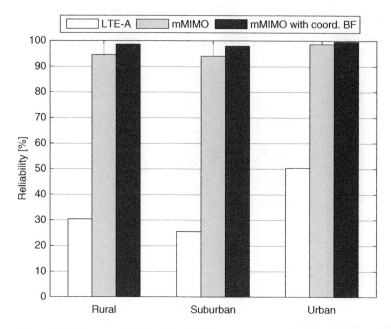

Figure 15-3. Coverage obtained for a 50 Mbps target user throughput in DL in the "wide area coverage" deployment scenario.

users' channel quality indication (CQI) feedback, similarly to non-MBSFN operation, and broadcast packets that are received with errors by some users are retransmitted only to such users, over dedicated unicast bearers using HARQ. This concept is evaluated using a transmission of a high definition video encoded at 17 Mbps, assuming a bandwidth of 20 MHz and 6 dedicated sub-frames for every radio frame made up by 10 sub-frames [25]. Without loss of generality, it is assumed that the single frequency network (SFN) is extended on a large scale. However, for each cell, only the closest rings of adjacent cells are able to boost the channel quality through a constructive signal. The others, instead, become progressively more interfering as the propagation delay exceeds the duration of the cyclic prefix. Results from system-level simulations are reported in Figures 15-4 and 15-5. They show that there is an optimal value for the MCS, because for lower values the throughput is reduced, and for higher values too much reception errors occur. The most advanced approach automatically selects the best MCS value without static or manual configuration. In addition, the packet loss rate is reduced compared to the baseline approach (for any MCS value) thanks to the HARQ retransmissions.

15.2.3.2 METIS-II

METIS-II has developed the overall 5G RAN design and provided the technical enablers needed for an efficient integration and use of the various 5G technologies and components. Additionally, METIS-II has provided the 5G collaboration framework within 5G PPP for a common evaluation of 5G RAN concepts, and prepared concerted action towards regulatory and standardisation bodies. A summary of the simulation KPIs is captured in Table 15-1 and followed by two exemplary evaluations of the METIS-II use cases.

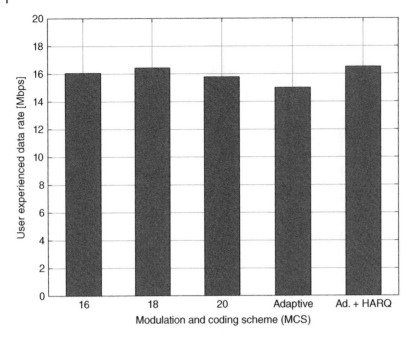

Figure 15-4. User experienced data rate for the "video broadcasting" scenario.

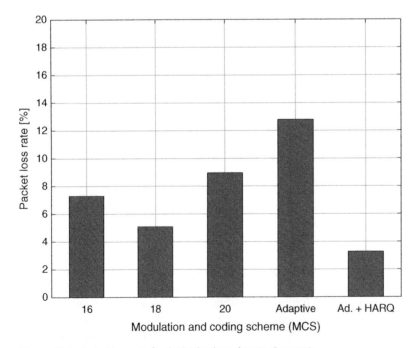

Figure 15-5. Packet loss rate for the "video broadcasting" scenario.

Table 15-1. Summary of simulation performance evaluation results from METIS-II [12].

Use case	KPI	Expected performance	Evaluated performance
Dense urban information society	Experienced user throughput	300 Mbps	>1 Gbps
Virtual reality office		Up to 5 Gbps	7.85 Gbps
Broadband access everywhere		50/25 Mbps for DL/UL	50/25 Mbps for DL/UL
Massive deployment of sensors and actuators	mMTC device density	>1 million/km^2	4 million/km^2
Connected cars	Reliability	99.999% at 50/1000 m for urban/highway	99.999% at 45/150 m for urban/highway with 20 MHz Requirement fulfilled with 40/100 MHz for urban/highway

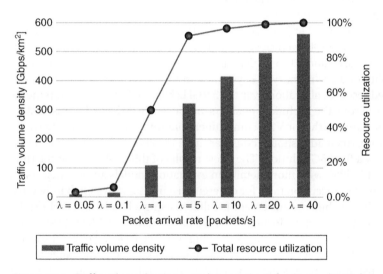

Figure 15-6. Traffic volume density vs. packet arrival rate for dense urban information society.

Dense Urban Information Society Use Case

In this use case, users exchange information with cloud servers and with other users, devices or sensors located in close vicinity, see also Section 2.5.1. A heterogeneous network of BSs deployed in a dense urban environment caters for connectivity requirements. Two important KPIs that are used to quantify 5G performance are network energy efficiency, covered in more detail in Section 15.3.4, and traffic volume density. The latter reflects the capability of the 5G system to handle massive traffic, which is expected especially in cities due to the large concentration of users in limited areas. Results presented in Figure 15-6 show the DL traffic volume density that may be supported by 5G for a given traffic load, represented by the packet arrival rate for individual users. It should be highlighted that these results were obtained for simulations limited to operations using 100 MHz only, mainly for simulation complexity reasons. In practice, bandwidths as high as 1 GHz can be expected in such deployments, leading to a potential improvement factor of at least 10, given that beyond a linear scaling of capacity with system bandwidth also pooling and multiplexing gains would be expected.

Connected Cars Use Case

This use case addresses the information exchange among vehicles and with the infrastructure to enable a safer and more efficient transportation and real-time remote computing for mobile terminals. The evaluations conducted in METIS-II focused on traffic safety and efficiency, where a reliability of 99.999% for the transmission of packets is required with a maximum E2E delay of 5 ms, considering certain communication ranges that depend on the mobility scenario. To assess this requirement, it is assumed that each vehicle periodically broadcasts packets of at least 1600 payload bytes with a repetition frequency of at least 5-10 Hz. Three main mobility environments are considered, namely an urban, rural and highway environment, with a maximum speed of 60 km/h, 120 km/h and 250 km/h, along with a required coverage range of 50 m, 500 m and 1 km, respectively. Further, three relevant scenarios are envisioned, namely a realistic urban scenario, earlier on already introduced as the so-called Madrid grid, a synthetic urban scenario, and a highway scenario [27]. The last two are based on scenarios defined by 3GPP in [20]. Concerning the density of vehicles, the specific values considered in the evaluation were 1000 vehicles/km^2 in the urban realistic scenario, 595 vehicles/km^2 in the urban synthetic scenario, and 10.25 vehicles per lane and km in the highway scenario. METIS-II has assessed the ability of a preliminary 5G system with direct vehicle-to-vehicle (V2V) communication and a centralized resource allocation over the 5.9 GHz band to fulfill the requirements of the connected cars use case. From the technical enablers proposed in METIS-II, this system considers the availability of large bandwidths for V2V communications, but no multi-antenna transmission scheme is used. Therefore, the reader should consider these results as a baseline evaluation that could be further improved by a final 5G solution. The latest results of this evaluation at the time of writing this book can be found in [28], with more details on evaluation models and results. As an example, Figure 15-7 shows the packet reception ratio for different distances and system bandwidths in the urban realistic scenario. The reliability requirement is fulfilled using a bandwidth of 40 MHz, while the system is already close to the target performance with a bandwidth of 20 MHz. Results are similar in the urban synthetic scenario. On the contrary, in the highway scenario, a large bandwidth of 100 MHz is needed to fulfil the reliability level required for a range of 1 km, while the range would be 600 m with 50 MHz. In conclusion, the results have shown that the requirements can be fulfilled with system bandwidths between 30 and 100 MHz, depending on the scenario. In addition, it seems feasible to reduce the needed bandwidth down to around 20 MHz in urban scenarios and 50 MHz in the highway scenario with a more advanced 5G system.

15.2.3.3 SPEED-5G

SPEED-5G's main objective was to investigate and develop technologies that address the well-known challenges with respect to the predicted growth in mobile connections and traffic volume by successfully addressing the lack of dynamic control across wireless network resources, so far leading to unbalanced spectrum loads and a perceived capacity bottleneck. Consequently, SPEED-5G has focused on resource management with the three degrees of freedom of densification, rationalized traffic allocation over heterogeneous wireless technologies, and a better load balancing across available spectrum.

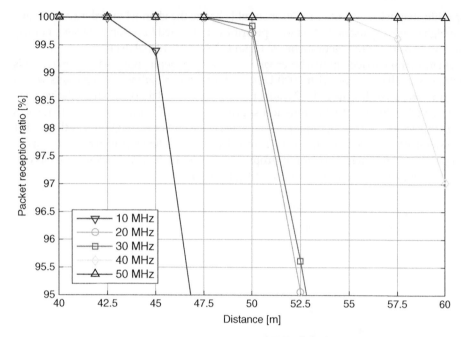

Figure 15-7. Packet reception ratio vs. distance in the Madrid Grid urban scenario.

In particular, SPEED-5G has focused on 4 different use cases: Massive IoT, broadband wireless, ultra-reliable communications, and high-speed mobility. The project has investigated various scenarios where capacity demands are the highest, but also where extended dynamic spectrum access (eDSA) [29] is expected to be most effective for exploiting co-operation across technologies and bands.

Future Dense Urban Use Case

One of the main solutions that have been investigated is in the context of broadband wireless with hierarchical management capabilities; that is, blending distributed and centralized management of ultra-dense multi-RAT and multi-band networks. In SPEED-5G, centralized management is used as a baseline and can be expanded with distributed management by moving management decisions related to RAT, spectrum or channel selection closer to the node level.

In order to obtain quality of service (QoS) and capacity expansion, operators and regulatory bodies are increasingly pursuing policy innovations based on the paradigm of shared spectrum, which allows spectrum bands that are under-utilized by primary owners to be exploited opportunistically by secondary users. Specifically, the solution captured below focuses on the Spectrum Access System (SAS) in the 3.5 GHz band, which consists of a hierarchical three-tier model: an incumbent with high priority, Priority Access Licenses (PAL) with medium priority, and General Authorized Access (GAA) with the lowest usage priority, see also Section 3.2. The higher priority users have a better utilization of channels compared to those at lowest priority. PAL and GAA users are controlled by the SAS and thus must register and check all of their operations in order to provide an interference-free environment to higher-tier users (i.e., incumbents). SAS is a monitoring system that checks whether a given user category can transmit over a specific channel, or whether the user should change the channel in order to avoid any interference to the higher priority users. In this way, priorities are formed between the users, namely high, medium and low priority for the incumbent, PAL and GAA users, respectively.

Performance evaluation was carried out in a heterogeneous network deployment with 19 macro BSs complemented with 285 small cells, and with 8000 UEs, each one downloading 2 MB packets with a different packet arrival rate, according to an FTP traffic model. Figure 15-8 illustrates the relative average DL throughput of a UE belonging to different access priorities. Specifically, it is shown that users (especially with higher priority, such as incumbents) can experience higher throughputs as the packet arrival rate increases, but after a certain point the PAL and GAA (i.e., the lower priority users) start to compete for radio resources and their throughput drops. Moreover, the relative packet transmission latency is better for higher priority users, as depicted in Figure 15-9.

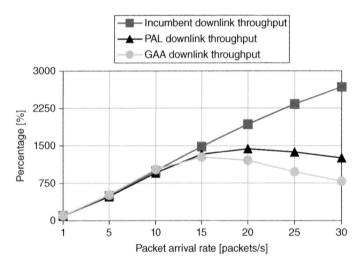

Figure 15-8. Relative increase of average DL throughput for different access priorities. Performance achieved at packet arrival rate of 1 packet/s (low load) is the baseline (100%).

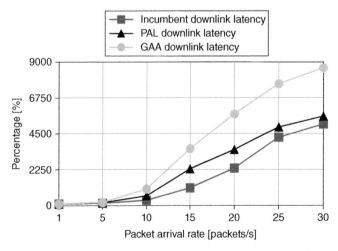

Figure 15-9. Relative increase of average packet transmission latency for different access priorities. Performance achieved at packet arrival rate of 1 packet/s (low load) is the baseline (100%).

15.3   Network Energy Efficiency

As stated previously, network energy efficiency is an important indicator for the deployed 5G networks, considering the growth of traffic and the number of users. In this section, we elaborate on why the energy efficiency has become so relevant recently, and we introduce energy efficiency metrics and methods of measurements. Finally, we propose a preliminary evaluation, based on simulations, of the energy efficiency of 5G versus the legacy radio systems.

15.3.1   Why is Network Energy Efficiency Important?

Network energy consumption translates to a substantial cost for operators. In mature markets, energy costs account for 10-15% of the total network operating expenditure (OPEX) and can reach up to 50% in developing markets with a high number of off-grid sites, or where only a poor-quality electricity grid is available [30].

Because of the rapidly increasing usage of mobile broadband connectivity, the largest network operators have recently reported a growth of 15-35% in their network energy consumption, the main reason being an increasing demand for mobile network coverage and capacity. Part of this growth is driven by the rise of the global mobile broadband subscriber base, which is expected to grow beyond six billion subscriptions globally in 2017, increasing by 10% per year, with a mobile broadband penetration likely hitting 100% by 2020. On top of this, each mobile broadband subscriber will use an average of 25-50% more data per year, resulting in an expected sevenfold increase of mobile data traffic between 2016 and 2021 [31].

In mature markets, where the number of subscribers is saturated, operator revenue has been flat. Users expect faster services with higher data rates, but are rarely willing to pay additional money. Operators must hence provide the growing data rates at constant cost. It is often claimed that energy efficiency was not considered in earlier mobile telecom generations, while in fact, it is the contrary: increasing energy efficiency has been one enabler for the rapid data traffic growth.

Energy cost is not the only driver to increase energy efficiency. In addition, global warming is a direct result of the greenhouse gas emissions caused by power consumption in general. The European Commission has correspondingly set three key objectives, known as the "20-20-20" targets: a 20% reduction in European Union (EU) greenhouse gas emissions from 1990 levels, an increase of the share of EU energy consumption produced from renewable resources to 20%, and a 20% improvement in the EU's energy efficiency [32]. Measuring the progress towards these goals is based on accurate data on energy consumption and emissions supported by several telecom power consumption and efficiency test standards developed by ITU-T, European Telecommunications Standards Institute (ETSI), the Alliance for Telecommunications Industry Solutions (ATIS), etc.

5G will introduce several new network features having a large impact on network energy consumption and efficiency. For instance, massive MIMO and antenna beam steering will be essential to increase link budget and compensate fading in particular for millimetre-wave carrier frequencies, and in general to increase spectral efficiency per area. However, the introduction of such new solutions can potentially increase power consumption. Moreover, IoT services involving a huge number of connected devices and increasing coverage requirements, as well as services related to ultra-high reliability and very high data rates and hence requiring simultaneous connection via multiple frequency bands, pose a serious challenge towards an improved energy efficiency.

15.3.2 Energy Efficiency Metrics and Models

To quantify network energy efficiency, the following metrics are often used [33]:

- **Energy per bit**, especially used in urban environments, where the planning of the network is usually capacity-constrained. Here, E stands for consumed energy in a given observation period measured at Medium Access Control (MAC) layer:

$$\lambda_I = \frac{E}{I} \quad \left[\text{J/bit} \right]$$

Under certain circumstances also throughput vs. power consumption [bps/W] can be applied as efficiency indicator.

- **Power per area unit**

$$\lambda_A = \frac{P}{A} \quad \left[\text{W/m}^2 \right],$$

typically applicable in suburban or rural environments, where the planning of the network is mainly constrained by the achieved coverage (A is the area coverage).

Both metrics are often applied by academic organizations, while for reporting and product evaluation purposes the inverse measures are usually used [34][35][37]:

- **Number of delivered bits per energy** [bit/J] is another common metric to assess equipment and operational network energy efficiency in some environmental standards from ITU-T, ETSI, etc.;
- **Coverage area per daily energy consumption** [m²/J] or [m²/Wh] is applied as energy efficiency parameter for operational mobile networks, while the previous metrics are often used also in simulated scenarios to estimate the energy efficiency without real measurements of the involved parameters.

In order to evaluate the energy efficiency of a network in a simulated scenario, power consumption models of the elements in the network are needed. One of the first widespread power consumption models was developed in the OPERA-net project [36], which was initiated in 2008 as one of the first international projects dealing specifically with mobile network energy efficiency. One of its key objectives was to develop metrics and KPIs for mobile network efficiency. Three operational sites (rural, sub-urban and urban) were selected and equipped with several power meters to allow detailed measurements of the power consumption of the different elements of the BSs' sites. Simultaneous temperature and load measurements (i.e., the amount of DL data per cell) allowed analyzing variations during the day and the correlation between power consumption, data rate and temperature. Based on these measurements, a BS model was created, which described the power consumption of different configurations (i.e., number of sectors, number of transmitters per sector, maximum installed radio frequency (RF) power and actual load), but which was at the same time simple enough to be directly applied in a network planning tool. This enabled simultaneous simulations of network capacity, coverage and power consumption for different network configurations. The power consumption of the BS was calculated as:

$$P_{BS} = \tau P_{P+} n P_{TRX} + \left(k_1 P_{RF1} + \ldots + k_n P_{RFn} \right)/c,$$

where, $k_1 \ldots k_n$ are load factors describing the fraction of available RF power transmitted in sectors $1 \ldots n$, with n being the number of installed sectors. τ denotes the increment of installed baseband processing capacity, P_P is the power consumption of the processing unit, while P_{TRX} describes the power consumption of the radio module, P_{RFn} represents the maximum RF output power of sector n, and c is a direct current (DC) to RF conversion slope parameter. This model is suited particularly to analyze the efficiency of different network configurations within a network planning tool, e.g., the effect of an increasing number of cells per macro site or the addition of small cells in a macro layer. The model requires the knowledge of practical BS power consumption parameters for different load levels, which can be derived from the BS manufacturer's data sheets. It should be noted that ETSI has created a standard to measure power consumption of BSs at different load levels [35].

Although a significant development effort was spent to decrease the power consumption as a function of the load in active mode, today's BSs show relative large and fixed power consumption level already in idle mode. BS sleep modes, where a single cell or even a complete BS is put in hibernation, play an important role to minimize network power consumption during low load levels and periods. However, the broadcast and pilot channel requirements of current radio systems limit the time when sleep modes can be activated.

To overcome this, the 5G system will introduce novel signaling channel approaches, and 5G BSs will be specifically designed to enter different sleep modes that are characterized by a different extent of functionality deactivation and consequently different extents of power savings and recovery times. For instance, it will be possible to switch fast to a sleep mode with medium power saving, allowing to switch back again fast, as opposed to hibernating the BS, involving a longer recovery time. Consequently, a new power model, which allows to model different levels of component (de-)activation and time-variant power consumption, is therefore needed.

This topic was in the past first investigated by the EARTH project [38], which addressed the global environmental challenge by studying and introducing effective solutions to lower energy consumption and increase energy efficiency of mobile broadband communication systems, without affecting users' perceived QoS and system capacity. Some interesting results from the project are collected in [39]. More specifically, some of the achievements from the EARTH project were architectures, network solutions and deployment strategies to achieve the goals of energy efficiency increase in future networks, as well as an in-deep analysis of the energy consumption sources in BSs. The project is for instance well known for the introduction of the so-called *power model* of a BS, representing a detailed investigation of the energy consumed in the different parts constituting a radio BS hardware [40].

The power model concept was carried on and refined by the GreenTouch consortium [41], which investigated energy efficiency aspects from 2010 to 2015. The ambitious goal of GreenTouch was to evolve the cellular network to ensure an increase of energy efficiency by a factor of 1000 comparing to the level in 2010. To achieve this goal, GreenTouch studied improvements in energy efficiency of mobile, fixed and core networks.

The most recent power model proposed for 5G in METIS-II [12] considers power consumption behaviour for various deployments, parameterization capability and flexibility. It allows describing the actual power consumption behavior of the whole BS under different deployment solutions and network statuses.

As it can be seen in Figure 15-10, a BS's instantaneous power consumption is basically proportional to the bandwidth load level λ with a constant power spectrum density $aPSD$. As the load level grows, the overall power consumption of the BS increases accordingly. *Pmax* is the maximum output power, while P_0 is the power consumption at the minimum non-zero output power

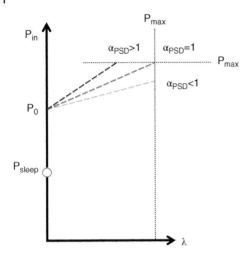

Figure 15-10. Illustration of power consumption behaviour of a BS with a constant power spectrum density.

due to load-independent operation. When the load is low, the BS can switch to micro-discontinuous transmission, which means that instead of continuous operation the BS is rapidly switched into sleep mode for a very short interval, as anticipated in the initial part of this section. During this time, the power consumption will further decrease to *Psleep*, which denotes BS power consumption in a sleep mode.

Note that the actual power consumption of the BS is tightly connected with the BS transmit power, or equivalently, to the power spectrum density ratio *αPSD*, which is defined as the ratio of the actual power spectrum density to the one with maximum transmit power multiplied with the total bandwidth. Based on Figure 15-10, the overall power consumption behaviour of a BS is calculated as:

$$P_{BS} = \begin{cases} n\left(P_0 + v_p P_{max} \lambda \alpha_{PSD} + P_1 \lambda\right), & 0 < \lambda < 1 \\ nP_{sleep}, & \lambda = 0 \end{cases},$$

where n is the number of sectors in the BS, v_p is the slope of the load dependent power consumption largely determined by the radio unit efficiency, and P_1 is baseband related power consumption.

The models described so far require relatively detailed knowledge of the BS, but at the same time omit some or all the site-specific factors. Practical network deployments include different variations of BSs: some are optimized for area coverage, others for high capacity; some are designed for extreme climate conditions and others for indoor use only. Specific models are developed for academic purposes to allow a simulation with additional environmental parameters, but without increasing the model complexity. All these models have in common that they focus on BSs only, without considering that what is really important for the operators is the power consumed in the whole site, including also all the other equipment necessary for the operation of the network, such as backhaul and fronthaul, see Chapter 7. The countless variants of sites with their specific needs and equipment are very difficult to be taken into account. The simplest way to take the site elements into account is to define a *site efficiency* as the average BS power consumption divided by the average total site power consumption (or alternatively the inverse, which is called the *power usage effectiveness*).

In this sense, the following sections describe two different sets of measurements of the energy efficiency. First, we consider the energy efficiency measured in laboratory environments, typically for the BSs and the most widely used equipment in the network, as detailed later. Secondly, we consider the measurement of the energy efficiency of a whole network in a live environment.

15.3.3 Energy Efficiency Metrics and Product Assessment In the Laboratory

As a first step to improve and measure the efficiency of the separate telecom equipment of a network, specific laboratory test standards have been developed for BSs, routers, etc. The standards allow assessing the efficiency for equipment in standalone mode under defined laboratory conditions.

In 2009, the ETSI Environmental Engineering (EE) Technical Committee published the specification on the energy efficiency of mobile radio BSs for Global System for Mobile Communications (GSM) and Wideband Code Division Multiple Access (WCDMA). This standard is regularly updated to cover the latest development in BS technology [35]. In 2013, the specification on the energy efficiency of routers and switch equipment for core, edge and access routers followed [42]. The defined metric is based on the so-called energy efficiency ratio of equipment and is defined as the throughput vs. power consumption [bps/W]. A set of weights is given to consider different load levels of the equipment as well.

In 2014, a metric and measurement method for the energy efficiency of mobile core network and radio access control equipment was specified [43].

Network performance including its power consumption can be modeled and simulated. However, simulations always depend on assumptions that might be different from real operating conditions. These simulations usually cover the telecom equipment, but it is rarely possible to include all the support equipment (like air-conditioning, security systems, lightning, networking equipment related to fronthaul or backhaul, etc.) installed at the different sites. Laboratory and field measurements are therefore unavoidable to measure the actual network performance.

Ultimately, because of the complex nature of telecom networks, product-based methods cannot describe the actual mobile network efficiency. Many features that are influencing network efficiency are not visible in a stand-alone test of the BS in a laboratory, but appear when considering the whole network. Such effects are particularly visible between different network generations and, of course, will have to be considered also when 5G equipment will be measured and evaluated with respect to previous generations.

15.3.4 Numeric Network Energy Efficiency Evaluation

The standardization activity for the evaluation of RAN energy efficiency started in 2012 with the publication of a technical report [44] that paved the way for the activities that followed and were summarized by the publication of the standard [35]. This standard, which has been also adopted as a recommendation by the ITU-T Study Group 5, presents metrics and methods to measure in a live environment the energy efficiency of mobile networks, including coexisting 2G, 3G and 4G systems.

The metrics proposed in this standard are twofold: there is the metric based on the ratio between throughput and energy consumed to deliver that throughput, and a metric based on coverage (i.e., the area covered by the network) and energy consumed by hardware providing this coverage, similar to what was described in Section 15.3.2, but applied to real networks. This duality is intended to cater for cases where the network is deployed for capacity purposes and those for coverage reasons mainly.

The method is based on a set of measurements (i.e., energy, data volume, coverage) made directly on field in the network under test, where the network is split into so-called "partial" networks, which are manageable in terms of number of requested measurements.

To extend the application of the method to wider networks, an extrapolation method is proposed. The specification is not yet applicable to 5G networks, but the activity to evaluate the extension of the specification to 5G networks is ongoing in order to suggest a set of additions and modifications to the existing specifications to cover also the new system.

Following the overall setup specified in the mentioned standards, in order to estimate the network efficiency of the 5G networks, a network energy efficiency process is proposed in [12]. This process captures the assessment of both rural and dense urban deployments, as well as a 24 hours timeframe, to account for the spatial and temporal fluctuations of traffic. This approach allows to compare 5G solutions aiming at coverage-limited (macro and rural BSs) and capacity-limited deployments (micro BSs and small cells). The main idea of the process is to prove that 5G can provide similar power consumption with respect to the traffic witnessed by early 4G deployments, even considering the massive traffic uptake expected in 2020 and beyond. The defined procedure consists of the following steps:

1) Calculate the expected traffic volume density for a 5G dense urban deployment and estimate corresponding packet inter-arrival time (IAT);
2) Scale the obtained IAT to account for different load levels of three periods calculated for traffic profiles proposed in [40];
3) Repeat steps 1 and 2 for rural 5G network deployments, considering different experienced user data rates;
4) Derive the total 5G radio network power consumption at a given load via simulations based on calculated IATs and load points;
5) Repeat steps 1-4 for baseline 4G deployments, considering 1000x lower traffic volume and different deployments of 4G versus 5G systems;
6) Integrate obtained results with network-specific weights (which can be different, e.g., from country to country) and compare 5G power consumption to 4G, to derive the overall energy efficiency improvement.

In [12], the network energy efficiency performance of the dense urban information society use case was evaluated, and the outcome is captured in Figures 15-11 and 15-12.

From Figure 15-11, it can be observed that higher 5G RAN energy efficiency performance is expected for higher traffic load levels, as more traffic can be delivered while the ratio of load-independent static power consumption could be reduced accordingly. In addition, advanced sleeping strategies can provide further performance gains, especially in low load scenarios. Figure 15-12 proves that when the load level is low, a high performance gain of 5G over 4G can be achieved, and similarly this gain increases as more advanced sleeping strategies are implemented. Even when the system load level is very high, the improvement is noticeable, which is mainly due to the introduction of small cells and more energy efficient hardware for 5G.

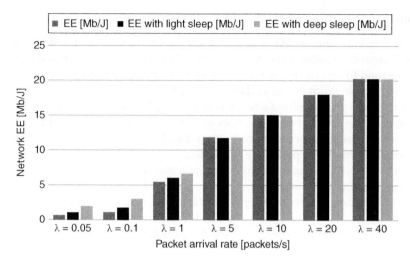

Figure 15-11. RAN energy efficiency for the dense urban information society use case.

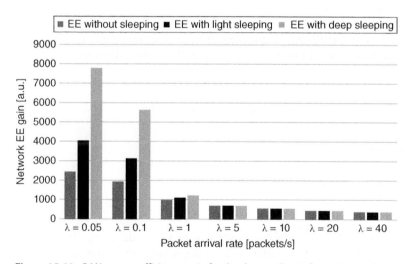

Figure 15-12. RAN energy efficiency gain for the dense urban information society use case over a baseline 4G deployment.

15.4 Techno-Economic Evaluation and Analysis of 5G Deployment

This section covers the economic assessment of the deployment of a new technology such as 5G both in terms of OPEX and CAPEX, by describing a methodology for this assessment and presenting the overall techno-economic evaluation and analysis.

15.4.1   Economic Assessment of New Technology Deployment in Mobile Networks

Before introducing a new technology such as 5G, a financial assessment to support the decision about its launch is needed. Economic studies must analyze rational economic criteria in order to make the deployment decision and to choose the best scenario for making the technology, being not only profitable in a long term, but also profitable in a relatively short period of time.

The economic analysis to judge the profitability of a new technology deployment is generally based on a free cash flow analysis: the difference between revenues and savings on one side and expenditures on the other side. The new technology is economically interesting for a company when the cash flow generated by income and/or savings overpasses the expenditures.

In practice, the net present value (NPV) is the most often used indicator to analyze the profitability of a new technology deployment. The NPV is the cumulative discounted free cash flow over the period of time under economic analysis. The technology is profitable when the NPV is positive.

The payback period is another frequently used indicator which measures the necessary time for the new technology to become profitable. It is the date when the cumulative discounted free cash flow becomes positive.

Generally speaking, an MNO can expect additional revenues generated by a new technology like 5G from new customer subscriptions, higher subscription fees and induced savings in the network operations, e.g., energy saving of new generation of equipment.

Expenditures are typically divided into capital expenditures (CAPEX) and operating expenditures (OPEX), where CAPEX include all expenditures that an operator invests for the initial setup of the network, while OPEX are recurrent expenditures to operate, run and maintain the network.

15.4.1.1   CAPEX

The main items constituting the CAPEX of a mobile network are related to:

- Radio spectrum licensing;
- Site building including site acquisition;
- Purchase of antennas and feeders;
- Purchase of radio access network equipment (e.g., BSs);
- Purchase of core network equipment;
- Transport network building.

In general, the site building in the CAPEX model covers site design, site engineering, site research, site acquisition, civil work, mast purchase and installation, housing, non-telecommunication equipment installation and commissioning, and the purchase of power supply equipment. The civil work includes concrete plinth and steel beam as foundation, support of mast and radio cabinet, fence and access path to the site, etc. The housing refers to purchase, delivery and installation of shelter, air conditioning, fire protection system and site adaptation. The electricity power equipment corresponds to electricity feeding and connection, purchase, delivery and installation of generator, fuel tank, DC power cabinet etc.

CAPEX differs for a site built on rooftop and a site built on a green field. A green field site is not situated in the existing architecture but built on the ground. A green field site's infrastructure comprises more elements, and it is often more expensive than a rooftop site. Moreover, two values of site building cost need to be considered, depending on whether it is an existing site shared with other previous generations of mobile technologies, or a totally new site dedicated for the new technology.

Finally, it should be highlighted that a major element of the CAPEX is the cost of transport network building, i.e., the costs of fronthaul and backhaul networks from the BS to the first aggregation point, for which either wireless or wireline technologies can be utilized, as detailed in Chapter 7.

15.4.1.2 OPEX

The main items constituting the OPEX are site OPEX, which may be shared among different cellular generations, energy cost (cf. Section 15.3), transport network OPEX, billing and sales cost. The site OPEX consist of site rent, labor cost and diverse maintenance costs, e.g. related to mast maintenance for macro sites, supplier maintenance, site security, site caretaking, etc. Labor cost is defined as the cost for the operator's internal staff working directly on the operation, maintenance and supervision of the network. The transport network OPEX reflects the rent paid to potential third parties for the rental of transport network infrastructure, e.g., optical fiber fronthaul or backhaul connectivity, see Chapter 7.

15.4.2 Methodology of 5G Deployment Assessment

The previous generation of cellular networks, i.e. 4G, was predominantly designed for mobile broadband users. In this context, provisioning the targeted mobile broadband experience in terms of coverage and capacity has been the main objective in the evolution of mobile networks. Usually, only the trade-off between the targeted mobile service experience on one side and the network CAPEX and OPEX on the other side was investigated. However, the increasing adoption of mMTC services brings new challenges to traditional cellular network signaling mechanisms and control plane system capacity. Therefore, even if 5G eMBB services can be assessed by making usage of classic methods, new techno-economic assessment methodology should be developed for mMTC services.

A methodology for 5G deployment assessment proposed in [45] consists of following steps:

- 5G traffic forecast;
- Estimation of 5G revenue;
- Dimensioning of 5G networks;
- Assessment of deployment scenarios;
- Techno-economic analysis.

For the **5G traffic forecast** it is assumed that eMBB traffic demand per area unit is equal to the average mobile data usage per user, times the number of users per area unit. Mobile data usage is the amount of data sent and received per user during one month. For the mMTC services, traffic forecast is based on the activity predicted per day and area, the number of devices per service, and the time the service needs to access the network and to transmit its payload.

Three classes of mMTC services have been defined in [45]. The indoor mMTC/IoT service class is foreseen for the stationary sensors deployed indoors. The outdoor services class represents sensors and actuators deployed outside, possibly involving mobility. The third class represents services that in all cases require device mobility.

Traffic profiles are defined for three cases: low load, baseline and high load, where the low and high load cases represent a load that is ten times lower or higher than the baseline, respectively. The values considered for eMBB and for mMTC are listed in Table 15-2.

Table 15-2. Parameters for eMBB and mMTC traffic profiles.

Service	Parameters	High Load	Baseline Load	Low Load
eMBB	Monthly traffic in GB/month/subscriber (heavy usage)	100	50	20
	Monthly traffic in GB/month/subscriber (medium usage)	30	15	8
	Monthly traffic in GB/month/subscriber (low usage)	4	2	1
	Share of total eMBB users [%] (heavy usage)	70	20	10
	Share of total eMBB users [%] (medium usage)	20	50	20
	Share of total eMBB users [%] (low usage)	10	30	70
mMTC	MTC services penetration rate [%]	68	48	28
	CAGR of MTC penetration [%]	39	29	19
	Device arrival window[1]	5	100	200
	Event cogeneration factor[2]	0.0028	0.00028	0.000072
	NB-IoT control plane usage rate per frame[3]	0.25	0.5	0.8

[1] defined as the number of devices that can arrive within that window.
[2] defining how many events are active and reserve some resource at given point of time.
[3] As NB-IoT is using the same resources for control and data transmission, this factor indicates how many resources can be occupied by the CP at a given point of time

To assess **5G revenues**, the average revenue per user or per unit (ARPU) for each device type is estimated. The estimation takes into account the pricing "tactics" of the operators, the evaluation of customers' willingness to pay, the traffic per customer, the 5G deployment schedules and the vertical market analysis.

IoT revenues assumptions take into account the development of mMTC and URLLC applications. In terms of technology, machine-to-machine (M2M) applications currently rely mainly on 2G/2.5G. By 2020, 4G will have a significant position, with estimated 65.2% compound annual growth rate (CAGR) at global level between 2015 and 2020. Commercial launches of connected and autonomous cars will stimulate the 5G take-off from 2021 onwards, as they require more accuracy and reactivity in data treatment, as detailed in Chapter 14.

mMTC services' ARPU is expected to be relatively low, as so-called low power wide area operators are currently setting the tariffs for the lower end of the market. Estimations of the ARPU from mMTC also take into account the trends observed from main mobile operators in Europe, South Korea and in the USA, and are depicted for the next years in Table 15-3. It should be kept in mind that mMTC is attractive for mobile operators due to the reduced churn of these subscriptions and the fact that customers pay for connectivity and not for the traffic volume.

For eMBB services, the legacy **dimensioning methods** can be used for 5G. For mMTC, a particular attention should be paid to event-driven services like smart grid, disaster management, earthquake or flood detection, etc., because they can cause an access crunch by triggering huge numbers of devices in a limited geographical area to send incident reports at the same time. In such conditions, the traffic and network access patterns will be different from those experienced in current human-centric service networks.

Table 15-3. ARPU estimates in EUR for EU 28 countries for 2020-2025 period.

Year	2020	2021	2022	2023	2024	2025
mMTC	5.0	5.0	4.0	3.5	3.0	2.5
URLLC	300	297	294	291	288	285

From mMTC perspective, the number of Physical Random Access CHannel (PRACH) slots required to achieve an appropriate access rate should be estimated first. PRACH dimensioning impacts only the uplink, while the maximum number of devices that attempt to connect within the retransmission period will also impact the downlink. In fact, more spectrum should be allocated to physical downlink control channels to meet required performance expectations.

Finally, **5G deployment scenarios** have to be assessed. Based on experience from previous cellular generations, the first step of 5G deployment is assumed to be a 5G roll-out in existing macro cell sites. The small cell deployment will follow the macro cell deployment when all macro cell frequency bands are overloaded. A maximum ratio of outdoor small cell number to a macro site is assumed, while the percentage of area covered by indoor small cells, the average number of floors and the small cell coverage surface have to be set as parameters in order to define the limit of indoor small cell density. If all macro and small cell radio resources are utilized, then the macro site densification is considered as the last technical option due to its high cost. Considered frequency bands are:

- Macro: 700 MHz FDD, 3500 MHz TDD and 2600 MHz TDD;
- Small cells: 2600 MHz TDD, 3500 MHz TDD and 30 GHz TDD.

The upper limit of an outdoor small cell capacity is estimated according to its spectrum efficiency, and its lower limit by the traffic in its coverage. The capacity of an indoor small cell is assumed to be equal to the traffic of an eMBB user for a home/office indoor small cell, and equal to its spectrum efficiency times its frequency bandwidth for a hotspot indoor small cell.

15.4.3 Techno-Economic Evaluation and Deployment Analysis Results

Following the assumptions and the methodology described in the previous subsection, a typical European dense urban area of $4 \, km^2$ is now considered for a 5G deployment techno-economic analysis, as detailed in [45]. Note that only eMBB and mMTC services are taken into account in this analysis.

Based on the methodology and traffic forecast described in Section 15.4.2, the analysis shows that macro cells will provide enough capacity during the first years of 5G roll-out. Nevertheless, after this period of time, a large number of small cells, especially for home/office indoor usage will be necessary to deliver the forecasted traffic. Comparing to other mobile technologies, small cells are of particular interest in 5G. The very big traffic volume per eMBB user in a dense urban information society will necessitate a high small cells deployment cost, which will take a much larger part of total CAPEX and OPEX than in the precedent mobile technology generations. Under the assumption that 5G eMBB ARPU remains the same as in precedent generations, or even decreases with time, the large amount of additional small cells will make it more challenging for MNOs to make 5G deployments profitable. Figure 15-13 represents the cumulative discounted cash flow of an MNO sharing a

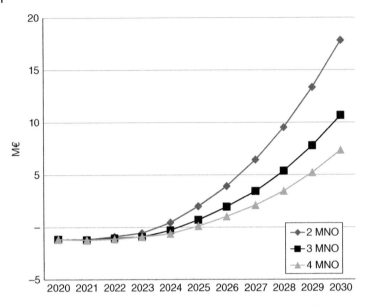

Figure 15-13. Cumulative discounted cash flow of an MNO with different numbers of MNOs in the area [45].

market with a different numbers of MNOs. The calculation has been made assuming each MNO has an equal share of the market. For the considered dense urban area, the cumulative discounted cash flow of an MNO will become positive several years after the beginning of the 5G deployment.

Only in isolated business districts where the population density during daytime is much higher than in an average European dense urban area, the macro cell densification will be required, mainly due to indoor traffic. Since it is practically impossible to densify the existing macro network in such areas, the indoor traffic offload by alternative radio solutions such as Wi-Fi should be envisioned. Regarding mMTC, it is noted that mMTC consumes little radio resources. Its incremental expenditure is in consequence very low. As a result, the mMTC contribution to MNO cash flow will be positive.

15.5 Summary

A fair evaluation of 5G performance, energy efficiency and techno-economic aspects is one of the major steps that have to be taken to answer key design questions for the new generation of cellular technology. This chapter has presented an evaluation framework for the qualitative and quantitative assessment of 5G. Comparing to 4G, such framework has to address new use cases, such as mMTC and URLLC, and also reflect the trend of going towards more diversified and heterogeneous deployments and operations in higher frequency regimes. First evaluation results indicate that 5G will bring significant gains in all generic use cases, e.g., handling an order of magnitude higher traffic densities in wide area eMBB scenarios, 10+ year operations for mMTC devices with a single battery, and 99.999% reliability for URLLC services.

5G is expected to bring a major improvement in energy efficiency of the network infrastructure, therefore this chapter has introduced some key projects and standards that aim at this goal. In order to assess energy efficiency correctly, proposed models cater for temporal and spatial variations of the traffic as well as for static and dynamic power consumption. An exemplary assessment that was done for dense urban deployments indicates that 5G is able to bring energy efficiency improvements that at least follow the traffic growth, i.e. allowing for a flat overall energy consumption over time.

Finally, this chapter has provided a techno-economic analysis of 5G. Based on traffic forecasts and ARPU predictions, a methodology has been developed and applied to a European dense urban area to analyze the 5G cash flow of a mobile network operator and identify the factors impacting it. It was concluded that for the considered dense urban area, the cumulative discounted cash flow of an MNO will become positive about 4-6 years after the beginning of the 5G deployment.

References

1 ITU-R M.2135, "Guidelines for evaluation of radio interface technologies for IMT-Advanced", 2008

2 ITU-R M.2134, "Requirements related to technical performance for IMT-Advanced radio interface(s)", 2008

3 3GPP TR 36.814, "Further advancements for E-UTRA physical layer aspects (Release 9)", Mar. 2010

4 3GPP TR 36.839, "Mobility enhancements in heterogeneous networks (Release 11)", Dec. 2012

5 Y. Srinivasan, J. Zhuang, L. Jalloul, R. Novak and J. Park, "IEEE 802.16 m Evaluation Methodology Document (EMD)", IEEE 802.16 m Broadband Wireless Access Working Group, IEEE 802.16 m-08/004r5, Jan. 2009

6 ITU-R WP 5D, M.2140, "Minimum requirements related to technical performance for IMT-2020 radio interface(s)", Nov. 2017

7 ITU-R WP 5D, M.2412, "Guidelines for the evaluation of the radio interface technologies for IMT-2020", Nov. 2017

8 ITU-R WP 5D, Document IMT-2020/2 Rev. 1, "Submission, evaluation process and consensus building for IMT-2020", Feb. 2017

9 3GPP TR 38.913, "Study on Scenarios and Requirements for Next Generation Access Technologies (Release 14)", March 2016

10 Y. Wang, J. Xu and L. Jiang, "Challenges of System-Level Simulations and Performance Evaluation for 5G Wireless Networks", IEEE Access, vol. 2, pp. 1553–1561, Dec. 2014

11 5G PPP FANTASTIC-5G project, Internal Report IR2.1, "Use cases, KPIs and requirements", Oct. 2015

12 5G-PPP White Paper, "Living document on use cases and performance evaluation", Version 1, April 2016

13 5G PPP METIS-II project, Deliverable D2.3, "Performance evaluation framework", Feb. 2017

14 5G PPP mmMAGIC project, Deliverable D1.1, "Use case characterization, KPIs and preferred suitable frequency ranges for future 5G systems between 6 GHz and 100 GHz", Nov. 2015

15 5G PPP SPEED-5G project, Deliverable D3.2, "SPEED-5G enhanced functional and system architecture, scenarios and performance evaluation metrics", June 2016

16 Aalto University et al., White Paper, "5G Channel Model for bands up to 100 GHz", Dec. 2015, see http://www.5gworkshops.com/5GCM.html

17 3GPP TR 36.873, "Study on 3D channel model for LTE (Release 12)", June 2015

18 3GPP TR 36.942, "Radio Frequency (RF) system scenarios (Release 13)", Jan. 2016

19 3GPP TR 36.931, "Radio Frequency (RF) requirements for LTE Pico Node B (Release 13)", Jan. 2016

20 3GPP TR 36.885, "Study on LTE-based V2X Services (Release 14)", June 2016

21 F. Schaich, B. Sayrac, S. Elayoubi, I.-P. Belikaidis, M. Caretti, A. Georgakopoulos, X. Gong, E. Kosmatos, H. Lin, P. Demestichas et al., "Fantastic-5 g: flexible air interface for scalable service delivery within wireless communication networks of the 5th generation", Transactions on Emerging Telecommunications Technologies, Wiley, vol. 27, no. 9, pp. 1216–1224, Sept. 2016

22 NGMN, White Paper, "NGMN 5G Initiative White Paper", Feb. 2015

23 A. Adhikary, J. Nam, J.-Y. Ahn, and G. Caire, "Joint spatial division and multiplexing: The large-scale array regime", IEEE Transactions on Information Theory, vol. 59, no. 10, pp. 6441–6463, Oct. 2013

24 M. Kurras, L. Thiele, and G. Caire, "Multi-Stage Beamforming for Interference Coordination in Massive MIMO Networks", Asilomar Conference on Signals, Systems and Computers, Nov. 2015

25 5G PPP FANTASTIC-5G project, Deliverable D2.1, "Air interface framework and specification of system level simulations", May 2016

26 B. Mouhouche and M. Al-Imari, "Optimization of Delivery Time in Broadcast with Acknowledgement and Partial Retransmission", IEEE International Symposium on Broadband Multimedia Systems and Broadcasting (BMSB 2016), June 2016

27 5G PPP METIS-II project, Deliverable D2.1, "Preliminary performance evaluation framework", Jan. 2016

28 D. Martín-Sacristán, C. Herranz and J. F. Monserrat, "Traffic Safety in the METIS-II 5G Connected Cars Use Case: Technology Enablers and Baseline Evaluation", European Conference on Networks and Communications (EuCNC 2017), June 2017c

29 I.-P. Belikaidis, A. Georgakopoulos, P. Demestichas, B. Miscopein, M. Filo, S. Vahid, B. Okyere and M. Fitch, "Multi-RAT Dynamic Spectrum Access for 5G Heterogeneous Networks: The SPEED-5G Approach", IEEE Wireless Communications Magazine, Special Issue on "Dynamic Spectrum Management for 5G", July 2017

30 Nokia, "Flatten network energy consumption technology vision 2020", White Paper, Aug. 2015

31 Cisco, "Cisco Visual Networking Index: Global Mobile Data Traffic Forecast Update, 2016–2021 White Paper", Feb. 2017

32 Commission of the European communities, "Limiting Global Climate Change to 2 degrees Celsius. The way ahead for 2020 and beyond", Jan. 2007, see https://ec.europa.eu/clima/policies/strategies/2020_en

33 5G PPP METIS-II project, Deliverable D1.1, "Refined scenarios and requirements, consolidated use cases, and qualitative techno-economic feasibility assessment", Jan. 2016

34 ITU-T Recommendation L.1310, "Energy efficiency metrics and measurement methods for telecommunication equipment", Version 3.0, July 2017

35 ETSI ES 202 706, "Measurement method for power consumption and energy efficiency of wireless access network equipment", Version 1.4.1, Oct. 2014

36 CELTIC Opera-NET project, see https://www.celticplus.eu/project-operanet/

37 ETSI ES 203 228, "Environmental Engineering (EE); Assessment of mobile network energy efficiency", Version 1.2.1, Apr. 2017

38 FP7 EARTH project, see https://www.ict-earth.eu/

39 A. Galis and A. Gavrasm, "The Future Internet - Future Internet Assembly 2013: Validated Results and New Horizons", Springer, 2013

40 FP7 EARTH project, Deliverable D2.3, "Energy efficiency analysis of the reference systems, areas of improvements and target breakdown", Version 2, Jan. 2012

41 GreenTouch, White Paper, "Improving the nationwide energy efficiency in 2020 by more than a factor of 10000 in relation to the 2010 reference scenario", Aug. 2015

42 ETSI ES 203 136, "Environmental Engineering (EE); Measurement methods for energy efficiency of router and switch equipment", Version 1.1.1, May 2013

43 ETSI ES 201 554, "Environmental Engineering (EE); Measurement method for Energy efficiency of Mobile Core network and Radio Access Control equipment", Version 1.2.1, May 2014

44 ETSI TR 103 117, "Environmental Engineering (EE); Principles for Mobile Network level energy efficiency", Version 1.1.1, Nov. 2012

45 5G PPP METIS-II project, Deliverable D1.2, "Quantitative techno-economic feasibility assessment", June 2017

16

Implementation of Hardware and Software Platforms

Chia-Yu Chang[1,], Dario Sabella[2], David García-Roger[3], Dieter Ferling[4], Fredrik Tillman[5], Gian Michele Dell'Aera[6], Leonardo Gomes Baltar[2], Michael Färber[2], Miquel Payaró[7], Navid Nikaein[1], Pablo Serrano[8], Raymond Knopp[1], Sandra Roger[3], Sylvie Mayrargue[9] and Tapio Rautio[10]*

[1] EURECOM, France
[2] Intel, Germany
[3] Universitat Politècnica de València, Spain
[4] Nokia, Germany
[5] Ericsson, Sweden
[6] Telecom Italia, Italy
[7] CTTC/CERCA, Spain
[8] Universidad Carlos III de Madrid, Spain
[9] CEA-LETI, France
[10] VTT, Finland
*Note that the authors are listed in alphabetical order in this chapter

16.1   Introduction

The exponential growth in mobile data traffic, anticipated to reach a 1000-fold increase over the next decade, and the large diversity of applications - ranging from low bit-rate and low-power machine-to-machine (M2M) applications to highly interactive and high-resolution entertainment applications - impose a number of distinct technical requirements on hardware (HW) and software (SW) platforms, where all mobile communication and related functionalities are implemented and executed. Among the most prominent requirements, the following must be highlighted: the further improvement of quality of experience (e.g., capacity, latency, resilience) and the energy efficiency, as well as the scalability, modularity, and reconfigurability when multiple radio access technologies (RATs) are considered.

In order to deal with the increasing traffic demand and, especially, with the anticipated heterogeneity of traffic in 5th generation (5G) cellular systems, future networks will require an increase of HW versatility and the ability to operate with increasingly higher bandwidths, especially at millimetre-wave (mmWave) bands. Although the software-defined radio (SDR) paradigm and, in general, the

5G System Design: Architectural and Functional Considerations and Long Term Research, First Edition.
Edited by Patrick Marsch, Ömer Bulakçı, Olav Queseth and Mauro Boldi.
© 2018 John Wiley & Sons Ltd. Published 2018 by John Wiley & Sons Ltd.

digitally-assisted analog front-end have contributed in this direction, further improvements cannot be made without novel enhancements on the HW itself. Versatile and/or reconfigurable HW components and platforms, which are able to cope with all the functionalities needed and invoked by the SW domain, are thus required to deliver innovative, cost effective and efficient network elements and devices, in order to achieve a successful commercial exploitation and deployment of 5G.

In addition, the control and the management of the heterogeneous HW infrastructure and devices expected in 5G are of utmost importance. In particular, these should guarantee an effective (i.e., rapid, easy and dependable) service development and deployment, as well as the adaptability to very demanding and changing contexts of operation, while maintaining the Quality of Service (QoS) and Quality of Experience (QoE). Thus, programmability and dynamic reconfiguration through interface abstractions or uniform application programming interfaces (API) are required to enable a HW-agnostic operation.

Within the context presented above, this chapter addresses implementation challenges and a complexity analysis of 5G HW and SW platforms, and is structured as follows: The focus of Section 16.2 is on analog and mixed signal HW, whereas in Section 16.3 digital HW is addressed. The link between the HW and SW domains is established in Section 16.4, which deals with HW/SW function partitioning aspects. A set of functional requirements for SW platforms that can cope with the challenges outlined in the previous paragraph is described in Section 16.5. In Section 16.6, an example of platform implementation targeting a virtual radio access network (vRAN) or cloud radio access network (CRAN) architecture is provided, before the chapter is summarized in Section 16.7.

16.2   Solutions for Radio Frontend Implementation

16.2.1   Requirements on 5G Radio Frontends

The requirement of significantly increased radio bandwidth for mobile communication asks for exploiting additional spectral resources to the already used frequency bands, as covered in Section 3.4, and, consequently, impacts the components utilized for the air interface.

The operation in several radio bands defined in the 3rd Generation Partnership Project (3GPP) Long Term Evolution (LTE) standard [1] for carrier aggregation between 700 MHz and 3.6 GHz has been extended to the frequency range of 450 MHz to 6 GHz to include new available radio bands allowing for increased capacity in mobile systems. The concurrent operation in different radio bands increases the HW complexity when duplicating the individual transceiver chains to support each of the bands. Thus, solutions for multiple band transceivers help to decrease the HW complexity by reducing the number of components and have motivated the research on such topics.

An additional increase of operating bandwidth is achieved by exploiting higher frequency bands of centimetre and millimetre wavelength. There, several GHz of spectrum will become available for mobile communications. System architectures and air interfaces are currently being defined, and solutions for radio frontends are necessary.

Multi-antenna systems will be used in all frequency ranges to enhance the radio performance, as addressed in detail in Section 11.5. At lower frequencies, an increased number of multiple-input multiple-output (MIMO) streams is targeted to increase spectral efficiency by exploiting spatial multiplexing. Similar approaches are also considered for mmWave, where multiple antennas are required to enable mobile radio links at these frequencies by overcoming the particular radio propagation conditions through beamforming and the related antenna gain. These approaches, however, lead to HW complexity that scales (linearly or even non-linearly) with the number of antennas and MIMO streams. Solutions to decrease complexity are hence required to enable their implementation.

An efficient utilization of spectrum is mandatory due to its limited availability. Solutions for more efficient spectrum usage are necessary to increase the mobile data volume for a defined bandwidth. One considered option is to use a certain frequency band simultaneously for transmission and reception, referred to as full-duplex.

A further requirement on implemented HW components is to improve their energy efficiency with the goal of simplifying their integration, due to lower cooling requirements, and decreasing the carbon footprint of wireless communication systems, as detailed in Section 15.3. This requirement implies the utilization of semiconductor technologies with lower power consumption and techniques for energy efficient component operation. It is worth highlighting here again that if the 1000-fold increase in data rate foreseen in 5G is expected to dissipate the same energy as in 4G systems, this would require improving the energy efficiency per bit by the same factor of 1000, as also discussed in Section 2.3.

Concepts and solutions to meet the requirements for 5G radio frontends are presented in the following sections, which give insights on recent research, while more details are available in [2] and [3].

16.2.2 Multi-Band Transceivers

The focus of this section is on bands below 6 GHz. Some aspects related to mmWave bands can be found in Section 16.2.3.2, especially related to multi-antenna operation, which is key at these higher bands.

To decrease the HW complexity, the signals of multiple radio bands are ideally fed through a single transceiver chain. The multiplication of transceiver chains by the number of supported bands and the increase of the number of included components are avoided, reducing size, weight and cost. The approach is based on broadband or multi-band capability of components used for data and frequency conversion, amplification and filtering.

Newly developed radio frequency (RF) data converters facilitate the generation of signals directly at radio frequencies. This includes the digital-to-analog and frequency conversion. Operating at sampling rates between 9 and 15 Gsamples/s, they show a broadband performance of up to 2 GHz signal bandwidth positioned arbitrarily at up to 6 GHz carrier frequency. The high-speed serial data interface allows the separated transmission of signals for different radio bands enabling an efficient utilization of the interface. This advantage is supported by numerically controlled oscillators (NCO) together with up or down conversion functionalities included in broadband digital-to-analog converters (DAC) or analog-to-digital converters (ADC). In this way, data and frequency conversion are provided by a single component for multiple radio bands.

The implementation of the RF signal generation with newest samples of DACs operating at 12 Gsamples/s, (e.g., those available at [4],[5]), targets three-band operation at carrier frequencies around 2.6, 2.8, and 3.5 GHz, as shown in Figure 16-1. Six times 20 MHz signal carriers provide 120 MHz of aggregated operating bandwidth positioned as two adjacent carriers in the three different radio bands. The pre-distorted signals generated in the digital signal processor (DSP) to compensate nonlinear distortions in the power amplifier (PA) show five times larger bandwidth than the wanted signal, and are adapted to the bandwidth of the third and fifth order intermodulation distortions caused by amplifiers. Thus, signals of 600 MHz aggregated bandwidth will pass through the component, utilizing

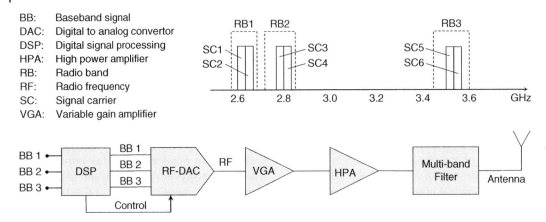

Figure 16-1. Concept of a three-band transmitter with an example of signal carrier positions.

mostly the capacity of the high-speed serial data link, which connects the DSP with the RF-DAC. It supports aggregated multiple complex input data streams up to a maximum complex data rate of 1.5 Gsamples/s. The three individual baseband signals accepted at the digital interface of the RF-DAC are individually configured in gain, converted to RF signals and positioned in the targeted radio bands. These signals feed the amplification stages of the transmitter as shown in Figure 16-1.

Broadband amplification is state-of-the-art at low power levels. For higher power, an increase in power efficiency of amplifiers is required to reduce the effort for component cooling and the power consumption of the system. The required energy efficiency stays in contradiction with the wide bandwidth, and a trade-off between output power, bandwidth and efficiency has to be considered. Accordingly, it is more appropriate to realize multi-band PAs for radio bands positioned in a restricted frequency range to limit the overall amplifier bandwidth, and implement separate transceiver chains for widely spaced bands.

To cover the three envisaged radio bands, a dual-band high-power amplifier is designed with a relative bandwidth of 10.9% (2.6 GHz – 2.9 GHz) and 5.7% (3.4 – 3.6 GHz) in a Doherty topology [6] with a peak output power of more than 53 dBm. It allows the operation in all three envisaged bands simultaneously. Considering the concurrent operation of six modulated carriers, an average transmission power of up to 38 dBm per carrier is achieved, corresponding to the power level of medium range base stations (BSs). This combination of bandwidth, power level and carrier frequencies is achieved in a Gallium Nitride (GaN) technology by realizing two transistor power elements, which are mounted and matched in ceramic packages. These are integrated with circuitries on printed circuit boards, resulting in a Doherty amplifier module.

In multi-band transceiver chains, signal filtering is required for each supported radio band. This demands solutions which allow for a closer integration of filters for different bands by enabling volume and weight reduction. Two concepts of resonators with multiple resonant frequencies are key building blocks for multi-band filters. One considers a coaxial resonator with two conductive posts of different size inside, which determine the two resonance frequencies of a dual-frequency resonator. This concept requires a minimum distance between the resonant frequencies as an inherent limitation, which prevents frequency ratios close to 1. The second approach provides full flexibility on the positioning of the resonant frequencies and allows optimizing the volume and resonator quality.

The two cavities are intertwined while the top surface of the lower one determines the lower surface, as an enlarged post of the upper one. The theoretical and experimental evaluation of these dual-band filter solutions, carried out for frequency bands of 2.6 and 3.5 GHz, prove their feasibility and demonstrate a volume reduction of up to 40% compared to the volume of two single-band filters taken individually and separated.

These two concepts of dual-band frequency resonators allow to be combined, resulting in a three-band resonator solution, anticipating significant benefit on volume and weight reduction. The presented concepts for multi-band filters show a variety of possibilities on the port implementations. Coaxial connectors at the input and output can be assigned to one, two or even three bands allowing a flexible adaptation to the transceiver architecture, especially to the number of amplifiers or antennas used to cover the envisaged radio bands.

The list of key building blocks for multi-band transceivers has to be completed by the antennas. Many activities are focusing on broadband solutions, but significant research includes also multi-band approaches at frequencies up to 6 GHz and covering radio bands which are further apart. Antenna topologies based on substrate integrated waveguide (SIW) technology [7][8], which allows for more than 1 GHz bandwidth, were proved for the range of 3.5 to 4.5 GHz and provide advantages on the integration environment. The shielded feeding line prevents coupling with adjacent components, and parasitic surface waves are minimized on the radiating element, which is realized as a slot in a conductor sheet. This eases a compact integration of the antenna in the system. Multi-band solutions also ask for different resonant structures or mechanisms realized as a radiating patch on a printed circuit board for the lower frequencies, and a radiating slot within the patch serving the upper frequencies. This allows a free positioning of the two radiation frequencies, but shows a restricted relative bandwidth around 1.5% to 3%, as evaluated for radio bands at 2.6 and 3.5 GHz.

The building blocks for multi-band transceivers promote carrier aggregation in radio frontends for increased capacity of mobile systems, as for instance addressed in Section 6.5. This is supported by the expected technology evolution on RF data converters with high sampling rates and decreased power consumption and by decreased fabrication costs of GaN based amplifiers, when moving from silicon carbide (SiC) to silicon (Si) as substrate material, which allows the fabrication on larger wafers.

16.2.3 Multi-Antenna Transceivers

In this section, concepts and solutions for multi-antenna transceivers are presented, which take into account in which frequency range they are going to operate. Section 16.2.3.1 considers the sub-6 GHz region, whereas Section 16.2.3.2 deals with the mmWave domain.

16.2.3.1 Solutions for Base Stations at Lower Frequencies

Approaches for compact multi-antenna system implementation address multiple signal generation and the chains to connect the antenna elements, including amplification and filtering, and resulting in a complete transmitter setup.

A compact solution for RF signal generation is achieved by integrating several transmit chains in a single field programmable gate array (FPGA) or application-specific integrated circuit (ASIC) as digital transmitters. Each chain comprises the generation of a binary pulse train, which includes the modulated RF waveform positioned at the targeted carrier frequency. The signal encoding is based on pulse-width modulation combined with delta-sigma modulation at sampling rates around 24.5 Gsamples/s, enabling high coding efficiency and a large spurious-free bandwidth [9]. The concept is proved with 8 chains integrated in an FPGA, whereby the number of chains is limited by the high-speed interfaces realized by the board implementation. It allows to position carriers of 5 or 10 MHz bandwidth between 0.9 and 4 GHz, meets the required signal performance, and provides a coding efficiency of 50% and 100 MHz of spurious-free bandwidth. Within this bandwidth, the mandatory requirements on spectral emission are met. If a spurious-free bandwidth of two times the radio band is provided, no additional filtering is required for spurious emissions and noise suppression beside the radio band (RB) filtering. These results show significant performance improvements compared to prior research results [9]. Further progression is expected when utilizing new FPGA families with a higher sampling rate and an increased number of high speed interfaces used as RF signal ports.

The RF digital signal provided by the FPGA has to be amplified to achieve the signal level targeted at the antenna input. Switched mode amplifiers allow to maximize efficiency as well as to utilize the digital modulation scheme available at the differential interface. Realized as an integrated circuit, it allows to design amplifiers with a form factor suited for mounting behind the antennas, adapted to the grid of the antenna array and enabling a compact system integration. Prototypes realized in GaN technology provide 38 dBm output power in the frequency band between 3.4 and 4.2 GHz [10]. The amplifier is followed by a filter to suppress the unwanted spurious emissions and noise signals. Figure 16-2 shows a schematic architecture of the discussed multi-antenna transmitter with digital transmitter, PA, filter and antenna as the mandatory parts.

Fibre-to-the-antenna (FTTA) links are a promising solution to connect large amounts of distributed antennas to a central BS. In the transmitter architecture, these links are placed at the RF interface between the central multi-chain component (digital transmitter at RF frequency) and the individual RF chains (amplifier, filter) close to the antennas, as depicted in Figure 16-2. Investigations with vertical-cavity surface-emitting lasers and electro-absorption modulators (EAMs) working as electrical-optical converters (E/O), and with photo-diodes and EAMs working as optical-electrical converters (O/E) show the potential of these low-cost components to be used in high data rate transmission systems [3].

16.2.3.2 Enablers for mmWave Transceivers

Concepts exploiting the large spectral resources available in mmWave bands ask for different building blocks to enable the envisaged radio interfaces. They have to support transceiver systems which include large antenna arrays for spatial directivity based on different beamforming schemes.

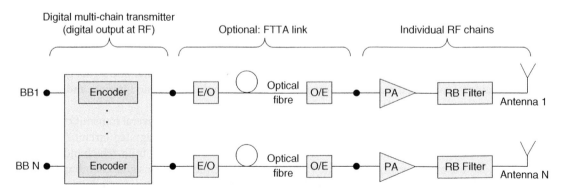

Figure 16-2. Architecture of a possible multi-antenna transmitter.

Antenna arrays combined with large bandwidths are a great challenge for HW implementation, and power consumption will essentially limit the design space. Analog or hybrid beamforming are currently considered as possible solutions to reduce the power consumption. However, these systems highly depend on the calibration of the analog components. Another major disadvantage is the large overhead associated with the alignment of transmit and receive beams of the BSs and mobile ends. Specifically, if a high gain is needed, the beamwidth has to be small, thus the acquisition and tracking of the optimal beam alignment in a dynamic environment is very challenging, as discussed also in Section 11.5. Alternatively, power consumption can be reduced by decreasing the resolution of the ADCs, see also Section 11.5.3. This solution has the advantage of keeping the flexibility of fully digital processing, while having reduced power consumption. This is especially advantageous for the case of multiple users transmitting to one access point.

Since hybrid beamforming is the dominating technology used for mmWave prototyping systems today, see also Section 11.5.2, low-resolution ADC systems should be compared to it. It was shown in [3], based on a theoretical evaluation and not considering the beam training overhead, that a low-resolution ADC based massive beamforming approach in many situations outperforms hybrid beamforming in terms of throughput and energy efficiency. Overall, this technology might be an interesting new candidate for future mmWave mobile broadband systems.

For providing local oscillator (LO) signals to transceivers for large antenna arrays, one may refrain from using a centralized LO at the high frequency, and instead consider an architecture based on distributed LO generation [2]. In this case, only the reference signal is common, which is much easier to distribute due to its lower frequency, and the approach has the advantage of phase noise suppression for signals combined at the antenna elements, as the phase noise is then uncorrelated. To support this approach, research was conducted on frequency generation for 28 and 60 GHz carrier frequencies, and it was shown that phase-locked loops (PLLs) with an 8 GHz tuning range are suitable for integration on transceiver chips in 28 nm silicon-on-insulator (SOI) complementary metal oxide semiconductor (CMOS) technology.

The development of amplifiers required for mmWave applications is determined by key performance indicators (KPIs) and parameters such as frequency, bandwidth, efficiency, output power, linearity, gain and cost. To address the latter, the cost-effective nanoscale bulk CMOS technology can be exploited. Its potential was demonstrated by investigations on three power amplifier circuits by playing with the trade-off between output power, bandwidth (BW) and power added efficiency (PAE) [2]. Here, the design was focused on efficiency, yielding 37.6% PAE combined with 16.8 dBm output power, and 3 GHz BW at 28 GHz operating frequency. An alternative solution provided 6 GHz BW at the same frequency at a similar output power of 16 dBm, but with a reduced PAE of 21%. With the third solution, the BW was further increased to 28 GHz between 29 and 57 GHz, the PAE maintained at 22%, but the output power was 13.4 dBm only. This shows the potential for designing amplifiers adapted to desired applications.

A further significant building block for multi-antenna systems considers antennas of small form factor and low cost. Planar antennas realized in SIW technology provide such advantages and ease the system integration due to their inherent shielding of the feeding lines also at mmWave frequencies [2]. Thus, the mounting of the amplifier on the antenna substrate allows a close integration of these building blocks, including a co-optimization of their interconnection with the advantage of reduced reflection of signal power at the antenna port and size reduction of this sub-system.

16.2.4 Full-Duplex Transceivers

The challenge of in-band full-duplex (IBFD) transceivers is to simultaneously transmit and receive at the same frequency band. To be able to listen and decode a low-power received signal while transmitting a high-power signal, it is necessary to mitigate the self-interference (SI) signal caused by leakages between the transmitter (Tx) and the receiver (Rx). To avoid placing the computational load of SI cancelation (SIC) on the user equipment (UE), IBFD is considered here only at the BS. The scenario considered is a frequency division duplex (FDD) or time division duplex (TDD) system, where a UE1 transmits to the BS on a certain frequency, while the BS simultaneously transmits to a UE2 on the same frequency. UE1 and UE2 are chosen to be spatially separated to avoid UE1 to interfere towards UE2. Moreover, a small cell scenario is considered, where there is a smaller difference between transmit and receive power than in large cell scenarios, thus requiring less isolation between transmitter and receiver, and therefore less SIC capability. Such transmission schemes can bring an important capacity gain compared to standard TDD [11] or FDD.

The SIC is done in several steps, as shown in Figure 16-3. First, a circulator with only one antenna can provide around 20 dB isolation between Tx and Rx ports. Alternatively, two antennas can also be used to further increase the isolation. As a second step, analog cancelers (RFSIC) must be used in order to avoid the saturation of the low noise amplifier (LNA), and to allow the ADC to convert the useful signal within the adapted power range. Finally, a digital SIC (DSIC) after the ADC can be applied to remove the remaining part of the SI. The receiver gain must be set in order not to saturate the ADC. An additional hybrid SIC (HSIC) may also be used between the RFSIC and the DSIC. It combines a duplication of the main Tx chain (the so-called auxiliary chain), with a digital linear filter whose role is to take into account the effect of the RFSIC, circulator, coupler, etc.

Figure 16-3 presents the architecture. It is based on a classical 2x2 MIMO transceiver. Only a few functions have been added to manage the swap between TDD MIMO and IBFD SISO.

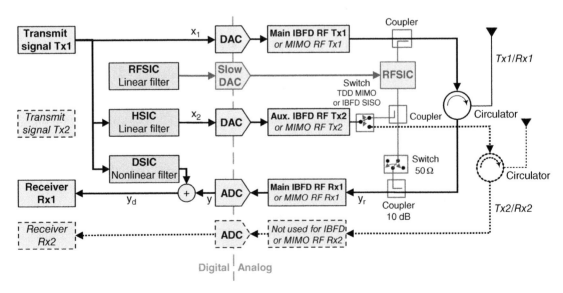

Figure 16-3. Architecture overview of the various stages of the IBFD SIC. The dashed parts are not used in the IBFD implementation.

RFSIC mitigates the strong part of the SI and has low tracking capability. It is based on a digitally controlled vector modulator component. The HSIC mitigates the remaining linear component of the SI. The HSIC can easily manage long delays and provide medium tracking capability. The DSIC cancels the nonlinear contribution of the SI. It is a fully adaptive filter working after the ADC.

Each canceller has been evaluated in a standalone way on real signals thanks to a test bench based on a SDR transceiver board. Globally, with less than 20 dB antenna subsystem isolation, both analog and digital cancelers together are able to reduce the SI signal by 85 dB over a 40 MHz bandwidth while receiving a useful signal. The main limitation of the current solution is the noise generated by the analog cancelers. Extra nonlinear distortions of the analog cancelers are also a limiting constraint. Nevertheless, the depicted solution can lead to 69% capacity gain on the downlink (DL) on a system level [11].

16.2.5 Techniques for the Enhancement of Power Amplifier Efficiency

As it has been pointed out above, low overall power consumption is a highly desirable feature, in particular because it can increase the battery life of user devices. In this context, poor power added efficiency (PAE, i.e. the ratio between the difference of the RF output power and the RF input power to the DC input power) of the RF PA is one of the primary factors that decrease the battery life. Accordingly, the following two techniques target to enhance the PAE in 5G systems.

16.2.5.1 Envelope Tracking for RF Power Amplifiers

In PAs for 2nd generation (2G) mobile radio, where the output signal has near-constant amplitude, a fixed supply has been used to power the PA achieving high PA efficiency. However, for 3G/4G devices where variable envelope signals are the norm, a fixed supply causes power to be wasted when a device transmits below its maximum output power. Envelope tracking (ET) is a well-known technique to improve PA efficiency in that case [3]. The principle behind ET is to continuously adjust the power supply voltage to follow the envelope of the transmitted signal in order to ensure the desired PA efficiency. This way, ET optimizes PA power consumption. Alternatively, ET can linearize a PA by reducing the amplitude-to-amplitude (AM-AM) distortion, which allows reducing out-of-band emissions or the transmission of higher order modulation schemes and, therefore, higher data rates. The main elements of the ET procedure comprise the detection of the RF magnitude at I/Q sub-sample rate and, accordingly, the update of the power supply voltage. It is of highest importance that this so called "envelope path" is well calibrated to support the targeted waveform and takes into account system imperfections such as the delay and gain offsets introduced by the circuit blocks in the signal path. Looking at 5G systems, the main challenges for ET performance compared to LTE/4G include the wider bandwidths (well above 20 MHz), the higher carrier frequency (>6 GHz) and the new waveforms used. A wider signal bandwidth has a direct impact on the susceptibility of the system to delay imbalances; doubling the bandwidth requires the delay mismatch to be halved. A higher signal bandwidth also has impact on the bandwidth of the envelope signal. Moreover, at higher carrier frequencies, PA memory effects become non-negligible and impact out-of-band emissions. In [3], a suitable PA model was adopted to evaluate the memory effects for ET and showed that distortion cannot be counteracted efficiently. Especially at higher frequency offsets, there is a mismatch between supply voltage and RF signal magnitude, which results in some higher-order distortion terms. While nearby spectral emissions are suppressed, further distant spectral emissions actually degrade when compared to a fixed supply voltage. Finally, waveforms such as filtered orthogonal frequency division multiplex (OFDM) can provide better frequency localization than LTE, and the advantage is that interference with other users is reduced due to much lower unwanted emissions. However, once PA nonlinearity is considered, much of the spectral properties are lost due to intermodulation products polluting nearby frequencies. All things considered, it is proven that fully dynamic ET may be challenging for 5G and less aggressive techniques may be preferred, like average power tracking (APT), where the PA supply voltage is adjusted based on the average power over a certain period.

16.2.5.2 Weighted Selective Mapping and Digital Pre-distortion Techniques for 5G Waveforms

5G waveform candidates such as Orthogonal Frequency Division Multiple Access (OFDMA) and filterbank multi-carrier (FBMC) are based on a multi-carrier approach. They are robust to the frequency-selective fading channel and can achieve a high data transmission. However, the major drawback of multi-carrier is a high peak-to-average power ratio (PAPR), and this drawback causes signal distortion and high energy consumption [12]. A PA is very sensitive w.r.t. its operational area, and hence a high PAPR causes a nonlinear operation problem and consequently signal distortion.

A new PAPR reduction technique named Weighted Selective Mapping Technique (WSLM) was developed and its complementary cumulative distribution function (CCDF) performances evaluated in OFDM and FBMC systems [3]. In addition, a memory polynomial digital pre-distortion (DPD) was used to compensate for nonlinear behaviour of a PA. It applies inverse distortion to the input signal of the PA in order to compensate the distortion generated by it. The performance of WSLM and DPD techniques was evaluated and about 2.5 dB gain was achieved under the given test configuration.

16.3 Solutions for Digital HW Implementation

16.3.1 Requirements on 5G Digital HW

Device chipset and platform suppliers will face the need to support an increasing number of RATs together with the trend toward a highly diverse set of device form factors, with cars, wearable and machine-type devices joining tablets and smartphones. This progressive increase in form factor diversity is driving multiple levels of platform support and capability, making transceiver complexity a key challenge for 5G devices. In addition, 5G devices are expected to integrate and interact with a multiplicity of sensors (e.g., those related to location and positioning, environmental conditions, image processing, etc.) that will provide context awareness to the communication and the deployed RAT, which will allow improvements in the efficiency of existing services, and help to provide new and more user-centric and personalized services.

A heterogeneous situation similar to that of 5G devices is also valid for 5G network elements. Since cost and flexibility of deployment will also be important factors, a shift towards SW-based implementations and virtualization technologies will be required, as detailed in Section 5.2.3. Ideally, these SW-based implementations should be as HW-agnostic as possible and, at the same time, supported by versatile, reconfigurable and flexible HW platforms that are able to cope with all the functionalities needed and invoked from the SW domain.

16.3.2 Complexity Analysis of the Individual Implementation of New Waveforms

An indicative way to indirectly quantify the implementation cost of digital HW is the analysis of the computation complexity of digital baseband processing blocks. The waveform generation and recovery are two of the basic digital signal processing blocks that need to be flexible and scalable in 5G system architectures due to the various possible use cases, scenarios and their requirements. In MIMO systems, for example, one waveform generation and recovery block is necessary for each antenna, and for carrier aggregation the same is true for each component carrier. In addition to that, if multiple subcarrier spacings are employed for different time-frequency resource blocks, separate waveform generation and recovery may need to be implemented for each of them.

In this subsection, the computational complexity of the waveforms proposed for 5G and beyond are analyzed by considering individual implementations, and in the next subsection, by a harmonized implementation of multiple waveforms as proposed in Section 11.4.4.

Multi-carrier (MC) and OFDM-based single carrier (SC) modulations considered for 5G and beyond include:

- Conventional cylic prefix OFDM (CP-OFDM) with windowing [13];
- Discrete Fourier transform (DFT)-spread-OFDM, also referred to as DFT-s-OFDM;
- Single carrier frequency division multiple access (SC-FDMA);
- Zero tail DFT-s-OFDM (ZT-DFT-s-OFDM) [14];
- FBMC with offset quadrature amplitude modulation (FBMC-OQAM) [15] and with quadrature amplitude modulation (FBMC-QAM) [16];
- Filtered multitone (FMT), a.k.a. pulse-shaped OFDM (P-OFDM) [17];
- Universal filtered multicarrier (UFMC or UF-OFDM);
- Cyclic convolution based FBMC, a.k.a. generalized frequency division multiplexing (GFDM).

In this section, the complexity is quantified and compared in terms of the total number of real multiplications and additions.

We consider the signal processing operations involved in the generation of the MC and SC signals, as well as the recovery of the subcarrier/subchannel/subband signals and equalization in the presence of multipath propagation. Here, we do not consider the operations involved in channel estimation or calculation of the equalizer coefficients. The first reason is because those signal processing tasks are not in the user data chain, which is the one that concentrates the processing burden, and the second is because of the many existing algorithms for those tasks, making the choice of one not trivial. Moreover, we assume that all systems are perfectly synchronized.

Because all waveforms are based on MC modulation, we assume that a total of N subcarriers are available out of which N_f are occupied with symbols. We will consider first the number of real-valued multiplications and additions to transmit one block of N_f symbols.

The transmitter (Tx) and receiver (Rx) of a CP-OFDM system are basically built with one single inverse fast Fourier transform (IFFT) and fast Fourier transform (FFT) and, possibly, a windowing operation. The SC waveforms cyclic-prefix-based DFT-s-OFDM and ZT-DFT-s-OFDM have the same complexity of CP-OFDM plus the DFT spreading part, which slightly increases the complexity. In the case of ZT-DFT-s-OFDM, the complexity is similar to cyclic-prefix-based DFT-s-OFDM, but the two main differences are that no cyclic prefix is attached and zero-valued headers and tails on a subchannel level are added.

One of the basic and common building blocks for all waveforms is the FFT/IFFT. The number of real multiplications and additions of an N-point FFT/IFFT using a split-radix algorithm are given by [18]:

$$
\begin{aligned}
Cm_{FFT}(N) &= N\big(log_2(N)-3\big)+4, \\
Ca_{FFT}(N) &= 3N\big(log_2(N)-1\big)+4.
\end{aligned}
\tag{16-1}
$$

Assuming an FBMC-OQAM system where the prototype filter has length KM, two approaches can be adopted for the generation and recovery of the MC signal: the polyphase-based and the frequency-spread-based structure.

We consider first the complexity of FBMC-OQAM implemented with a structure based on the polyphase decomposition of the prototype filter and using a direct form realization of the polyphase components (PPC) [18]. The Tx is composed of 3 steps after the OQAM modulation:

- Phase rotations to get linear phase filters in each subcarrier, i.e., $4N_f$ real multiplications;
- IFFT;
- Polyphase filtering followed by block overlapping of 50%, i.e. $4KN$ real multiplications.

At the receiver (Rx) side, similar operations in the inverted order are implemented, including one more step: polyphase filtering, FFT, multi-tap channel equalization per subcarrier with an equalizer of length L_{eq}, resulting in a complexity of $4N_fL_{eq}$, and the OQAM demodulation. The phase rotations at the Rx side can be embedded in the equalizer coefficients.

The second approach is a frequency domain filtering, also known as frequency spread based (FS)-FBMC-OQAM [19], featuring also a prototype of length KN and designed using the frequency sampling approach [20] with only $2(K-1)$ non-zero coefficients in the frequency domain. In this case, the structure changes drastically. The subcarrier signals have to be spread over $K+1$ frequency domain samples, and each of them multiplied by one of the prototype frequency domain coefficients. The complexity of the frequency domain filtering encompasses $8N_f(K-1)$ multiplications. The overlapping parts in frequency domain are all added and then transformed with an IFFT of size KN. Finally, the first $M/2$ samples of a given block of KN IFFT outputs are added to the last $M/2$ samples of the previous block to generate the serialized time domain signal. At the Rx side, the inverse operations are done. In addition to the FFT at the Rx, a frequency domain filtering with $16N_f(K-1)$ multiplications is also performed.

The FMT/P-OFDM waveforms are similar to the FBMC/OQAM case, but there is neither OQAM nor a 2-stage up-sampling. Nevertheless, for the calculation of the number of operations per sample, we need to consider the lower sampling rate.

The UFMC system can be parametrized between two extremes: on one end, the whole CP-OFDM signal is filtered by one filter to reduce the out-of-band radiation. At the other end, each or a minimum number of resource blocks is transformed with the IFFT and filtered with its own filter. In a UFMC system with maximum granularity, N_B resource blocks each with M subcarriers require N_B FFTs of size N, where each of them has only NN_B non-zero inputs.

The modulation is performed in the following steps: First, the signal of each subband is spread over the whole symbol length and transformed into frequency domain. Then, the filtering is performed in frequency domain, and the sum of all subbands is converted into time domain [21].

Instead of filtering and then transforming, a non-matched filtering is applied in frequency domain [22]. The Rx then has 3 steps:

- Windowing in the time domain;
- FFT transformation of size N_uM with zero padding and half of the outputs thrown away;
- Frequency domain filtering and equalization.

The GFDM modulation scheme is based on circularly convolving each subcarrier in a block of data with a filter kernel. In contrast to OFDM, a cyclic prefix is added per block and not per symbol [23][24]. Since a circular convolution can be calculated as a multiplication of two vectors in frequency domain, then both transmitter and receiver can be efficiently implemented using the FFT.

Out of a total number of subcarriers N only N_f are used. M_B symbols per subcarrier are combined to form one transmission block. In total, N_fM_B data symbols can be transmitted per block. The prototype filter is designed to overlap with N_a adjacent subcarriers, and it is typically chosen to be an RRC filter with $N_a = 2$.

As described in [23], excluding the trivial operations like reordering, the following signal processing tasks need to be performed at the transceiver:

- Transformation of the data signal of each subcarrier into frequency domain;
- Filtering in frequency domain;
- Transformation of the signal into time domain.

The details of the corresponding receiver are described in [24]. It is important to mention that since the subcarriers are overlapping, it is necessary to cancel this interference to achieve a sufficient performance. In [24], the authors use the detected symbols to subtract the interference to adjacent subcarriers in an iterative fashion. For a constellation as large as 64QAM it was shown that $J = 8$ iterations are sufficient. The receiver can be divided into the following signal processing tasks:

- Transformation of the signal into the frequency domain;
- Channel equalization;
- Filtering in the frequency domain;
- Iterative interference cancelation.

Let us now consider a numerical evaluation of the individual waveform implementations. For that, we calculate here only the complexity at the base station. We utilize the complexity formulas presented in detail in [25]. The metric utilized is the number of multiplications and additions normalized by the number of QAM symbols transmitted across the used subcarriers. We assume a similar overhead in terms of training or reference signals for all waveforms.

Moreover, we consider here four 3GPP New Radio (NR) wideband parameters [26]:

1) NR downlink with BS wideband and UE narrowband allocation (UE: 1 MHz, BS: 100 MHz)
2) NR downlink with wideband allocation (100 MHz)
3) NR uplink with BS wideband and UE narrowband allocation (UE: 1 MHz BS: 100 MHz)
4) NR uplink with wideband allocation (100 MHz)

In scenarios 1 and 3, six resource blocks are allocated for each UE, and 85 mobiles are served simultaneously. In scenarios 2 and 4, only one UE is served, and all 6600 available subcarriers are allocated to it.

The parameters have been chosen as given in Table 16-1.

Table 16-1. Parameters for the numerical complexity analysis.

• CP-OFDM	• FFT size: $N = 8192$
	• Number of active subcarriers: $N_f = 6600$ (wideband), 6120 (narrowband MS, wideband BS)
	• Cyclic prefix length: $L_{CP} = 576$
	• Number of RB: $N_{min}^{RB} = 6$ (narrowband) or $N_{max}^{RB} = 550$ (wideband)
• DFT-S-OFDM/	• Size of the resource blocks: $M = 12$
• ZT-DFT-s-OFDM	• Size of the small FFTs: $MN_{min}^{RB} = 72$ (narrowband) or $MN_{max}^{RB} = 6600$ (wideband)
• FBMC	• Time domain overlapping factor $K = 4$
	• Number of equalizer taps/subcarrier: $L_{eq} = 3$ (polyphase)
	• Frequency sampling: $L_{eq} = K$
• UFMC	• Number of subcarriers/RB: $M = 12$
	• Filter length: $L = L_{CP} + 1$
	• Frequency oversampling factor: $N_u = 2$
	• Size of narrowband (NB) FFT: $N_{NB} = 64$
• GFDM	• Number of symbols/subcarrier: $M_B = 4$
	• Number of Overlap subcarriers: $N_a = 2$
	• Number of SIC iterations: $J = 8$

In Figure 16-4, the BS complexity results related to the different waveforms for the different scenarios are shown.

16.3.3   Complexity Analysis of a Multi-Waveform Harmonized Implementation

A flexible implementation integrating multiple waveforms in a single harmonized solution can be very useful to select the waveform that better matches a particular communication scenario and to reduce implementation costs. For this particular purpose, a generic MC waveform implementation was presented in Section 11.4.4. In such a tuneable waveform implementation, by selectively enabling or disabling particular blocks, one out of six different MC waveforms could be generated. The waveforms of interest include classical CP-OFDM, windowed OFDM (W-OFDM), P-OFDM and SC-FDMA or ZT-DFTs-OFDM. Furthermore, the framework is also valid to represent waveforms of the FBMC family such as FBMC-QAM and FBMC-OQAM.

Firstly, focusing on the transmitters for the independent waveforms, it can be noted that they all include the following blocks: one for serial-to-parallel data conversion, one for QAM constellation mapping, a subcarrier mapping and pilot insertion block, a block for MC modulation through FFT and, before sending the data to the channel, one parallel-to-serial data conversion block. However, recall that there are some extra modules needed for specific waveforms, which should be considered within the harmonized implementation:

- **DFT spreading/de-spreading**: This block is intended to perform the spreading/de-spreading operations necessary for the ZT-DFT-s-OFDM waveform transmission/reception. It has the same cost as an FFT of size M;

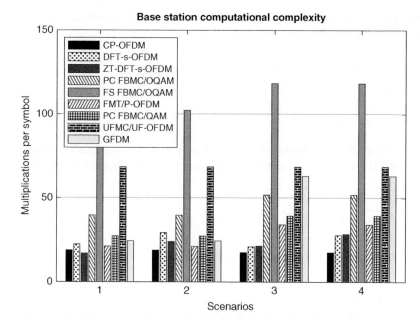

Figure 16-4. BS computational complexity in terms of number of real-valued multiplications per Tx/Rx symbol – NR wideband.

- **OQAM pre-processing/post-processing**: The FBMC-OQAM waveform needs a specific module for complex to real conversion of QAM symbols, up-sampling, and time staggering. At the receiver, these operations obviously need to be reverted;
- **Reconstructing IFFT block**: Because of its importance for reducing the complexity of FBMC-OQAM, we propose the simultaneous calculation of the FFTs of size N for the two real-valued inputs involved, via a single FFT of size N with complex-valued inputs, as shown in [27]. After performing the IFFT, the additional total cost of reconstructing each individual FFT for both two real-valued inputs is $8N$ multiplications and $4N$ additions;
- **Polyphase network (PPN)**: The implementations of all the waveforms that include filtering or windowing in time domain (W-OFDM, P-OFDM, FBMC-QAM and FBMC-OQAM) include a PPN, where the input data is convolved with a filter. For the complexity evaluation in the next section, it will be assumed that the prototype filter used in the PPN has real coefficients, i.e., it is symmetric in frequency domain, the most common assumption in waveform implementations.

16.3.3.1 Complexity Evaluation

The harmonized implementation at the minimum complexity requires: one FFT block of size N (common to all waveforms); one block to rebuild the FFT of two real-valued inputs (required for FBMC-OQAM); two blocks of real-valued PPN filtering; and one block for DFT spreading of size M, necessary for ZT-DFT-s-OFDM. Considering the complexity analysis carried out in the previous section for new waveforms, the aggregated complexity *cost* in terms of multiplications, additions, and floating point operations per second (flops) is [28]

$$Cm_{HARM}(N, K) = N \log_2 N + 4NK + 5N + M \log_2 M - 3M + 8,$$
$$Ca_{HARM}(N, K) = 3N \log_2 N + 4NK - 3N + 3M \log_2 M - 3M + 8,$$
$$Cf_{HARM}(N, K) = 4N \log_2 N + 8KN + 2N + 4M \log_2 M - 6M + 16.$$

(16-2)

The complexity cost of a solution where all the waveforms are drawn together as standalone implementations is found by adding the number of multiplications, additions, or flops of all of them. The total result is

$$Cm_{NOHARM}(N, K) = 6N \log_2 N + 8NK - 8N + M \log_2 M - 3M + 28,$$
$$Ca_{NOHARM}(N, K) = 18N \log_2 N + 8NK - 22N + 3M \log_2 M - 3M + 28,$$
$$Cf_{NOHARM}(N, K) = 24N \log_2 N + 16NK - 30N + 4M \log_2 M - 6M + 56.$$

(16-3)

Assuming the typical values $K = 4$ and $M = 12$ and a number of subcarriers ranging from 16 to 4096, Figure 16-5 and Figure 16-6 show the complexity in terms of multiplications, additions and flops of the harmonized and non-harmonized implementations for different values of N. From basic calculations based on the values in the figures, it can be observed that the typical savings due to harmonized implementation range between 60–75%.

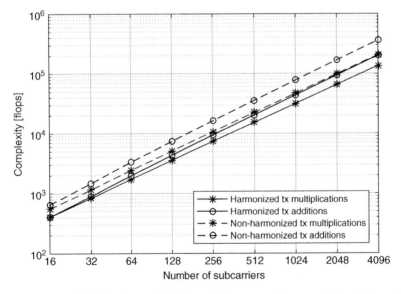

Figure 16-5. Multiplications and additions of the proposed harmonized and the non-harmonized implementation of the six waveforms.

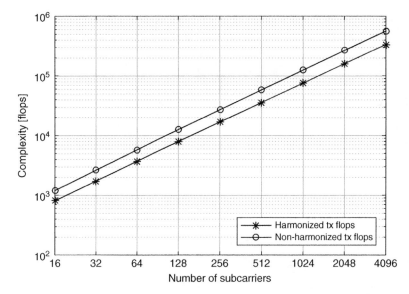

Figure 16-6. Number of flops of the proposed harmonized and the non-harmonized implementation of the six waveforms.

16.3.3.2 Multi-Processor Baseband Architectures for 5G Network Elements

Current baseband processing architectures, such as the ones required to implement the waveforms whose complexity has been analyzed in the previous sections, typically rely on a combination of general purpose processor, DSP processors and dedicated HW due to stringent performance and power-efficiency requirements. The flexibility and programmability of these solutions featuring fixed partitioning is typically limited since the DSPs are not efficient in general purpose processing. Further, general purpose processors are not efficient in dedicated tasks, and it is impossible to use HW accelerators for anything other than their dedicated tasks. While it is not possible to entirely remove these limitations, with a smart enough architecture it is possible to have a fully programmable and flexible computing solution for 5G baseband computation, which can be suitable for implementation in network elements, for example.

REPLICA is an architectural framework (or family of chip multi-processors) aimed to solve the performance and programmability bottlenecks of current multi-core processors [29]. The performance comes from a scalable latency hiding (a technique to eliminate delays caused by the shared memory system) combined with a low-cost inter-thread synchronization mechanism and efficient exploitation of low-level parallelism. Programmability is achieved via support for architecture-independent abstraction, strict memory consistency and synchronous, more deterministic execution than in current alternatives. A REPLICA chip multi-processor (CMP) consists of P processors, M_S shared memory modules, M_l local memory modules, P instruction memory modules and a high-bandwidth network for connecting the elements, as shown in Figure 16-7. REPLICA processors are T_p-way multi-threaded to support latency hiding of the shared memory system, where T_p is a design-time parameter chosen so that the latency of typical memory accesses gets hidden.

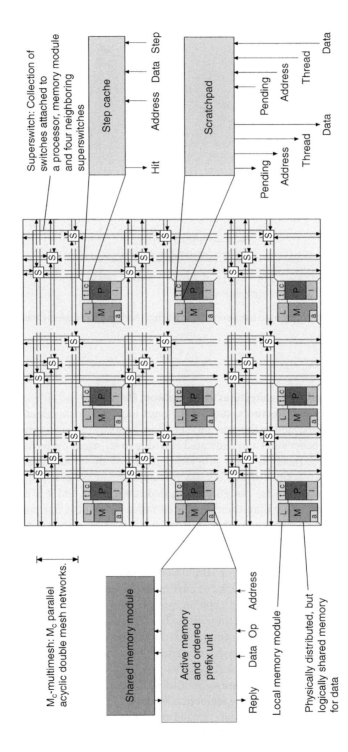

Figure 16-7. REPLICA CMP [29] with P processors, M_c-way multi-mesh network and P active memory modules (P=processor core, I=instruction memory module, t=scratchpad, c=step cache, a=active memory unit, M=shared memory module, L=local memory module and S=switch).

They feature a chained very long instruction word organization of functional units to support execution of dependent sub-instructions within an instruction and have a $T_p + 1$ stage pipeline to avoid pipeline delays that affect current processors [30].

Applying the REPLICA architecture framework for 5G baseband processing requires architectural improvements to fulfil performance requirements with affordable energy efficiency. Modifications to the threading system have been introduced, and an architectural unit to speed up the FFT computation has been added. In terms of the silicon area and power consumption, this solution looks affordable with respect to commercial general purpose processor alternatives [25].

16.3.4 Channel Decoder Implementations for 5G

5G systems should also consider advanced forward error correction (FEC) technologies such as advanced Turbo-codes, low-density parity check (LDPC) codes and Polar codes. From the decoder standpoint, such FEC codes are often decoded using iterative decoding strategies, relying on maximum a posteriori (MAP) or similar algorithms. 3GPP has moved forward on the selection of the coding scheme for eMBB, and LDPC and Polar codes are most likely to be considered for the data and control signals, respectively. As the data rates in 5G will be very high also on the terminal side, there is a need to develop high-speed channel decoders that are also energy and area efficient. However, for backward compatibility with LTE, it is also of high interest to investigate flexible decoder architectures able to address Polar, LDPC or Turbo codes.

16.3.4.1 Message-Passing Decoding for LDPC Codes

LDPC codes are a class of error correction codes known to closely approach the Shannon limit under iterative message-passing (MP) decoding algorithms [31]. MP architectures are composed of processing units that perform the desired computations and pass messages to each other. The way such architecture applies to LDPC decoding is closely related to the bipartite graph representation of LDPC codes [32]. It comprises two types of nodes, known as variable-nodes (VNs) and check-nodes (CNs), corresponding respectively to coded bits and parity-check equations. Accordingly, an LDPC decoder comprises two types of processing units, namely VNUs and CNUs, which exchange messages according to the structure of the bipartite graph.

Two layered decoder architectures for quasi-cyclic (QC)-LDPC codes have been designed targeting high data throughput and an efficient use of HW resources. High throughput is achieved by either pipelining the datapath or increasing the HW parallelism in the architecture. Both architecture variants may accommodate two different decoding kernels. The first decoding kernel corresponds to the conventional Min-Sum (MS) decoding, while the second corresponds to the Non-Surjective Finite Alphabet Iterative Decoder (NS-FAID), which has been proposed in [33], and it was later shown that NS-FAIDs can provide different trade-offs between HW complexity and decoding performance for both regular and irregular LDPC codes [25].

The layered decoder architectures with both MS and NS-FAID decoding kernels were evaluated in terms of throughput and resource consumption, for both regular and irregular QC-LDPC codes. Implementation results of the proposed architecture for both FPGA and ASIC platforms show improvements in throughput compared to state-of-the-art implementations. For the ASIC implementation of 65 nm CMOS technology, an area-normalized useful throughput of 2293 Mbps/mm^2 was achieved while the comparable state-of-the-art implementations achieve a performance between 106 Mbps/mm^2 and 1257 Mbps/mm^2 [25], It has also been shown that NS-FAIDs allow significant improvements in terms of both throughput and HW resource consumption, as compared to the conventional MS solution, while also improving the error correction performance.

16.3.4.2 Turbo Decoder Design Optimized for mMTC

For backward compatibility with LTE, it is of high interest to investigate flexible decoder structures able to address Polar, LDPC or Turbo codes. The focus here is on iterative decoders based on MAP or related measures such max-log MAP. The latter is an approximate low-complexity algorithm with some degradation in performance when compared to a log-MAP or MAP algorithm.

In order to meet very high data rates, typical implementations consider parallelized architectures, in which several MAP instances are used simultaneously to speed up the overall processing. On the other hand, having several parallel instances of such modules increases both the size (and thus cost) and the power consumption of the solution. Such approach could, therefore, be challenged when addressing low-end devices (e.g., connected objects in a Smart Cities use case), for which cost and power efficiency become key characteristics. A flexible Turbo decoder has been designed that is able to adapt dynamically to the different services expected for 5G [25]. By configuring the number of active MAP engines, the number of iterations and number of warm-up stages, it is possible to configure the decoder to fulfill latency, throughput, power consumption and performance requirements as needed.

16.4 Flexible HW/SW Partitioning Solutions for 5G

Mobile wireless systems in 5G are envisioned to offer increased performance on top of flexible heterogeneous devices while reducing the overall energy consumption by efficient virtualization and coordination technologies. In order to reduce the number of physical network elements, new systems will need to incorporate virtualization mechanisms, which are able to cope with the increased network demands without affecting timing constraints and communication overhead. New implementation techniques must be introduced in order to further increase reusability, flexibility and performance along with a reduction of energy consumption at the same time. Some solutions that introduce static and customized functions virtualization able to fully virtualize a network element and offer significant power reduction are already available. In the case of RAN network nodes, existing virtualization technologies perform the entire digital baseband processing part in SW platforms and utilize SDR platforms connected to remote radio heads (RRHs) for amplification and transmission purposes, as for instance discussed in Section 6.7. Some implementations are also able to simulate full LTE networks. The most notable among them are the LENA ns-3 [34], which provides full SW implementation of virtual LTE networks, the OpenAirInterface [35], OpenLTE [36] and srsLTE [37].

16.4.1 Architecture for Supporting MAC/PHY Cross-Layer Reconfiguration

In this section, we describe an architecture for wireless terminals able to support advanced Medium Access Control (MAC) and physical layer (PHY) reconfigurations. The design has been motivated by the need to improve the terminal capabilities for exploiting context information and adapting the PHY to the application requirements and network operating conditions. Following this need, PHY-layer improvements can be divided into two main groups: i) those aiming at supporting new PHY primitives, devised to configure the central frequency and the bandwidth of the transceiver, in order to optimize the spectrum utilization and minimize the interference generated to other coexisting links; ii) those aiming at supporting new PHY measurements, in order to characterize the network environment.

The terminal is based on an extension of the so-called wireless MAC processor architecture [25], according to which the terminal transceiver is driven by a programmable state machine which specifies the MAC protocol logic and the configuration of the transceiver and receiver processing chain. The prototype has been implemented on the WARP v3 research platform, by exploiting the implementation of a complete Institute of Electrical and Electronics Engineers (IEEE) 802.11 MAC/PHY legacy stack, already available for this platform. For this reason, the design has been organized in terms of incremental extensions of the IEEE 802.11 implementation. While the generic executor of MAC/PHY state machines was implemented at the firmware level with minimal modifications of the original firmware (devised to take into account the possibility to specify chains of multiple actions), the implementation of advanced PHY layer primitives required to work on the IP cores and HW blocks of the platform [38].

The developed architecture has been applied to evaluate the benefits of dynamic bandwidth adaptation and innovative signaling mechanisms based on tones. Bandwidth adaptation is more powerful than simple adaptive modulation schemes, because it can increase robustness by identifying the most suitable spectrum portion to be utilized for transmissions. Adaptive modulation can still be utilized on top of bandwidth adaptation. Although the idea of bandwidth-dynamic adaptation for wireless links is not new and especially relevant in the cognitive network scenario, the proposed solution is promising, since it does not require an out-of-band or in-band signaling channel to be used for negotiating the bandwidth to be utilized.

16.4.2 Cognitive Dynamic HW/SW Partitioning Algorithm

The static HW/SW partitioning methods may generally lead to reduced energy consumption, but they are not reusable and not reconfigurable and thus limiting network upgrades and resource allocation. Also, the processing power cannot be shared among nodes, which can lead to load unbalancing and, thus, a decrease in efficiency. Full virtualization is not always available as the devices of the underlying network might not be able of performing such a task, or virtualization might be limited according to available physical and computational resources. Within this context, the cognitive and dynamic HW/SW partitioning algorithm's task is to provide the best HW/SW partitioning of the 5G network functions for both device and network elements. The algorithm takes the required KPIs for a given scenario and the available HW/SW resources and optimizes for high flexibility, configurability, performance and/or energy efficiency [38].

In [38], the cognitive and dynamic HW/SW partitioning algorithm has been tested considering a dynamic hotspot use case, resulting in the LTE enhanced Node-B (eNB) PHY layer reconfiguration according to pre-specified measured KPIs and optimization goals that lead to higher flexibility, lower execution time and energy consumption reduction. The cognitive and dynamic HW/SW partitioning offers extended flexibility and re-configurability inside a network node, while also ensuring the KPI requirements and further applying an optimized functional partitioning towards user-defined KPIs. In terms of reconfiguration, the proposed solution performs its optimization operation in less than 17 ms for common scenarios.

16.5 Implementation of SW Platforms

For 5G systems, it is of utmost importance that the control and the management of heterogeneous HW infrastructure and devices expected are done in a way that guarantees: (i) an effective (i.e., rapid, easy and dependable) service development and deployment, (ii) the ability to adapt to very demanding and changing contexts of operation, and (iii) Quality of Service (QoS) and Quality of Experience (QoE). To ease this burden while efficiently making use of the available resources, the use of two enabling technologies is required: on one hand, the deployment of scalable, flexible and multi-RAT SW networks, and on the other hand, introducing a HW-agnostic operation through the use of programmability and dynamic reconfiguration via interface abstractions and uniform application programming interfaces (APIs). With the above, changes in context (as for instance users' demands, interference, availability of resources, etc.) as identified by sensors will trigger changes in the operation of radio technologies, but at an affordable control and management overhead.

In the last years, the need to introduce this reconfigurability in the wireless networks has been identified as a key requirement to efficiently deploy and maintain high-performance wireless deployments [39], where many end-devices can be densely deployed and may be equipped with different radio communication technologies (Wi-Fi, ZigBee, Bluetooth, 2G/3G/4G, etc.) to cope with different application requirements in terms of data rate, latency, reliability and energy consumption [40]. Different wireless technologies will all have to work with limited spectral resources, and it is therefore very important to deploy a suitable protocol stack with proper configuration settings that adapts to changing wireless and application contexts. Reconfiguration involves the ability to change protocol parameters, a network protocol, e.g., the MAC layer, and the radio operation mode (e.g., the modulation and coding scheme, channel frequency, transmitted power, etc.) [41][42].

In order to support the above vision, the selection of a suitable protocol stack and the corresponding configuration parameters should happen in an autonomous way without the involvement of radio or network experts, otherwise the complexity of the network precludes its scalability. In the area of cognitive radio networks, a lot of attention is paid to programmable SW communication architectures [43][44][45][46][47][48][49]. These architectures focus either on a single-radio technology or sophisticated SDR platforms. Single-radio platforms are cost efficient, but offer limited radio flexibility. SDR platforms, on the contrary, are very powerful, but expensive and consume a lot of energy. Both types of platform are expected to coexist in future dense wireless scenarios, although their difference in complexity will result in different levels of adoption. Furthermore, while these platforms provide the ability to adapt the operation of the wireless interfaces to the estimated environment conditions, they do not provide a complete wireless architecture including all required functionalities (apart from re-configuration). In what follows, we describe the functionality that is envisioned to leverage this re-configurability to optimize performance in 5G systems.

16.5.1 Functional Modules

Versatile wireless interfaces are a key requirement for future 5G systems, as these enable to adapt to the estimated context conditions. However, to take advantage of this versatility, the interfaces have to be provided with the adequate functionality to properly estimate network conditions, and to react accordingly in the most appropriate manner, which may involve altering the operation of other wireless interfaces. To satisfy the requirements discussed above in 5G systems, the architecture of future wireless interfaces have to be designed building on three key concepts, namely, flexibility, reconfigurability and monitoring. This required functionality can be divided into three different areas: the operation of a single device, i.e. intra-node enhancements, the coordination of multiple devices, i.e. inter-node functionality, and the control and optimization of the network, i.e. intelligent programs to optimize performance.

For the single node operation, the most relevant concept is flexibility: to cope with the high variability of 5G application requirements and network topologies, 5G technologies will support advanced reconfiguration capabilities at both PHY and MAC levels. Architectures are needed which move from the traditional approach of a "one-size-fits-all" MAC/PHY protocol stack to an innovative paradigm of on-the-fly configuration of context-specific stacks, see also Section 6.4, implementing abstracted versions of wireless primitives that sit between the current ossified stacks and the fully-programmable SDR solutions.

The control and optimization of the network deals with the optimization of network performance, and should be based on an information-centric operation and exploit the re-configurability and monitoring features. This requires functionality such as a monitoring library that provides access to the data collected by sensors and monitoring agents throughout the network, which should feed two modules that extract network-wide performance and context estimations to trigger different types of optimizations. For instance, performance degradations or context variations may trigger the activation of additional network elements or switching to a radio access technology that is more robust. Moreover, we also envision the need of a global scheduler, coordinating the operation of the BSs, enabling the C-RAN and vRAN vision as described next, and a service scheduler that coordinates the optimization of all these elements, to enable a smooth operation that precludes conflicts.

Finally, to connect these two areas, a third one is needed: multi-node coordination. Here, we envision functionalities such as: a monitoring service which gathers data from the local monitoring agents of the multiple devices (including sensors), processes their information, and forwards it to the optimization modules; and modules for functional re-composition, which stitch together SW and HW functions and abstract the changing of the operation of multiple devices to the performance optimizers, hence providing a technology-agnostic re-configurability service to the intelligent programs.

As an example of field-use of the proposed architecture, context-based system requirements could trigger a network-wide partitioning and reconfiguration of HW-accelerated and SW baseband functions tailored for different traffic service delivery and QoE needs. Out of the different available options, the flexible partitioning could result in placing i) the gNB layer-2 and above together with the EPC functions in the cloud, ii) the gNB layer-2 and above functions at a local MEC server or distributed unit (DU) or iii) move most of the gNB stack to the cloud and transform the remainder of the gNB into a remote radio head (RRH). This dynamic split and reconfiguration of baseband functions is complementary to the partitioning of the gNB protocol stack, either at stack or algorithm level, e.g., MAC-PHY, Radio Link Control (RLC), Packet Data Convergence Protocol (PDCP), etc., that was promoted recently by key industry actors in [50]. See also details on related RAN split options as considered by 3GPP in Section 6.6, and related deployment options in Section 6.7.

16.5.2 SW Platform Solutions for Prototyping 5G Systems

We will now review some of the most relevant existing SW platforms that can support a vision such as the one described before. Several efforts have been done recently related to prototyping mobile networks in SW, with most of the platforms exploiting and supporting the GNU Radio development suite [51] and the Ettus Research USRP SDR platforms [52][53]. One prominent product is Amari LTE 100, a fully SW-based LTE base station commercialized by Amarisoft [54] that offers a complete out-of-the-box solution for students and researchers. While being a relevant and promising product with an outstanding computational efficiency, its closed license makes it unsuitable for MAC scheduling algorithms optimization and other advanced research fields.

The most popular open-source LTE SDR SW available for testbeds today are Eurecom's OpenAirInterface (OAI) [35] and openLTE [55]. OAI provides a standard-compliant implementation of a subset of Release 10 LTE, including key elements of the network such as UE, eNB, Mobility Management Entity (MME), Home Subscriber Server (HSS), serving gateway (S-GW) and packet gateway (P-GW) on standard Linux-based computing equipment (Intel x86 PC architectures). The SW can be used in conjunction with standard RF laboratory equipment available in many labs (i.e., National Instruments/Ettus USRP and PXIe platforms). Although OAI's implementation is very complete and provides relatively good performance, the code structure is complex and difficult to customize, plus the implementation of the CN functionality does not provide a very stable performance (although the pace of updates has increased recently).

openLTE is an open-source implementation of LTE specifications. The project includes a C library, Octave code for testing DL and UL Physical Random Access CHannel (PRACH) functionalities, GNU Radio applications for DL functionalities, both simulated and using HW platforms, and a simple implementation of an eNB using Universal Software Radio Peripheral (USRP). It runs with the Ettus Research B2x0 USRP and provides eNB, MME and HSS functionalities. openLTE code is well organized, documented and easy to customize or modify. However, it is incomplete and many features are still unstable or under development. Furthermore, it does not provide a UE, limiting the testbed capabilities in terms of instrumentation and measurement.

A relatively recent library is srsLTE [37], an open-source platform for LTE experimentation designed for maximum modularity and code reuse and fully compliant with LTE Release 8, which can be considered as the evolution of the libLTE [56], an initial attempt that suffered from lack of functionality and poorly documented code. Although the platform provides a complete UE, this is not the case for the implementation of the eNB and core network, which are available for purchase, while the open version of the eNB lacks functionality such as Hybrid Automatic Repeat reQuest (HARQ).

Following the above overview, the OAI project by Eurecom [35] appears to be the most complete and flexible LTE SDR platform to date [57][58], and an implementation example will be described in the following sections targeting a vRAN/CRAN architecture.

16.6 Implementation Example: vRAN/C-RAN Architecture in OAI

As the wide adoption of the cloud computing concept is progressing, the currently distributed radio access network (D-RAN) architecture is expected to evolve toward a cloud/centralized radio access network (C-RAN) architecture that stands out as a promising solution for 5G. In C-RAN architecture, the original BS is decoupled into centralized baseband units (BBU) and the remote radio unit (RRU) at the network edge. These centralized BBUs can be pooled and used as shared resources, offering statistical multiplexing gains and energy efficiency. Further, C-RAN facilitates advanced coordinated multi-point (CoMP) processing and satisfies the stringent synchronization constraints of CoMP [59]. Finally, the BBU/RRU network functions can be implemented on commodity HW and executed on a virtualized environment, i.e. a virtualized RAN (vRAN), further benefiting from network softwarization and network function virtualization (NFV) concepts.

Despite its appeal, one key obstacle in the adoption of C-RAN are the excessive capacity requirements on the fronthaul (FH) link that provides BBU and RRU interconnections [60], as quantified for an example scenario in Section 6.6.2. To relax these requirements, the concept of C-RAN has been revisited, and a more flexible distribution of baseband functionalities between the RRU and BBU is considered in [61] and also discussed in 3GPP. Rather than offloading all baseband processing to the BBU, it is possible to keep a subset of this processing at the RRU, which is the case for the 3GPP split options 7-x detailed in Section 6.2.2. This concept is also known as *flexible centralization* that splits the functions between RRU and BBU. Nevertheless, flexible centralization has two main drawbacks related to the initially envisioned benefits of C-RAN: (a) complex and expensive RRUs, and (b) reduction of the opportunities for multiplexing gains and coordinated processing. In consequence, flexible centralization is a trade-off between what is gained in terms of FH requirements and what is lost in terms of C-RAN features.

As a consequence, a three-tier architecture is envisioned with the RRU at the network edge, a DU at some aggregation points and a central unit (CU) [62], see also Section 6.7. The low-level functional split between RRU and DU can maintain the low-cost RRUs, whereas the high-level functional split between DU and CU still provides multiplexing gain and supports advanced coordination schemes.

16.6.1 Overall Architecture

To support the envisioned three-tier architecture, OAI is evolving from an isolated monolithic BS concept (i.e., LTE eNB) to a more flexible and disaggregated BS spanning multiple modules. The overall architecture supports multi-RAT, namely (e)LTE, NR and NB-IoT as shown in Figure 16-8. The envisioned RRU supports low-level L1 functionalities, while more signal processing is concentrated at the DU, when using low-latency FH links. Otherwise, more functions have to be distributed to RRUs, for example the whole PHY layer may be placed at the RRU if 3GPP split option 6 is applied. Further, the CU covers a larger area than the DU and possesses a centralized PDCP functionality to allocate the user plane among different RATs within its coverage area. The same CU also hosts the control functions of the core network. All these three-tier components can be further managed by the global orchestrator under different deployment topologies. Please also refer to Section 6.7 for further details.

16.6.2 Deployment Topology

Since the C-RAN concept first appeared, a single FH link per RRU that connects all the way directly to the BBU has normally been assumed. However, it is expected that the FH network will evolve to a more complex multi-hop mesh network topology that requires switching and aggregation [63], see also Chapter 7. Hence, we focus on a multi-segment FH network topology in Figure 16-9. The considered topology can support a generic mesh deployment of the FH network that can be shared with other RRU traffic flows. The RRU gateways are the multiplex/de-multiplex point in the network and can be utilized to transport not only the C-RAN traffic, but also other traffic flows. Further, the BBU can be pooled and distributed centrally in the cloud, or at the network edge in some aggregated points. Figure 16-9 also shows how the considered C-RAN network can be mapped to the three-tier architecture stated before (i.e., RRU, DU, CU) and the ones provided by Next Generation Fronthaul Interface (NGFI), such as the RRU, the radio aggregation unit (RAU) and radio cloud center (RCC) in [64], or

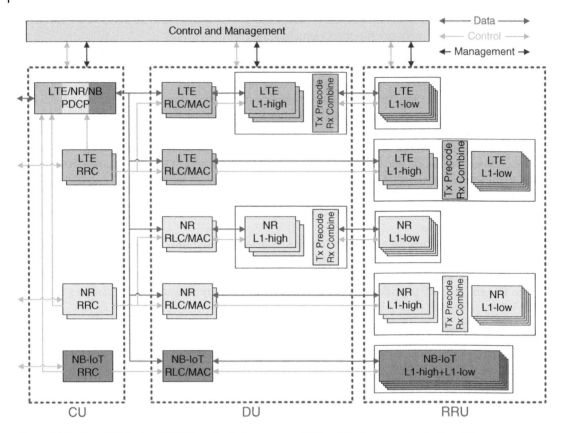

Figure 16-8. OpenAirInterface (OAI) three-tier heterogeneous RAN architecture, see also Section 6.7.

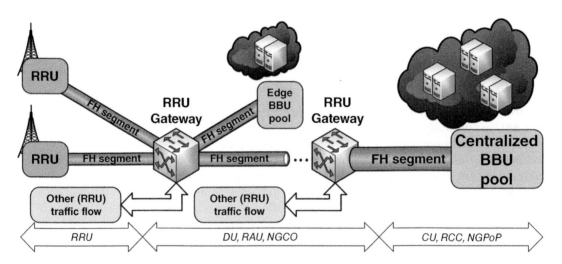

Figure 16-9. Considered C-RAN network topology.

the telecom industry, which considered a RRU, new generation central office (NGCO), and a next generation point of presence (NGPoP). Another key characteristic is how the information between the RRU and BBU is transported over the FH link. A number of FH transmission protocols are already used in the field, such as CPRI [65], OBSAI [66] and ORI [67]. However, as stressed before, these techniques consider carrying raw I/Q samples that can be utilized in the FH segment between the first and second tier of the architecture in Figure 16-9. In light of the flexible centralization concept, different types of information are transported over the FH link based on the functional split between RRU and BBU, such as the 3GPP split options 7-x that are described in detail in Section 6.6.2. Given the extensive adoption of Ethernet in remote clouds, data centers and the core network, the Radio over Ethernet (RoE) [68] approach is a generic, cost-effective and off-the-shelf alternative for FH link traffic transport.

16.6.3 Performance Results

We evaluate the C-RAN implementation using OAI [35], a SW-based LTE/LTE-A system implementation spanning the full 3GPP protocol stack with a third-party EPC, and commercial off-the-shelf (COTS) UE and USRP B210 SDR. In the following, we consider two C-RAN network deployments: (a) no RRU gateway (i.e., 1 hop between RRU/DU) and (b) a single RRU gateway (i.e., 2 hops between RRU/DU) with each FH link being made up of a 3-meter cable. In simple, there is only 1 RRU and 1 DU in the considered C-RAN network using split A or split B defined in [69] with 5 MHz/10 MHz radio bandwidth. In the following, some important KPIs related to C-RAN are presented.

16.6.3.1 FH-related KPIs: FH Link Throughput

The FH link throughput together with the theoretical rate of both 5 MHz and 10 MHz cases is shown in Figure 16-10. First, the RAW transmission throughput only has little overhead (between 3 to 4 Mbps) compared with the theoretical rates. The UDP transportation shows little further overhead (up to 3 Mbps) compared with RAW transmission. Moreover, the applied a-law compression scheme reaches almost 50% reduction in the FH link throughput. Further, using split B shows a gain of 43.8% in terms of FH link throughput reduction compared with split A via moving the FFT/IFFT operation to RRU, as considered in 3GPP split option 7-1. This reduction ratio is similar to that presented in Section 6.2.2, and also close to the analysis in [60] that shows 45.3% of throughput reduction. We can also observe that the throughput is scaling with the channel bandwidth and hence the predicted FH capacity when using 20 MHz channel bandwidth without any compression scheme will be around 1 Gbps.

16.6.3.2 FH-related KPIs: Round-trip Time of the FH and RF Front-end

These KPIs measure the round-trip latency between the FH link and RF devices. Such metrics are crucial to evaluate the possible RRU-DU functional split, and for instance NGMN adopts 250 μs as the maximum one-way FH latency [70], and SCF categorizes the one-way FH latency from 250 μs to the millisecond level to evaluate the applicable RRU-DU split [71]. Here, we use the difference of timestamps at RRU side to measure both round-trip times (RTTs). The RTT of the FH is measured as the time elapsed at the RRU between the start of sending the receiver data samples and the end of reading the corresponding transmitter data samples from the DU on the FH link. In detail, this RTT of the FH is made up of 5 components: (a) compress time, (b) FH write time, (c) FH link RTT, (d) FH read time, and (e) decompress time. Moreover, the RTT of

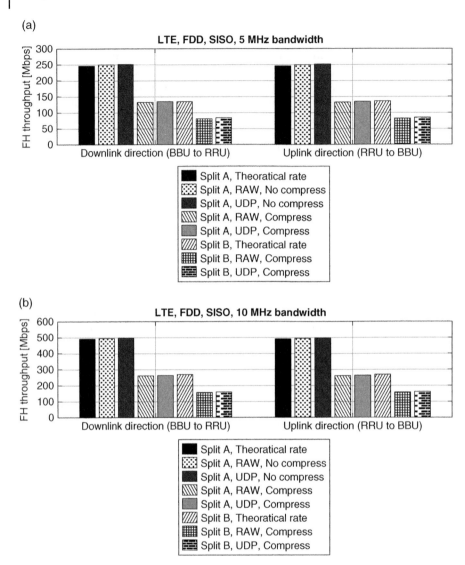

Figure 16-10. FH throughput needed for 5 MHz and 10 MHz bandwidth.

the RF front-end is defined as the time elapsed from the reading of data samples of RF devices until the writing of samples to RF devices, which includes the RTT of the FH.

Based on aforementioned definitions, the results are shown in Figure 16-11. The reduction of the RTT in the FH is proportional to the reduction of throughput (by a factor of 2) when applying the a-law compression which comes at the cost of the extra processing time for compression and decompression (see Table 16-2) but with much fewer FH link read/write time (see Table 16-2), which confirms the benefit of the compression in the FH network. In addition, it can be seen from Figure 16-11(b) that the average RTT of the RF front-end remains comparable for the 5 MHz case even without the

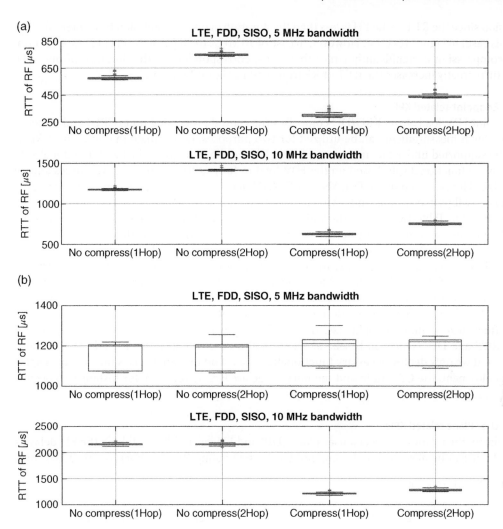

Figure 16-11. RTT of the FH and RF for 5 MHz/10 MHz bandwidth.

Table 16-2. Mean time for FH link read/write and compression/decompression.

Bandwidth	Compression	Compress time [µs]	FH write time [µs]	FH read time [µs]	Decompress time [µs]
5 MHz	No	-	69.19	224.37	-
	Yes	16.53	68.80	53.13	23.43
10 MHz	No	-	72.35	469.53	-
	Yes	31.70	71.95	170.13	35.93

compression since the RTT of the FH is less than the duration of one transmission time interval (TTI, 1 ms in this case) in Figure 16-11(a). However, in case of 10 MHz, the RTT of the FH in Figure 16-11(a) without compression is significantly larger than 1 ms, corresponding to the duration of one TTI, which in turn greatly increases the RTT of RF front-end in Figure 16-11(b).

16.6.3.3 Endpoint-related KPIs

The end-point HW load at the RRU/DU comprises the central processing unit (CPU) and memory utilization. Such measurement can be utilized for two purposes: (a) estimate the number of RRUs that can be supported under the limited number of DUs in the pool given the fixed functional split or (b) dynamic functional split based on the HW load to fully utilize all available resources among RRU and DU. Hence, we compare D-RAN and C-RAN deployments given the constant user traffic of 15 Mbps/30 Mbps in downlink and 5 Mbps/10 Mbps in uplink for 5 MHz/10 MHz bandwidth. The results are listed in Table 16-3, where the CPU utilization ratio is the percentage of CPU processing time of the process, and the memory usage is measured based on the proportional set size (PSS) in KBytes. RRU, DU and eNB are deployed in a hex-core CPU each with Intel i7 Sandy Bridge architecture in 3.2 GHz. As a result, 2 CPU cores are required to deploy the proposed RRU for 10 MHz bandwidth for split A and 3 cores for split B. Moreover, the sum of CPU resources required by RRU and DU is slightly higher than the one required by the eNB deployment. In addition, the memory usage at RRU and DU does not have large differences since our C-RAN deployment is considered to support the fully flexible function split, i.e., the baseband processing can be dynamically allocated between RRU and DU. Hence, all baseband functionalities are still at both RRU and DU. We also observe that if we only deploy necessary functionalities based on the functional split, i.e., use a split-specific deployment, the CPU ratio and memory usage are largely reduced, for instance, to occupy 5% of CPU ratio and 16 KBytes of memory usage for RRU in split A.

16.6.3.4 User Plane related KPIs

We first clarify the relations between user plane (UP) and FH-related KPIs. Taking the FH delay as an example, it can be absorbed and compensated by scheduling the transmission ahead of time,

Table 16-3. HW load of RRU/DU.

Bandwidth	Endpoint	Split	CPU ratio	Memory usage [kBytes]
5 MHz	eNB		40.15%	1002019
	RRU	A	16.76%	917486
	DU	A	26.57%	918794
	RRU	B	24.19%	917478
	DU	B	22.71%	917174
10 MHz	eNB		65.02%	1195059
	RRU	A	29.23%	1107382
	DU	A	45.70%	1180126
	RRU	B	41.12%	1107374
	DU	B	32.40%	1124989

which in turn reduces the total transmitter and receiver processing time to provide extra FH transportation time. However, such shortened processing time might not be enough for some processing (e.g., Turbo decoder) in some cases [72], and it can cause an extra UP delay due to retransmission. To evaluate the UP KPIs, we use 15 Mbps/30 Mbps user traffic for 5 MHz/10 MHz in downlink direction separately.

To characterize the UP QoS from the user perspective, we measure the delay jitter and good-put. First, the delay jitter is shown in Figure 16-12. Due to the extra route from the user to the gateway that leads to less available transmitter/receiver processing time, the jitter of the C-RAN deployment is larger than the one in a legacy D-RAN, and the jitter of the case with two hops is larger than the one in a one-hop case. Further, the measured good-put at application level is shown in Figure 16-13, and seven different C-RAN deployments (A1 to A5, B1 and B2 in Table 16-4) are considered to be compared to a legacy eNB/D-RAN deployment. Note that the good-put is the application-level successful throughput that is significant from the user perspective. We observe that these deployment scenarios show almost the same good-put variation as the D-RAN one; that is to say, the experienced good-put will be maintained among different RAN deployments.

For the UP delay, we measure the application RTT over the default radio bearer. It not only considers the impact of the FH link, but also the transmitter and receiver processing time at the RRU and DU of both downlink and uplink directions. Hence, we use the ping utility with a 8192 bytes packet size and 0.2 s inter-departure time to characterize the delay in Figure 16-14. We observe that the average RTT in a

(a)

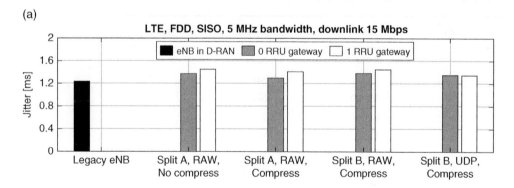

(b)

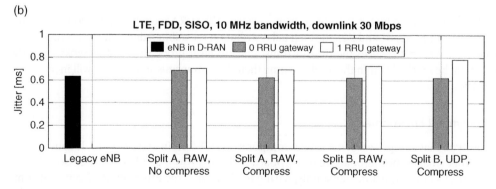

Figure 16-12. User plane packet delay jitter for 5 MHz/10 MHz bandwidth.

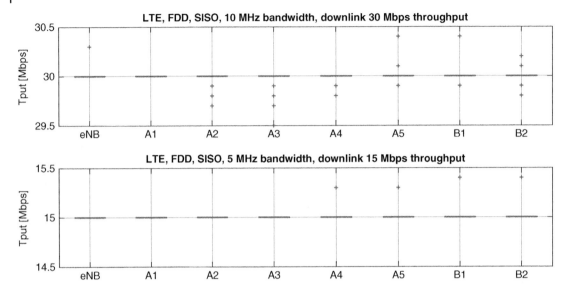

Figure 16-13. User plane good-put of several deployment scenarios.

Table 16-4. Parameters for the different C-RAN deployments.

Mode	Split	Protocol	Compression	Hop count
A1	A	RAW	No	1
A2	A	RAW	No	2
A3	A	UDP	No	2
A4	A	RAW	Yes	2
A5	A	UDP	Yes	2
B1	B	RAW	Yes	2
B2	B	UDP	Yes	2

C-RAN deployment is a little higher than the one in a legacy eNB/D-RAN deployment. However, in the two-hop cases (i.e., A2 to A5, B1 and B2), the long tail distribution is exhibited due to the extra route from the user to the gateway that reduces the time or transmitter/receiver processing. In this sense, once the processing cannot be finished within the available time then the re-transmission scheme will increase the user plane delay.

16.6.4 Deployment Environment

An RRU prototype that comprises all necessary components is shown in Figure 16-15. It contains the Pico-ITX or other smaller motherboard (e.g., UpBoard with Intel Atom quad-core processor), Power over Ethernet (PoE+) to support power supply, wiring for 1 Gigabit/sec Ethernet, RF front-end components (PA, LNA, Switch), 10 MHz/PPS frequency synchronization cable, baseband-to-RF radio unit (e.g., USRP B200-mini) and RF front-end circuits.

(a)

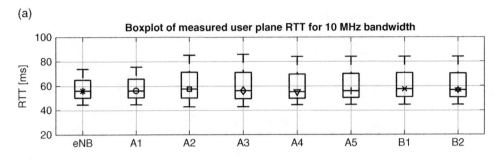

(b)

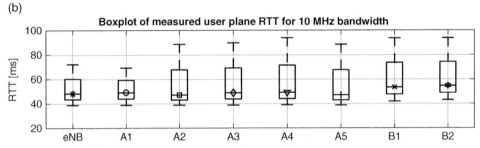

Figure 16-14. User plane packet RTT for 5 MHz and 10 MHz BW.

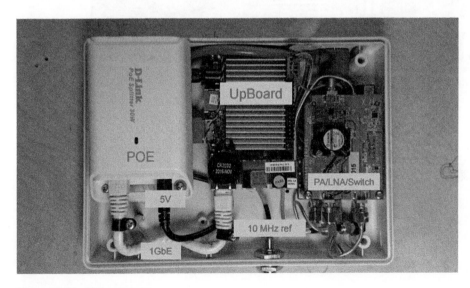

Figure 16-15. RRU prototype.

In Figure 16-16, the physical deployment of an indoor scenario is shown. Two distribution switches are placed and connected to the aggregation switch. As an extension, the planned outdoor deployment using a large number of RRUs (i.e., up to 64 RRU elements) and higher FH capacity is depicted in Figure 16-17. This deployment showcases the coordination of the indoor and outdoor network segments and the orchestration on a larger scale.

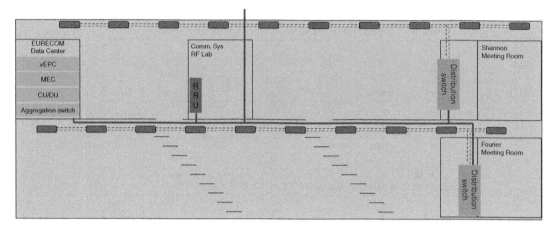

Figure 16-16. Indoor deployment floorplan.

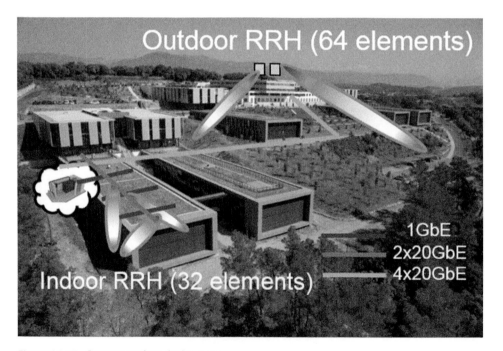

Figure 16-17. Campus outdoor deployment.

16.7 Summary

In this chapter, a number of implementation challenges that future 5G HW and SW platforms will face have been presented, and some preliminary solutions in different technology domains have been provided. In particular, Section 16.2 has provided an overview on challenges and solutions regarding the analog HW and mixed-signal stages, whereas in Section 16.3 digital HW has been addressed with a strong emphasis on waveform implementation complexity. Section 16.4 has focused on HW/SW function splitting, and a set of functional requirements for SW platforms has been described in Section 16.5 together with a summary of different existing SW platforms which could support these. As an example of one of such platforms, in Section 16.6, an implementation targeting a vRAN/CRAN architecture has been provided.

References

1 ETSI TS 136 104, "LTE; Evolved Universal Terrestrial Radio Access E-UTRA); Base Station (BS) radio transmission and reception (3GPP TS 36.104 version 12. Release 12)", V12.5.0, Oct. 2014

2 5G PPP Flex5Gware project, Deliverable D2.2, "Analogue components for high-performing and versatile 5G RF front-ends", May 2017

3 5G PPP Flex5Gware project, Deliverable D3.2, "Mixed-signal strategies for 5G: final architecture and design for flexibility, power and spectral efficiency", May 2017

4 Analog Devices, AD9172, see http://www.analog.com/media/en/technical-documentation/data-sheets/AD9172.pdf

5 Vadatech, AMC589, see http://www.vadatech.com/media/AMC589_AMC589_Datasheet.pdf

6 W.H. Doherty, "A new high efficiency power amplifier for modulated waves", Proceedings of the Institute of Radio Engineers, vol. 24, no. 9, pp. 1163–1182, Sept. 1936

7 M. Bozzi, A. Georgiadis and K. Wu, "Review of substrate-integrated waveguide circuits and antennas", in IET Microwaves, Ant. & Prop., vol. 5, no. 8, pp. 909–920, June 2011

8 S. Agneessens and H. Rogier, "Compact Half Diamond Dual-Band Textile HMSIW On-Body Antenna", in IEEE Trans. on Antennas and Propagation, vol. 62, no. 5, May 2014

9 D. Markert, Y. Xin, H. Heimpel and G. Fischer, "An All-Digital, Single- Bit RF Transmitter for Massive MIMO", IEEE Transactions on Circuits and Systems, vol. 64, no. 3, pp. 696–704, Dec. 2016

10 5G PPP Flex5Gware project, Deliverable D6.2, "Final PoC evaluation in Flex5Gware", June 2017

11 Goyal, S. et al., "Improving Small Cell Capacity with Common-Carrier Full Duplex Radios", IEEE International Conference on Communications (ICC 2014), June 2014

12 E. Costa, M. Midro and S. Pupolin, "Impact of amplifier nonlinearities on OFDM transmission system performance", IEEE Comm. Letters, vol. 3, pp.37–39, Feb. 1999

13 P. Achaichia, M. L. Bot and P. Siohan, "Windowed OFDM versus OFDM/OQAM: A transmission capacity comparison in the HomePlug AV context", IEEE Intl. Symp. on Power Line Communications and Its Applications (ISPLC 2011), Apr. 2011

14 G. Berardinelli, F. M. L. Tavares, T. B. Sørensen, P. Mogensen and K. Pajukoski, "On the potential of zero-tail DFT-spread-OFDM in 5G networks", IEEE Vehicular Technology Conference (VTC Fall 2014), Sept. 2014

15 P. Siohan, C. Siclet and N. Lacaille, "Analysis and design of OFDM/OQAM systems based on filterbank theory", IEEE Transactions on Signal Processing, vol. 50, no. 5, pages 1170–1183, 2002

16 C. Kim, Y. H. Yun, K. Kim and J. Y. Seol, "Introduction to QAM-FBMC: From waveform optimization to system design", IEEE Comm. Magazine, vol. 54, no. 11, pages 66–73, 2016

17 Z. Zhao, M. Schellmann, Q. Wang, X. Gong, R. Boehnke and W. Xu, "Pulse shaped OFDM for asynchronous uplink access", 49th Asilomar Conference on Signals, Systems and Computers, Nov. 2015

18 L.G. Baltar, F. Schaich, M. Renfors and J.A. Nossek, "Computational complexity analysis of advanced physical layers based on multicarrier modulation", Future Network & Mobile Summit (FutureNetw 2011), June 2011

19 M. Bellanger et al., "FBMC physical layer: a primer", PHYDYAS, Jan. 2010

20 M. Bellanger, "Specification and design of a prototype filter for filter bank based multicarrier transmission", IEEE International Conference on Acoustics, Speech, and Signal Processing, May 2001

21 T. Wild and F. Schaich, "A Reduced Complexity Transmitter for UF-OFDM", Vehicular Technology Conference (VTC Spring 2015), May 2015

22 Y. Chen, F. Schaich and T. Wild, "Multiple Access and Waveforms for 5G: IDMA and Universal Filtered MultiCarrier", Vehicular Technology Conference (VTC Spring 2014), May 2014

23 N. Michailow, I. Gaspar, S. Krone, M. Lentmaier and G. Fettweis, "Generalized frequency division multiplexing: Analysis of an alternative multi-carrier technique for next generation cellular systems", International Symposium on Wireless Communication Systems (ISWCS), Aug. 2012

24 I. Gaspar, N. Michailow, A. Navarro, E. Ohlmer, S. Krone and G. Fettweis, "Low Complexity GFDM Receiver Based on Sparse Frequency Domain Processing", Vehicular Technology Conference (VTC Spring 2013), June 2013

25 5G PPP Flex5Gware project, Deliverable D4.2, "Final report on HW architectures", May 2017

26 3GPP TS 38.802, "Study on New Radio Access Technology - Physical Layer Aspects", March 2017

27 E. O. Brigham, "The fast Fourier transform", Prentice-Hall Inc., 1998

28 5G PPP METIS-II project, Deliverable D4.2, "Final air interface harmonization and user plane design", April 2017

29 M. Forsell and J. Roivainen, "REPLICA T7-16-128 - A 2048-threaded 16-core 7-FU chained VLIW chip multiprocessor", Special session on Multicore, Manycore and Distributed systems at the 48th Asilomar Conference on Signals, Systems, and Computers, Nov. 2014

30 M. Forsell, J. Roivainen and V. Leppänen, "Prototyping the MBTAC processor for the REPLICA CMP", Advances in Parallel and Distributed Computational Models (APDCM'14) in conjunction with the IEEE International Parallel and Distributed Processing Symposium (IPDPS'14), May 2014

31 T. Richardson, A. Shokrollahi and R. Urbanke, "Design of capacity-approaching irregular low-density parity-check codes", IEEE Transactions on Information Theory, vol. 47, no. 2, pp. 619–637, 2001

32 R. Tanner, "A recursive approach to low complexity codes", IEEE Transactions on Information Theory, vol. 27, no. 5, pp. 533–547, 1981

33 Truong Nguyen-Ly, Khoa Le, V. Savin, D. Declercq, F. Ghaffariy, and O. Boncalo, "Non-Surjective Finite Alphabet Iterative Decoders", IEEE International Conference on Communications (ICC 2016), May 2016

34 LTE-EPC Network Simulator (LENA), see http://iptechwiki.cttc.es/LTE-EPC Network Simulator (LENA)

35 N. Nikaein, R. Knopp, F. Kaltenberger, L. Gauthier, C. Bonnet, D. Nussbaum and R. Ghaddab, "OpenAirInterface 4G: an open LTE network in a PC", 2013

36 Q. Zheng, H. Du, J. Li, W. Zhang and Q. Li, "Open-LTE: An Open LTE simulator for mobile video streaming", IEEE International Conference on Multimedia and Expo Workshops (ICMEW 2014), July 2014

37 I. Gomez-Miguelez, A. Garcia-Saavedra, P. D. Sutton, P. Serrano, C. Cano and D. J. Leith, "srsLTE: An Open-Source Platform for LTE Evolution and Experimentation", Feb. 2016

38 5G PPP Flex5Gware project, Deliverable D4.2, "Final report on HW architectures", May 2017

39 C. J. Bernardos et al., "An Architecture for Software Defined Wireless Networking", IEEE Wireless Communications, vol. 21, no. 6, pp. 52—61, June 2014

40 C. Donato et al., "An OpenFlow Architecture for Energy Aware Traffic Engineering in Mobile Networks", IEEE Network, vol.29, no.4, pp. 54–60, July-August 2015

41 P. Demestichas, G. Dimitrakopoulos, J. Strassner and D. Bourse, "Introducing Reconfigurability and Cognitive Networks Concepts in the Wireless World", IEEE Vehicular Technology Magazine, 2006

42 S. Hong, J. Mehlman and S. Katti, "Picasso: Flexible RF and Spectrum Slicing,", ACM SIGCOMM, Aug. 2012

43 M. Mueck, et al., "ETSI Reconfigurable Radio Systems: Status and Future Directions on Software Defined Radio and Cognitive Radio Standards", IEEE Communications Magazine, vol. 48, no. 9, Sept. 2010

44 C. R. Aguayo González, C. B. Dietrich and J. H. Reed, "Understanding the Software Communications Architecture", IEEE Communications Magazine, vol. 47, no. 9, Sep. 2009

45 J. Bard and V. J. Kovarik Jr., "Software Defined Radio: The Software Communications Architecture", John Wiley & Sons Ltd., 2007

46 P. De Mil, B. Jooris, L. Tytgat, J. Hoebeke, I. Moerman and P. Demeester, "SnapMac: A generic MAC/PHY architecture enabling flexible MAC design", Ad Hoc Networks, 2014

47 I. Tinnirello, G. Bianchi, P. Gallo, D. Garlisi, F. Giuliano and F. Gringoli, "Wireless MAC Processors: Programming MAC Protocols on Commodity Hardware", IEEE INFOCOM, Mar. 2012

48 G. Bianchi, P. Gallo, D. Garlisi, F. Giuliano, F. Gringoli and I. Tinnirello, "MAClets: Active MAC Protocols over Hard-Coded Devices", ACM CONEXT, Dec. 2012

49 P. D. Sutton et al., "Iris: An Architecture for Cognitive Radio Networking Testbeds", IEEE Communications Magazine, vol. 48, no. 9, Sept. 2010

50 Small Cell Forum, "Virtualization for small cells: Overview", see http://www.scf.io/en/documents/106__Virtualization_for_small_ cells_Overview.php

51 GNU Radio, "GNU Radio: The Free & Open Software Radio Ecosystem", see http://gnuradio.org

52 Ettus Research, see http://www.ettus.com/

53 M. Abirami, V. Hariharan, M. Sruthi, R. Gandhiraj and K. Soman, "Exploiting GNU Radio and USRP: An Economical Test Bed for Real Time Communication Systems", IEEE ICCCNT, July 2013

54 Amarisoft, "Amari LTE 100 - Software LTE base station on PC", see http://www.amarisoft.com/index.php?p=amarilte

55 B. Wojtowicz, "openLTE: An open source 3GPP LTE implementation", last update Sept. 2014, see http://sourceforge.net/projects/openlte/

56 I. Gomez, "libLTE: Open source 3GPP LTE library", last update Oct 2013, see http://sourceforge.net/projects/liblte/

57 R. Wang, Y. Peng, H. Qu, W. Li, H. Zhao and B. Wu, "OpenAirInterface-An Effective Emulation Platform for LTE and LTE-Advanced", IEEE ICUFN, 2014

58 H. Anouar, C. Bonnet, D. Câmara, F. Filali and R. Knopp, "An Overview of OpenAirInterface Wireless Network Emulation Methodology", ACM SIGMETRICS Perform. Eval. Rev., 2008

59 A. Checko et al., "Cloud RAN for mobile networks - a technology overview", IEEE Communications Surveys & Tutorials, 2014

60 C.-Y. Chang et al., "Impact of packetization and functional split on C-RAN fronthaul performance", IEEE International Conference on Communications, (ICC 2016), May 2016

61 D. Wübben et al., "Benefits and impact of cloud computing on 5G signal processing: Flexible centralization through cloud-RAN", IEEE Signal Processing Magazine, vol 31, no. 6, pp. 35–44, Oct. 2014

62 I. Chih-Lin, "RAN revolution with NGFI (xhaul) for 5G", Optical Fiber Communications Conference, March 2017

63 C.-Y. Chang et al., "Impact of packetization and scheduling on C-RAN fronthaul performance", IEEE Global Communications Conference (GLOBECOM 2016), Dec. 2016

64 Y. Zhiling et al., White Paper, "White paper of next generation fronthaul interface v1.0", China Mobile Research Institute, Tech. Rep., 2015

65 CPRI, "Interface specification", v7.0, 2015

66 OBSAI, "BTS system reference document", v2.0, 2006

67 ETSI GS ORI 001, "Open Radio equipment Interface (ORI); requirements for ORI", Oct. 2014

68 IEEE, Technical Report, "1904.3 task force: Standard for radio over Ethernet encapsulations and mapping", 2015

69 C.-Y. Chang et al., "FlexCRAN: A Flexible Functional Split Framework over Ethernet Fronthaul in Cloud-RAN", IEEE International Conference on Communications (ICC 2017), May 2017

70 NGMN, Technical Report, "Further study on critical C-RAN technologies", 2015

71 Small Cell Forum, Technical Report, "Small cell virtualization functional splits and use cases", 2015

72 N. Nikaein, "Processing radio access network functions in the Cloud: Critical issues and modeling", Intl. Workshop on Mobile Cloud Computing and Services, Sept. 2015

17

Standardization, Trials, and Early Commercialization

Terje Tjelta[1], Olav Queseth[2], Didier Bourse[3], Yves Bellego[4], Raffaele de Peppe[5], Hisham Elshaer[6], Frederic Pujol[7], Chris Pearson[8], Chen Xiaobei[9], Takehiro Nakamura[10], Akira Matsunaga[11], Hitoshi Yoshino[12], Yukihiko Okumura[13], Dong Ku Kim[14], Jinhyo Park[15] and Hong Beom Jeon[16]

[1] Telenor, Norway
[2] Ericsson, Sweden
[3] Nokia, France
[4] Orange, France
[5] Telecom Italia, Italy
[6] Vodafone, UK
[7] iDATE, France
[8] 5GAmericas, US
[9] IMT-2020 PG, China
[10] NTT DOCOMO, Japan
[11] KDDI, Japan
[12] Softbank, Japan
[13] NTT DOCOMO, Japan
[14] 5G Forum, South Korea
[15] SK Telecom, South Korea
[16] KT, South Korea

17.1 Introduction

With 5th generation (5G) cellular systems, both traditional services from earlier cellular generations and new services will be provided, as underlined in Chapter 2. The network will become very fast and provide quick responses, very large individual user peak data rates, and much higher total capacity per area unit. The ability to handle large numbers of things and provide ultra-reliable connections will open up many new applications, beneficial services for the society, and business opportunities.

The time for a commercial launch comes after 5G technology is available in the form of internationally agreed standardized solutions. In preparation for roll-out and also the evaluation of options, a number of technology and service trials are taking place.

This chapter initially presents the roadmap of the expected standardization activities towards a full 5G system design in Section 17.2. It identifies the main standardization bodies and timelines for a two-phase approach, where the first phase focuses only on a sub-set of use cases and technology

5G System Design: Architectural and Functional Considerations and Long Term Research, First Edition.
Edited by Patrick Marsch, Ömer Bulakçı, Olav Queseth and Mauro Boldi.
© 2018 John Wiley & Sons Ltd. Published 2018 by John Wiley & Sons Ltd.

solutions, while the second phase reflects the full 5G standard addressing all main services types introduced in Section 2.2. Then, the chapter covers trials and early commercialization plans in the three regions Europe, Americas and Asia in Section 17.3. A number of technical and service-oriented tests have been performed and are planned, and an early commercial launch will happen as soon as the first phase standard solution is agreed and equipment is available. The chapter is finally summarized in Section 17.4.

17.2   Standardization Roadmap

17.2.1   3GPP New Radio

The purpose of standards for telecom systems is to ensure interoperability between equipment from different manufacturers. The most prominent part is the interface between the terminal and the network, which is loosely known as the air interface, see Chapter 11. In addition, there are a number of other interfaces that ensure interoperability between various equipment in the network, e.g., the interface between the radio access network (RAN) and the core network (CN), covered for instance in Section 5.3.2. The 3rd Generation Partnership Project (3GPP) is the organization developing the specifications for NR as well as for Long Term Evolution (LTE), High-Speed Packet Access (HSPA), Wideband Code Division Multiple Access (WCDMA), and the Global System for Mobile Communications (GSM). It is a partnership between the seven standards organizations that develop standards for telecommunications. The organizational partners of 3GPP are ARIB, ATIS, CCSA, ETSI, TSDSI, TTA, and TTC [1]. The reports and specifications developed by 3GPP are then used as the basis when the organizational partners develop standards applicable in the respective regions. In practice, the standards developed by the respective organizations tend to follow very closely the specifications developed by 3GPP. This ensures interoperability, so that equipment can be used globally. The exact differences between the specifications created by 3GPP and the ones from the organizational partners of 3GPP are beyond the scope of this book, but some examples include the support for frequency bands that are only available in regional specifications.

When looking at the work of 3GPP from the outside, it may seem very orderly with a sequence of releases following each other in very clean manner. However, when one looks into the process in more detail, it can be seen that there are many things happening in parallel, and that the timing is varying in different parts of the standards. To get a better understanding of the process and understand some of the timings, an overview of the process is given in the following. For more details, the interested reader is referred to [2][3].

The work on developing a specification is done in a number of stages. In the first stage, a feasibility study is done to study how a feature or function can be developed, and to obtain a rough understanding of how the specification would be done. This first step is also known as "Stage 0". Formally, the work is done in a study item (SI), and the outcome of the study is described in a technical report (TR). If the outcome of the SI is satisfactory, it is agreed to start a work item (WI) that will define the detailed specifications for that functionality. The work is divided into stages, where Stage 1 refers to the service description from a user's point of view. Stage 2 makes a logical analysis and defines architecture, functions and information flow. Stage 3 defines the concrete implementation and protocols. The interfaces are described using Abstract Syntax Notation One (ASN.1), and there are usually a few extra months to complete this description after all other work in Stage 3 is done.

The functionalities of 3GPP specifications are described in parallel releases. A release provides implementers with a stable set of functionalities once the release is frozen. A release is considered to be started when the first WI or SI targeting that release is approved. This is a somewhat arbitrary point in time, since discussions about the content of a release have usually taken place prior to that. The work on developing the specifications progresses successively until the release is frozen. The freeze is done in three phases, and different parts of the specifications are frozen at different times. E.g., the test specifications are frozen later than the definition of the functions themselves. After a release is frozen, no new functionality can be added, but the specifications are updated on a regular basis to correct errors. Eventually, the release is closed, which indicates that the release is no longer maintained. A new release is typically started every 1-2 years, and the duration of the development is 2-4 years.

One of the main 5G components developed by 3GPP is the New Radio (NR) specification. It will describe the new 5G air interface and the required functions for interfacing the CN and other 3GPP air interfaces.

For NR, the relevant releases are Release 14, 15 and 16. Release 14 started in 14Q3 and ended in mid-2017, see Figure 17-1. In Rel. 14, a number of preliminary activities were done to prepare for the specification of 5G. For instance, one study was completed mid-2016 to develop propagation models for spectrum above 6 GHz [4], as detailed in Chapter 4. Another study was done on scenarios and requirements for 5G and concluded at the end of 2016 [5]. In addition, a feasibility study was done of the NR air interface itself. This study was concluded in March 2017 and generated a number of reports covering all aspects of the new air interface. An overview can be found in [6].

Release 15 will contain the specifications for the first phase of 5G. A Rel. 15 WI is defined to generate the specifications. It started in mid-2016 and is expected to end in the second half of 2018. The process for this specification is somewhat unusual with two "drops" to match the deployment plans of the operators. The first drop of the standard was largely completed at the end of 2017 and contains specifications of all functions necessary to enable non-stand-alone operation. In non-stand-alone operation,

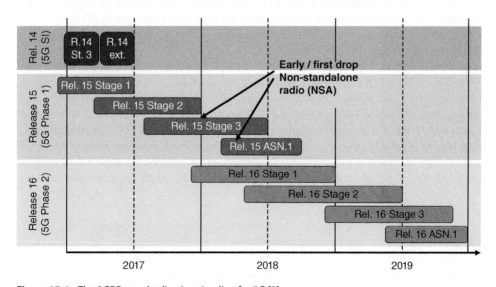

Figure 17-1. The 3GPP standardization timeline for 5G [2].

the NR carriers are used in combination with LTE carriers in a dual connectivity manner, as detailed in Sections 5.5 and 6.5. This allows postponing the development of some functions and procedures, such as initial access, and speeds up the development of the standard. The functions for standalone operation will take another 6 months to develop, and the complete specification of NR is expected to end in the second half of 2018. In addition, a number of feasibility studies are carried out for functionality that will be specified in Rel. 16.

Rel. 16 will begin in 2018 and last until 2019 with the specification of the second phase of 5G. The content of Rel. 16 is not clear at the time of writing, but it is expected that the main focus will be on functions needed for massive machine-type communications (mMTC) and ultra-reliabile and low-latency communications (URLLC) type of services, as described in Section 2.2.

17.2.2 IMT-2020

In parallel to the standardization efforts in 3GPP, there are activities ongoing in the International Telecommunication Union (ITU) to define the criteria a system must fulfil in order to be classified as an International Mobile Communications 2020 (IMT-2020) system, see also Section 15.2.1.

IMT-2020 is the third instance of the IMT family of technologies, where the predecessors are IMT-2000 and IMT-Advanced. There are several benefits for a technology to meet the IMT requirements, but the most prominent is that spectrum is identified by the World Radio Conference (WRC) for IMT systems, as discussed in Chapter 3, meaning that access to spectrum is facilitated. The second reason is that the IMT-2020 recognition is ratified by ITU-R, i.e., the radio leg of ITU, being a United Nations (UN) agency bringing together almost 200 administrations as well as several hundreds of other members from industry, international organizations as well as academia.

In September 2015, ITU issued the recommendation on "IMT Vision - Framework and overall objectives of the future development of IMT for 2020 and beyond" [7]. This document describes the rationale, scenarios and use cases that IMT-2020 systems should be able to fulfil, and outlines the capabilities of the IMT-2020 systems. This recommendation sets the scope for the continued work in ITU.

After WRC-15, the work has continued with the definition of the performance requirements that an IMT-2020 compliant technology must meet. In February 2017, the technical performance requirements were finalized by ITU-R Work Package 5D (WP5D), and in mid-2017 the deliverables on evaluation criteria and methodologies were finalized and are anticipated to be adopted in November 17. In October 2017, a workshop was held to kick off the proposal submission and subsequent evaluation work. In 2018-2019, technology proposals will be submitted and evaluated, and if they meet the requirements, they will be approved as IMT-2020 technologies, as shown in Figure 17-2. Obviously, not only the technologies developed by 3GPP can be part of IMT-2020 (3GPP decided that both NR and enhanced LTE (eLTE) will be submitted to the ITU-R process), but also other technologies that meet the IMT-2020 requirements will be approved.

17.2.3 3GPP eLTE

5G is not equivalent to the NR air interface, as already stressed in Chapter 1, and the term should not be limited to the air interface only. In fact, from a 3GPP perspective, 5G is generally considered

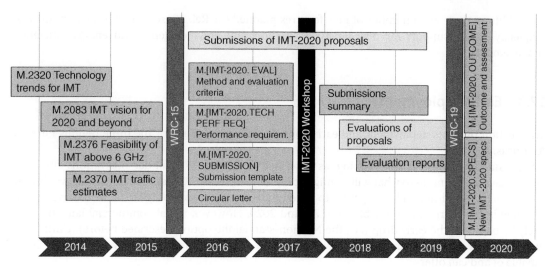

Figure 17-2. The ITU-R IMT-2020 (5G) timeline [8].

as a system that comprises the NR currently under development, as well as an evolution of LTE. The same applies for the CN, where a new NextGen core is developed together with the Evolved Packet Core (EPC).

In parallel to the NR activities in 3GPP, there has been ongoing work to add features to LTE so that some of the 5G use cases can already be fulfilled using the LTE air interface. The first LTE air interface release was Rel. 8, which was frozen in the beginning of 2009. In the six releases since, the original LTE specification has been enhanced with a number of functions and features, and there are still plans for enhancements in the ongoing and coming releases.

One of the areas where LTE has been enhanced is the ability to address mMTC use cases. Two features achieving this are known as MTC [9] and narrowband Internet of Things (NB-IoT) [10]. The headline capabilities of these are enhanced coverage, support for a very large number of devices and reduced energy consumption. This has been done by allowing for simpler user equipments (UEs) which only support half duplex, by allowing longer sleep times for UEs, and by reducing the transmission bandwidth of UEs down to 3.75 kHz.

To cater for the need of vehicular use cases, LTE has been enhanced with modes for vehicle-to-vehicle (V2V) communications [11], based on direct communication between UEs known as the LTE sidelink. For Rel. 15, work is planned to support advanced vehicle-to-anything (V2X) services [12], i.e. vehicle platooning, extended sensors beyond a single vehicle, or semi or fully automated driving and remote driving, as detailed in Chapter 14. In addition, activities are ongoing to investigate new security challenges, especially for the case of operation without direct communication with the network.

LTE has also evolved to handle low-latency and URLLC services. A number of suggested changes for handling low-latency communications have been devised [13], and some of them were introduced in Rel. 14. This work is continued, and in Rel.15 shorter transmit time intervals (TTIs) than the legacy 1 ms in LTE will be introduced. Work will also be done to support high reliability, i.e., to achieve error probabilities below 10^{-5} for small data packets within a latency of 1 ms.

In addition, there are a number of new features planned for Rel. 15, such as the support for new frequency bands, support for new UE types, architecture enhancements, and energy efficiency enhancements.

17.3   Early Deployments

This section presents deployment and early commercialization plans for the three regions Europe, Americas, and Asia.

Since standardization is still ongoing, as described in Section 17.2, all demonstrations currently taking place are pointing to what will or might become part of the 5G standard. It is conceivable that the commercial roll-out will only get up to speed once products according to the new standard are available in the market, presumably from around 2020. However, initial commercial launches will likely be based on the early drop (i.e., the NR non-standalone option described before). Until then, there will be a number of trials and early deployments to gain experience and validate what 5G is capable of and can offer.

17.3.1   Early Deployment in Europe

Several European network operators have shown interest in early testing of 5G technology. This is typically done through tests of selected 5G capabilities, such as very high data throughput or very low latency, and by partnering with equipment vendors to gain experience within some 5G technology areas for potential future exploration. The large test activity shows that 5G is considered very important for the future development.

The European community has discussed, and is still discussing 5G trials and commercial deployment. Some will be private initiatives such as between a network operator and equipment vendor, or among a few entities and under closed collaboration. Other trials may involve public entities together with industry such as initiatives in support of smart cities. Yet others include international cooperation. An overview issued by the 5G Infrastructure Association (5G IA) [14], is provided in Figure 17-3.

The trials roadmap has been worked out by the European Trials working group (WG) coordinated by the 5G IA. The roadmap expands the work initiated by the industry and the European Commission (EC) in the context of the 5G Manifesto [15] and the 5G Action Plan (5GAP) [16] and follows the definition of the roadmap strategy introduced at Mobile World Congress (MWC) 2017 [17].

The main objectives of the roadmap are to:

- Support global European leadership in 5G technology, 5G networks deployment and profitable 5G business;
- Validate benefits of 5G to vertical sectors, public sector, businesses and consumers;
- Initiate a clear path to successful and timely 5G deployment;
- Expand commercial trials and demonstrations as well as national initiatives.

Most of the roadmap implementation is and will be covered by the industry on a private basis, with part of this implementation supported by the EC through the 5G Infrastructure Public Private Partnership (PPP) Phase 3, the EC 5G Investment Fund, and by member states through specific national programs.

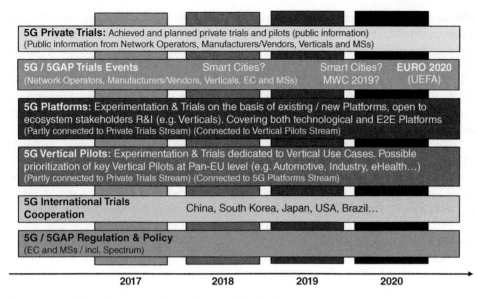

Figure 17-3. 5G Pan-European trials roadmap strategy [14].

Building Europe's 5G Readiness Through 5G Private Trials

Europe is home to an increasing number of 5G private experimentation and trials (pre-commercial and commercial), involving a multiplicity of stakeholders and notably network operators, manufacturers/vendors and some vertical actors. Several major network operators in Europe have already announced first demonstrations and experimentation results achieved and planned on demonstrating specific 5G features, bilaterally with one single manufacturer/vendor or bilaterally with multiple manufacturers/vendors. The main target of the current trials is to demonstrate high data rates and low latency. In 2017, there were only a few 5G private trials including vertical stakeholders. Trials in 2016-2017 have been focused on enabling technologies related to the radio interface (e.g., demonstrating high throughput, millimeter-waves and other new large spectrum bands, or antenna technologies, etc.), the network architecture (e.g., virtualization, cloudification, network slicing, edge computing, etc.) and the introduction of new technologies dedicated to specific use cases (e.g., technologies for IoT, automotive, etc.). It is foreseen that, when the maturity level of 5G features increases, vertical stakeholders will be more directly involved in trials. Some of the 5G trials announced include joint work on experimentation platforms that could become open to new ecosystems, in order to develop 5G applications and services in the context of the digital transformation of vertical industries.

Before the first 5G release, technology demonstrations and trials will be implemented in different countries, driven by the location of the different network operators. These demonstrations and trials will be done partly independently of the status of standardization, though network operators may prioritize features well advanced in standardization, to demonstrate and validate early the new 5G capabilities as well as foster an ecosystem around these new 5G capabilities. Consequently, the running trials and demonstrations are already today building concrete know-how and readiness of Europe to benefit from an early 5G launch when the standards will be fully stable

The 5G readiness will be further consolidated during and after 2018, when European stakeholders will move to agree on detailed trial specifications (e.g., use cases, scenarios, interfaces, agreements to transfer use cases across trial networks, etc.) valid for Pan-European trials, largely based on standard-compliant systems. These trials will take advantage of the first 5G release of the 3GPP Standard (Rel. 15 Stage 3 - June 2018) and use some of the additional frequency spectrum proposed for identification in WRC-19, enabling to demonstrate the full performance capabilities of 5G in terms of capacity and speed. These trials will aim to demonstrate wider interoperability and support for vertical use cases in order to validate new business models.

The 5G IA Trials WG targets a tight connection with the Next Generation Mobile Networks (NGMN) 5G Trial & Testing Initiative (5GTTI) launched in June 2016. 5GTTI enables a global collaboration on testing activities, consolidating contributions and reports on industry progress, and testing future 5G use cases with industry stakeholders, in particular from vertical industries. It involves testing technology building blocks, proofs of concept, testing of key interfaces for ensuring interoperability, and pre-commercial network trials to visualize 5G capabilities and advantages on a system close to real network operation.

At the international level, multiple standalone 5G network trials will be in progress by 2018. Hence, it is targeted to run interoperability trials in coordination with other main regions. In order to enforce the coordination with other regions, the bi- and multi-lateral agreements of the 5G IA with other 5G initiatives in China, Japan, Korea, the United States and Brazil will be leveraged.

5G for Vertical Industries - Building the Case Through Pilots

After 5G technologies have reached a maturity level beyond pure laboratory experiments, trials and pilots at local as well as Pan-European levels will be targeted to accelerate the deployment of these new technologies. The main goals of such trials are the validation of 5G technologies in multi-vendor and multi-user environments. Verifying the stability and advantages in the environments will be an important step in moving 5G technologies from the research and development (R&D) laboratories into the market. In addition, such trials will serve to validate the technology as well as to identify further features to serve the needs of specific industry segments and user groups. A secondary and equally important goal is to increase the understanding of the new possibilities 5G technologies offer in industry segments which are less information and communication technology (ICT) focused and thereby, to help jump-start 5G adoption across a broad range of industries.

Following the 5G Manifesto, eleven different industry segments are considered by the 5G IA Trials WG. Common and new functions utilized by almost all vertical industries are Targeted Virtual Networks, enhanced privacy and security, and IoT enabler functions. Beyond these, different verticals require functions such as localization techniques, cloud and edge computing support, or heterogeneous network access. While understanding the benefits of 5G for different industries is a crucial point for a successful deployment, the 5G IA will assure that an umbrella perspective across all verticals will be developed to avoid siloes and a harmful segmentation of the 5G trials.

The ambition and the reality of demonstration coming from vertical trials have a strong dependency on the underlying platform capabilities. The platform interoperability issues are considered from at least two different angles:

- Verticals should benefit from meaningful common and standard interfaces such as application programming interfaces (APIs), which represent 5G services offerings well. This will allow discovering, triggering, negotiating and controlling the platform capabilities;

- Since many verticals should involve several platforms to demonstrate end-to-end (E2E) capabilities, interoperability between platforms is becoming a must.

In addition, platforms should serve key performance indicator (KPI) validation and benchmarking considering the vertical mission, and they should be replicable and deployable where the vertical resides.

Beyond functional aspects, different verticals have different levels of 5G readiness. In some verticals, large industry alliances or consortia exist, such as the European Automotive Telecom Alliance (EATA) and the 5G Automotive Association (5GAA) in the automotive sector, or the Industrial Internet Consortium (IIC) in the manufacturing industry. These are well positioned to partner with the 5G industry to establish, drive and shape 5G adoption within their vertical domains. From that perspective, Connected Cars are considered as one of the prioritized use cases for a strategic 5G European roadmap targeting vertical use cases.

There are several other vertical segments such as eHealth, the various Smart X areas (where X can stand for cities, buildings, transport, agriculture, energy, etc.), and public safety. Relevant sector organizations, such as the Personal Connected Health Alliance (PCHA) for the eHealth sector or other organizations, as far as available, will be invited to participate in this initiative. Planned 5G pilots at high visibility public events such as media and entertainment services for the EURO 2020 or 5G services for the lighthouse Smart City (including various other Smart X sectors) will deliver awareness for end users and get them interested in these new services. An important part of such trials will be the understanding of business aspects to support the setup of the 5G infrastructure. However, in some sectors like public safety or eHealth, business models may be of secondary importance, as the value to society is the main focus.

5G for UEFA EURO 2020 – An Opportunity for a Pan-European Showcase Event

In order for 5G to be truly successful, high profile trial(s), accessible to large public audiences, are planned. The target flagship event has to get widespread media attention and serve as a milestone for industry, governments and the general public in demonstrating that 5G is coming now and is beneficial for individuals and the society.

The Union of European Football Associations (UEFA) EURO 2020 football championships will be played in 13 different cities in Europe (Glasgow, Dublin, Copenhagen, Budapest, Bucharest, Brussels, Bilbao, Amsterdam, Saint Petersburg, Rome, Munich, Baku and London). This makes the EURO 2020 an excellent opportunity for a 5G Pan-European trial, also because of the media attention it will get. The timing of EURO 2020 in summer 2020, just before the 2020 Olympics in Japan, fits well with the EC 5GAP.

The proposal is that the EURO 2020 acts as the "launching event" for 5G in Europe with a number of 5G services that will be trialed around the EURO 2020 football cup. Three different types of trial services are proposed:

- For stadiums, around stadiums and in fan zones, 5G augmented and virtual reality applications related to EURO 2020 or football in general can provide ways to entertain fans before, during and after games, including immersive experiences around the competition. These services could be available in hosting cities but also in other cities;
- The EURO 2020 will be the opportunity to demonstrate services related to automated transportation around the stadiums and relevant transport routes. Scenarios include transportation to and from airports, and automated vehicles for the transport of officials, staff and supporters;

- Public safety authorities present for security around the stadiums could benefit from advanced public safety services. For example, augmented reality can be used to visually mark persons of interest based on facial recognition. Tailored services (e.g., access security, person localization, etc.) will create a significant improvement in safety and security.

Content and services will be shared across the different sites, which will be connected by different means towards the provision of services beyond the capabilities of current technologies. In each of the cities, a consortium is needed of local governments, playing stadiums, operators, infrastructure vendors and application providers. The City of Amsterdam, with the Amsterdam Arena as EURO 2020 host stadium, publicly announced at the MWC 2017 in Barcelona that they are committed to participate in 5G trials and to get as many playing cities on board. For each playing city, an agreement with (at least one) operator is targeted to ensure that there will be a 5G coverage on which the intended 5G services can be trialed. Sufficient spectrum (at least 100 MHz per operator across the 3.4-3.8 GHz band, and several hundreds of MHz in the 26 GHz band) is targeted for availability from the related member states, in order to allow demonstrating the full performance capabilities of 5G. The trial services will be developed together with the local partners, as for instance public safety trials will need collaboration with local governments. Where possible, a replication of 5G trial services across multiple cities should be aimed for. A Pan-European steering committee will ensure a consistent coordination of trial objectives and implementation, including also the marketing and communications aspects of this profile event across Europe and the world.

Many of the trial services are not only relevant for EURO 2020, but are also related to Smart City type applications. A relation with the UEFA will be established in order to investigate the rights and constraints associated with using EURO 2020 as a flagship event and to ensure that the trials add to the success of UEFA as well. A win-win-win perspective between 5G Infrastructure PPP, EC, member states, related cities and other stakeholders is sought. Specific actions may be partly funded by the EC in the context of 5G PPP phase 3 and by member states on national level.

5G Trials Cities

In complement to the 5G private trials under development and the 5G for UEFA EURO 2020 flagship event definition, specific cities in Europe already announced their plans to become 5G Trials Cities, at the forefront of 5G experimentation and trials. The different involved stakeholders come together to enable societal infrastructure benefits to the public. 5G will clearly be part of future cities, and to conduct relevant trials is a way to ensure the development of the best feasible 5G solutions. A non-exhaustive list of 5G Trials Cities is: Amsterdam, Bari, Berlin, Espoo, L'Aquila, London, Madrid, Malaga, Matera, Milan, Oulu, Prato, Stockholm, Tallinn and Turin. In addition, there are also 5G Research & Innovation Programs running in several member states, including the development of specific labs, experimentation and trials platforms. These platforms, being generally anchored in specific labs and cities (before their replication), contribute to the 5G momentum in specific countries and cities. Some of the 13 cities where the EURO 2020 competition will be organized already work on the possible 5G demonstrations and showcases. It is also anticipated that member states will communicate before the end of 2017, in the context of the 5GAP, the information on their 5G Pioneering City (or multiple cities) where 5G will be deployed in 2020. These different actions clearly create a strong momentum on 5G from cities and countries perspectives.

Under the 5G Infrastructure PPP initiative, a "5G City Challenge" will be organized as a call for interest towards interested cities prepared to sign a 5G charter and aiming at supporting cooperation among the cities involved in 5G experimentation and trials, e.g. for the possible development of best practices and sharing of lessons learnt. The number of collaborations is foreseen to grow as the various trial activities in the cities are maturing. In connection to the current European Smart Cities developments, a Charter or Alliance of 5G Trial Cities can be developed in that context.

Here again, the availability of sufficient amounts of spectrum, as discussed in Chapter 3, will be sought to deliver the full benefits of the 5G city trials.

17.3.2 Early Deployment in Americas

Just when it seems like mobile data usage can't get any bigger, it does. In 2016 alone, it grew 63% world-wide, and 18-fold over the past five years, according to Cisco's Visual Networking Index [18]. Over the next five years, mobile data will have a compound annual growth rate (CAGR) of 47%. The end of that period also is when commercial 5G networks should be widely available across North America - just in time to provide the speeds and spectral efficiency necessary to accommodate that demand. It won't be easy, but operators and regulators are already making major strides toward that goal.

Part of the challenge is spectrum, which is in chronically short supply in the U.S. and the rest of the Americas. This was a challenge for 3G and 4G, but 5G aims to tackle the problem by supporting far more bands, from 600 MHz to potentially as high as 95 GHz. This broad selection also will enable 5G to support a wider range of applications and deployment scenarios than its predecessors.

Pioneering New Spectrum Opportunities
The United States Federal Communications Commission (FCC) laid the foundation for this and other 5G trials in October 2014, when it issued a notice of inquiry (NOI) to explore the use of cellular in bands above 24 GHz. This NOI had global implications because in the process, the FCC became the world's first regulatory body to formally initiate proceedings for 5G spectrum. This is one example of why the rest of the world is so focused on the Americas: It's home to much of the pioneering work on 5G, including spectrum.

The FCC achieved another milestone in July 2016, when it announced that 28 GHz, 37 GHz and 39 GHz licensees "will have the flexibility to provide any fixed or mobile service that is consistent with their spectrum allocation [19]. This breaks with the recent past in which licensees were limited to only single use licenses in these bands." These three bands are under study for WRC-19. The FCC ruling helps that work providing real-world insights into how 5G can be used in millimeter-wave (mmWave) spectrum.

Although cellular systems traditionally have used only licensed spectrum, some 5G applications will use unlicensed bands, too, as also discussed in Section 3.2. That could mean aggregating licensed and unlicensed frequencies to create a connection fast enough to support a bandwidth-intensive application, or using only unlicensed spectrum by itself. Either way, making unlicensed bands available for cellular is one of many important elements for alleviating the chronic spectrum shortage.

The FCC's October 2015 notice of proposed rulemaking (NPRM) includes the use of the 64-71 GHz unlicensed band, while other commission initiatives are exploring unlicensed services in bands above 95 GHz. These high frequencies could be a particularly good fit for residential applications such as video and broadband because they can provide high bandwidth, while their limited range reduces the risk of one household's system interfering with those of its neighbors.

This example also shows how 5G in unlicensed mmWave spectrum could be used to deliver services that currently are offered only over copper or fiber, thus giving consumers more options.

Therefore, by laying the regulatory foundation for those new business models, the FCC also is giving regulators and operators in other countries an opportunity to see how they fare in the U.S. market.

The top four U.S. National Operators Show What's Possible

Many U.S. operators have been trialing 5G technology for over a year. Some noteworthy examples are:

- **AT&T** has worked on 5G technology for several years, for instance performing 5G trials with vendor partners in 2016. Additionally, in 2017, AT&T began trials of its DIRECTV NOW video service over 5G to residential customers. "We expect to further advance our 5G learnings – especially in how fixed wireless mmWave technology handles heavy video traffic," AT&T said. Additionally, AT&T announced the acquisition of FiberTower in 2017 for its 24 GHz and 39 GHz spectrum rights. Furthermore, in 2017, AT&T announced plans for its "5G Evolution" branded service [20];
- **Sprint** has licenses for more than 100 MHz of 2.5 GHz spectrum. In May 2017, Sprint announced that it is working with SoftBank and Qualcomm to develop 5G technologies for use at 2.5 GHz, with the goal of offering commercial services and devices by late 2019 [21]. This work is noteworthy from a global perspective, as Sprint CTO John Saw explained [22]: "The Next Generation Mobile Networks Forum recently released their position on 5G spectrum and recognized 2.5 GHz as one of the recommended bands for sub-6 GHz 5G. This is also why in March this year, it was decided at the 3GPP plenary meeting that 2.5 GHz would be included as a sub-6 GHz band for the 3GPP 5G standard. At this important meeting, 3GPP also agreed to provide an early NR release that includes 2.5 GHz." Additionally, in 2016, Sprint with vendor partners conducted 5G trials at two U.S. locations of the Copa América Centenario soccer tournament;
- **T-Mobile**'s 2012 acquisition of Metro PCS included licenses for 28 GHz and 39 GHz. T-Mobile's 28 GHz trials with Ericsson have shown that 5G is capable of delivering unprecedented throughput [23]. "The innovative demonstration showed that it was possible to achieve connection throughput of over 12 Gbps for 5G downloads and ultra-low latency connections of less than 2 ms," T-Mobile said. "The 5G trial occurred on August 3, 2016, and included test demonstrations of two-directional beam steering, operation of multiple simultaneous 4K video streams and the first test of a voice call connecting between 4G and 5G networks." Also, in May 2017, T-Mobile announced plans to deploy 5G at 600 MHz beginning in 2019, with nationwide coverage by 2020 [24];
- Since early 2016, **Verizon** has been working with its 5G Technology Forum partners—which include Apple, Cisco, Ericsson, Intel, Nokia, Qualcomm and Samsung—on a variety of fixed and mobile applications and use cases. The tests explore antenna technologies such as massive multiple-input multiple-output (mMIMO), and enormous bandwidths of up to 1 GHz [25]. "Throughput [was] in the multiple gigabits per second range," Verizon said. "Latency was measured in the millisecond range across varied distances, delivering superb video quality." In February 2017, Verizon finalized a deal to lease 24 GHz and 39 GHz spectrum from NextLink Wireless, with an option to buy the company [26].

The above examples should not be considered a comprehensive list of all trials occurring in the U.S. Besides operators and vendors, many universities also are testing 5G. For example, New York University's (NYU) Tandon School of Engineering trialed 5G in New York City. This test is noteworthy because it showed that the usage of mmWave spectrum is viable even in dense-urban environments. NYU determined that a 200 m cell radius is "very feasible" using only 1 W of transmit power, and that a range of 450 m or more is possible when combining antenna beams [27].

Although 5G is a global technology, there will be a myriad of national and regional differences in terms of spectrum, use cases and roll-outs. The U.S. and the rest of the Americas are providing the world with deep, actionable insights into what's possible.

17.3.3 Early Deployment in Asia

17.3.3.1 China

The 5G Trials in China include two phases, Technology R&D Trial (Phase 1) and Product R&D Trial (Phase 2). The target of phase 1 is to promote 5G core technology research and development in China, to verify 5G technology schemes and to support global unified 5G standards. Then, the phase 2 aims to validate the 5G product. More specifically,

- **Technology R&D Trial (Phase 1)** is led by the IMT-2020 (5G) Promotion Group and conducted from 2016 to 2018;
- **Product R&D Trial (Phase 2)** is considered as the 5G scale trial in China as well, and will be led by Chinese operators and conducted from 2018 to 2020.

In January 2016, the Ministry of Industry and Information Technology (MIIT) in China has launched the 5G Technology R&D Trial. The three steps of this trial are as follows:

- **Step 1 (Key technology trial)**: implemented from January 2016 to September 2016, focusing on the test of 5G key technology prototypes, the evaluation of function and performance, and the promotion of standards consensus building;
- **Step 2 (Technology scheme trial)**: implemented from June 2016 to December 2017, targeting to test the performance of a single base station (BS) and to verify the functionality and performance of different vendors' 5G technology schemes;
- **Step 3 (System trial)**: Implemented from June 2017 to October 2018, targeting to verify the networking and interconnection performance of the 5G system, and demonstrate the 5G typical services.

In September 2016, the Step 1 trials were completed and the testing results were released. Some key technologies of 5G have been identified: Massive MIMO is the key technology to provide superior spectrum efficiency in 5G, as covered in detail in Section 11.3. With 100 MHz bandwidth at 3.5 GHz, the bandwidth efficiency is close to 30 bit/s/Hz, which is 2-3 times that of 4G. For high frequency communication, the tests have been conducted on 15 GHz, 28 GHz and 73 GHz. The peak data rate with 1 GHz bandwidth at 28 GHz can reach more than 20 Gbps. The outdoor coverage range at 15 GHz is 350 m. Meanwhile, a number of other key technologies such as new waveforms, new multi-technology and polar codes have been tested. The test results show the advantage of these potential key technologies of 5G.

In November 2016, China released the test specifications of Step 2 trials and then initialized the test. The Step 2 trials will be conducted based on the unified test platform, requirements and spectrum. This step will be the key focus of the IMT-2020 (5G) Promotion Group in 2017 as well. Currently, China is setting up the lab and field test environment, to support an E2E RF test, function test and performance test. Now, China has set up the world's largest 5G field trial in Huairou district of Beijing. 30 sites were planned, which will support the requirements of network performance test for 6 equipment vendors, including Huawei, Ericsson, ZTE, Datang, Nokia, and Samsung. China has also invited test instrument manufacturers and chipset vendors to join the field trial, and therefore the whole industry can work together to build a 5G industry ecosystem.

In addition, the China 5G technology R&D trial is open to foreign enterprises, and invites both domestic and international major operators, equipment vendors, chip companies, metering & instrument enterprises as well as research institutions to participate. 21 companies have taken part in the trial for the time being. They work together to formulate the testing specifications and carry out the test. The final target is to achieve commercial deployment of 5G networks in the year of 2020.

Moreover, the integration of 5G and vehicle networks is considered as a key focus of China. Currently, several provinces and cities including Shanghai, Zhejiang, Beijing-Hebei, Chongqing, Jilin, Hubei and Wuxi have been supported by MIIT and planned as demonstration areas or testing sites to explore and promote intelligent connected vehicles based on broadband wireless communication networks.

Network Evolution Plan of Operators in China
The Chinese operators pay a lot of attention to 5G, and they make great efforts on 5G R&D and standardization. Currently, the largest operator in China, CMCC, has announced its initial strategy of 5G network evolution. The other two operators in China, China Telecom and China Unicom, are working on their 5G pre-commercial plans intensively as well.

According to CMCC's workplan towards the enhanced mobile broadband (eMBB) scenario, the network will keep evolving based on the current technology, while the new air interface will be introduced in 2020. In addition, 2017 is the first year for IoT deployment, where both NB-IoT and eMTC networks will be rolled out. Moreover, the pre-commercial launch of LTE V2X is planned in 2019.

For the new radio eMBB, CMCC has also developed its own 5G trial plan. With the completion of key technology component validation by the end of 2016, CMCC will carry out a proof-of-concept system field trial with 7 sites per city in 2017. Then a pre-commercial trial with 20 sites per city will be launched in 2018, where multi-vendor inter-operation and E2E testing will be the main task in this phase. In 2019, a large-scale pre-commercial trial with more than 100 sites per city will be carried out, targeting the commercial launch of more than 10,000 5G sites in 2020.

17.3.3.2 Japan
Nowadays, the use of radio waves is expanding into a wide variety of fields that are essential for our daily lives, including not only communication entities and networks such as mobile phones and wireless LANs, but also robotics, medicine and the environment. It is also expected that the Internet of Things (IoT), where everything is connected to a network, will soon be in full swing.

On the other hand, there are demands for the provision of new mobile services that can handle the increased traffic levels resulting from the evolution of mobile broadband networks. In response to this situation, the Ministry of Internal Affairs and Communications has been promoting efforts aimed at delivering 5G mobile communication systems in time for the Tokyo 2020 Olympic and Paralympic Games. These efforts include joint R&D projects between industry, academia and government institutions with the aim of delivering eMBB while supporting mMTC, international cooperation involving stronger ties with governments and 5G promotion organizations worldwide, comprehensive trials that are scheduled to begin in 2017, and discussions of frequency allocations for 5G networks.

In the following, the efforts made to realize 5G are introduced, particularly regarding the Radio Policy 2020 Study Group that was opened by the Senior Vice-Minister for Internal Affairs and Communications between January and July 2016, and regarding the 5G System Field Trials Plan in Japan.

Final Report of the Radio Policy 2020 Study Group

The Radio Policy 2020 Study Group was opened by the Senior Vice-Minister for Internal Affairs and Communications between January and July 2016, and studied promotion policies aimed at the realization of new mobile services including 5G and next-generation Intelligent Transport Systems (ITS).

In its final report of July 2016, 5G is positioned as the ICT platform for the IoT era in which the circulation of vast quantities of diverse kinds of data will take place in the cloud. The report presents nine fields of use, three projects and nine promotion models, and concludes that the promotion of these projects will accelerate efforts to realize 5G by 2020. This article summarizes the 5G final report of the Radio Policy 2020 Study Group.

Up to 4G, services were centered around the distribution of information to devices such as smartphones and tablets. On the other hand, 5G not only provides eMBB, but also facilitates URLLC and mMTC to large numbers of sensors and terminals, and is expected to be used in a much wider diversity of fields including automobiles. Achieving horizontal development in the following nine fields is vital for the spread of 5G:

1) Sports (fitness, etc.)
2) Entertainment (games, tourism, etc.)
3) Offices/workplaces
4) Medical care (healthcare, nursing)
5) Smart houses/daily life (daily necessities, communications, etc.)
6) Retail (financing, payments)
7) Agriculture, forestry and fisheries
8) Smart cities/smart areas
9) Traffic (passenger transport, freight distribution, etc.)

To further develop these nine fields, cooperation with diverse utilization fields is essential from the stage of R&D and verification before a system is introduced, rather than going ahead with horizontal deployment from the stage where 5G is implemented. Therefore, to achieve the early deployment of 5G in these nine fields, based on the main requirements of 5G (eMBB, mMTC, and URLLC), it was shown that three projects should be promoted: ultra-broadband, wireless IoT, and next-generation ITS, as shown in Figure 17-4.

With regard to these three projects, three specific "promotion models" were shown simultaneously for each project, as depicted in Figure 17-5. For the advancement of future projects, it is essential to construct an implementation system involving not only communication providers and the vendors of communication equipment, but also the users of this technology, and to consider the specific contents of these projects.

In the promotion of these projects, actions were taken from the following four viewpoints:

1) Strategic R&D and verification: It is important to promote R&D focused on key technologies, and to conduct R&D and verification based on strategic cooperation with other countries as in European joint studies;
2) Environmental improvements aimed at business expansion: Systems and maintenance should be studied at the same time as R&D, and frequencies should be reserved by taking factors such as international harmonization and ease of social implementation into consideration;

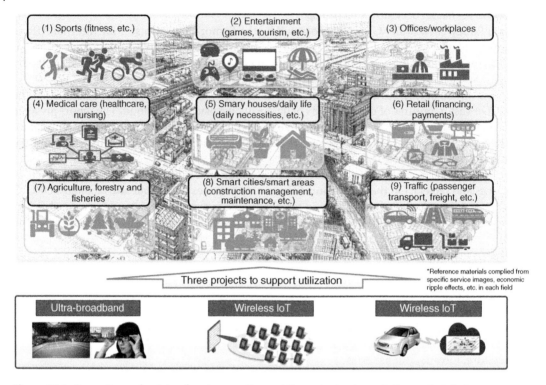

Figure 17-4. Promotion and uptake of next-generation mobile service implementation projects in Japan.

3) Deployment in regions that contribute to regional stimulation, etc.: Open test beds that allow a wide diversity of stakeholders to get involved have been set up not only in Tokyo but in all regions, and such measures have contributed to regional revitalization and local creation;

4) National standardization and international development: Ongoing strategic national standardization and international development will be promoted in partnership with Fifth Generation Mobile Communication Promotion Forum (5GMF) stakeholders and other related individuals from industry, academia and government, and the international expansion of a comprehensive system that links technologies and services.

Based on feedback from users, it is also expected that promotion efforts coupled with a fast "plan, do, check, adjust" (PDCA) cycle will help with the resolution of issues facing society and with the creation of new value in the 5G era.

5G System Field Trials Plan

In Japan, in addition to research and development by each operator, vendor, research institution, university, etc., the 5G research and development projects promoted by the Ministry of Internal affairs and Communications, Japan (MIC) are started with the plan for four years from fiscal year (FY) 2015. These 5G R&D projects include research on key technologies for 5G features, such as

Ultra-broadband

Ultra-fast simultaneous delivery

Wireless communication that is faster than 4G and accommodates more simultaneous usres

Stadium, etc.

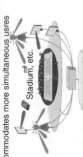

Wireless realism

Low-latency trasmission of high-definition video data (e.g., 4k/8k), and using VR technology to achieve a high sense of realism

Wireless VR

Rumor League

High performance image sensors

Acquiring big data by monitoring with mechanical "eyes" that exceed the capabilities of the human eye

Smart office/factory

Wireless IoT

Wireless network fusion

Realization of smart systems with optimal management of diverse wireless networks by unified deposits

Smart city

Large numbers of simultaneous connections

Implementing compact, inexpensive low-power wireless terminals that can achieve reliable wireless communication even when densely packed together invery large numbers

Wireless platform

Platforms that safely and quickly manage, analyze and utilize large quantities of data collected on wireless terminals

Terminal/robots   Server   Terminal/robots

Security robot

Next Generation ITS

Next Generation "Connected Cars"

Creation of new ITS business and services by sharing and utilizing data from vehicles that are permanently connected to the network

Connected car

Ultra-low latency inter-vehicle communication

Safely platooning vehicles by low-latency inter-vehicle communication

Ultra-fast communication for high-speed mobile services

Stress-free high-speed wireless communication, even in Bullet trains and other fast-moving vehicles

Linear cell

Optical fiber

| Sport | Entertainment | Office/workplace | Medical care | Smart houses/ daily life | Retail | Agriculture, forestry and fisheries | Smart cities & areas | Traffic |

Figure 17-5. Nine promotion models considered in Japan.

ultra-high-speed and low-latency transmission, large capacity, massive connection, and collaborative research with Europe. Utilizing the outcome of the 5G R&D projects, from FY 2017, the 5G system field trials also promoted by the MIC are started with a plan for three years. These trials are held in Tokyo and local areas in various application-specific fields for social implementation of 5G, and aim to include contributions to international standardization activities by establishing an open environment where companies and universities around the world can participate, as shown in Figure 17-6.

In the 5G system field trials, multiple test environments are assumed as follows:

- Urban micro-cell or Urban macro-cell,
- Suburban macro-cell or Rural macro-cell, and
- Indoor hotspot.

Also, the following three key capabilities will be proved with the trials:

- eMBB (10 Gbps peak data rate),
- mMTC (1 million connected devices/km2), and
- URLLC (1 ms over-the-air latency).

Table 17-1 provides an overview of trials in 2017 by six organizations (mobile operators and research institutes). It shows trial goals, technology targets, and locations. Each trial is considered to contain a demonstration test assuming various characteristic use cases. Examples including ultra-high-definition and high-presence video streaming, Smart City for safety and security, and telemedicine services are shown in Figure 17-7.

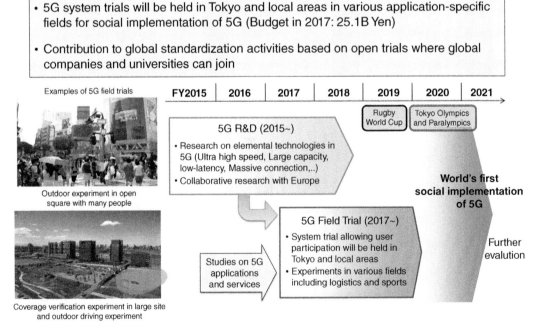

Figure 17-6. 5G R&D and system field trials in Japan.

17.3.3.3   South Korea

Introduction

Korea is one of the countries with the most mature and advanced LTE market and technology. In 2015, Korea has already reached an LTE penetration of 95.7% and was ranked by OpenSignals [28] as the country with the highest LTE penetration.

Given this early matured LTE market and technology, Korea had a relatively early start on 5G research and evolution. The global industry consensus is to bring 5G into the market in 2020, with phase 1 5G standardization. Korea, however, is now strongly driving the development of 5G mobile technologies with the ambitious launching timeline starting at 2018.

In 2013, the Giga-Korea project was launched in the collaboration by Ministry of Science, Information and Future Planning (MSIP) with participation of multiple other related ministries for the development of Gbps mobile broadband connectivity by 2020, where the R&D roadmap is to not just upgrade 5G mobile communication network infrastructure, but also pursue shared growth together with industries in all contents, platforms, and devices. It focuses on R&D for the first five years, and continues for service feasibility trials until 2020.

MSIP announced the 5G Mobile Strategy 1.0 in January 2014, which aimed at early 5G commercialization by 2020. It derived joint R&D projects of industry, government institutes, and academia

Table 17-1. Overview of 5G field trials in Japan in 2017.

	Responsible organization	Main partners	Trial overview	Main trial locations	Technology target
I	NTT DOCOMO	• TOBU TOWER SKYTREE • ALSOK • Wakayama Pref.	• Ultra HD video streaming • Advanced city security • Remote medical services	• Tokyo • Wakayama	eMBB • 5 Gbps/UE (10 Gbps/BS)
II	NTT Communications	• Tobu Railways • Infocity	• Entertainment for high mobility transportation	• Tochigi • Shizuoka	eMBB • 2 Gbps at high mobility environment
III	KDDI	• Obayashi Corp. • NEC	• Remote operation for construction machinery	• Saitama	URLLC • low latency communication within 1 ms (over the air)
IV	ATR	• Naha City • Keikyu Railways	• Entertainment in stadium • Ultra HD video streaming	• Okinawa • Tokyo(HND)	eMBB • 5 Gbps/UE (10 Gbps/BS)
V	Softbank	• Advanced Smart Mobility Co., Ltd. • SBDrive Corp.	• Platooning vehicles and remote operation for trucks	• Yamaguchi	URLLC • low latency communication within 1ms (over the air)
VI	NICT		• Logistics • Smart office	• Hokkaido • Osaka	mMTC • 1 million devices/km^2

to realize eMBB, mMTC and URLLC. It also includes international collaboration in R&D, spectrum, and policy development with governments and global 5G promotion organizations. The 5G Mobile Strategy 2.0 was also announced in December 2016, emphasizing the 5G convergence service with other industries and ecosystem building by 2022. MSIP and the 5G Forum published the 5G Service Roadmap 2020 [29] in 2016. It investigates the latest issues, forecasts future society, derives potential

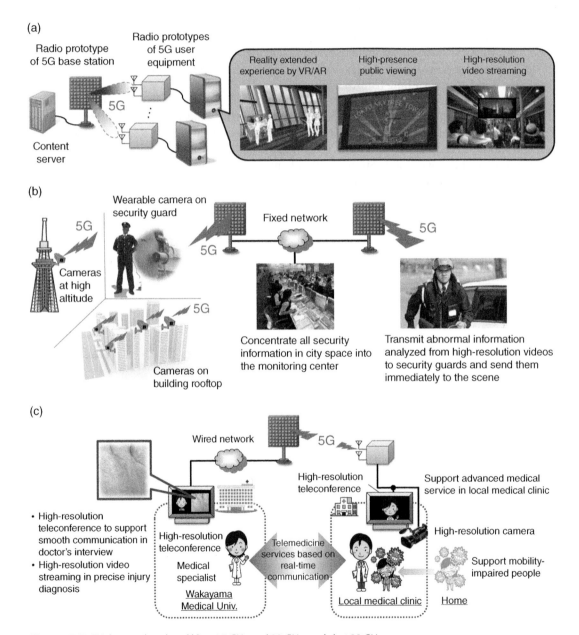

Figure 17-7. Trial examples, a) and b) at 4.5 GHz and 28 GHz, and c) at 28 GHz.

service fields and suggests feasible options (i.e., necessary technologies, cooperation measures, government policies, etc.), and finally addresses five key vertical services that Korea is interested in throughout the time period 2018 to 2022, as shown in Figure 17-8.

Currently, three task forces in the 5G Forum with MSIP are working on developing details on service scenarios, business models, and network architectures for each of the five key services. Results will be published in November 2017, based on which 5G vertical service feasibility trials will be scheduled from early 2018. Especially the country's two biggest telecommunication operators are actively leading their industry in building 5G infrastructures. The vision for 5G is a lifestyle transformer in the upcoming years [30].

Network Architecture

The 5G network architecture may be described with three main layers, a service enablement, All-IT infrastructure, and hyper-connected radio:

1) The service enablement layer understands both the service requirements and the underlying system capabilities. It can dynamically gather necessary system resources to create diverse services on-demand;
2) The All-IT infrastructure is the layer that abstracts physical system resources into logical (virtualized) system resources, and provides the abstracted system resources to the service enablement layer. The All-IT infrastructure is composed of commodity hardware and open software and heavily leverages network function virtualization (NFV) and software defined networking (SDN) technology, as detailed in Section 10.2;
3) Hyper-connected radio is the layer composed of multiple radio access technologies, providing radio connectivity satisfying bandwidth, latency and density requirements of 5G to the All-IT infrastructure layer.

A unified 5G network supports seamless operations across many different types of wired and wireless networks. The unified access architecture allows sharing legacy optical-fiber infrastructure with both wireless and wireline services. Furthermore, gateway functionality for both wireless and wired services can be merged into a single gateway at the network edge, enabling unified wireline and wireless services across different access technologies. These new features help to quickly develop new business opportunities, adapting to diverse market demands, such as the demand for near-zero latency.

Trials 2015-2017

SK Telecom has been actively conducting 5G pilots and trials for harmonized and timely 5G development and commercialization, since industry-wide harmonization is as critical as facilitated commercialization. One of the noticeable activities was the opening of the "5G Global Innovations Centre" in October 2015. During the opening ceremony, a number of 5G key enabling technologies were assessed and demonstrated, including mmWave radio technology with low latency, massive MIMO, virtualized RAN and core network, and network slicing.

Globally, in 2016, collaborations between the mobile network industry and other vertical markets were initiated fast. As part of strategic collaboration efforts with new vertical markets, SK Telecom has formed a partnership with BMW and Ericsson to develop and demonstrate 5G connected cars with a number of relevant V2X use cases, including:

- 28 GHz 5G radio beamforming, fast beam-tracking with high mobility;
- Video recognition based safety features, enabled by mobile edge computing;

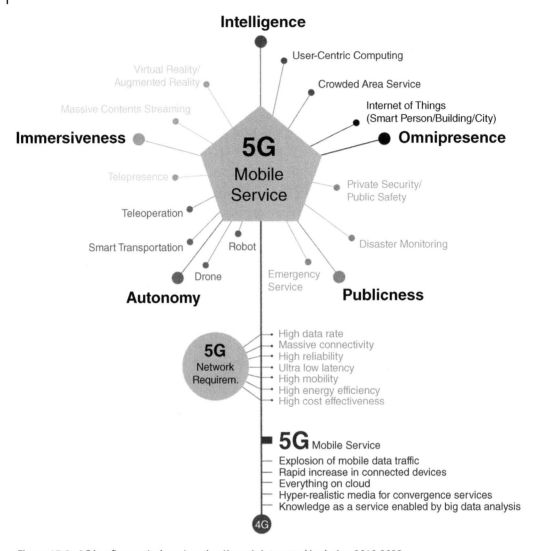

Figure 17-8. 5G key five vertical services that Korea is interested in during 2018-2022.

- Drone Helper;
- 4K UHD Streaming and 360 VR Live Streaming;
- UHD Live Conference;
- 5G Experience Bus with AR and VR.

In addition, network slice interworking technology in supporting vertical markets at a global scale became important as the new 5G emerging vertical services are designed and developed at a global scale. These efforts towards 5G commercialization are well recognized by the global industry, which is visible by the fact that both T5 - the world's 1st 5G connected car - and an outdoor mmWave handover trial have received the MWC Shanghai AMO Award (June 2017, Best Innovation for LTE

T5–World's 1ˢᵗ 5G Connected Car ('16.11.15)   World's Fastest 5G Connected Car ('17.2.07)

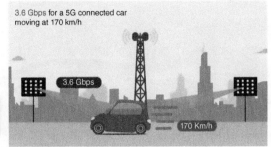

5G Connected Car with 28 GHz 5G Infra   Achieved 3.6 Gbps speed at 28 GHz band
                                                 for 5G Connected Car moving at 170 km/h

Figure 17-9. SK Telecom's T5: The world's 1st 5G Connected Cars.

to 5G Evolution), Telecom Asia Award (May 2017, Most Innovative Partnership Strategy) and GTB Award (May 2017, Infrastructure Innovation), see Figure 17-9. Other global awards include 5G & LTE Asia awards (June 2016, 5G Trials and Development & Best 5G R&D) and the RANNY award (June 2016, Best 5G Initiative).

In 2017, SK Telecom has partnered with Deutsche Telecom and Ericsson to setup an inter-continent 5G testbed, and successfully developed and demonstrated federated network slicing technology for the first time. The 5G R&D, pilot, and trial activities performed since 2015 are summarized in Table 17-2.

Korea Telecom (KT) is proceeding to commercialize 5G in 2019 by deploying a 5G testbed for early verification of 5G key technologies, developing 5G technical specifications, and launching 5G

Table 17-2. Summarized list of trials by SK Telecom.

Year	Activity
2017	3.5 GHz 5G E2E system trial with Samsung and Nokia (June 2017)
	5G trial services in SK Happy Dream Park (see Figure 17-10)
	Federated network slicing by SK Telecom, DT, and Ericsson (Feb. 2017, World's 1st)
	3.6 Gbps inside a vehicle moving at 170 km (Feb. 2017, World's fastest)
2016	5G connected car by SK Telecom, BMW, and Ericsson (Nov. 2016, World's 1st)
	5G mobility and handover performance test (Sep. 2016, World's 1st)
	5G radio ray tracing and deployment analysis (Aug. 2016)
	5G trial specification finalization and outdoor field trial (Mar. 2016, World's 1st)
	World's fastest speed with 25 Gbps (Feb. 2016)
2015	5G 19.1 Gbps with centimeter-wave 5G radio (Oct. 2015)
	Global Innovations Centre Opening (Oct. 2015)
	19.1 Gbps with centimeter-wave 5G radio (Oct. 2015)
	5G network slicing (Oct. 2015, World's 1st)
	In-band full-duplex demonstration (May 2015, World's 1st)
	5G Robot Demo with 7.55 Gbps millimeter-wave 5G system (Mar. 2015, World's 1st)

Figure 17-10. SK Telecom's Happy Dream Park Baseball stadium.

trial services alongside the PyeongChang 2018 Olympics. Since opening the KT 5G R&D Center in July 2015, KT has been actively working on 5G technologies, along with many world's first trials including massive MIMO, unified wired and wireless access, mmWave cell planning, 5G multi-user MIMO (MU-MIMO), 5G first call, and 5G self-driving car.

For 5G Trials in 2018, KT developed PeongChang 5G specifications – world's first 5G common specifications co-working with Ericsson, Intel, Nokia, Samsung, and Qualcomm. KT released 5G physical layer specifications in Mar. 2016, and full 5G specifications including the higher layers (L2/L3) in June 2016 [32]. These specifications target 20 Gbps using 800 MHz bandwidth, and KT is primarily considering the 28 GHz spectrum band for 5G trial services.

Figure 17-11. KT's 5G Trial Sites at Ganghwamun and PyeongChang.

Based on 5G PyeongChang specifications, KT developed 5G stations and devices, and accordingly built up an outdoor 5G trial network around Gwanghwamun Square in central Seoul, a crowded area with a large floating population and several tall buildings, as shown in Figure 17-11. They also set up the trial network in Gangnam area, another downtown in Seoul, and PyeongChang, a hosting city of the 2018 Winter Olympics, and finished the field tests including massive-MIMO/beamforming, beam-tracking, and handover. Since 2016, KT have showcased diverse 5G use cases for the 2018

Winter Olympics, mainly focusing on hyper-realistic media services on top of KT's 5G trial networks as follows:

- **"Sync view"** enables real-time 360-degree VR videos from the player's view, making spectators feel as if they were actually with the bobsleigh crew during the race (PyeongChang Alpensia Sliding Center, October 2016);
- **"Omni point view"** provides multiple video streaming instances at the same time with cameras staged at various positions around the cross-country ski course (2017 Cross Country World Cup, March 2017);
- **"Interactive time slice"** delivers 3D videos captured at diverse angles on athletes, which enables interactive selection of angle that people wants to watch (Four Continents Figure Skating Championships games, February 2017);
- **"Hologram live"** supports real-time interactive hologram communications, making people feel closer to athletes on the interview. KT demonstrated the live hologram interview between Seoul and PyeongChang, which are 200 km apart (March 2016).

Furthermore, during the Olympic test event in March 2017, KT had a self-driving 5G bus in operation in the area of PyeongChang, which is the world's first self-driving car supporting 5G connectivity for delivering both tremendous driving-related information and hyper-realistic mobile media.

Trials 2018 and Beyond then Followed by Commercialization

KT, as an official telecommunication partner of 2018 PyeongChang Winter Olympics, is aiming at launching a 5G trial service in time before the event begins, and ultimately at commercial 5G services in 2019 [31].

KT is planning to provide 5G trial services during the PyeongChang 2018 Olympics, mainly in the Olympic venues and nearby areas. A 5G network is also deployed in targeted areas within Seoul metropolitan area, KTX (Korea Train eXpress), and Incheon International Airport, as shown in Figure 17-12. 5G-enabled smartphone and IoT modules are ready for Olympic staff, athletes and

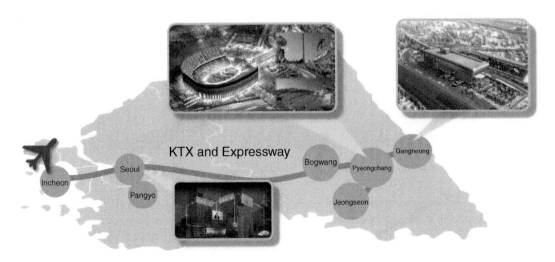

Figure 17-12. KT's 5G deployment plan for the 2018 Winter Olympics.

visitors to experience real 5G services, 2 years ahead of the commercialization of 5G. Even after the Olympic Games, 5G trial systems will be moved and applied to several test sites to verify the diverse 5G use cases, and KT's 5G specifications to be tested during the trial have been contributed to the global 5G standardization and applied to the subsequent 5G commercialization phase.

While mainly considering 28 GHz for 5G trials in 2018, KT plans for commercial 5G services using both 3.5 GHz for a nationwide area and 28 GHz for urban hotspots in 2019. Starting from the heavily crowded areas of major cities, KT will quickly extend 5G coverage to nationwide based on its optical-fiber infrastructure, and deploy the world's first nationwide 5G network.

SK Telecom, along with AT&T, KT, NTT DoCoMo, Vodafone, Ericsson and others, led the 3GPP 5G NR acceleration schedule by introducing an intermediate milestone for an early completion of non-standalone 5G in March 2017, as mentioned in Section 17.2. Aligned with this accelerated 3GPP timeline on 5G, SK Telecom plans to commercialize 5G in Korea. In June 2017, as preparation towards 5G commercialization, SK Telecom completed a 3.5 GHz 5G E2E system trial and thereby secured both 3.5 GHz and 28 GHz 5G systems. Beside the key enabling technologies on the RAN, CN capabilities such as network slicing and mobile edge computing will also be part of SK Telecom's 5G commercialization.

17.4 Summary

This chapter has presented the standardization roadmap towards 5G with its main features along with timelines and the main contributing organizations. The process of developing the new 5G radio interface involves development and proposals, their evaluation and adoption. Particularly, the roles of 3GPP and ITU-R have been highlighted.

The chapter has further presented trial activities and roadmaps in the three regions Europe, Americas, and Asia. It has detailed early findings utilizing pre-standard equipment, and plans under realistic environments such as big sport events to test the new standardized solutions as soon as these arrive. Finally, some commercialization plans have been outlined as well.

In conclusion, the chapter has given a complete picture of the roadmap towards 5G in terms of standardization and activities leading to early deployment and commercialization, showing that there is a great interest in the new technology capabilities.

References

1 3rd Generation Partnership Project (3GPP), see http://www.3gpp.org/about-3gpp/about-3gpp
2 3GPP release overview, see http://www.3gpp.org/specifications/releases
3 3GPP TR 21.900, "Technical Specification Group Services and System Aspects; Technical Specification Group working methods, Release 14", V14.0.0, March 2017
4 3GPP TR 38.900, "Technical Specification Group Radio Access Network; Study on channel model for frequency spectrum above 6 GHz, Release 14", V14.3.1, July 2017
5 3GPP TR 38.913, "Technical Specification Group Radio Access Network; Study on Scenarios and Requirements for Next Generation Access Technologies, Release 14", V14.2.0, March 2017
6 3GPP TR 38.912, "Technical Specification Group Radio Access Network; Study on New Radio (NR) access technology, Release 14", V14.0.0, March 2017

7 ITU-R Recommendation M.2083, "IMT Vision – Framework and overall objectives of the future development of IMT for 2020 and beyond", Sept. 2015

8 ITU-R WP5D Contribution 836, "Report on the twentieth meeting of Working Party 5D (Geneva, 15-22 October 2014)", Chapter 2, Oct. 2014

9 3GPP TR 36.888, "Technical Specification Group Radio Access Network; Study on provision of low-cost Machine-Type Communications (MTC) User Equipments (UEs) based on LTE, Release 12", V12.0.0, June 2013

10 3GPP TR 36.802, "Technical Specification Group Radio Access Network; Evolved Universal Terrestrial Radio Access (E-UTRA); NB-IOT; Technical Report for BS and UE radio transmission and reception, Release 13", V13.0.0, June 2016

11 3GPP TR 36.885, "Technical Specification Group Radio Access Network; Study on LTE-based V2X Services, Release 14", V14.0.0, June 2016

12 3GPP TR 22.886, "Technical Specification Group Services and System Aspects; Study on enhancement of 3GPP Support for 5G V2X Services, Release 15", V15.1.0, March 2017

13 3GPP TR 36.881, "Technical Specification Group Radio Access Network; Study on latency reduction techniques for LTE, Release 14", V14.0.0, June 2016

14 5G Infrastructure Association, "5G Pan-European trials roadmap strategy", February 2017, see https://5g-ppp.eu/wp-content/uploads/2017/05/5GInfraPPP-Trials-Roadmap-Strategy_Short_28-February-2017.pdf

15 European Commission and European industry, "5G Manifesto", July 2016, see http://ec.europa.eu/newsroom/dae/document.cfm?action=display&doc_id=16579

16 European Commission, 5G Action Plan, Sept. 2016, see http://ec.europa.eu/newsroom/dae/document.cfm?doc_id=17131

17 5G Infrastructure Association, "Roadmap Strategy", Febr. 2017, see https://5g-ppp.eu/wp-content/uploads/2017/05/5GInfraPPP_TrialsWG_Roadmap_Version1.0.pdf

18 CISCO, "Visual Networking Index: Global Mobile Data Traffic Forecast Update, 2016–2021 White Paper", March 2017, see http://www.cisco.com/c/en/us/solutions/collateral/service-provider/visual-networking-index-vni/mobile-white-paper-c11-520862.html

19 FCC, "Report and order and further notice of proposed rulemaking", FCC 16-89, July 2016, see https://apps.fcc.gov/edocs_public/attachmatch/FCC-16-89A1.pdf

20 AT&T, "AT&T Details 5G Evolution", Jan. 2017, see http://www.prnewswire.com/news-releases/att-details-5g-evolution-300385196.html

21 Sprint, "Qualcomm, SoftBank and Sprint Announce Collaboration on 2.5 GHz 5G", May 2017, see http://newsroom.sprint.com/news-releases/qualcomm-softbank-and-sprint-announce-collaboration-on-25-ghz-5g.htm

22 J. Saw, Sprint, "CTO Blog: Sprint, Qualcomm Technologies and SoftBank Accelerate 5G for 2.5 GHz", May 2017, see http://newsroom.sprint.com/blogs/sprint-perspectives/sprint-qualcomm-technologies-and-softbank-accelerate-5g-for-25-ghz.htm

23 Ericsson, "T-Mobile and Ericsson achieve over 12 Gbps on 5G connection", Sept. 2016, see https://www.ericsson.com/en/press-releases/2016/9/2043477-t-mobile-and-ericsson-achieve-over-12-gbps-on-5g-connection

24 T-Mobile, "T-Mobile Announces Plans for Real Nationwide Mobile 5G", May 2017, see https://newsroom.t-mobile.com/news-and-blogs/nationwide-5g.htm

25 Verizon, "Verizon 5G trials driving ecosystem towards rapid commercialization", Feb. 2016, see http://www.verizon.com/about/news/verizon-5g-trials-driving-ecosystem-towards-rapid-commercialization

26 Verizon, "Verizon completes purchase of XO Communications' fiber business", Feb. 2017, see http://www.prnewswire.com/news-releases/verizon-completes-purchase-of-xo-communications-fiber-business-300400440.html

27 T. Rappaport, New York University University Tandon School of Engineering, "Spectrum Frontiers: The New World of Millimeter-Wave Mobile Communication", March 2016, see https://transition.fcc.gov/oet/5G/Workshop/Keynote Rappaport NYU.pdf

28 OpenSignals, "State of LTE", 2016, see https://opensignal.com/reports/2016/11/state-of-lte

29 MSIP and 5G Forum, "5G Service Roadmap 2020", 2015, see https://www.5gforum.org/home

30 SK Telecom, see https://developers.sktelecom.com/resource/document/#

31 Korea Telecom, "KT 5G Master plan", Sept. 2015, see http://www.kt.com/eng/biz/kt5g_01.jsp

32 Korea Telecom, „KT 5G SIG", June 2016, see http://www.kt.com/eng/biz/kt5g_02.jsp

Index

5G System Design: Architectural and Functional Considerations and Long Term Research, First Edition.
Edited by Patrick Marsch, Ömer Bulakçı, Olav Queseth and Mauro Boldi.
© 2018 John Wiley & Sons Ltd. Published 2018 by John Wiley & Sons Ltd.